Systems Theory and Practice in the Knowledge Age

Systems Theory and Practice in the Knowledge Age

Edited by

Gillian Ragsdell
Daune West
University of Paisley
Paisley, Scotland

and

Jennifer Wilby
University of York
York, England

Kluwer Academic / Plenum Publishers
New York, Boston, Dordrecht, London, Moscow

Proceedings of the Seventh International Conference of the UK Systems Society on Systems Theory and Practice in the Knowledge Age, held July 7–10, 2002 in York, United Kingdom

ISBN: 0-306-47247-3

233 Spring Street, New York, New York 10013

http://www.wkap.nl/

10 9 8 7 6 5 4 3 2 1

A C.I.P. record for this book is available from the Library of Congress

Printed in the United States of America

PREFACE

Welcome to the proceedings of the Seventh International Conference of the UK Systems Society being held at York University, United Kingdom from July 7th to 10th, 2002. It is a pleasure to be able to share with you this collection of papers that have been contributed by systems thinkers from around the world. As with previous UKSS conferences, the aim of this conference is to encourage debate and promote development of pertinent issues in systems theory and practice. In current times where the focus has moved from 'information' to 'knowledge' and where 'knowledge management', 'knowledge assets' and so on, have become part of everyday speak, it seemed fitting to offer a conference title of 'Systems Theory and Practice in the Knowledge Age'.

In keeping with another tradition of previous conferences, the UKSS Conference 2002 Committee decided to compile a collection of delegates' papers before the event as a platform from which to launch discussions in York. Ideas presented in the following papers will, undoubtedly, be developed during the dialogue generated at the conference and new papers will emerge. In his abstract for his plenary at this conference, Professor Peter Checkland throws down the gauntlet to systems thinking and its relevance in the knowledge age with the following statement:

> ***"30 Years In The Systems Movement: Disappointments I Have Known and Hopes for the Future***
> Springing from a lunchtime conversation at an American University, the Systems Movement is now nearly 50 years old. If we look back at its progress as objectively as we can, we have to acknowledge its relative failure. Today's Systems Movement has: a toe-hold in academe; modest institutions and journals; a great propensity to talk to itself; and virtually no effect on the 'influentials' who shape social change.
> This is ironic given the potential relevance of systemic thinking in 'the knowledge age', the focus of this conference.
> The presentation will illustrate and seek to explain the 'failure' of the Movement and to suggest a pattern which would give hope for the future."

The fifty four papers that are included in this volume go some way to facing that challenge. The papers are organised under ten streams that will be mirrored in the conference programme: systems theory, systems methodologies, knowledge management, environmental management, critical systems thinking, social systems design, business, information systems, modelling, and student presentations. Not only is it interesting to note the favoured streams, but it is interesting to note the origins of the fifty four papers that have been included in this book - there are papers from authors in eleven different countries. This book stands as an indication of the international systems movement. It is also worth commenting that there are papers from both practitioners and academics, and that there are a significant number of papers from research students. All of these features are promising signs. The number of new contributors, (i.e. those who have not participated in the UKSS proceedings before), gives equally positive signals about the future of the systems movement.

While each paper was double blind reviewed by respected people in the systems field, it is fair to say that the editors were keen to uphold the authors' arguments. The role of the editors was therefore more concerned with checking the formatting, the quality of presentation and so on, in order to ensure a high-quality book rather than to impose a particular view of systems theory and practice. The diversity of view points offered here will be a trigger to stimulating discussion at the conference and to a fruitful synthesis of ideas.

In conclusion, we would like to extend thanks to the authors and to the anonymous reviewers for their contributions of time and commitment to this project. We look forward to joining with you in York where we anticipate many enjoyable and meaningful discussions on the topics you have introduced here. Internationally eminent speakers such as Professors Peter Checkland, Mike Jackson, Max Boisot, Norman Ball and John Mingers have already accepted invitations to speak in plenary sessions. The scene is set for a memorable time.

Finally, we would like to extend our sincere thanks to the University of Paisley, in particular to the Division of Computing and Information Systems and its staff, for the financial support, practical help, and moral support that you have given us in the organising of the conference.

Gillian Ragsdell
Daune West
Jennifer Wilby
December 2001

Table of Contents

Systems Theory and Practice in the Knowledge Age

Section One

Systems Theory

EMERGENT OR JUST UNEXPECTED?

Joyce Fortune and Diana White

Centre for Complexity and Change
The Open University
Walton Hall
Milton Keynes
MK7 6AA
J.Fortune@open.ac.uk

INTRODUCTION

Discussion within the systems movement about the nature of 'emergence' dates back at least as far as Ashby (1956; 1968) and Bertalanffy (1968) but the concept remains elusive. Over twenty years ago Checkland seemed to provide a more tangible way of viewing it when he linked it firmly to hierarchy and defined it by reference to different hierarchical levels of organization where each particular level is 'characterised by emergent properties which do not exist at the lower [less complex] level' (Checkland, 1981). Since then, however, the term has been given a whole new set of meanings. These almost always draw attention to the notion of interaction as a cause of emergence but they also emphasise changes over time and concepts such as unpredictability and counter-intuitiveness which are not specifically linked to the systems vocabulary. For example, Casti (1997) states:

> Emergence is the overall system behaviour that comes out of the interaction of many participants. It is *behaviour that cannot be predicted or even envisioned* from a knowledge of what each component of a system does in isolation.

Hebel (1999) provides the following definition:

> An emergent property of any system is the result of a collective interaction of components, although it is often an unintended consequence. It can be singular or multiple and have positive and negative effects.

Harrington *et al.* (1999) provide a 'down to earth' version aimed at quality management practitioners:

> It is the emergent properties that give a system life, and it is both the nature of the elements and their interrelationships that give rise to the emergent

> properties. ... Which emergent properties are useful depends on the nature of the system. But here is the "dark side of the force": not every emergent property is useful – some are either wasteful or dangerous. Some systems do unexpected things, show surprising behavior, or result in unintended consequences.

These definitions emphasise interaction but none of them contain the notion of hierarchy. Are they, therefore, talking about something different from the concept used by Checkland (op cit) and others who shared his view such as Young (1964), Wilson (1984) and Bergandi (2000)? And if they are not, is the emergent property or behaviour referred to something that can really be illuminated through use of the concept of 'system'?

This paper questions the value of labelling unpredictable and 'unenvisionable' behaviour as emergent. Using a phenomenon widely reported in the risk literature it suggests that some behaviours and properties which could easily be cited as evidence of emergence may owe nothing to the concept of hierarchy and have explanations which lie wholly outside the field of systems.

UNEXPECTED OUTCOMES OF SAFETY MEASURES

Many safety measures such as the introduction of design improvements and greater regulation fail to live up to expectations even when there is almost unanimous agreement that they are highly relevant and appropriate. For example, a review by Adams (1988) of accident statistics across the world found that in spite of the introduction of numerous safety measures the number of accidents had not declined. Research by Peltzman (1975) found that traffic safety regulations did not affect the total number (drivers, passengers and pedestrians taken together) of accident fatalities. A study by Hurtwood (1968) reported that fewer accidents took place in adventure playgrounds than in those containing ordinary (safer) fixed equipment.

A particularly well-researched example is a study commissioned by the German Federal Ministry of Transport into the effectiveness of anti-lock braking systems (ABS) in reducing traffic accidents. This study, undertaken by Aschenbrenner and Biehl (1994) with the help of a Munich taxicab company, found that drivers 'utilized ABS to their advantage, but no improvement in the accident loss per time unit of exposure to traffic could be seen' (Wilde, 1994). It is this example that will be used in this paper. It represents behaviour which was not "predicted or even envisioned from a knowledge of what each component of a system does in isolation" and which appears to have "come out of the interaction of many participants" and thus, on the face of it, meets Casti's definition of emergence.

Figure 1 shows a systems map of the situation Aschenbrenner and Biehl addressed. Cabs with and without ABS but identical in all other respects were assigned randomly to drivers from a Munich taxicab firm. The cabs were observed over a period of three years. In addition, 10 cabs with ABS and 10 non-ABS cabs were installed with instruments to measure acceleration and deceleration. Trained observers were used to evaluate the cab drivers' behaviour but the drivers were not aware when their behaviour was being observed. Unlike the drivers, the observers did not know whether they were in cabs with or without ABS.

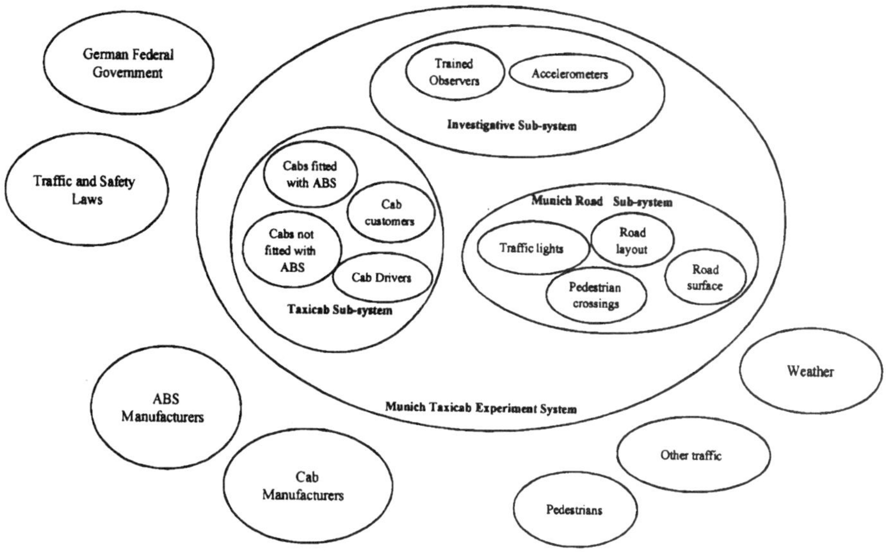

Figure 1. Systems Map of the Munich Taxicab Experiment System

Findings showed that:

> Drivers of cabs with ABS made sharper turns in curves, were less accurate in their lane-holding behaviour, proceeded at a shorter forward sight distance, made more poorly adjusted merging manoeuvres and created more "traffic conflicts". ... [Furthermore,] among a total of 747 accidents incurred by the company's taxis during that period, the involvement rate of the ABS vehicles was not lower, but slightly *higher*, although not significantly so in a statistical sense (Wilde, 1994).

This paper will now look at a number of explanations for this 'emergent' behaviour. All except the last have been prompted by a study of the risk and accident literature.

POSSIBLE EXPLANATIONS FOR THE UNEXPECTED OUTCOME

Explanation 1: Human Error

The cause that is most commonly cited in explanations for accidents is 'human error'. However, there are strong arguments that this term seldom provides sufficient explanation unless one accepts that people will always, at random, make mistakes because they are only human.

Explanation 2: Lulling

It has been suggested that safety measures may have a 'lulling effect', leading people to develop a false sense of security because they overestimate the improvements the safety measures will bring. (See Viscusi and Cavallo (1994)). For example, drivers may take less care and drive faster if their cars are fitted with air bags (Viscusi and Cavallo (op cit)) and 'pedestrians and cyclists wearing fluorescent or reflective clothing may cycle or walk in places and in ways in which they would not otherwise if they felt invisible to motorists' (Hillman *et al.*, 1990).

Explanation 3: Crossover Effect

There is also evidence that changes in the safety of one product can lead individuals to modify their behaviour with respect to an entire class of similar products. For example, Viscusi (1984) found that after the introduction of safety caps on medicine bottles parents became increasingly prone to leaving the bottles lying around and soon adopted a lax storage policy for all medicines regardless of whether their containers had safety caps. Similarly, tests with a new 'child resistant' cigarette lighter conducted by Viscusi and Cavallo (1994) found that parents took fewer precautions against a child tampering with the 'safer' lighter than with a conventional model.

Explanation 4: Reduction in Learning

It has also been suggested that safety measures prevent people learning directly from their own mistakes. For instance: during an emergency stop the restraining effect of seat belts prevents motorists from appreciating the full effect of hitting their heads on the windscreen; and children who fall off a slide on a well-padded playground floor will not learn to appreciate the dangers of falling from a height (Adams, 1988). It follows from this that if safety measures are introduced the opportunity for learning will be reduced and this may be translated into an increase in accidents which negates the safety gains.

Explanation 5: Risk Compensation

Another possible explanation for safety measures failing to achieve their objectives is 'risk compensation'. This theory evolved from Wilde's work on driver behaviour and traffic accidents (see, for example, Wilde (1976; 1982a; 1982b)) in which he proposed the theory of risk homeostasis. Adams (1988), although accepting that Wilde's model was un-testable, described the theory as plausible and built on it to develop the theory of risk compensation (Adams, 1995).

Adams argues that everyone has a propensity to take risks and that this propensity is influenced by the potential rewards of the risk taking. In addition, perceptions of risk are influenced by experience of past accidents. Risk taking decisions therefore represent a balancing act in which perceptions of risk are weighed against propensity to take risks. Adams further contends that every individual has a level of risk with which he or she feels comfortable. (He likens this to the setting of a thermostat.) Therefore, any safety intervention that does not affect the setting of the 'thermostat' is likely to be frustrated by behavioural responses that reassert the level of risk with which that person was originally content.

A number of authors agree that safety interventions which decrease the feeling of risk act as motivators which result in behavioural adjustments that re-establish a comfortable (target) level of risk. (See, for example, Saunders and Wheeler (1991), and Shanteu (1992)). However, the literature also provides ammunition with which to challenge the notion of risk compensation. Numerous studies into the introduction of seat belts laws found no evidence of risk compensation. (See for example, Lund and Zador (1984), Rock (1993), and Shannon and Szatmari (1994)). McKenna (1985) also queries the conditions that underlie any compensating behaviour, suggesting, for example, that it is more likely that behaviour modification would occur from conspicuous safety measures that affect the user (for example studded tyres) than from measures that are not obvious and do not affect the user (for example penetration resistant windscreens).

Explanation 6: Cultural Theory

It has been argued (Cutter, 1993), that perception of risk is value-laden and culturally constructed so an individual (who may be the designer, implementer or user of a safety measure) will perceive risk in a way that is coloured by his/her own 'world view' (See also Thompson (1980)). Douglas and Wildavsky (1982) have proposed that risk decisions are influenced by an individual's morals, values, wealth and power and that the 'value' of risk has been passed down through the generations by ancestors who took risks in order to survive. They also suggest that this 'value' is taught at school and influenced by history, fiction, cultural heroes and the media with the result that competitiveness becomes socially desirable and cultural demands 'compel' individuals to take risks in order to gain self-respect and acceptance in society. However, in highly industrialised societies level of risk is reduced by the introduction of numerous safety measures so people deliberately take risks. At the extreme this may involve participating in pursuits such as sky diving and mountain climbing but in relation to negating safety measures it may be achieved by, for example, driving faster.

Explanation 7: Hierarchy

The system depicted in Figure 1 clearly has properties at the system level which do not exist at the sub-system level and properties at the sub-system level which do not exist at component level. For example, the Munich road sub-system is not able to transport passengers. However, observations such as these are no more profound than saying a pile of components from which a car could be made cannot be driven from A to B. They do nothing to explain the unexpected outcome.

CONCLUSION

This paper has drawn attention to two fundamentally different definitions of emergence: one which is closely related to hierarchical level; and one which arises from interaction but manifests itself mainly as unexpected properties or behaviour. It has also described a situation where the outcome of change has been more than unexpected – it has been counter-intuitive – and put forward a range of explanations for that outcome.

The authors would argue that almost all the explanations are plausible except for two: human error; and hierarchy. All the rest rely on interaction, usually between the people involved and the safety measures taken, but they are not concerned with hierarchy. Thus the situation exhibits emergence according to one form of the definition of emergence but not the other. Perhaps it is time to start to draw a more robust and more explicit distinction between the two.

REFERENCES

Adams, J.G.U. (1988). "Risk Homeostasis and the purpose of safety regulations", *Ergonomics*, 31(4):407-428.

Adams, J.G.U. (1995). *Risk*, UCL Press Ltd, London.

Aschenbrenner, M., and Biehl, B. (1994). "Improved safety through improved technical measures? Empirical studies regarding risk compensation processes in relation to anti-lock braking systems", in *Challenges to accident prevention: The issue of risk compensation behaviour*, (R. M. Trimpop and G.J.S. Wilde, eds.), Styx Publications, Groningen, the Netherlands.

Ashby W.R. (1956). *An Introduction to Cybernetics*, Chapman and Hall Ltd., London.

Ashby W.R. (1968). "Principles of the self-organizing system", in *Modern Systems Research for the Behavioral Scientist*, (W. Buckley, ed.) Aldine Publishing Company, Chicago.

Bergandi, D. (2000). "Eco-cybernetics: the ecology and cybernetics of missing emergences", *Kybernetes*, 29(7/8):928-942.

Bertalanffy, L. Von, (1968). *General System Theory. Foundations, Development, Applications*, Allen Lane. London.

Casti, J.L. (1997). *Would Be Worlds*, John Wiley and Sons, New York.

Checkland, P. (1981). *Systems Thinking, Systems Practice*, John Wiley and Sons, Chichester.

Cutter, S. (1993). *Living with Risk: Geography of Technological Hazards*, Edward Arnold, Sevenoaks.

Douglas, M., and Wildavsky, A. (1982). *Risk and Culture: an Essay on the Selection of Technological and Environmental Dangers*, University of California Press, Berkeley.

Harrington, H.J., Carr, J.J., and Reid, R.P. (1999). "What's this "systems" stuff, anyhow?", *The TQM Magazine*, 11(1):54-57.

Hebel, M. (1999). "World-views as the emergent property of human value systems", *Systems Research and Behavioral Science*, 16:253-261.

Hillman, M., Adams, J. and, Whitelegg, J. (1990). *One False Move - A study of Children's Independent Mobility*, Policy Studies Institute, London.

Hurtwood, Lady Allen of (1968). *Planning for Play*, Thames and Hudson, London.

Lund, A.K., and Zador, P. (1984). "Mandatory belt use and driver risk taking", *Risk Analysis*, 4(1):41-53.

McKenna, F.P. (1985). "Evidence and assumptions relevant to risk homeostasis", *Ergonomics*, 28(11):1539-1541.

Peltzman, S. (1975). "The Effects of automobile safety regulations", *Journal of Political Economy*, 83(4):677-725.

Rock, S.M. (1993). "Risk compensation and the Illinois seat belt use law", *Accident Analysis and Prevention*, 25(5):537-544.

Saunders, R. and, Wheeler, T. (1991). *Handbook of Safety Management*, Pitmans Publishing, London.

Shannon, H.S. and, Szatmari, P. (1994). "Seat belt legislation and risk homeostasis: further analysis of the British data", *Accident Analysis and Prevention*, 26(6):803-805.

Shanteau, J. (1992). "Decision making under risk: applications to insurance purchasing", *Advances in Consumer Research*, 19:177-181.

Thompson, M. (1980). "Aesthetics of risk: culture or context", in How Safe is Safe Enough?, (R.C. Schwing, and W.A. Albers, eds.), Plenum Press, New York.

Viscusi. W.K. (1984). "The lulling effect: the impact of child-resistant packaging on aspirin and analgesic ingestions", *American Economic Review*, 74(2):324-327.

Viscusi, W.K. and Cavallo, G.O. (1994). "The effect of product safety regulation on safety precautions", *Risk Analysis*, 14(6):917-930.

Young, O.R. (1964). *General Systems: Yearbook of the Society for General Systems Research*, 9:61.

Wilde, J.S.G. (1976). "The risk compensation theory of accident causation and its practical consequences for accident prevention", paper presented to the Annual Meeting of the Osterreichische Gesellschaft fur Unfallchirurgies, Salsburg.

Wilde, J.S.G. (1982a). "The theory of risk homeostasis: implications for safety and health", *Risk Analysis*, 2(4):209-225.

Wilde, J.S.G. (1982b). "Critical issues in risk homeostasis theory", *Risk Analysis*, 2(4):249-258.

Wilde, J.S.G. (1994). *Target Risk*, PDE Publications, Toronto.

Wilson, B. (1984). *Systems: Concepts, Methodologies and Applications*, John Wiley and Sons, Chichester.

SAWU BONA: SYSTEMS THEORY IN DESIGN

Johann van der Merwe

Faculty of the Built Environment and Design
Cape Technikon, South Africa
johannv@ctech.ac.za

INTRODUCTION

Systems theory and practice in design will achieve nothing of lasting worth if not predicated on human values and human understanding – if the social system we all have to deal with is not allowed to either affirm or deny that which we are engaged in. The basis of that which we are engaged in happens to be the raw (social) material, if you will, of all systems theory and practice, or at least those aspects that have a direct influence on the discipline of design, hence on design communication, design research and on the teaching of design theory and practice. That basis is the social structure from within which design draws its reality and reason, and, I would presume, that same basis would also serve to authenticate systems theory and practice. The correspondences between design theory and systems theory – and more specifically action learning - points to one of the strengths of both, namely adaptability in the face of different cultural conceptions of reality.

Yet there are differences in cultural ways of seeing that could shed new light on our own specific ways of seeing and understanding. One such aspect of another culture is used to introduce the main text of *The Fifth Discipline Fieldbook* (Senge, *et al.*, 1994:3). It is the simple yet wide-ranging Zulu greeting, *Sawu bona*! This culturally specific greeting is more of an acknowledgement, however, since it means *I see you*. Instead of wishing you an extrinsic *good morning*, by saying *Sawu bona* I bestow upon you cultural and intrinsic existence. Systems theory in/and design can only be this acknowledgement of *I see you*, this reciprocal acknowledgement between systems thinking/design and the social bases of everyday reality, while the traditional reply *Sikhona* (I am here), conveys the proper response to this acknowledgement, namely that now the initial positions have been established, we may start to talk to one another. Von Foerster (Waters, 1999) says when we use language it is an invitation to dialogue, an invitation to dance – "Togetherness is the point in a dialogue." Senge *et al.* (1994:4) voiced this same sentiment: "You could argue that we invoke each other's potential by our willingness to see the essence of each other." To von Foerster and Glanville (2000), both second order

Edited by Ragsdell *et al.*, Kluwer Academic/Plenum Publishers, New York, 2002.

cyberneticists, this *willingness* revolves around conversation, the humanness of the interaction leading to unfeigned communication. Von Foerster's example of a conversational dance is quite apt, since a true dialogue needs the willing interaction of both sides, and as in a dance, "when we are talking with each other, we ... invent what we both wish the other would invent with me" (Waters, 1999). In this respect Glanville (1999) also introduces the concepts of generosity and creativity, and argues that agreement and understanding requires both, but that this would fail "unless we are prepared to look well on what our conversational partner offers us." *Sawu bona*, indeed.

This notion of reciprocity is at the heart of design (theory, research, practice) and systems theory in design. As individuals we already live within a social system of constructed reality. Within this field we go about our business, dependent largely upon tacit knowledge, which is itself already a system *of* thought, within a larger system of social behaviour, *for* thought. This work will therefore examine some of the ways a designer may view and make use of systems theory, as influenced by the sociological and phenomenological models of experience.

SYSTEMS INTO ACTION

In design understanding, the whole is more important than the parts. Learning what our perceptions may mean involves not a knowledge of different viewpoints so much as experiencing the interaction of these viewpoints in a continuous and collective field of activity wherein significant relations are exposed through focusing on asking questions instead of providing definitive answers. This fluid dialogue between what is known and what is not known is a socialisation process that includes tacit knowledge transfer from person to person. From lower-order object and action recognition people can move to higher-order cognitive skills acquisition in this field wherein knowledge is interchangeable and capable of transformation via creative chaos and diversity. The transformative power of abstract thinking leads to the real-world application of concepts, in order to make sense of real experiences and their meanings (Lundberg, 1999; Pasca and Pietroni, 2000; Schuck, 1996).

It is precisely because we have to deal with the real world that systems thinking and practice becomes a designed tool for thought and action. The world is an unformulated space filled with Rittelish wicked problems. When designers speak of systems theory we really mean soft systems methodology that deals with real-world problems that can only be described as ill-defined, obscure, and where problems present themselves through exhibiting a distinct mismatch between existing situations and preferred ones (here my paraphrasing interweaves Peter Checkland's and Herbert Simon's views on systems thinking and design thinking respectively: what both are essentially saying is that systems theory and design has this purpose – to change existing situations into preferred ones). Soft systems investigate models of social reality that are far from *given*, because this alternative phenomenological and ever-changing world is endowed with meaning by social actors who create and recreate their perceived reality. When speaking of a social *system*, therefore, the intention is not to claim it as *systemic*, precisely because it is not a given that can be investigated using hard systems methodology, nor is it a new version of functionalism. Social dynamics cannot be described unambiguously as goal orientated, thus its problems are not as easily identified as in natural science systems (Checkland, 1981). Hard systems methodology does not allow for complexity and uncertainty, which observational point is necessary in dealing with problems that follow a pattern of mutability because of their complex social structure.

Systems theory is in agreement with action learning in that both make use of a specific learning paradigm in which questions are more important than answers, and the social learning process provides a mutually enriching environment. Paradoxically this environment involves the dynamics of recognised ignorance, or the knowledge that people know too little about problem situations. According to Checkland (1981) this will appeal to all disciplines that study social interactions from a vantage point inside the system, and it is here that action learning demonstrates that when we start to learn from each other, under these circumstances, we discover that nobody really knows the answers (Revans, 1985). To be in a state of anomalous ignorance is like being in the sea with nothing to drink. Since anomalous means deviating from the norm or the expected, and finding yourself in a situation of plenty but useless for complex problem solving purposes, start with the known and familiar, since only through this gateway can you glimpse the usefulness of the unexpected and find information to the unknown that can combat a state of knowing ignorance.

It is important to understand that we are more culturally bound than we realise, and our language use often determines the way we deal with information and knowledge acquisition. Anything outside the norm, and a social system is also a normative system, requires a different mindset and an extension to what is culturally known. The problem is thus one of formulating questions, in this case, in order to determine the "properties of something which ... is not yet culturally known" (Eco, 1976:249), yet it is possible to "construct the unknown or unknowable in terms which are familiar" (George, 1979:129). The familiar is the vast stock of social knowledge that everyone has access to, and hence may share as existing knowledge. Human systems are structured, hence are fluid structures that require constant maintenance. Restructuring brings innovation and renewal, and a social system may find its new configuration, not specifically in any particular property of any one constituent part, but in the new perspective gained by the system as a whole if it allows this *new* property (which is usually a familiar and always-been-there property) to influence the direction the system takes in its development, a development towards the as-yet-unknown.

DECONSTRUCTING THE SYSTEM

One solution that systems thinking shares with action learning is team work, and especially the dynamics of compromise (Newport, 1998), wherein the notion of multiple perspectives is put to the test (a foretaste of the so-called *unknown* or not-quite-so-familiar). Deconstructing the system means breaking free of old habits of thought, and means deconstructing (aka seriously questioning the basis for knowledge) the paradigmatic models that use exemplars (tradition) and normative reasoning to shape our world (Bermudez, 1996). Since we rarely break away from this scenario, by ourselves, we need the dynamics of group compromise to break the frame. Again, *ignorance* enters the equation, but in the sense that so-called unqualified people, that are ignorant about your field or speciality and thus outside your frame of reference, may provide just the type of fresh insight and approach necessary to help you give up your present mode of searching for solutions (Braddick and Casey, 1996; Revans, 1980).

Braddick and Casey's (1996:237) notion of the set adviser's role in action learning is wonderfully deconstructive in nature: "to encourage iconoclastic smashing of frames by encouraging meta-questions about how the project is being done. These meta-questions have the power to jump the problem" from one level of reasoning to a next higher and more insightful level. How does one *jump the problem*? Deconstructively breaking the frame takes courage when you see yourself cutting the towrope that connects the dinghy you are in from the mother ship. The mistake we make is in thinking that the object we are looking at, the mother ship, is the social stock of knowledge we depend on for survival. It is not. Our social matrix lies beneath our feet, or as in this example, the ocean we are drifting upon, which will guide our every move, which forms, at the same time, the past, the present, and the future. The idea remains the same: stop looking in the direction of the so-called problem (because there is nothing, literally, yet, in the future), and focus on everything and anything that is at hand to help you towards a possible solution. Abstract thinking means defocusing, and then refocusing, rather, on general principles and theories instead of on specific instances; it means focussing on the internal structures and features of social contexts.

Dewey (1916a) would have called the mother ship our mental habits and emotional characters regulating our behaviour. The social support base you use for your systems methodology geared towards discovery, is at once an investigative tool and a pattern of behaviour leading to ways of seeing and eventually to ways of knowing. Your choice becomes a method of action and interaction that regulates the conditions necessary to solution-finding – but this is only possible through action that has intelligent direction. Your behavioural action takes its direction from the surrounding social and contextual issues by yielding to its inherent forces. Society is the *standing effective agency* that directs this activity of discovery, "the social medium in which an individual lives, moves and has his being," and an intelligent direction thus depends on yielding to this interaction by taking cognisance of the value of social conditions and observed relations, and consequently planning and executing action in view of this knowledge (Dewey 1916b; 1933). This journey or method of discovery must depend on the every day working logic of social reality, yet design cannot depend on so-called *normal* logic, since this term usually indicates the way things have always been done, traditionally and safely.

Thus the design problem involves a *logical gap*, and a journey of discovery is concentrated mainly on crossing this logical gap (or asking meta-questions to *jump the problem*), allowing us to "gain a foothold at another shore of reality" (Polanyi 1962:123). If we intelligently use our social environment the "hidden aspects of the unknown" can be seen as simply the known data turned pattern, clues and pointers to the so-called unknown solution (Polanyi 1962:120–128). Being creatures of habit should, in fact, be a help and not a hindrance, since finding *the new*, or *insight* into the unknown, is often a matter of finding a *stimulus pattern* that shares structural qualities with the problem being investigated, the unshared pattern structures helping to reintegrate (recognize) the unknown parts of the problem (Dreistadt, in Langley and Jones 1988: 185). Using these patterned habits of intelligent application allows action learning to take place, but only if there is a *purposive tension* at work between the perceived (everyday) structure of the problem and the looked-for new knowledge, between a stage of perplexity and a stage of finding your way (Polanyi 1962: 123). This cognitive dissonance (Duit and Confrey 1996: 83-84) is quite close to what Schon (1987) means by "indeterminate zones of practice – uncertainty, situations of confusion and messiness where you don't know what the problem is." In searching for design solutions you should not look at the object but at the process. Designers should use a paradoxical way of seeing turned into a way of knowing, and enquire about that which they do not know by looking at that which they do know, only not directly *at* but *through* what they know. According to Ohlsson (Langley and Jones 1988: 181) unsolved problems create (logical) gaps "between one's current state and the goal state." Under tension restructuring occurs, and this means looking differently through the descriptive problem space for a problem which demands a different way of looking, of seeing the problem and its environmental context. Systems thinking requires a look at the whole scenario, which may equate with this description space, and demands a restructuring of the familiar, of the normative, to find the gaps that exist between the known and the as-yet-unknown. For this one needs the conditions of ignorance that action learning requires in order to ask searching questions that can unbalance the existing structures. Design practitioners and design researchers have been doing exactly that – finding new ways of looking at things. The next section reports on some of the leading issues that are being debated in contemporary design research, and it is notable that the main language use is couched in a systems thinking mode.

PERSPECTIVES FROM THE REAL WORLD

The practice of design and consequently researching design has become anything but a straightforward exercise in linear problem solving. Increasingly, design tasks are being described as ill defined, the reason being that society itself faces ever increasing complex situations through greater change. These levels of complexity mean that design researchers need to adopt a different approach to design knowledge that will enhance its contemporary content value, especially since many new tasks have to be designed as part of the project. Researchers must acquire the skills necessary in order to deal with the notion of complexity and its related areas of uncertainty, systems theory and multidimensional knowledge (Hohmann and Jonas, 2000; Manzini, 2000; Pizzocaro, 2000; van der Merwe, 2000). Design needs to understand the interaction between itself, as design entity focused on the understanding of product function, and society at large. This complex interconnectivity between researchers and the dynamics of social objectives cannot be fully experienced or understood from within the confines of traditional discipline parameters. The pace of contextual change brings with it the challenge of constant interaction with the design environment by refocusing design thinking from product to process. The role of shifting social variables that cannot be planned for becomes part of the research structural plan, and once having understood this factor the researcher may turn uncertainty into a stochastic process. This process, that incorporates an element of probability or chance with intentionality, and illustrating the perennial struggle between tradition and change, admits that research creativity moves along a continuum between reason and imagination, admits that chance really is a fine thing in researchable terms. Systems thinking, alone, accommodates the unthinkable, and complexity becomes a collective experience acting as a new gateway to reality (Hohmann & Jonas, 2000; Margolin, 2000; Pizzocaro, 2000; van der Merwe, 2000).

In order to adapt their working methods to social reality, design researchers will have to substantially revise their interpretive frameworks, or mental models of reality, by means of the holistic structure of systems thinking, since this theory facilitates the change necessary to review assumptions about the world. The notion of complexity highlights the uses of multidimensional knowledge, and the conceptual tools inherent in this concept allow the questioning of established design theories and practices. The activity of design appears much like a flexible network, affording design theory and practices multiple viewpoints that allow perspective shifts and the free circulation of ideas. In the new knowledge economy design takes on the form of a co-producer of knowledge and meaning, a systematic activity looking to social co-producers to complete the extended frameworks of understanding – a continuous and innovative conversation in which the new meaning can emerge (Manzini, 2000; Pizzocaro, 2000; van der Merwe, 2000).

Designers and design researchers will thus learn the quality of real-time adaptability while work is in progress, by learning how to be open to contextual signals from outside the immediate confines of the research project, inputs which have a substantial bearing on the research outcome. In this respect it is suggested that specific questions asked at the outset of the research project may mislead the researcher / designer, while inaccurate questions and methods, which may seem somewhat vague and likely to generate inaccurate answers, may in fact help redefine the questions and methods used (Dominoni and Trabucco, 2000).

CONCLUSION

Herbert Simon wrote "Few engineers and composers... can carry on a mutually rewarding conversation about the content of each other's professional work. What I am suggesting is that they *can* carry on such a conversation about design, can begin to share their experiences of the creative, professional design process" (in Cross, 2000). Our systems of language very often determine how we see the world, depending on what the observer is prepared to see and, crucially, depending on the observational position. Those very same systems of communication can be used to bring people together, to get them talking to one another in the spirit of *Sawu bona*! Design is one such system that can facilitate the process of investigation. Most people find the assimilation of new knowledge difficult to handle without some form of visual structure. By using the term *visual* I do not mean to suggest that only what can be seen can be fully understood, but that design methodology as a vehicle for thought has a way of structuring even abstract material, and a method of logical organization that uses a type of visual intelligence that can be articulated via various methodological practices and theories, including systems theory and practice.

I would like to suggest that design thinking – systems theory for practice – is a natural method towards systems awareness. Design, by its very nature, is geared towards the holistic whole while recognising and working with the constituent parts. Design is a reciprocal conversation with objects, events and, above all, with people. A system of design as action learning – design thinking – is a way to deal with change in a world of ambiguity and complex uncertainty.

REFERENCES

Bermudez, J. (1996). *On Paradigms & Avant Garde: Peeking into the Architectural Mind*, URL:www.arch.utah.edu/people/faculty/julio/parad.htm

Braddick, B., and Casey, D. (1996). "Developing The Forgotten Army: Learning and the Top Manager," in *How Organizations Learn*, (K. Starkey, ed.), International Thomson Business Press, London.

Checkland, P. (1981). *Systems Thinking, Systems Practice*, John Wiley, Chichester.

Cross, N. (2000). "Designerly Ways of Knowing: Design Discipline versus Design Science," in *Design (plus) Research*. Proceedings of the Politecnico di Milano conference. May 18-20, 2000. (S. Pizzocaro, A. Arruda and D. de Moraes, eds.), Politecnico di Milano, Milan.

Dewey, J. (1916a). *Democracy and Education*. URL:www.ilt.columbia.edu/academic/texts/dewey/d_e/chapter2.html

Dewey, J. (1916b). *Democracy and Education*. URL:www.ilt.columbia.edu/academic/texts/"ewey/d_e/chapter3.html

Dewey, J. (1933). *The Quest for Certainty*. URL:www.marxist.org/reference/subject/philosophy/works/us/dewey.htm

Dominoni, A., and Trabucco, F. (2000). "Project Research: When the Project is Inherent to the Research," in *Design (plus) Research*. Proceedings of the Politecnico di Milano conference. May 18-20, 2000. (S. Pizzocaro, A. Arruda and D. de Moraes, eds.), Politecnico di Milano, Milan.

Duit, R., and Confrey, J. (1996). "Reorganizing the Curriculum and Teaching to Improve Learning in Science and Mathematics," in *Improving Teaching and Learning in Science and Mathematics*, (D. F. Treagust, R. Duit and B. J. Fraser, eds.), Teachers College Press, New York.

Eco, U. (1976). *A Theory of Semiotics*, Indiana University Press, Bloomington.

George, F. (1979). *Man the Machine*, Granada, London.

Glanville, R. (1999). "A (Cybernetic) Musing: Thinking the New Millennium," in *Cybernetics and Human Knowing*. 6(4):65-71. URL: www.imprint-academic.demon.co.uk/C&HK/vol6/v6-4ind.html

Glanville, R. (2000). "A (Cybernetic) Musing: The State of Cybernetics," in *Cybernetics and Human Knowing*. 7(2-3):151-159. URL: www.imprint-academic.demon.co.uk/C&HK/vol7/v7-2-3ind.html

Hohmann, P., and Jonas, W. (2000). "Design Scenario Building: an Integrative Process Model for Projective Design Tasks," in *Design (plus) Research*. Proceedings of the Politecnico di Milano conference. May 18-20, 2000. (S. Pizzocaro, A. Arruda and D. de Moraes, eds.), Politecnico di Milano, Milan.

Langley, P., and Jones, R. (1988). "A Computational Model of Scientific Thought," in *The nature of Creativity*, (R.J. Sternberg, ed.), Cambridge University Press, Cambridge.

Lundberg, B.G. (1999). *The academic organisation: an approach to knowledge management*. URL: www.informatik.uni-bonn.de/~prosec/ECSCW-XMWS/

Manzini, E. (2000). "Design Research: Reasons and Possibilities," in *Design (plus) Research*. Proceedings of the Politecnico di Milano conference. May 18-20, 2000. (S. Pizzocaro, A. Arruda and D. de Moraes, eds.), Politecnico di Milano, Milan.

Margolin, V. (2000). "Building a Design Research Community," in *Design (plus) Research*. Proceedings of the Politecnico di Milano conference. May 18-20, 2000. (S. Pizzocaro, A. Arruda and D. de Moraes, eds.), Politecnico di Milano, Milan.

Newport, R. (1998). *Private Interview*, Birmingham.

Pasca, V., and Pietroni, L. (2000). "History of Design and Design Research: Toward a Multilinear History of Design," in *Design (plus) Research.* Proceedings of the Politecnico di Milano conference. May 18-20, 2000. (S. Pizzocaro, A. Arruda and D. de Moraes, eds.), Politecnico di Milano, Milan.

Pizzocaro, S. (2000). "Research, Theory and Design Culture: A Knowledge Growing within Complexity," in *Design (plus) Research.* Proceedings of the Politecnico di Milano conference. May 18-20, 2000. (S. Pizzocaro, A. Arruda and D. de Moraes, eds.), Politecnico di Milano, Milan.

Polanyi, M. (1962). *Personal Knowledge: Towards a Post-Critical Philosophy*, Routledge and Kegan Paul, London.

Revans, R.W. (1980). *Action Learning: New Techniques for Management.* Blonde and Briggs, London.

Revans, R.W. (1985). "Action Learning: Its Origins and nature," in *Action Learning in Practice*, (M. Pedler, ed.), Gower, Aldershot.

Schon, D. (1987). *Donald Schon's Presentation 'Educating the Reflective Practitioner.'* URL:euphrates.Stanford.edu/other/schon87.htm

Schuck, G. (1996). "Intelligent Technology, Intelligent Workers," in *How Organizations Learn*, (K. Starkey, ed.), International Thomson Business Press, London.

Senge, P., Roberts, C., Ross, R.B., Smith, B.J., and Kleiner, A. (1994). *The Fifth Discipline Fieldbook*, Nicholas Brealey, London.

Van der Merwe, J. (2000). "The Innovative Principle of a Design Language," in *Design (plus) Research.* Proceedings of the Politecnico di Milano conference. May 18-20, 2000. (S. Pizzocaro, A. Arruda and D. de Moraes, eds.), Politecnico di Milano, Milan.

Waters, C. (1999). "Invitation to Dance – A Conversation with Heinz von Foerster," in *Cybernetics and Human Knowing.* 6(4):81-84.
URL: www.imprint-academic.demon.co.uk/C&HK/vol6/v6-4ind.html

MAPPING THE STAGES OF THE DEVELOPMENT OF MANAGEMENT SCIENCES METHODOLOGIES IN THE UK: FROM SINGLE SOLVING METHODS TO MULTIMETHODOLOGY

Alberto Páucar-Cáceres

The Business School Manchester Metropolitan University
Aytoun Building, Aytoun Street
Manchester M1 3GH, England
a.paucar@mmu.ac.uk

INTRODUCTION

In this paper, we shall trace the development in the UK of some of the main Management Sciences Methodologies (MSM) since the 1940s to the present time. Checkland's first map of the systems movement showing the influences of the different methodologies is briefly described and a framework containing two dimensions: (a) time; and (b) development of paradigms in management sciences is then introduced and four stages/groups of methods/methodologies are identified: (i) problem-solving methods (1940-1960); (ii) improving-situation methodologies (1970-80); (iii) intervention-empowering emancipatory systems methodologies (1980-90); and (iv) multi-methodological and pluralism systems practice (1990-). The main features of a sample of these methodologies are outlined. Conclusions from the grouping proposed are drawn.

MAPPING THE SYSTEMS MOVEMENT

One of the first to map the development of the so-called systems movement (defined as a loose amalgamation of people who claim to use systems ideas in tackling real-world situations) was Checkland (1976). His map makes explicit the difference between the hard and soft paradigms and depicts the influences received by the then emerging soft systems methodology. Although this map was extensively used by systems practitioners and was very useful in the 80s for making clear the difference between soft and hard approaches, it was however a snapshot of the development of systems thinking and the systems movement up to that time and it does not include the later developments in systems thinking in the 80s and 90s.

Systems Theory and Practice in the Knowledge Age
Edited by Ragsdell *et al.*, Kluwer Academic/Plenum Publishers, New York, 2002.

The other important attempt in tracing the development of the systems movement came from Jackson and his systems of systems methodologies developed in the late 80s (Jackson, 1991). However this is, arguably, not a map of methodologies but a grid that relate the methodologies to their social science underpinnings. It does not show the influences received or exerted by the different methodologies in the late 80s. His account nevertheless takes into account more recent developments in the systems field and has also been used extensively during the last years. More recently, Mabey and Mayon-White (1993) have constructed a map including methodologies developed since the late 50s until the early 90s. Although this is a useful historical account of the development of systems methodologies in both sides of the Atlantic, the account is strongly biased towards the methodologies developed by the Open University systems group in the UK. This paper suggests a general framework to understand the historical development of the systems movement in the UK. This framework helps to situate the main systems methodologies developed over the last 60 years; based on this framework, the paper proposes a classification into the four general groups or stages and are discussed in the next section.

MANAGEMENT SCIENCES PARADIGMS OVER TIME

Using time and the paradigms developed in the systems movement over the last 60 years, a framework to show the developments of the systems movement in the UK was constructed as shown in Fig. 1. It depicts the emergence and development of the main Systems Methodologies in the UK, showing the major direct and indirect influences between them over the last six decades. The map depicts four main paradigms. The emergence of hard approaches is located in the late 50s and it has been associated with the developments of operational research (OR) in the UK and the USA and with the developments of systems engineering/systems analysis in the USA. During the 60s and 70s, a number of soft systems thinking methodologies emerged in the UK amongst them the more influential were Checkland's soft systems methodology (Checkland 1981; Checkland and Scholes, 1990) and cognitive mapping developed by Eden *et al.* (1983). In the late 80s and 90s, Critical Systems Thinking (CST) became prominent in the UK when 'total systems intervention' developed by Flood and Jackson (1991) embraced the CST commitments in systems practice. The other methodology that follows a critical approach is Ulrich's critical heuristics. Finally, the more recent debate in OR and systems communities in the UK is around the use of methodologies in combination and acknowledging various paradigms. We call this stage, Pluralistic/Multi-paradigmatic thinking.

These major methodological developments together with the main theoretical influences of the four paradigms over time are depicted in Fig. 1: (a) Problem solving methods (1940-1960); (b) Situation-improving methodologies (1970-1980); (c) Intervention-empowering emancipatory methodologies; and (d) Pluralistic/multimethodology practice (1990-). A brief description of a sample of these major methodologies follows in the next sections.

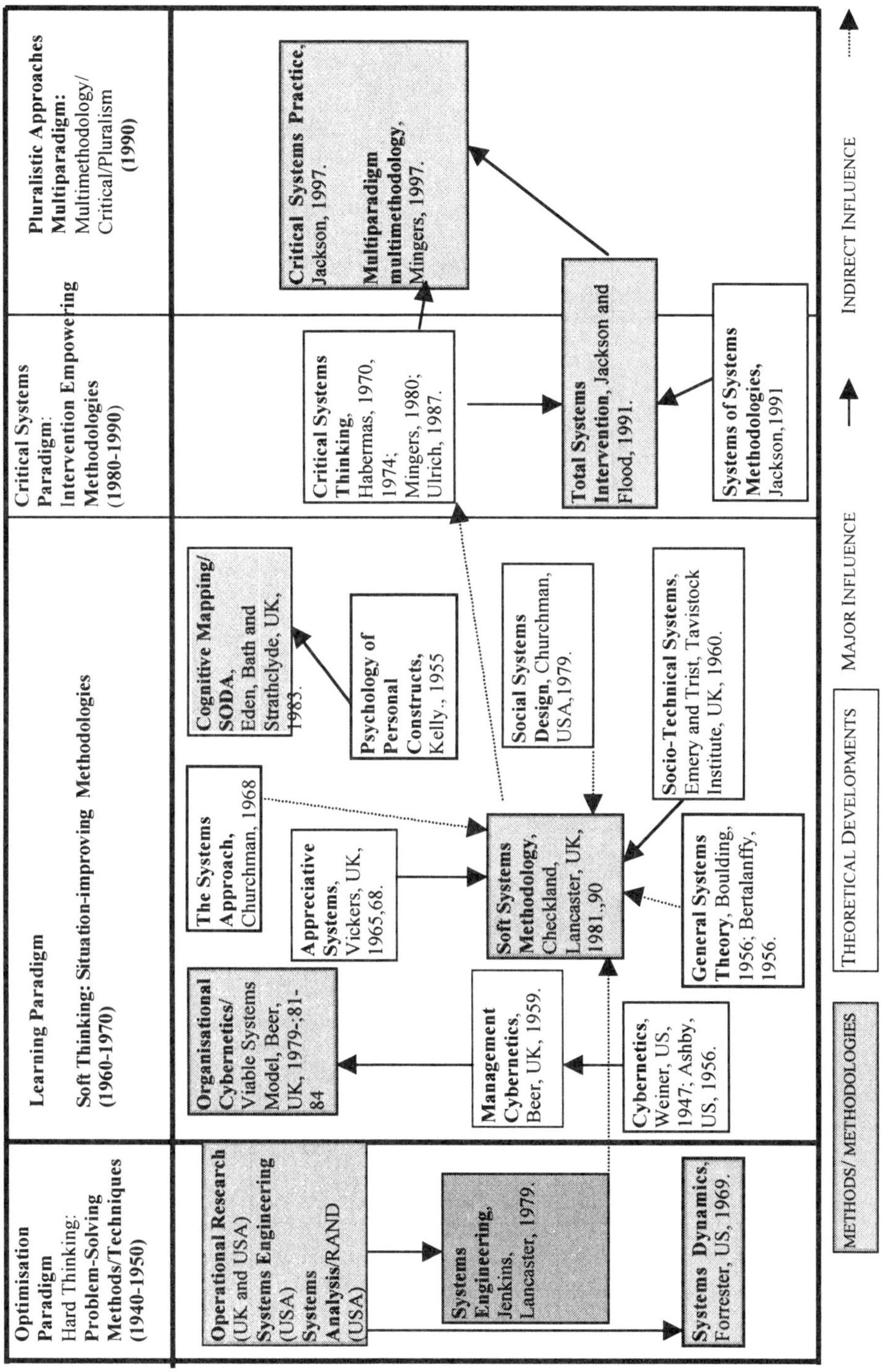

Fig. 1. The Systems Movement 1940/50–90. Development of major Mgt Science methodologies in the UK: Major and Indirect Influences.

Problem-Solving Methods: 1940-1960

The term 'Hard Systems Thinking' and its use in managerial situations were first identified by Checkland, who originally included system engineering and computer systems analysis into the category of hard systems. He identified another group under the umbrella of aid to decision making in which the RAND type of systems analysis is placed. Operational Research/management science can, arguably, be regarded as a type of decision-making aids.

Systems Engineering

The term Systems Engineering (SE) was first used by the scientist and engineers of the Bell Telephone Corporation in the United States in the 1940s. Although the methodological approaches of SE were used profusely after the war in the UK it was in the 50s that the Bell Corporation group refined the methodology and applied to a range of problems. It was conceived as a systematic approach to model building and model optimisation to meet a specific need. Checkland regards SE as an important influence on the development of SSM.

Systems Analysis

Together with the developments of Systems Engineering, the post war period witnessed the increasing use of a methodological approach called Systems Analysis. This approach has been associated largely with the work of the RAND Corporation and its work for the Department of Defence of the United States. Briefly this approach can be defined as a technique which incorporates the use of systems ideas and methods or techniques in organisational and government decision-making, Quade (1963). Since its early applications, Systems Analysis has became more associated with Computer Systems and although its applications to more general managerial problems is still present, methodologically has not evolved very much from its beginnings at the RAND Corporation.

Operational Research / Management Sciences

It is well known that OR started as an effort to improve military operations during the Second World War. It started in the UK from where it expanded to the United States. After the war the civilian application of OR become more a more popular in both sides of the Atlantic. Although linked to the hard approach paradigm and heavily criticised in the last decades OR has absorbed the criticism and been very influential on the systems movement and developed its own soft approaches that have been practiced under the umbrella of 'soft' OR.

Systems Dynamics

Systems Dynamics (SD) originally known as Industrial Dynamics is a creation of Jay Forrester in the 1960s in the Massachusetts Institute of Technology, (Forrester, 1961). SD is mainly a methodology which uses theory of information feedback and control in order to evaluate businesses. The basic idea underpinning this approach is that any complex situation can be described in terms of elements and flows; flows being the relationships between the elements. The main focus of the methodology is the structure (elements) and processes (flows) between the elements. This description constitutes the dynamic behaviour of the system. Essentially, SD aims to predict the behaviour of a system, it relies heavily on the use of a model which aims to contain the intricacies of the complex structure and the multiple feedback loops that link each element of the structure. SD has been seen as an outsider in the SM and perhaps most of its practitioners have been situated on the hard end of the systems approach. However, during the 90s, a number of SD and systems commentators have been making bridges between systems dynamics and the general developments of systems thinking, Senge, (1990); and Lane and Oliva (1994) amongst others.

Situation-improving Methodologies: 1970-1980

Over the last 50 years systems thinking in the UK has developed rapidly, particularly in two areas: the theoretical development of systems ideas; and the practical application of systems thinking to real-world situations. Within the latter area, soft systems thinking challenges the goal-seeking model followed by the hard approach and advocates the conscious use of 'systems' as mental constructs for perceiving a problematic situation with the view of bringing not solutions but improvement and learning from the intervention. Three methodologies are briefly described here: Checkland's Soft Systems Methodology, Beer's Viable Systems Model and Eden's cognitive mapping.

Checkland's Soft Systems Methodology

Peter Checkland's Soft Systems Methodology (SSM) is the most developed Systems Methodology in terms of its theoretical premises and philosophical underpinnings. It is also one of the most widely used in the UK and in other parts of the world (Mingers, and Taylor, (1992); Ledington, *et al*, (1997)). During the 1970s, Checkland and his colleagues at Lancaster University questioned the use of hard systems thinking to real-world situations and started to test a new methodology that shifted the systemicity from the real world to the process of enquiry itself. SSM articulates a learning process which takes the form of an enquiry process in a situation that people are concerned. This process leads to action in a never ending learning cycle: once the action is taken, a new situation with new characteristics arises and the learning process starts again.

Organisational Cybernetics

Beer has made enormous efforts over the years to persuade management science to move out from the traditional use of the company organisation chart as a model of the organisation, Beer (1966, 1979, 1981). For him a system is viable if it is capable of responding to complex environmental changes. He proposes the use of a Viable System Model (VSM) which is a model of the organisational features of any viable system. He argues that the human body, controlled by its nervous system is the most known control system (neurocybernetic systems containing five-level hierarchy of systems) from which we can develop cybernetic principles that can be applied to any organisational system. VSM can be used to diagnose whether an organisation is built according to cybernetic principles.

Cognitive Mapping

Developed by Eden and colleagues (Eden, et al, 1983, Eden 1989), cognitive mapping is a technique that stems from cognitive psychology. Eden's concern is to use cognitive maps to understand the way people construct their thoughts. The methodology to accomplish this by interviewing people and try to sketch out a 'map' or model of their thinking by understanding the concepts they use and by trying to comprehend the way these concepts are linked together. The approach is accompanied by a method called SODA (Strategic Options Development and Analysis) which aims to support the process of shared appreciation of the different ways people involved perceive the situation. SODA aims to encourage the consultant to develop two types of skills: (a) to be a faciltator of the *process*; and (b) to be able to construct a model and analyse the *content* of the views of each member. SODA shares some views with SSM in the sense that that it has its foundation on subjectivism because assumes that each member in the group hold his or her personal subjective view of the 'real' problem. SODA manages the process and the content of the intervention, (Eden, 1989).

Intervention-empowering Emancipatory Methodologies: 1980-1990

During the 80s, a new set of methodologies based on a Critical Systems Thinking (CST) appeared in the UK systems movement. It can be argued that the main feature of these approaches is that they try to empower the actors in the intervention. Two of the main approaches based on CST are briefly outlined here.

Total Systems Intervention

Total Systems Intervention (TSI) is a relatively new development in management science and has come to the management science arena mainly through the work of Jackson and Flood in Hull/Humberside in the late 80s and 90s. Flood and Jackson (1991), define Total Systems Intervention (TSI) as *a new approach to planning, designing, "problem solving" and evaluation* .The main philosophy behind TSI is 'critical systems thinking' and it stands on: (1) Complementarism. TSI argues that systems studies must use different management methodologies from different paradigms in a complementary way: alternative theoretical positions should be respected; (2) Social awareness. In any intervention, the practitioner(s) needs to be aware of the sociological consequences that a particular methodology entails; and (3) Human well-being and emancipation. This is based on the work developed by Habermas on human cognitive interests.

Critical Systems Heuristics

Developed by Ulrich (1987) Critical Heuristics of Social Planning, this approach aims to equip planners and citizens concerned with a method which will reveal the 'normative content' of any proposed plan. The method will aim to make the planners' intentions transparent to the people involved and will equip the latter with arguments to test the planners 'boundary judgements'. The philosophy behind CSH is Kant's conceptualisation of these three words: (a) *critical* approach implies that planners should make transparent to themselves and others the normative content of the planning design; (b) *system* in the sense that CSH will try to capture the totality of relevant conditions of the problematical situation; and (c) *heuristics* is the process of uncovering deceptions and messy situations in order to help planners and people concerned. Ulrich and his CSH contribution to management sciences has been greatly acknowledged as a major development of critical systems thinking, particularly to the emancipatory strand.

Multi-paradigmatic Thinking: 1990-

During the 90s there has been a great debate in the systems community around issues concerning the use of more that one methodology (combinations of them or parts of them) when intervening in complex situations. The general term of multimethodology has been coined to group this practice. The theoretical and practical problems of multimethodology are still being debated strongly nowadays within the UK Systems Society. One of the main problems facing the practice of *multi*methodology is that the practitioner needs to 'migrate' from one paradigmatic stance to another in order to fully understand (being conversant) and apply it. This section will discuss the two more developed current approaches to multimethodology: (a) pluralism/complementarism as initiated by Flood and Jackson and lately developed into 'coherent pluralism' by Jackson; and (b) multi-paradigm multimethodology/Critical pluralism developed by Mingers. Jackson's approach is briefly outlined in the next section.

Complementarism/Pluralism/Critical Systems Practice

The history of multimethodology in the UK is, in many respects, closely related to the development of CST and the methodological pluralism proposed mainly by Jackson and Flood at Hull and Humberside Universities. The Systems of Systems Methodologies, Jackson (1992) stated clearly that a pluralistic approach in management sciences was needed and disregarded isolationism, imperialist and pragmatist tendency within the systems movement.

Jackson's pluralism was originally called 'complementarism', which was introduced as part of the TSI approach. In a recent paper, Jackson (1999) explains the reasons behind his reverting to the word pluralism and states his position in what he calls 'coherent pluralism' in management sciences. In this later paper Jackson defines pluralism as:

> ...a response to many methods, models and techniques (tools), and methodologies, developed by management scientists... [t]he point of pluralist thinking and practice is to make the best use of the tools and methodologies by using them in a way that allows us to continually improve them (do research) and, at the same time, improve our ability to tackle diverse and difficult problem situations, (Jackson, 1999, pp. 17).

Multi-paradigm Multi-methodology/Critical Pluralism

This area has been developed mainly by Mingers (1997a, 1999) who argues that CSTand TSI are only one particular form of multimethodology and takes the view that any intervention should gain benefits from being approached with a variety of management science methodologies in what he calls 'strong pluralism' arguing that agent(s) that is person(s) intervening in the situation would benefit if the intervention is tackled using a 'blend of methodologies'. The following arguments favour an application of a multiplicity of methodologies: (1) any situation is in itself complex that not a single methodology can claim to be able to tackle it completely, rather we should pay attention to three aspects involved in any intervention: material, social and personal. Some methodologies will bring more enlightenment to some of the three aspects; (2) an intervention is not a discrete event but continuous and therefore some methodologies are more suitable to certain phases of the intervention. We should not disregard the possibility of combining methodological stages, methods or tools from different methodologies serving to different paradigms; and (3) finally there are practical reasons in favour multiparadigm multimethodology: many systems practitioners have already started to practice it. Mingers provides numerous examples supporting his claim and uses five dimensions to characterise the different types of multimethodology practice: (a) one/more methodologies; (b) single/multi paradigm; (c) same/different intervention; (d) whole/part methodology; and (e) imperialist/mixed (Mingers, 1997b).

CONCLUSIONS

1. The UK systems movement has been a main contributor to the systems movement worldwide over the last six decades. The development in management sciences methodologies can be seen as a consequence of these developments. This paper has traced these developments over the last 60 years and mapped the main management science methodologies developed in the UK in those years.
2. A paradigmatic shift widely acknowledged within the systems movement, has occurred in management sciences practice. From the optimisation paradigm of the 50s where solving problem methods flourish to the learning paradigm epitomised by soft systems methodologies of the 70s and then to the critical systems paradigm of the 80s where both the optimisation and learning paradigm were questioned.
3. Four stages of the development of MSM have been proposed to further understand the evolution of MSM: (a) problem solving methods (1950-1970); (b) situation-improving methodologies (1970-1980; (c) intervention empowering emancipatory methodologies (1980-1990); and (d) multiparadigmatic/pluralistic approaches (1990-). From the literature reviewed it is evident that during the first stage of these developments, the US management science practice was mainly involved whereas in the last three stages, the UK systems movement has been the major contributor.
4. Systems thinkers from the USA (mainly Ackoff and Churchman) have made important contributions to support the theory behind many methodologies that later emerged in the UK. This contribution is depicted in Fig 1. It can be seen that an increasing number have of methodological approaches have been developed in the UK in the last two decades. The pragmatism embedded in British culture seems to have been evident in transforming theoretical developments into methodologies born and tested in practice.
5. The four stages proposed also highlight the MSM changes in the approach used to tackle a managerial situation: from method (1950-70) to methodology (1970-90) to methodological intervention (1980-90) and finally to the broader philosophical approach embedded in the multiparadigmatic practice of the 90s to date.
6. Soft methodologies and in particular Checkland's SSM is the most established methodology since the 70s. Critical systems thinking approaches, mainly under the banner of Jackson and Flood have developed Total Systems Intervention (TSI) which has been in practice since the 80s. Two approaches that take a multi-

paradigmatic/pluralistic stand have been developed in the 90s: Mingers' critical pluralism and Jackson's coherent pluralism.

7. Further research is necessary to assess the influence/importance of the main systems thinkers/authors in the development of the systems movement in the UK.

REFERENCES

Beer, S. (1966). *Decision and Control,* Wiley, Chichester.

Beer, S. (1979). *The Heart of the Enterprise,* Wiley, Chichester.

Beer, S. (1981). *Brain of the Firm,* Wiley, Chichester.

Checkland, P. B. (1976). "Science and the Systems Paradigm," *International Journal of General Systems* 3(2): 127-134.

Checkland, P. B. (1981). *Systems Thinking, Systems Practice*, Wiley, Chichester.

Checkland, P. B. and Scholes, J. (1990). *Soft Systems Methodology in Action*, Wiley, Winchester.

Eden, C., Jones, S. and Sims, D. (1983). *Messing About in Problems,* Pergamon, Oxford.

Eden, C. (1989). "Using cognitive mapping for strategic options development and analysis," in *Rational Analysis for a Problematic World,* (J. Rosenhead, ed.),Wiley, Chichester.

Flood, R. L. and Jackson, M. C. (1991). *Creative Problem Solving: Total Systems Intervention*, Wiley, Chichester.

Forrester, J. (1961). *Industrial Dynamics,* MIT Press, Cambridge, MA.

Jackson, M. C. (1991). *Systems Methodology for the Management Sciences*, Plenum, New York.

Jackson, M. C. (1997). "Pluralism in Systems Thinking and Practice", in *Multimethodology*, (J. Mingers and A. Gill, eds.), Wiley, Chichester.

Jackson, M. C. (1999). "Towards coherent pluralism in management science", *Journal of the Operational Research Society,* 50: 12-22.

Lane, D. and Oliva, R. (1994). "The Greater Whole: towards a synthesis of SD and SSM", in *1994 International Systems Dynamics Conference,* Stirling, UK.

Ledington, P. and Donaldson J. (1997). "Soft OR and management practice: a study of the adoption of SSM," *Journal of Operational Research Society,* 48:229-240.

Mabey, C. and Mayon-White, B. (eds.) (1993). *Managing Change*, Paul Chapman Publishing, London.

Mingers, J. and Taylor, S. (1992). "The Use of Soft Systems Methodology in Practice", *Journal of Operational Research,* 43(4): 321-332.

Mingers, J. (1997a). "Towards critical pluralism", in *Multimethodology*, (J. Mingers and A. Gill, eds.), Wiley, Chichester.

Mingers, J. (1997b). "Multi-paradigm Multimethodology," in *Multimethodology*, (J. Mingers and A. Gill, eds.), Wiley, Chichester.

Mingers, J. (1999). "A Comparative Characterisation of Management Sciences Methodologies," in: *Systemist,* 21(2): 81-92.

Quade, E. (1963). "Military Systems Analysis", in *Systems Analysis,* (S.L. Optner, ed., 1965) Penguin, Harmondsworth.

Senge, P. (1990). *TheFifth Discipline, the Art and Practice of Learning Organization,* Century Business, London.

Ulrich, W. (1987). "Critical Heuristics of Social Systems Design," *European Journal of Operational Research*, 31:276-83.

RELEVANCE AND RIGOUR, THEORY AND PRACTICE IN SYSTEMS

Stephen K. Probert

Computing and Information Systems Management Group
Cranfield University
RMCS Shrivenham
Swindon SN6 8LA
United Kingdom
s.k.probert@rmcs.cranfield.ac.uk

INTRODUCTION

This paper describes work utilising some of the philosophical concepts employed in contemporary epistemology; these concepts have been utilised to develop a framework for understanding the different approaches needed for research and practice in systems analysis and design. The paper concentrates on the differing needs for *relevance* and *rigour* in research and practice, and the paper re-conceptualises these concepts using the traditional philosophical distinctions between *universals* and *particulars* (this argument is fully explained). The conclusion drawn is that *research* can be of two forms. Firstly, it can be concerned with developing interesting and (possibly) insightful generalisations, or *abstractions*, (relevance). Secondly, it can be concerned with developing well-grounded but strongly-bounded *particularities* (rigour).

As Loucopoulos and Karakostas (1995, p4) argue, "Information systems are entering a new phase, moving beyond the traditional automation of routine organisational processes and towards the assisting of critical tactical and strategic enterprise processes. Development of such systems needs to concentrate on organisational aspects, delivering systems that are closer to the culture of organisations and the wishes of individuals." Clearly, such demands are difficult to achieve in practice. Over the past twenty years or so there has been a considerable growth of interest in IS development methodologies. Also, unsurprisingly, (given the above remarks) there has been a steady growth in interest in softer methodological approaches – more attuned to the social needs of the organisation than the technical aspects of IS development. One result of the growth of interest in (and use of) methodologies has led to a new set of problems for IS practitioners; two of which will be considered herein. Let us call the first problem the *problem of (IS) methodological rigour*. Simply stated, this problem concerns the *degree*

Edited by Ragsdell *et al.*, Kluwer Academic/Plenum Publishers, New York, 2002.

to which a IS practitioner needs to adhere to the prescriptions of the chosen methodology, i.e. should he/she use a *cookbook* or a *toolkit* approach. Academically, it is worth qualifying this problem somewhat, as different methodologies – indeed different *versions* of the same methodology – have tended to give different prescriptions in (precisely) this respect. Fidler and Rogerson (1996, p269) neatly capture this notion with the phrase, "The rule prescriptiveness of the methodology". It will be argued below that, regardless of the rule prescriptiveness of the methodology, the extent to which a IS practitioner *follows* the precise prescriptions of a (highly rule-prescriptive) methodology remains a matter of *choice* for the IS practitioner. Furthermore, there is considerable evidence to support the view that such choices are frequently made, e.g. as Jayaratna concludes:

> "The structure, steps, models, values and philosophy of the methodology-in-action may very well be different from either those explicitly outlined in the methodology (creators' rationale) or those that were interpreted and changed by the methodology users in the context of their own 'mental constructs' (methodology users' rationale) before intervention. This is because a methodology has to match the dynamic nature of the situation and the interactions of the methodology users in the situation". (Jayaratna 1994, p. 229)

As the practical choice as to whether to use a methodology in a cookbook or a toolkit manner remains largely open, this paper will focus on this issue. The second problem concerns that of methodology choice and the movement between methodologies (within the same IS project). Let us call the second problem the *problem of IS methodology relevance.* A considerable literature has built-up in recent years to help practitioners with this problem (e.g. Jayaratna 1994), and this paper will not revisit this territory. However, what will be argued is that the *problem of (IS) methodological rigour* and the *problem of IS methodology relevance* are epistemologically related; they are "two sides of the same coin". Consequently, whilst the main focus of this paper concerns the *problem of (IS) methodological rigour*, many of the arguments may help to inform the debate on the *problem of IS methodology relevance*.

Although the paper utilises (some aspects of) *contemporary epistemology*, the requisite concepts and arguments will only be introduced where appropriate. This is a broad area of philosophy, and – in a short paper such as this – it has been necessary to be selective in this respect. However, by keeping the philosophical discussion to a minimum, it is hoped that relevance can be emphasised without a serious loss of philosophical rigour! Although the terms 'relevance' and 'rigour' are used herein they are not used in quite the same sense as they are used in the *IS research* debate concerning relevance and rigour. I take the *research* debate to be one concerned with the comparative merits of research leading to interesting and (possibly) insightful generalisations, or *abstractions*, (relevance) - versus research leading to well-grounded but strongly-bounded *particularities* (rigour). It is tautological to say it, but that debate relates to *academic research* (termed 'theory' in Figure 1 below). This paper is concerned with *IS development* (termed 'practice' in Figure 1 below); here relevance relates to *appropriateness* in a *particular* situation (e.g. a Hospital in England, a Power Plant in the Ukraine, a DotCom in California, etc.). Here, rigour relates to broad *principles* that apply in *any* (IS development) situation, i.e. they are generalised *abstractions*. Using the more traditional philosophical term 'universals', rather than 'abstractions' or 'generalisations', these notions can be summarised in the following table with the examples given:

	Relevance	Rigour
Theory	*Universals* - S-curve model of growth for multi-channel shopping.	*Particulars* - % of families in Yorkshire with internet access and/or WAP phones.
Practice	*Particulars* - "No Purchase Orders without a P.O. number."	*Universals* - "Verified with both Accounts *and* Purchasing?"

Figure 1. Rigour, Relevance and IS Theory and Practice

CONTEMPORARY EPISTEMOLOGY AND IS METHODOLOGY RELEVANCE

In contemporary epistemology, an important distinction can be made between our *criteria for justification* for our belief-sets and prescriptions relating to the *conduct of inquiry* (for adding to our "stock" of beliefs).

Criteria for Justification

Our criteria for justification will often be based on some notion that what we believe, we believe for "good reasons". A crude example would be the dictum that "seeing is believing". Whilst not adequate in all situations (such as during a Magic show), seeing something (x) occurring generally provides a better justification for believing (that x occurred) than e.g. *hearsay*. In contemporary epistemology, this would be termed a *foundational* criterion for justification – as it is based on isolated occurrences, which we would claim to be fairly certain about. We might use some such phrase as "I believe x occurred because I saw it happening." in order to *justify* our belief that x occurred. However, another approach to justification is termed *coherentism*. This approach is holistic, in that it requires that all our beliefs "cohere", i.e. that they do not contradict each other. We might use some such phrase as, "Everything I know about macroeconomics tells me that we have not abolished the business cycle." in order to justify our belief that (e.g.) after a sustained period of economic growth an economic downturn was inevitable. Such a belief would not be justified on any "direct" *foundational* evidence, but rather by the totality of our beliefs (and previous foundational evidence) pertaining to macroeconomics.

In IS analysis, beliefs are generally justified by foundational arguments, but we can (and, I would argue, should) use techniques such as cross-referencing between different models (etc.) to check that our beliefs – about how a particular organisational IS functions – are correct. In practice, it is often necessary to obtain further foundational evidence (i.e. "go back to the users") if our belief sets do not cohere. A (somewhat crude) example would be: suppose that Accounts had told us that "No goods are ever ordered unless a Purchase Order had been raised", whereas Purchasing had told us that "Sometimes goods are obtained without a Purchase Order being raised." *Both* beliefs would be *foundationally* justified – but they are not *coherent*. Further investigation would be required until we could reach a belief-set that was justified both foundationally and coherently. At any rate, IS methodologies should emphasise a rational, mature approach to the justification of beliefs about an IS:

> "The goal of inquiry is substantial, significant, illuminating truth; the concept of justification is specifically focussed on security, on the likelihood of beliefs being true. Hence my claim that truth-indicative is what criteria of justification need to be to be good... [But] Even if our criteria of justification *are* truth-indicative, to reach the conclusion that our beliefs are mostly true would require the further assumption that our beliefs are mostly justified. But people have many beliefs in which they are not justified, or are justified to only a very modest degree. Superstition, wishful thinking, self-deception, jumping to conclusions, and so forth, are not, after all, so rare". (Haack 1993, p. 203)

In a sense, the very point of IS methodologies is to provide adequate, rational, defensible *justification* for the nature, scope and functioning (etc.) systems that are developed as a result of using an IS methodology. As Jayaratna put it, "Methodologies exist to help us in our reasoning. They attempt to raise our conscious thinking, to make us question the rationale of our planned action and to guide us in the transforming of situations." (Jayaratna 1994, p. xii). Furthermore, there is a welter of evidence to support the view that correctly understanding the operation of the current system and the users' requirements is crucial to the development of successful information systems. The models developed for the design of a new IS will need to be both (internally) coherent and (foundationally) justified by the users' requirements.

Conduct of Inquiry

Precisely how analysts come to arrive at such models is a different matter; not less important, but less rigidly definable and more open to variations – such that the varieties of organisational circumstances, in which a particular IS is to be developed, can be adequately catered for. Interestingly, Haack argues that this principle is true for all human inquiry in general:

> ... "[I]t is doubtful whether it is possible to give *rules* - as opposed to *guidelines*, the application of which requires judgement or discretion – for conducting inquiry... the 'conduct of inquiry' project is likely to be more hospitable to pluralism, for there may well be different, equally good, ways of proceeding in inquiry – indeed it may well be that the best thing is for different inquirers to proceed differently; whereas pluralism with respect to criteria of justification ... is not possible". (Haack 1993, p. 204)

Consequently, it is no real surprise that both hard and soft IS methodologies have similarities at the level of (how to go about) rational justification – lots of interaction with the users and such like, *and* no surprise that they have radically different recommendations for the conduct of inquiry - different models with different interpretations as to the purpose of modelling etc. Also, on this basis, new ideas for the conduct of inquiry (i.e. new IS development methodologies) can be cautiously welcomed *providing* that it can be shown that their criteria for the justification of their (intermediate and final) products adequately meet the criteria proposed (or, rather, endorsed) herein. Interestingly, in *this* respect there is sufficient common epistemological ground between hard and soft approaches to allow movement between hard and soft approaches, as the organisational circumstances dictate. A systems analyst may choose whichever approach he or she sees fit, and/or consult some texts to decide which methodology to choose in the circumstances pertaining, (e.g. Hirschheim et al. 1995) without committing themselves to a radically different *epistemological* basis – at least insofar as rational *justification* is concerned.

Conduct of Inquiry Versus Criteria of Justification

Haack developed a table to indicate the desirable features of "projects of devising guidelines for the conduct of inquiry" and for "the project of explicating / ratifying criteria of justification". These are summarised below (adapted from Haack 1993):

Table 1. Goal of Inquiry: Substantial, Significant Truths

"conduct of inquiry"	*"criteria of justification"*
More hospitable to pluralism	Oriented to truth
More recalcitrant to precision	Focused on security of belief
Guidelines, not rules	Focused on likelihood of belief
Requires discretion, good epistemic character	Focused on truth-indicativeness of belief
Social dimension important	

These considerations can help us to evaluate IS methodologies in the following way. For any particular IS methodology, we may simply substitute Haack's *conduct of inquiry* guidelines, with exception of the requirement (1), which (I would suggest) should be replaced with, "that any particular methodology – and especially a new methodology - should be sufficiently novel and distinct from other methodology with respect to guidelines (2) through (5) to warrant our attention and interest". Moreover, we may ask how *any* IS methodology meets Haack's *criteria of justification*; for if it fails to do so then, I conclude, its use should not be advocated. Furthermore, if one examines the *conduct of inquiry* guidelines one can easily provide a theoretical argument concerning the undesirability of slavishly applying an IS methodology in a "cookbook" manner. Not only should movement between hard and soft methodologies be encouraged by the above analysis, but "cookbook" uses should be actively discouraged. The appeal of a "cookbook" approach rests on the inability to make an important distinction between how to *conduct an inquiry* and what the *criteria of justification* for the products of a IS development project are to be; consequently it violates the freedom a IS practitioner needs to adapt his or her processes of inquiry to the actual needs of the project:

> "... [C]oncerns about justification are focused on one dimension, specifically, of the goal of inquiry ... This is not to suggest that the two kinds of epistemological project here distinguished are unrelated ... It is only to insist that, though related, they are distinct. But it is the distinctness of the two projects that needs emphasis here, because they have frequently been run together." (Haack 1993, p. 203-204)

In IS development methodologies (and, in particular, in their "theoretical underpinnings"), the consequences of "running these projects together" will result in a failure to discern the *real* differences between using a methodology in "toolkit" mode and in "cookbook" mode. A cookbook approach, similar to that described in Wastell (1996), will not result in improved *justification* for the models developed. A toolkit approach is not epistemically sloppy; rather it can now be seen as being (generally) epistemically desirable. Methodologies such as Soft Systems Methodology (Checkland and Scholes 1990; Checkland and Holwell 1998) have always recognised as much. Of course, there *will* be appropriate cases where a near-cookbook approach to methodology-use can sensibly be advocated. An inexperienced practitioner, with little or no opportunity to call on the resources of more experienced IS practitioners, may well do better with a cookbook approach – rather than no methodological approach at all. Moreover, there will always be phases / stages / tasks in a project where a near cookbook approach is the only practical approach available, e.g. if a IS practitioner decides to use a technique like normalization. However, even here, the "depth" to which a practitioner may "delve down" may be a matter of practical significance and choice:

> The normalization process is often described in terms of stages known as *first, second, third, fourth* and *fifth normal forms* (1NF-5NF)... fifth normal form deals with a rather unusual situation known as join dependency which is of little practical significance. (Howe, 1989, p. 87)

So, generally speaking, the *practical significance* of methodological prescriptions should determine which aspects of a methodology - even of near-cookbook process within a methodology – are actually performed.

CONCLUSION

In choosing a methodology, *relevance* should determine the *conduct of inquiry*; in using methodologies, *rigour* will stem from the existence of adequate *justification* for the requirements elicited by the practitioners. However, it can be concluded that practitioners should be given (by project managers, etc.) considerable freedom to choose approaches to IS development that – from their perspective - suit the pertaining organisational circumstances. Furthermore, it may often be necessary for practitioners to move between (broadly-based) hard and soft, and near-cookbook and near-*ad hoc* approaches in many IS development projects – as the practicalities dictate. Iteration between phases and cross referencing should be encouraged, at least as far as the time-constraints of a particular IS development project allow. Finally, it should be noted that whilst this paper has gone some way to providing a gone some way to providing a coherent framework for interpreting practical action in IS development, Jayaratna's advice to the practitioner, "Not to hand over his or her thought processes to be directed by any external person, model, methodology or framework, including the one advocated in this [1994] book. Methodology users must become responsible for their thinking and actions." (Jayaratna 1994, p. xiii) is pertinent here, as his comments apply to this paper also.

REFERENCES

Checkland, P., and Holwell, S. (1998). *Information, Systems and Information Systems – Making Sense of the Field,* Wiley, Chichester.

Checkland, P., and Scholes, J. (1990). *Soft Systems Methodology in Action*, Wiley, Chichester.

Fidler, C., and Rogerson, S. (1996). *Strategic Management Support Systems*, Pitman, London.

Haack, S. (1993). *Evidence and Inquiry*, Blackwell, Oxford.

Hirschheim R, Klein, H. K., and Lyytinen, K. (1995). *Information Systems Development and Data Modelling*, Cambridge University Press, Cambridge.

Howe, D. R. (1989). *Data Analysis for Data Base Design*, Edward Arnold, London.

Jayaratna, N. (1994). *Understanding and Evaluating Methodologies*, McGraw-Hill, Maidenhead.

Loucopoulos, P., and Karakostas, V. (1995). *System Requirements Engineering*, McGraw-Hill, Maidenhead.

Wastell, D. G. (1996). "The Fetish of Technique: Methodology as a Social Defence," *Information Systems Journal*, 6: 25-40.

VICKERS' CONCEPT OF RELATIONSHIP-MAINTAINING AND THE NATURE OF REGULATION

Daune West

Computing and Information Systems
University of Paisley
West-ci0@paisley.ac.uk

INTRODUCTION

In their text, *Client-led Design*, Stowell and West (1994) suggested that Vickers' concept of relationship-maintaining might provide a basis of a concept of organisation that was more appropriate to interpretive IS work than the 'traditional' goal-seeking view of organisation. Whilst they may have gone one step further than Walsham (1993) in trying to make explicit the model of organisation their work implied, they failed to develop the model beyond a description of the nature of appreciation and its application in IS work. More recently, Checkland and Holwell (1998) have set down what they refer to as the "conventional wisdom model of 'an organisation' in much of the IS literature" (p82), which is based upon the pursuit of organisational goals, and compare this with what they propose as "a richer model of the concept of 'an organisation' than the 'conventional wisdom' [model]" (ibid., p83). Their 'conventional wisdom' model is based upon the premise that organizations are taken to exist as entities in the 'real world'; that is, they *are* treated as "a conscious, person-like entity capable of unitary purposeful action" (ibid., p82). In this model, organizational members are considered to understand and share the organization's aims and objectives, and consequently, work within the organization to attain these objectives. Objective attainment is commonly expressed through the concepts and language of 'goal-seeking'. Checkland and Holwell state that "Philosophically, this is a positivistic model, sociologically, a functionalist one" (ibid., p82).

In comparison, their 'richer model' challenges the concept of organization as a single objective entity. Instead, in this model we recognize that organizations, whatever they may be, are complex phenomena, created out of a myriad of intention for purposeful action. However, they recognize that it may be useful to treat organizations *as if* they were single entities in order to facilitate discussion about them. But they are careful to point out that "Acknowledging that this is a linguistic convention rather than an account of 'what is the case' makes it possible to include a richer account of what goes on under

Edited by Ragsdell *et al.*, Kluwer Academic/Plenum Publishers, New York, 2002.

the umbrella of that convention" (ibid., p82). A second challenge to the 'conventional wisdom' model is that in their 'richer model' the pursuit of collective objectives in a rational manner is developed further by introducing the idea that there are many conceptualizations of the organisation, and consequently, many objectives which may be pursued by organizational members at the same time. The management of these different personal objectives means that the concept of organisational decision-making being a consensus-seeking activity is unrealistic. Instead, organisational members need to "seek *accommodations* between conflicting interests upon which action can be based" (ibid., p84). For them, this accommodation-seeking activity is consistent with Vickers' concept of managing relations. In comparison to the 'conventional wisdom' model, this richer model is "Philosophically ...a phenomenological model, sociologically an interpretive one" (ibid., p84).

Checkland and Holwell's description of the 'conventional wisdom' model and their 'richer' model of organisation explains the philosophical and sociological differences between goal-seeking and relationship-maintaining but in practical terms their model is disappointing since it tells us what is to be done but offers little advice or guidance about how these concepts can be put into use to help IS practitioners and researchers improve their understanding of organizations and the information systems that may support their activities. In the following sections goal-seeking and relationship-maintaining are compared and a different way of representing relationship-maintaining is offered which is felt to be more accessible to researchers and practitioners.

VICKERS' CRITICISMS OF 'GOAL-SEEKING'

For Vickers the concept of goal-seeking, which has lain at the heart of popular organizational models for many years, is poverty stricken. He criticized goal-seeking on three main grounds:

1. goals are considered to be once-and-for-all states to be attained and their establishment and pursuit equates to organizational planning and control. Once these goals are attained then new goals need to be considered and set.
2. goal-seeking encourages the simplification of a complex situation by the process of reducing multiple objectives to a single goal.
3. the need to measure goal attainment has led to "effectiveness" becoming the single most important criteria for judging the appropriateness of alternatives to achieve the goal (1970, p116).

In place of goal-seeking Vickers suggests a way of conceptualizing, planning and regulating a system over time through the concept of relationship-maintaining. The difference between goal-seeking and relationship-maintaining is an important but subtle one; they both strive to 'manage' a situation but go about it in very different ways based upon different assumptions about human decision-making and action. With goal-seeking the assumptions are that we can describe the single purpose of any action and that the action has a start and finish point and, hence, can be assessed in terms of the success of its achievement. Success, too, is assumed to be a 'known' in the situation and a suitable measurement can be devised and practised. Figure 1 illustrates the process of goal-seeking through the mechanism of a traditional classical feedback control model.

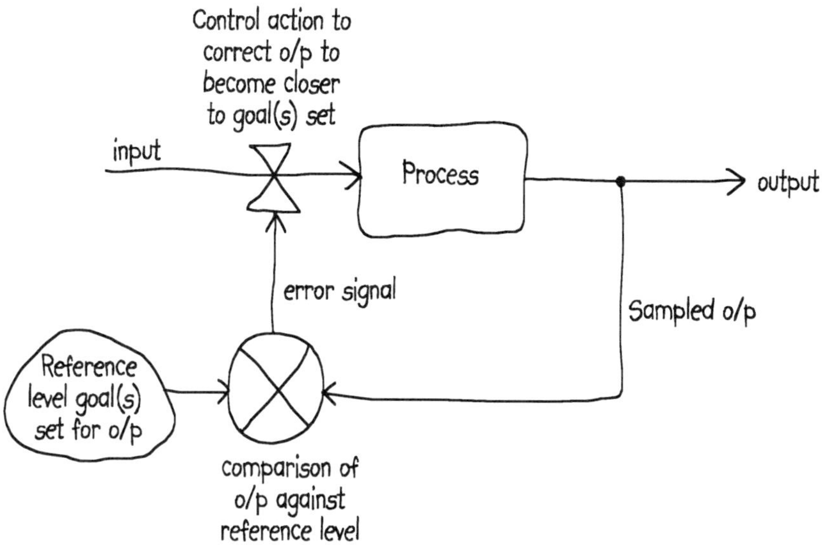

Figure 1. A Classical Feedback Control Model to Illustrate the Concept of Goal-seeking

For example, our goal might be to gain a particular academic qualification and we will know that we have succeeded when we are awarded that qualification: at this point we will know that we have fulfilled our goal. Vickers challenges the usefulness and meaningfulness of such an action-planning and action-regulating device by pointing out that our fulfilment of the goal is, in itself, somewhat pointless in that its achievement signifies the end of the regulating action: the qualification is achieved…and then what? How does the achievement of this qualification fit in with the wider picture of the reasons for wanting, or needing, the qualification? The qualification, as an object of the fulfilled goal, is not important in itself but it is important within a wider web, or context, of related desires. Vickers explains the difference between 'objects' as our focal point and the idea of 'objects in their context of relations', the view that he favours:

> "The objects of our desires and aversions are not objects but relations. No one 'wants an apple'. He may want to eat it, sell it, paint it, admire it, conceivably even merely to possess it – a common type of continuing relation – in any case to establish or change some relation with it. *The goals we seek are changes in our relations or in our opportunities for relating; but the bulk of our activity consists in the 'relating' itself*"(1965, p33).

VICKERS' CONCEPT OF RELATIONSHIP-MAINTAINING

Vickers did not dismiss the concept of 'goal' completely. Instead he argued its importance only as part of a wider whole, that is, within the context of some desired relationship. Let's go back to our somewhat simplistic example to help illustrate the idea of the importance of relations:

> The reason why I want to gain a higher academic qualification is because its attainment will, I believe, make me more eligible for promotion in the workplace, which would mean a higher salary and I feel this would provide my family with a better standard of living. The qualification sought is likely to take a minimum of six years to complete if I wish to maintain my current employment and income. I will be expected to commit myself to at least two evenings a week attending classes and perhaps another two evenings and one day each weekend to personal study and assessment. I have a young family and this commitment to the pursuit of the qualification means that I will spend considerably less time with them. I know I cannot replace this 'lost time' later. The extra work load will cause considerable pressure and may affect my relationships at home and at work and yet I may also feel a great sense of achievement if I am successful in my studies. My family's quality of life and my relationship with them is likely to be affected for the next six years – is this acceptable or not?

What was first expressed as a single goal can now be viewed as a complex web of relations - all of which I would like to maintain in the 'best' possible way I can. As Vickers explained, when discussing policy making, the successful management of any human activity system, whether business organization or one's personal life,

> "does not consist in prescribing one goal or even a series of goals; but in regulating a system over time in such a way as to optimize the realization of many conflicting relations without wrecking the *system* in the process" (my italics) (1970, p116).

When we make decisions about how to act we need to take into consideration the implications of all our actions and, as the diagram in Figure 2 illustrates, the relations likely to be effected surrounding a single goal are many and complex.

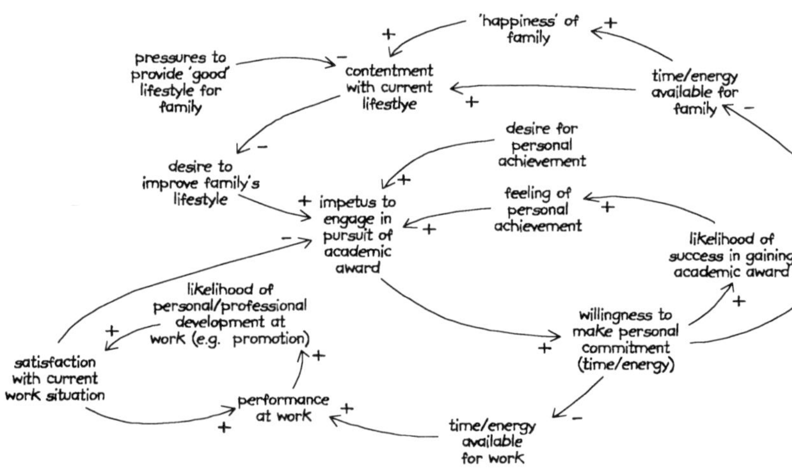

Figure 2. Causal Loop Diagram to Illustrate the Example

For Vickers, management, or regulation, is about trying to effect stability in a situation by encouraging and supporting desired relationships and avoiding or minimizing the effect of undesirable ones:

> "The meaning of stability is likely to remain obscured in Western cultures until they rediscover the fact that life consists in experiencing relations, rather than in seeking goals or 'ends'. The intrinsic confusion about means and ends arises from the fact that no end can ever be more than a means, if an end is equated with a goal. To get the job or marry the girl is indifferently an end, a means and a goal; it is an opportunity for a new relationship. But the object of the exercise is to do the job and live with the girl; to sustain through time a relationship which needs no further justification, because it is or is expected to be satisfying in itself" (1970, p128).

As we decide which relations we wish to maintain, avoid or sacrifice for the sake of others, we are making judgements about what we see and value and these judgments are continually being formed and reformed as, through experience, we re-evaluate the relations with which we are dealing. New experiences and ideas make us more susceptible to certain events and ideas that inform our thinking and action and change our attitudes and beliefs. In such a way we learn. Some relations disappear whilst new ones come into focus. We adapt and re-evaluate the context of our "goal" continually. Vickers described the result of this continuous process of learning as "appreciation" and its process of development an "appreciative system" (1965).

CHECKLAND AND CASAR'S DIAGRAMMATIC DESCRIPTION OF AN "APPRECIATIVE SYSTEM"

Checkland and Casar (1986) produced a useful diagrammatic illustration of Vickers' description of an "appreciative system" which is reproduced in Figure 3.

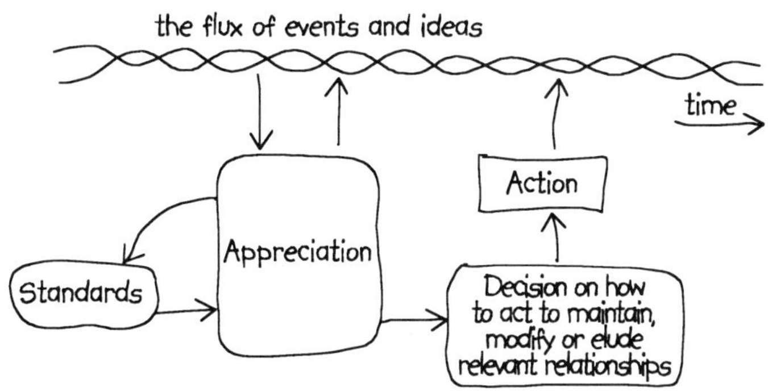

Figure 3. Checkland and Casar's Diagrammatic Representation of Vickers' Notion of an "Appreciative System" (Source: Checkland and Casar, 1986, Figure 3)

In Checkland and Casar's model the source of what we referred to above as the 'complex web of relations' in a situation is represented by what they call the "interacting flux of events and ideas", or *Lebenswelt* (1986, p5). For Vickers, decision-making (within any particular context) is conditioned:

> "... not only by the concrete situation in which it is taken but also by the sequence of past decisions; and that their new decisions in their turn will influence future decisions not only by their effect on the history of event but also by the precedents which they set and the changes which they make in the way decision makers in the future will see, interpret and respond to event, a separate development which for the moment I will label the history of ideas. Thus human history is a two stranded rope; the history of events and the history of ideas develop in intimate relation with each other yet each according to its own logic and its own time scale; and each conditions both its own future and the future of the other" (1965, p15).

Our past understanding of events and ideas provide us with our own "appreciative setting", that is, a "temporary expression, these readinesses to see, to value and to respond to situations in familiar ways which, while they last, exclude the power to see other possibilities" (ibid., p54). Our appreciative setting 'allows' our interest in certain events and ideas which we judge in terms of reality and value:

> "An appreciation involves making judgements of 'fact' about the state of the system', both internally and in its external relations. I will call these reality judgements. These include judgements about what the state will be or might be on various hypotheses as well as judgements of what it is and has been. They may thus be actual or hypothetical, past , present or future. It also involves making judgements about the significance of these facts to the appreciator or to the body for whom the appreciation is made. These judgements I will call value judgements. Reality judgements and value judgements are inseparable constituents of appreciation.... The relation between judgements of fact and of value is close and mutual; for facts are relevant only in relation to some judgement of value and judgements of value are operative only in relation to some configuration of fact" (1965, p40).

Over time our judgements of reality and value build up to provide a 'filter' through which further experiences of events and ideas are "appreciated". This "filter", which is continually open to development from these new experiences, can be seen as a form of dynamic regulating standard.

RE-REPRESENTING THE APPRECIATIVE SYSTEM BY MEANS OF A 'CONTROL MODEL'

The example given in Figure 2 above, hopefully, serves to illustrate why Vickers considered the maintenance of relations to be a richer and more useful concept than goal-seeking when thinking about decision-making and the regulation of action. The theory of 'maintaining relations' is explained but our difficulty now lies in putting this theory into practice. Checkland and Casar's model is invaluable in helping us to conceptualize Vickers' work and, as they claim, "can provide an epistemology for describing the social process in which human beings deliberate and act", (Checkland and Casar, 1986, p10). However, if the idea of relationship-maintenance is to be taught and used in organizational analysis and information systems work then we need to be able to provide a description of it that can be understood and the differences between it and existing models can be clearly illustrated. We used a traditional classical feedback control model above to demonstrate the principles of goal seeking and it seems possible to use this same type of popular model to explain Vickers' notion of regulation through relationship-maintenance but with some important differences. In Figure 4 we use the components of Checkland and Casar's representation of an appreciative system and superimpose them onto a classical feedback control model. We can see how the self-regulative action of appreciation means that we need to re-think the boundaries of the 'system'. For example, in the goal-seeking model, the 'reference level', or standards, are considered as being set from outside the system: even if they are altered as a result of the system's controller being unable to correct any deviant action, the 'goal' needs to be reset. By comparison, with relationship-maintenance the standards are within the system and are continually being reformed and adjusted in reaction to the past history, current experience and future expectations of the system itself. The controller, or regulator is clearly within the system boundary at all times (he/she does not stand outside the system to consider the systems performance objectively and reset the system's performance criteria), watching and judging its action.

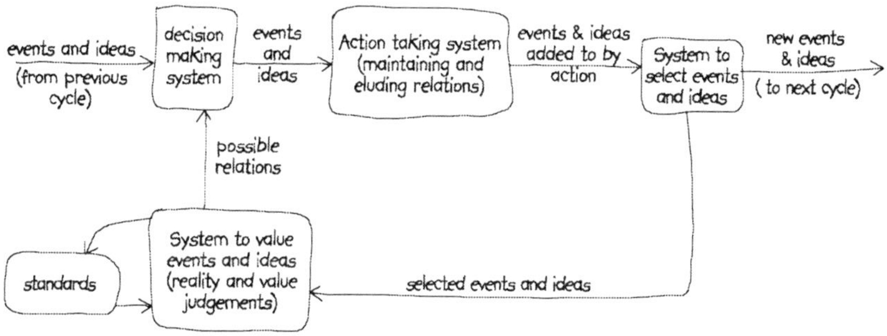

Figure 4. The Concept of Regulation through Relationship-maintenance Illustrated through an Adapted Classical Feedback Control Model

The major difference between the two notions of 'control' illustrated in Figures 1 and 4 is that in Figure 1 the course, which is set as the desired course (the standards or the reference level), is provided from outside of the system that is being controlled. In Figure 4, which illustrates the nature of the regulation involved in relationship-maintenance, the standards are developed out of the previous actions of the system itself: this difference concerning the establishment of standards or criteria for selecting courses of action "are not given from outside. They are themselves internally generated by the previous history of the system itself and its interactions with its environment" (Checkland and Holwell, 1998, p47). Vickers explains the difference between the conventional engineering model of regulation and the institutional model with which he is concerned:

> "The difference between the engineering model ... and the institutional model which I am concerned to develop lies primarily in the fact that in institutional behaviour the concepts of what relations should be regarded as regulable, the

> standards by which they should be regulated, and the ways of reconciling the inconsistent demands which they generate, are neither constant nor given but are themselves a function of the process which they are supposed to govern" (1965, p41).

CONCLUSIONS

The idea of relationship-maintenance lies at the heart of Vickers' notion of appreciation as a description of the process of continuous learning surrounding human decision-making and action. The idea of relationship-maintenance encourages a context-dependent view of organizational activity built upon organizational members' conceptualization of 'the organization' and its aims and objectives. The management of the organization becomes the management of the varied and complex relationships that are considered to exist for different organizational members. Such a view necessitates a way of describing the way in which *accommodations* are sought and regulated. Vickers' concept of relationship-maintenance seems to be a useful way of describing this regulating process. If the [theoretical] concept of relationship-maintenance is to be useful and to be made relevant and available to IS practitioners and researchers then it needs to be presented in a way which sets out its components in a manner which is easy to understand. It is suggested that by illustrating relationship-maintenance through an adapted classical feedback control model its use might be facilitated more readily than if it remains embedded within generic, theoretical process models like that offered in Figure 2 above or in the descriptive but largely implicit forms provided by researchers such as Walsham (1993) and Stowell and West (1994).

REFERENCES

Checkland, P.B. and Casar, A. (1986). "Vickers' concept of an appreciative system: a systemic account," *Journal of Applied Systems Analysis*, 13:3-17.

Checkland, P.B. and Holwell, S. (1998). *Information, Systems and Information Systems: Making Sense of the Field*, John Wiley and Sons, Chichester.

Stowell, F.A. and West, D. (1994). *Client-led Design: A Systemic Approach to Information Systems Definition*, McGraw-Hill, Maidenhead.

Vickers, G. (1965). *The Art of Judgement: A Study of Policy Making*, Chapman and Hall, London.

Vickers, G. (1970). *Freedom in a Rocking Boat: Changing Values in an Unstable Society*, Penguin Books Ltd, Harmondsworth.

Walsham, G. (1993). *Interpreting Information Systems in Organizations*, John Wiley and Sons, Chichester.

Section Two

Systems Practice

CONNECTING SOFT SYSTEMS THINKING WITH PROJECT MANAGEMENT PRACTICE: AN ORGANIZATIONAL CHANGE CASE STUDY

Kerry Costello, Lynn Crawford, Lesley Bentley and Julien Pollack

Project Management and Economics Program
University of Technology, Sydney
PO Box 123, Broadway NSW Australia 2007
Lynn.Crawford@uts.edu.au

INTRODUCTION

This paper reflects on the experience of a university research team working in collaboration with an emergency services agency undertaking significant organisational change. The team, from the Project Management and Economics Program at the University of Technology, Sydney (UTS) applied learning gained with other industry partners to support the New South Wales (NSW) Rural Fire Service in implementation of an innovative project management approach to meet its strategic reform agenda. Participation in this and other partnerships with government agencies, has led to re-examination of project management theory and practice by the research team.

BACKGROUND

Public sector organizations worldwide are moving towards project-based management. This is occurring against the background of the "new public management ", a trend for reform based on business management models. New technologies, new work practices and new ways of conceptualising organizational change have come with the "new public management". Within this environment:

> "The values informing public programs are invariably diverse and contested. Future issues are often hard to anticipate. Trade-offs and overlap between policy areas abound. New issues cross departmental and agency boundaries in unpredictable ways. Stakeholder environments are invariably multifaceted and complex." (Marsh, 1999, p65)

Edited by Ragsdell *et al.*, Kluwer Academic/Plenum Publishers, New York, 2002.

Also within this environment, governments are promoting the use of project management and formal project management methodologies (Cabinet Office, UK, 2000, p21). Underpinning these methodologies are industry standards such as the (USA) Project Management Body of Knowledge (Project Management Institute, 2000), and the (UK) Association for Project Management Body of Knowledge (Dixon, 2000). These standards were developed in the context of essentially “hard” projects in the construction, engineering, defence and aerospace industries. In "hard" projects both the goals to be achieved and the methods to be used are well developed (Turner and Cochrane, 1993). Problems in applying "hard" project management practice to "soft" projects have prompted rethinking of standards and practices, with a number of writers and researchers turning to systems theory for possible enlightenment (Neal, 1995; Rodrigues and Bowers, 1996). The UTS team noted apparent linkages between soft systems thinking and areas of project management involving "soft" projects through the common concern with ill-defined problem situations (Crawford and Costello, 2000). A research collaboration with the NSW Rural Fire Service (RFS) provided an opportunity to look at theory and practice within a "real work" context.

THEORY AND PRACTICE IN PROJECT MANAGEMENT

A review of the first ten years of the International Journal of Project Management in 1995 found project management practice quite well defined, but that papers reflected “a field that is very practice based, and concerned with the integration of information and experiences rather than being highly analytical or theoretical" (Betts and Lansley, 1995, p215), a position that has been recently confirmed (Themistocleous and Wearne, 2000).

Possible scenarios by which more theoretical and model-based contributions might emerge include application from other fields (Betts and Lansley, 1995, p215). Here a, parallel may be drawn between project management and information systems (IS) as "applied disciplines". It "is generally accepted that information systems is essentially a pluralistic field, founded on knowledge from many other well-established source disciplines" (Avison and Myers, 1995, p43). Issues being debated in IS reverberate in project management, including status as an academic discipline and need for a unifying theory. Within IS, soft systems thinking has been widely and successfully applied (Bennetts *et al.* 2000). Project management has continued, however, to be grounded in hard systems approaches. Further, breakdown into nine knowledge areas (Integration, Scope, Time, Cost, Quality, Human Resources, Communications, Risk and Procurement) (Project Management Institute, 2000) has created a tendency to lose sight of the systemic relationships that were well understood in the earlier years of project management.

PROJECT MANAGEMENT RESEARCH AT UTS

As project management is not recognized as a distinct academic discipline, research interest has been difficult to attract. Academics in the field have tended to come from engineering, although more recently those from other fields have shown interest. This has given rise to formation of the International Research Network for Organising by Projects (IRNOP) that has held a bi-annual conference since 1994.

Research at UTS has been concentrating on practitioner competence and has been developing profiles or benchmarks of project management knowledge and use of practices. Another major area of research has been directed at enhancing the understanding of complex, interdependent "soft" projects.

Yeo has highlighted the divergence of systems thinking and project management from their common origins in systems analysis and systems engineering and suggested a need for re-integration, making particular reference to Soft Systems Methodology (SSM) (Yeo, 1993, p116). In looking at the principles and application of systems thinking, the research team noted the benefits of multi-methodologies in "real world" applications (Midgley, 1997; p251). Also, it was found that different "levels" of project management were operating within the organizations concerned and were translating, according to the level of integration involved, into a hierarchy descending from portfolios of programs, to programs of projects, to individual projects (Payne and Turner, 1999, p57).

CONNECTING SSM WITH PROJECT MANAGEMENT PRACTICE

Initial exploration of SSM brought out its adaptability at the conceptual level as well as the variation occurring in practice (Holwell, 2000), including in its application to organizational change initiatives (Gregory and Midgley, 2000; Ragsdell, 2000).

Over time, the research team's conceptualisation of the possible connections between SSM and project management were transformed. This may be likened to progression from Mode-1 to Mode-2 applications of SSM. The former is sequential and methodology-driven, while the latter can be used as an internalised, situation-driven model for sense-making (Checkland, 1999, pA36; Checkland and Holwell, 1998, p164)). The knowledge creation process associated with Mode-1 was seen to be discipline based, hierarchical and academically driven while that for Mode-2 was seen to be practice-based, trans-disciplinary and socially distributed. Examples of models produced at the time are set out in Figures 1 and 2. Those in Figure 1 were produced at an early stage and involved linking organisational change processes and the seven-stage model of SSM to the project management lifecycle (Stretton, 2000).

Figure 2 is a model produced during a later iteration of the process (Crawford and Costello, 2000, p317). It involved using SSM for structuring thinking about the program / project and its environment and then to guide the application of project management tools and techniques. The basic model used was the "Processes for Organisational Meanings" (POM) model (Checkland and Holwell, 1998, p106), extended to include organizational structure and project management practice. The learning acquired through the development and testing of models attempting to connect SSM as a sense-making framework and project management practices and tools was re-examined in the Rural Fire Service change program.

RURAL FIRE SERVICE CASE STUDY

The NSW Rural Fire Service (RFS) is a community-based organisation founded on the dedication of volunteers, some 70,000 in total. Maintaining the sense of community ownership through these volunteers is essential for the RFS to function.

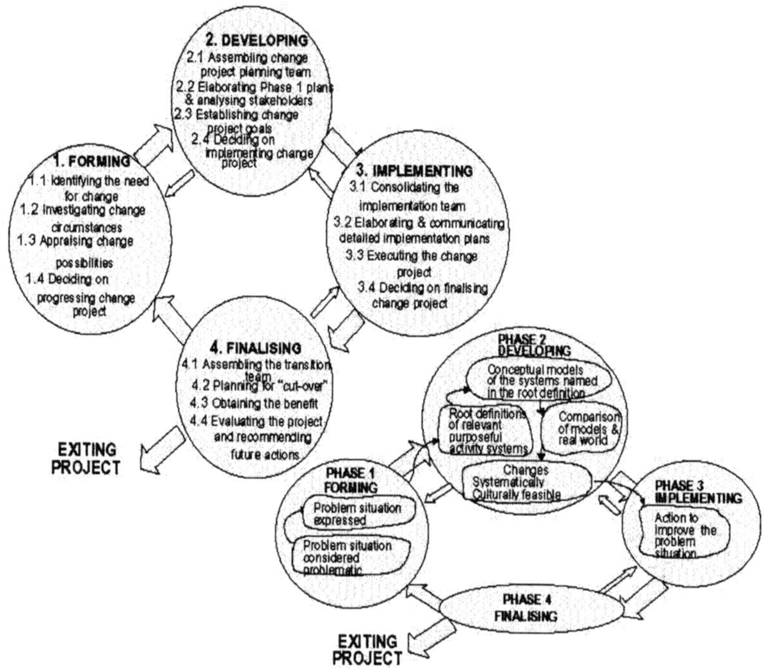

Figure 1. Examples of Models Linking the Project Management Lifecycle with Organisational Change Processes and SSM (Stretton, 2000, pp6-7)

Perceived Real-World Problem Situation

Following devastating fires in NSW during the 1990's, a number of inquiries were held into their cause and the effectiveness of the RFS response. Although major reform was undertaken in 1997, dual accountability of fire control staff remained (i.e. to the RFS in operational matters and to local government councils in administrative matters). On 7 June 2000, the Minister for Emergency Services announced that fire control staff would be transferred from local to state employment from 2001, thus effectively unifying the RFS. In undertaking this program, the RFS also recognised an opportunity to address change in the wider context of government reforms taking place in NSW. In effect the change program required review and reconstruction of key RFS administrative processes within a 12-month timeframe without any interruption to core business.

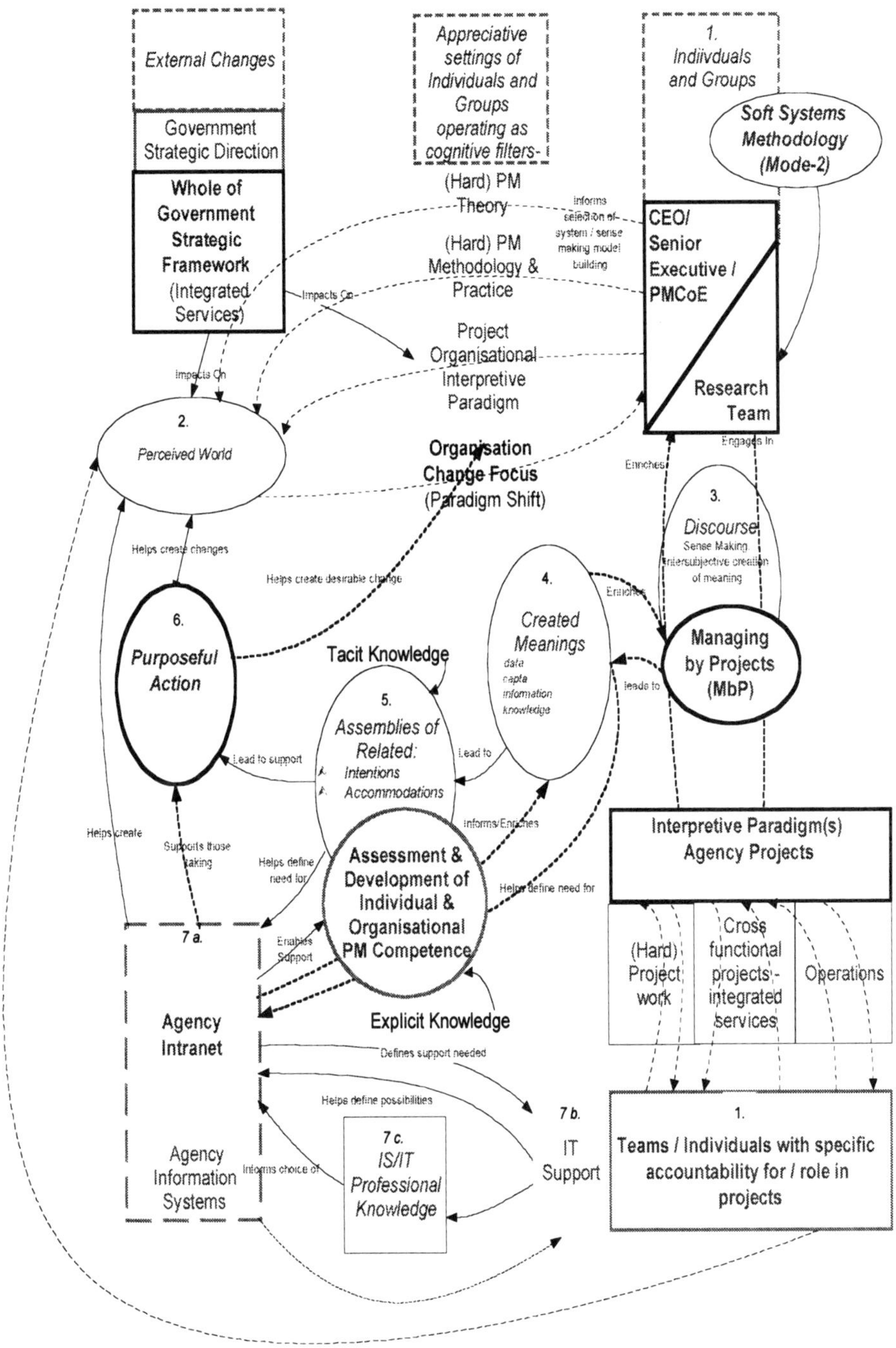

Figure 2. An Example of an Adapted "Processes for Organisational Meanings" (POM) Model (Checkland and Holwell, 1998, p106) for NSW Department of Public Works and Services (Crawford and Costello, 2000, p318)

Process for Tackling the Situation

A strategic approach was developed in facilitated sessions of RFS staff and UTS research team members. At the outset, it was recognized that the change process needed to be tackled as a whole system and scoping of the work was undertaken within a "portfolio of programs" structure. This provided a structure for the top-level deliverables to be progressively "unpacked" into program and project areas. In a facilitated process, ideas were compared and vigorously debated by participants to identify common objectives, realistic solutions and necessary actions

The debates were structured by the facilitator around a project brief template developed by the UTS research team and specifically tailored for the RFS. The template incorporated the three levels of program / project management i.e. integrative (portfolio), strategic (program) and tactical (project). All components of the brief including objectives, deliverables and risk assessment were designed to cascade from portfolio to project. Outputs (deliverables) identified during the process were allocated to program teams that took responsibility for their completion and formed smaller project teams. Eight program teams were established: Communications Change Management; Service Accommodation; Industrial Review; Workforce Planning and Organisational Structure Review; Financial Program Review; Legislation, Service Standards and Policy Review; Information Management; and Strategic Business Planning Review. These teams were overlaid in a matrix on top of the existing RFS hierarchy. Thus, whilst managers of the program teams were still required to perform their normal functional duties they were also expected to prioritise work to achieve delivery of the change initiatives. In this they were supported by a Project Office. As well as facilitating the planning of programs, Project Office staff captured key information, monitored progress, prepared reports, communicated program status to stakeholders and undertook risk management.

People Involved in the Process

Details of the change management structure and the people involved, including the composition of the project teams that were formed are at http://www.bushfire.nsw.gov.au. Within the structure, the "portfolio team" became known as the Program Managers Working Group (PMWG) and included the Sponsor, Project Manager, Program Managers and support staff. During facilitated sessions, the PMWG team reached agreement on a range of portfolio objectives and identified the key deliverables. The PMWG then met weekly to review progress, exchange ideas, discuss the status of dependencies and integrate the deliverables within and across programs. Project managers were recruited through Expressions of Interest advertised throughout the RFS. Program teams included internal and external stakeholders

The PMWG reported to the Organisational Change and Continuous Improvement Steering Committee (OCCISC) through comprehensive monthly status reports. OCCISC comprised senior executives and members of key internal and external stakeholder groups and met monthly to review the portfolio status and endorse findings.

The Combination of Situation, Process, and People

Translated into the framework of three of the nine project processes, the following is an example of some key outcomes:

Communications: An online project management system was developed for communication of program information to the Sponsor, Senior Executive and the PMWG. This enabled strategic information, including status, to be captured, stored, monitored, communicated, reported and archived by the Project Office.
Human Resources: A project governance structure was established to set scope, and cost/timeline expectations, to ensure appropriate volunteer and stakeholder input and to resolve complex process management issues.
Quality: A holistic approach was taken to total quality management. Quality, in accordance with quality standards (ISO 10006) was defined in terms of ensuring outcomes were fit for the purpose, provided value for money and satisfied volunteers.

The successful outcome of the RFS change program was recognized by an Award by the Australian Institute of Project Management. One requirement was demonstration that all nine project elements (Project Management Institute, 2000) had been addressed.

Project management practice includes a process for transferring good practice and for avoiding mistakes in the form of "Lessons Learned". A typical report format is given in PRINCE 2 (http://www.ogc.gov.uk/prince/index.htm). In undertaking this process for the RFS, it was found that while generic quality standards were met this did not capture the full richness of the experience. Regard was had to project management success stories, described by Laufer and Hoffman (2000) as the best way to convey meaningful knowledge. Themes emerging from the stories provided by RFS practitioners supported a view that successful projects require strong leadership as well as strong management; are founded in team work; that plans need to be flexible enough to support unavoidable changes; and that direction and a focus on goals is more important than detailed schedules.

CONCLUSION

Checkland refers to "the interaction between ideas and lived experience, between theory and practice" and in particular to "the ongoing cyclic process in which ideas and experience (theory and practice) continuously create each other" (Checkland, 2000, p800). This was borne out in the Rural Fire Service / University of Technology, Sydney experience. Here, the systems concepts applied, as well as the tools developed, in association with project management techniques, provided a structure for debate to be conducted, ideas to be contributed, insights to be gained, challenges to be appreciated, plans to be negotiated and action to proceed towards a successful conclusion within "traditional" project management time, cost and quality parameters.

REFERENCES

Avison, D.E. and Myers, M.D. (1995). "Information systems and anthropology: an anthropological perspective on IT and organizational culture," *Information Technology & People*, 8(3):43-56.

Bennetts, P.D.C, Wood-Harper, A.T. and Mills, S. (2000). "A holistic approach to the management of information systems development – a view using soft systems and multiple viewpoints," *Systemic Practice and Action Research*, 13(2):189-205.

Betts, M. and Lansley, P. (1995). "International Journal of Project Management: a review of the first ten years," *International Journal of Project Management*, 13 (4):207-217.

Cabinet Office, UK. (2000). *Successful IT: Modernising Government in Action* http://www.e-envoy.gov.uk/publications/reports_index.htm

Checkland, P. (2000). "The Emergent Properties of SSM in Use: A Symposium by Reflective Practitioners," *Systemic Practice and Action Research*, 13(6):799-823.

Checkland, P. (1999). "Soft Systems Methodology: a 30-year retrospective," in *Soft Systems Methodology in Action*, (P. Checkland, and J. Scholes, eds.). John Wiley and Sons, Chichester.

Checkland, P. and Howell, S. (1998). *Information, systems and information systems - making sense of the field*, John Wiley and Sons, Chichester.

Crawford, L. and Costello, K. (2000). "Towards a Transferable Methodology for Managing Strategic Change by Projects," in *IRNOP IV Conference - Paradoxes of Project Collaboration in the Global Economy: Interdependence, Complexity and Ambiguity*, (L. Crawford, and C. F. Clarke, eds). University of Technology, Sydney, Sydney, Australia:

Dixon, M. (2000). *APM Project Management Body of Knowledge*, Fourth edition. Association for Project Management, Peterborough, England.

Gregory, W.J. and Midgley, G. (2000). "Planning for disaster: developing a multi-agency counselling service," *Journal of the Operational Research Society*, 51(3):278-290.

Holwell, S. (2000). "Soft Systems Methodology: Other Voices," *Systemic Practice and Action Research*, 13(6):773-798.

Laufer, A. and Hoffman, E. (2000). *Project Management Success Stories: Lesson of Project Leaders*, John Wiley and Sons, New York.

Marsh, I. (1999). "Program Strategy and Coalition Building as Facets of New Public Management," *Australian Journal of Public Administration*, 58(4):54-67.

Midgley, G. (1997). "Mixing Methods: Developing Systemic Intervention," in *Multimethodology: The Theory and Practice of Combining Management Science Methodologies*, (J. Mingers and A. Gill, eds.). John Wiley and Sons, Chichester.

Neal, R.A. (1995). "Project definition: the soft-systems approach," *International Journal of Project Management*, 13(1):5-9.

Payne, J.H. and Turner, J.R. (1999). "Company-wide project management: the planning and control of programmes of projects of different type." *International Journal of Project Management*, 17(1):55-59.

Project Management Institute (2000). *A Guide to the Project Management Body of Knowledge*, Project Management Institute, Newtown Square, PA.

Ragsdell, G. (2000). "Engineering a paradigm shift? An holistic approach to organisational change management," *Journal of Organizational Change Management*, 13(2):104-120.

Rodrigues, A. and Bowers, J. (1996). "The role of system dynamics in project management," *International Journal of Project Management*, 14(4):213-220.

Stretton, A. (2000). "An Investigation of Connections Between Organisational Change and Project Management." in *IRNOP IV Conference - Paradoxes of Project Collaboration in the Global Economy: Interdependence, Complexity and Ambiguity*, (L. Crawford and Crawford, L. and C. F. Clarke, eds.). UTS, Sydney, Australia.

Themistocleous, G. and Wearne, S.H. (2000). "Project management topic coverage in journals," *International Journal of Project Management*, 18(1):7-11.

Turner, J.R. and Cochrane, R.A. (1993). "Goals-and-methods matrix: coping with projects with ill defined goals and/or methods of achieving them," *International Journal of Project Management*, 11(2): 93-101.

Yeo, K.T. (1993). "Systems thinking and project management - time to reunite," *International Journal of Project Management*, 11(2): 111-117.

CYBERNETICS/SYSTEMS APPROACH TO TRANSFORMATION OF THE TOWEL PRODUCTION FIRMS IN OSAKA AREA, JAPAN

Takayuki Koga

Osaka Chamber of Commerce & Industry
Osaka City, Japan
BYG02143@nifty.ne.jp

INTRODUCTION

The aim of this research is to present recommendations regarding transformation of the towel industry in Osaka area so that it can grow concurrently with increasing importation of towel products. For this purpose, I undertook a cybernetics/systems approach to analysing characteristics of the industry and its current problems through a field survey. The concept of the "system" is defined as follows: an organisation that has emerged from the combination and interaction of multiple factors, existing independently of the environment. This concept can be applied not only to living organisms but also to situations where a certain industry grows in a certain area. If individual firms making up a certain industry are disappearing, we can say that the industry is having trouble in its function as a system to adapt itself to changes in the environment. The towel industry in Osaka, which is the object of this research, according to our records began in 1887, and expanded constantly until 1991, except for a period of regulation during the Second World War. 1991 was a turning point for the industry, after which the number of firms and level of production began decreasing. In February 2001, a petition was filed for implementation of the textile safeguard in conformity with WTO agreements.

In this research, priority is placed on how to solve a problem actually occurring in society, rather than making a pure theoretical analysis. Therefore, the methodology and models used in this research must be ones already verified as applicable to actual problems. Here, as a methodology for analysing phenomena, VSM (organisation cybernetics that Beer proposed through a deductive study of cybernetics/systems theory), and SSM (epistemology that Chechland substantiated using the system thinking) was used. I have previously used VSM for analysing the evolving process and structure of Seven-Eleven Japan Co., Ltd., a company in the convenience store industry. Also, I have used SSM for analysis of the Yamato Transport Co., Ltd., a company that changed its business line from maker's production goods transactions to the Takkyubin business

Edited by Ragsdell *et al.*, Kluwer Academic/Plenum Publishers, New York, 2002.

(door-to-door delivery service). From these experiences, I have concluded that these system models are effective for analysing the phenomena in the real world.

Systemic Transformation System

Checkland (1981) presented the basic model of SSM. This model represents a circular learning system to acquire knowledge. It is also a transformation system leading to necessary actions to improve the situation, by comparing the model with the actual status it has perceived. When we have a theme of problem solving of a corporate organisation or an industrial structure, we can get a concrete milestone by applying VSM as an ideal conceptual model for comparison. From such a standpoint, I have created a systemic transformation system by combining VSM and SSM, shown in Figure 1. In this transformation system, SSM serves as a prime mover to promote mental activities, and VSM serves as a basic model that reflects the actual organisational structure. Human beings learn ideal types and methodologies, and try to achieve a higher level by improving themselves. This produces individuality, a precious value that cannot be replaced by anything else.

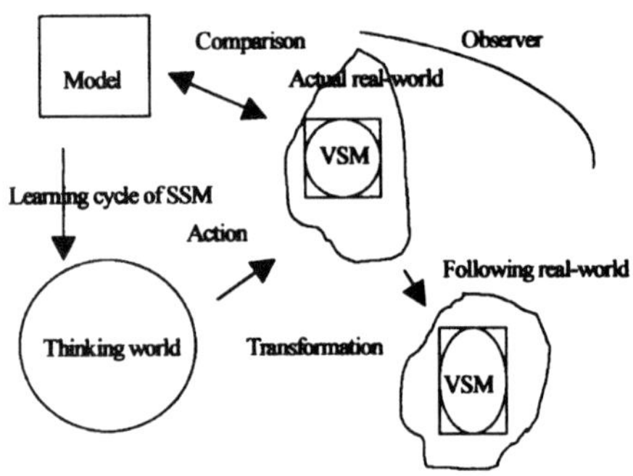

Figure 1. The Transformation Systems by SSM with VSM Incorporated

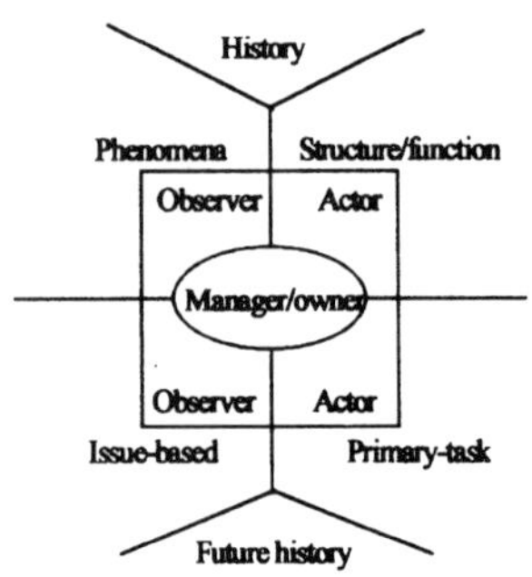

Figure 2. Transform from the Process of SSM to Questionnaire

I have produced a table of relevant systems and chosen categories after converting the process of SSM shown in Figure 2 for this survey. This figure is useful for making a comprehensive analysis of the problem and deriving a basic definition regarding the system conversion. I have prepared the following categories: "History" for a historical viewpoint, "Phenomenon" to identify changes and problems, "Structure/Function" to identify their role in society and decisive power in a political context, "Issue-based" from the standpoint of values and standards, "Primary-task" to determine concrete strategies and schemes, and "Future history" for their potential. Items in such categories as the history of the industry, current phenomenon and point of argument present findings and knowledge of an observer, while structure/function, task and prospects present the viewpoint of an actor.

Towel Industry in the Osaka Area

The towel industry split from the cotton industry and developed independently. The course of development was never easy, with excessive competition caused by an increased number of firms, industrial combination forced by wartime regulations, and a change in the target from overseas countries to the domestic market. Despite the harsh environment, the towel industry constantly grew until 1985, the turning point after which the number of firms started decreasing. In 1991, the level of production also started decreasing. The number of firms is currently 222, down from its peak of 613. On February 26, 2001, the Japan Towel Industry Association filed a petition with the Ministry of Economy Trade and Industry to ask the government to invoke a textile safeguard and surveys on the influence of imported goods on the towel industry in Japan were begun in compliance with the WTO agreement.

Hypothesis about Industry-Relevant Systems

A producing centre is an aggregation comprising of a number of small-scale firms. It is not an integrated single business entity. As top-down decision-making cannot be performed in such a system, a mechanism of self-organisation works to determine its direction, with each constituent manager's activities interacting. The source of the industry's leadership lies in the managers, and their outlook on the world and managerial activities determine the direction of the firms' future. If we assume that various phenomena in the social economy are caused by a natural mechanism, the rise and fall of an industry is an incidental result of natural selection, and when an industry starts declining, we cannot stop its flow. If, on the other hand, we acknowledge the effectiveness of rational decision-making based on reason, we can allow the industry to achieve a higher level by facing the challenges offered by the environment. Selections that are rationally made are not always correct. In the course of natural selection, however, we tend to get caught in the trap of the local maximum to avoid decline in performance, while rational selection allows an intentional transition toward the global maximum.

In order to apply a universal proposition that an industry grows from the local maximum to the global maximum to an individual case, we have, at first, to know characteristics specific to the producing centre. In July 2000, I conducted a business diagnosis of two firms located in the producing centre, and obtained some preparatory knowledge from interviews with the managers. The knowledge thus obtained allowed me to formulate the following hypothesis on the industry-relevant system.

1. Managers of the producing centre don't manage their business based on deductive business analysis or casual sequences. They make their decisions and take actions by comparing changes in the actual environment and formation of the industry and that this is the reason the industry was able to grow for over 100 years. Survey results also demonstrated that managers face up to changes in their business environment and possess interactions with ideal models. This management technique resembles the methodology of soft systems. If this is true, managers and researchers can take the same system approach.
2. Through an enhanced empirical study of managers, an industrial system has been formed in which a number of independent small firms divide the production process amongst themselves, in order to avoid risks. This system could cope with fluctuations in towel demand, and achieved cost effectiveness.
3. The reason for the decrease in the number of firms is that an increased amount of imported towels has resulted in excess supply, lowering transaction prices to the point that further production becomes impossible. This is one of the reasons for requesting implementation of the WTO safeguards regarding textiles.
4. Conventional division of labour doesn't have requisite variety that can deal with changes in the distribution structure brought about by the development of digital information technology, as well as disruption of the equilibrium of production costs caused by globalisation of free trade activities. It is necessary to design a new type of producing centre system that amplifies requisite variety.

Survey of Managers' Consciousness and Relevant Systems

As a field survey, I interviewed several managers in January and February 2001 to learn about their awareness and industry relevant systems. I have chosen the firms from the list of executives of the Osaka Towel Industry Association and the 1998 business directory. I asked 15 managers from these firms for an interview and was able to meet 9 of them to conduct a hearing survey.

The results of the survey illustrate an image of a sound manager. They believe in economic rules that they have learned from their own experiences, rather than economic theories written in books. They learn such economic rules by properly perceiving the status, and by comparing it with their acquired economic rules. This management technique is very close to the definition of system thinking, and 8 of 9 interviewees chose "System Thinking" in the consciousness survey. Managers respect such economic rules as "keeping business within its proper scale", "customer loyalty", and "efficient business" that have been inherited from previous generations. They also respect such enlightened ideas as "happiness of the staff involved", "flexible creativity", and "use of information technology". It is suspected that these attitudes of managers have worked effectively for the potential to overcome challenges.

Analysis of Survey Results

The field survey verified the hypothesis about characteristics and problems of the industry by using relevant systems. The following 4 major conclusions were drawn:

First, in the consciences survey managers chose system thinking as the decision-making process. Managers are involved in daily transactions and production processes while bearing managerial responsibilities, therefore they need to deal with all tasks immediately and appropriately. Under such circumstances, it is essential to make a skilful assessment of the managerial situation centring on human beings, while data analysis plays only a supplemental role. This fact indicates that managers and researchers can share the same system approach, and the characterisation of this fact as practical knowledge is a great step forward. System thinking, however, is not understood as SSM by the managers, but rather just what they have in their mind.

Second, the industry system has focused on the production process from purchasing cotton yarns to selling products to wholesalers, and has not involved the wholesale function within the system. This is because the production centre was located next to Osaka, a place of great consumption, and this enabled firms to have a close relationship with the wholesalers. The scale of firms has remained small, because they had learned that with their independent business style, expansion of scale could be accompanied by risks. Accumulation of production by many small firms achieved economies of scale by dividing work amongst themselves and resulted in improved productivity. A consecutive, large-lot production had advantages in making towels, and that led to formation of a division of labour system. That is, each firm specialised in either beaming, weaving, bleaching/dyeing, sewing, or the packaging process. Firms specialised in weaving played a main role in guaranteeing and selling products to wholesalers. The division of labour production system by small-scale firms was also effective in absorbing fluctuations in demand of towels, although fluctuations were comparatively small.

Third, the industry is affected by changes in distribution structure due to the development of Information Technology. Direct transaction by retailers has created unstable relation between production lines and wholesalers. Evaluating the low prices, distributors changed their strategies and began importing towels from overseas. Joint venture companies investing in China have no other choice but to follow the nation's policy to export 70% of their production. Increases in number of imported goods led to excessive stock, which became one of the factors causing a lowering in transaction prices. Low transaction prices made reproduction difficult, and created an environment that restricts the domestic industry's activities.

Fourth, current changes in the managerial environment are far beyond the industry's capability, and it is impossible to control such issues as international price differentials due to differences in cost, changes in distribution structure brought about by IT development, and the start of foreign investment and overseas production. That is to say, the towel industry does not have the requisite variety to deal with the changed environment.

Root Definition of System Transformation

Analysis of industry-relevant systems indicates two important points. First, in order to amplify the industry's requisite variety, it is necessary to establish a distribution system that involves the wholesale function and links production lines and retail business directly. In order to provide the wholesale function within the system, its economy has to be large enough to absorb its cost. This contradicts the conventional system that has produced most of the advantages associated with small-scale business. It is necessary to plan a new system to achieve economies of scale, while respecting the independent sprits of the managers.

Second, managers regard the industry to be as important as themselves personally, and are making every effort to improve the managerial environment. In the industry, such managerial activities are seen as getting into direct retail business, investing in distribution warehouses that enforce inventory functions, or in production facilities for prospective goods, developing differentiated products, and making a co-operative group with other firms sharing the same concepts. A change in the structure is going on, with firms parting from the aggregation of large number of elements of the same quality toward making groups by their own characteristics. This phenomenon suggests that the industry is ready to be transformed. The above examination leads to a root definition of a transformation system of an industry.

> "A system that converts division of labo ur into a co -ordinate supply system", which, "by integrating firms while keeping their independence and by including a system that co-ordinate such functions as order-taking, distribution, inventory, and wholesale ...", "amplifies requisite variety " "so that the industry can grow in balance with imported goods "

This root definition describes a strategy level that sets the direction of survival and growth of the industry. This industry transformation system has input of "division of labour production structure by many independent firms" and output of "a common supply system made of integrated independent firms, which include a distribution co-ordinate function" with a transformation process in between. The transformation system can be described with CATWOE elements as follows:

C : Manager, Consumer
A : Manager
T : Transformation from division of labour to a co-ordinate supply system
W : Rational decision-making enables an escape from the trap of the local maximum
O : Manager / Owner
E : Change in distribution structure due to IT development
Globalisation of free economic activities

From Division of Labour to a Co-ordinate Supply System

As a conceptual model that embodies the root definition derived from the industry-relevant systems, I suggest a co-ordinate supply system based on VSM structure. The nature of the problem is that the industry system does not have the requisite variety to deal with complexity, generated by the interaction between rapid IT development and globalisation of free trade. It is this lack of requisite variety that prevents the industry from intervening by changing the distribution structure.

One measure to amplify the requisite variety of the industry system is to provide the wholesale function linked to retail business within the system. Individual firms, however, do not have an economy large enough to include the distribution function. Economies of scale are necessary to ensure economical efficiency. A concrete measure addressing this problem is establishing a common supply system of recursive structure, which unifies independent firms and provides a co-ordination function. The basic conceptual model of the co-ordinate supply system is VSM. The concept of the co-ordinate supply system is different from that of a co-operative or trust. In the co-ordinate supply system, individual firms produce items, which are integrated and distributed in a co-ordinated process. As a similar business form, I analysed the course of development and structure of Seven-Eleven Japan, that developed a convenience store chain, and demonstrated the presence of VSM. Figure 3 shows the common supply system with a distribution co-ordination function, based on VSM as a conceptual model.

The co-ordinate supply system has the following characteristics. In System 1, managers take responsibility for all the economic activities of their firm. Therefore, the instruction function of the headquarters restricted, and managers need management skills making use of cybernetic indirect control. When System 1 falls into financial difficulties, a special committee is set up after receiving request from System 1, and takes charge of restructuring the system by discussing the situation with financial institutions. As low cost is a requirement of the co-ordinate supply system, a digital information network should be extensively utilised. Supply chain management using network and POS systems are already in actual use, and priority is now placed on the skill to design a realistic architecture and establishment of the system. The essence of transformation is that an information system is provided first, to which the industrial environment is to be adapted to in a flexible way.

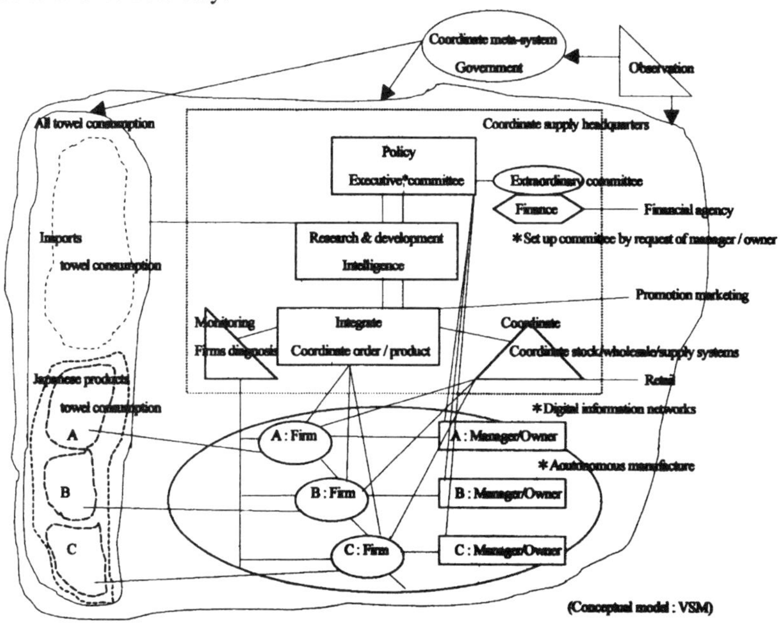

Figure 3. Co-ordinate Supply Systems of the Towel Production Firms

SUMMARY

In this document, I used a cybernetics/systems approach to perform research and analyse the problems in the towel industry in the Osaka area, where the number of firms is decreasing. As a conclusion, I suggest a transformation in the industry's structure from division of labour style to a co-ordinate supply system. Considering the qualities of managers as well as potential lying within their sense of community, it is highly likely that a co-ordinate supply system could be embodied. The industry, however, must overcome the actual situation of a diminishing local maximum, achieved by rational decision-making, and then move towards the global maximum. For this purpose, accommodation using a cybernetics/systems approach and establishment of a networked management organisation are essential.

Sincere effort and sound management philosophy of the managers played a significant role in keeping the Osaka towel industry successful for over 100 years. However the world challenges us with greater influences and neutralises the effect of our efforts. These difficult challenges can however stimulate reasoning of human beings, which, in turn, can become a motivation for greater development.

What is happening in the towel industry in Osaka reflects the reality of the social economy that Japan now faces. The role of the government as a meta-system will become more significant in adjusting the level of sacrifice and welfare in a world of free trade. It is not useful to make abstract generalisations by legislation or reductive measures based on casual relationships. What is required is providing support to the people working in an industry, who wish to contribute to its development by a comprehensive and realistic approach. Simultaneously, we have to establish moderate business relations from a global standpoint, reflecting that Japanese industries, with their high productivity, must have had a great influence on various overseas manufacturers.

REFERENCES

Beer, S. (1979). *The Heart of Enterprise*, John Wiley & Sons, Chichester.

Beer, S. (1981). *Brain of the Firm,* 2nd ed., John Wiley & Sons, Chichester.

Bertlanffy, L. (1968). *General System Theory*, George Braziller.

Checkland, P. (1981). *Systems Thinking, Systems Practice*, John Wiley & Sons, Chichester.

Checkland, P. and Scholes, J. (1990). *Soft Systems Methodology in Action*, John Wiley & Sons.

Espejo, R., Schuhmann, W., Schwaninger, M. and Bilello, U. (1996). *Organizational Transformation and Learning*, John Wiley & Sons, Chichester.

Flood, R.L. and Jackson, M.C. (1990). *Creative Problem Solving*, John Wiley & Sons, Chichester.

Wiener, N. (1961). *Cybernetics,* 2nd ed., The M.I.T. press.

Kusutarou, I. and Shouiti, O. (1935). *Japan Towel Industry History*, Japanese Towel Industry Assn.

Osaka Towel Industry Compilation Comm. (1966). *Osaka Towel Industry History*, Osaka Towel Assn.

Osaka Towel 100-Year Commemoration Compilation Committee. (1986). *Osaka Towel industry History (100 -Year)*, Osaka Towel Association.

Kyoichi, K. and Hiroshi, D. (1997). *Inquiry of System Intellect 1st*, Nikkagiren.

Takayuki, K. (1998). "Autonomy Network Organisation Approach to Systemic Management," *Spring 1998, Proceedings of Research Convention of JASMIN*, 219 - 222.

Takayuki, K. (1998). *Systemic Management*, Asahi Culture Centre.

Takayuki, K. (2000). "Autonomic System Model Appearance from Continual Problem-Solving with Digital Network Systems," *The 6th Society Information System Research Symposium Proceedings*, 23-28.

Takayuki, K. (2000). "On the Evolution of Systems and Undecidable Things," *Spring 2000, Proceedings of Research Convention of JASMIN,* 392-395.

TOWARDS A CONTINGENCY FRAMEWORK FOR INTERVENTION IN THE CONTEXT OF A SYSTEMS-BASED APPROACH

Eugene H. Melan

School of Management
Marist College
Poughkeepsie, New York 12601, USA
Eugene.Melan@Marist.edu

INTRODUCTION

In the practice of applying systems-based problem solving and learning approaches to a client system, the presence of mediating factors governing the outcome of the intervention is often not realized or may be assumed but not accounted for. Current theories of intervention are lacking a framework to enable a practitioner, experienced or novice, to understand and cope with various and oftentimes complex factors governing the success of the intervention. It is the purpose of this paper to examine the primary contingency factors involved in an intervention while employing a systems-based methodology such as SSM.

The origins and development of contingency theory and its current status will be reviewed. Its success in application to matters concerning organization suggests consideration in examining the factors underlying the intervention process between the practitioner as an interventionist and the system served or client system. The use of a method or methodology as an instrument of intervention, it will be argued, serves to give an additional level of complexity to the process and in itself possesses contingencies both general and method-specific.

A contingency model of intervention is proposed in the context of an SSM application and is intended to provide a practitioner with a schema for addressing the primary contingencies that need to be understood prior to engagement.

CONTINGENCY THEORY: ORIGINS AND CURRENT STATUS

During the first half of the last century the prevalent thinking of management theorists at the time was that there are certain general management principles that are applicable to all types of organizations. For example, one of the prevailing theories, Scientific Management, proposed a "one best way" of managing an organization (Taylor, 1947). In the 1950's, Burns and Stalker, (1961) and Woodward (1965) performed landmark studies of the organizational structure of various firms in the UK, which negated this view. In the Woodward research, for example, there was a realization early in the study that there was little correlation between management practice and organization size or efficiency. Woodward concluded

> The classical school of thought on organization appears to be inadequate not only because it fails to provide a direct and simple basis for relating organizational structure and business success, but also because its concepts and abstractions do not facilitate the work of those who plan organizational structure. Woodward, 1965, p 124.

This work on structural contingencies, in addition to that of Burns and Stalker (1961), Lawrence and Lorsch (1967), Thompson (1967), and (1967) laid the foundations of contingency theory.

In the following three decades, these early studies on structural contingencies were elaborated upon (for example, Fry & Slocum (1984) and Galbraith (1977)) and, more importantly, research expanded into other areas of management. Research on situational factors relating to effective leadership of work groups (Fiedler, 1967), organization culture (Child, 1981), innovation (Damanpour, 1996) and managerial work (Mintzberg, 1973) have been reported in management and organization literature. A large body of research on contingency theory and its application exists. One of the primary teachings of modern contingency theory is that individual and group behaviour in social situations is contextual (Greenberg and Baron, 1995).

INTERVENTION: A CONTINGENCY VIEW

Since intervention is a social process (Argyris, (1970), Beckhard, (1969), Eden, (1982), Harvey and Brown, (1982)) it can be argued that intervention possesses contextual factors and therefore, a contingency approach to intervention merits consideration.

There are several views of intervention that can be taken. Checkland and Holwell (1998) and Lewis (1994) view intervention in term of two interacting systems: the system served (problem content system) residing in an organization and the server system. Leavitt (1964) sees the system served as a complex sociotechnical system composed of people, technology, task and structure. Further, responsibility and authority for the development and maintenance of the system served lies with management whose work is contingent upon environmental task, operational and situational factors (Mintzberg, 1973). Intervention may also be viewed as a social process involving interventionists in interaction with the actors of the client system with an objective of alleviating the problem situation. The activities of the intervention process are dependent upon the basic assumptions underlying the process. For example, Argyris sees the process as ending with developing a client commitment to a choice among alternatives (1970). Checkland views it as a learning process (1999). Schein (1969) seeks to provide insight, and both Harvey and Brown (1982) and Ormerod (1995) view the process as achieving change.

Consequently, the process itself is contingent upon the steps used (reflecting the views of the practitioner) and its objectives as well as underlying assumptions. Intervention, then, can be taken as contextual and, therefore a contingency approach is suggested.

TOOLS OF THE INTERVENTION: METHODS/METHODOLOGIES USED

Many, if not all interventions, involve the use of tools, techniques, methods or methodologies ranging from simple to complex. In interventions employing systems-based methods or methodologies their classification is frequently done according to whether the problem situation is 'hard' or 'soft' reflecting a reductionist or an interpretive approach to addressing the problem. Of interest here is the employment of methods/methodologies in soft problematic situations, particularly Soft Systems Methodology (SSM) (Checkland, 1981; Checkland and Scholes, 1990; Wilson, 1990). SSM was chosen for several reasons: its popularity of use, particularly in the UK, its relative richness both in primary literature and in research performed and, importantly, access to practitioners of SSM, both novice and experienced. These reasons provided the basis for determining the contingency factors associated with a systems-based method employed in the intervention.

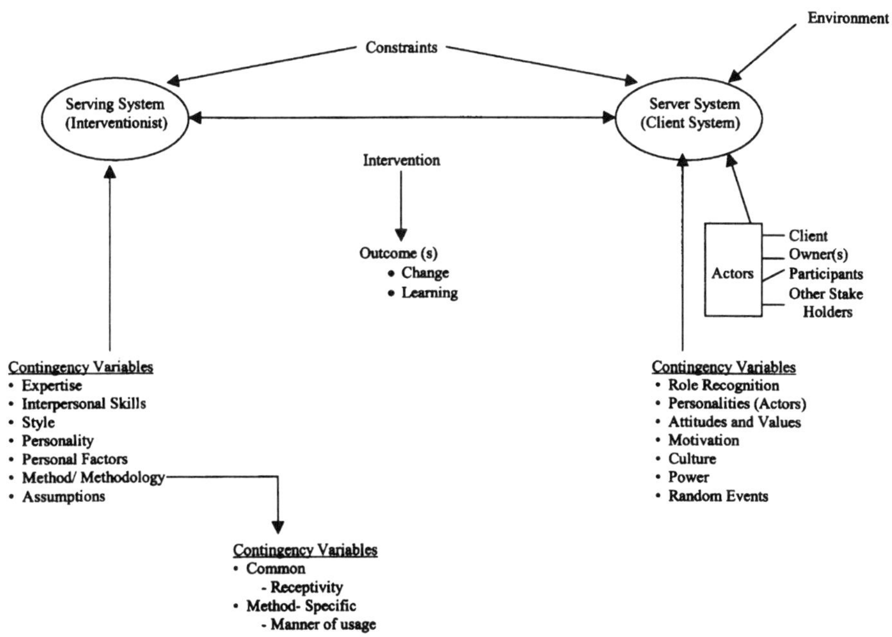

Figure 1. A Proposed Contingency Model of Intervention

CONTINGENCY FACTORS OF THE INTERVENTION SYSTEM

One can view intervention systemically in terms of two primary components or subsystems in interaction with one another: the interventionist and the client system. It is an interactive two-linked social system characterized by a 'problem solving' system and a 'problem context' system (Checkland, 1981) or more recently by a 'serving' system and a 'served' system (Checkland and Holwell, 1998), see Figure 1. The interaction between the two subsystems is subject to constraints such as resources, time, ethics and information. This system is purposeful as well as transient. It exists to provide an outcome, which is a product of the interaction. Once the outcome is achieved, the contract between the interventionist and the client is fulfilled and the intervention effectively ended.

Each of the two subsystems possess variables which are contingency factors that may affect the outcome of this interactive relationship. Based on content analysis of relevant literature, surveys, interviews of both novice and experienced practitioners, and the author's experience as a practitioner, the primary variables of the serving system affecting the outcome of the intervention are suggested to be:

- Expertise. The experience and knowledge of the individual in the methodology used in conducting the intervention.
- Interpersonal skills. The social skills of the practitioner in dealing with actors of the client system.
- Style. The compatibility of the intervention style or approach used by the practitioner (Tichy, 1974).
- Personality. The personality traits of the interventionist reflected in behaviour perceived by actors of the client system (such as introversion/extroversion).
- Personal factors. Economic, academic, or reputational factors that may affect behaviour and conduct of the interventionist.
- Method/Methodology. How the intervention method or methodology is used with actors of the client system and how it is communicated may affect the receptivity of the method and the ultimate success of the outcome. Some methods rely heavily on accuracy of the elicited inputs of the actors, others require participation of the actors and problem owners. For SSM, such factors as the jargon of the methodology, and difficulties with its epistomology have affected its use (Mingers and Taylor, 1992).
- Assumptions. Assumptions underlying the intervention as well as the methods used may affect the process and outcome as well. In SSM, for example, the outcome of the intervention is "action to improve the problem situation" if one assumes the use of either the seven-stage or the two-stream model (Checkland and Scholes, 1990). However, if one were to use the LUMAS model, learning is the assumed outcome (Checkland, 1999).

The client subsystem is the second component of the intervention system. This subsystem can be viewed as a set or sets of actor constituencies composed of the client, owner(s), participants and other stakeholders of the organization. The primary variables of this system that may influence the outcome of the intervention are:

- Role recognition. It is important for the interventionist to understand the roles of the various actors and their interrelationships in this social subsystem. Of particular interest is determining the respective roles and relationships of client, problem owner(s), the key actors and other stakeholders that may potentially affect the outcome. Knowledge of the roles played by the various actors during the decision making process is especially important in view of its potential effect on outcome.

- Personalities. Personality may influence the behaviour of individuals and groups, which may in turn, affect the interactions with the client system and consequently influence the outcome of the intervention.
- Attitudes and values of the actors. These are closely linked and may influence behaviour manifest during the intervention process. A negative attitude, for example, may result in uncooperative behaviour towards a consultant who is perceived as an instrument for downsizing the organization.
- Motivation is an underlying factor governing the behaviour of individuals and groups in organizations and may be either intrinsic or extrinsic or both. In the context of intervention, motivation refers to the willingness of the actor to participate in the process and relies on the actor's perceived self interest. The actor's perception of the benefits of intervention, the existence of peer and group social pressures and power may influence the actor's behaviour towards the process and consequently its outcome.
- Culture. Organization culture is a well-known factor influencing the behaviour of individuals and groups. It is a variable of the client subsystem that can influence the intervention process, interactions with the interventionist, receptivity towards the method used and the subsequent outcome of the intervention. The significance of culture in the practice of SSM is evident in its two-stream version where the first stream, the 'culture stream' takes into account analysis of the intervention, the 'social' system, and the 'political' system (Checkland and Scholes, 1990). Although not explicitly defined, cultural analysis is implied in the social and political systems analysis. The second stream refers to the logic-based stream comprising some of the traditional stages of SSM.
- Power. Power as manifested in organizations, like culture, is another significant variable that may govern the outcome of the intervention. Power underlies decision-making and consequently may govern the course of the intervention as well as its outcome. Power, covert or implied, may provide an element of coercion in the conduct of the actors during the intervention process and therefore bias the outcome.
- Random events. The intervention process, as with other matters of organization, is subject to unforeseen events that may affect its progress and outcome. Personnel changes may affect the stakeholder profile, key actors; and problem owners may change. Funding changes may affect the scope of the intervention as well as changes in views of the new problem owners. Determining organizational stability is only a partial answer to assessing change.

CONTINGENCY FACTORS OF METHOD

We have, thus far, discussed factors affecting the intervention process and its outcome in terms of two subsystems: the serving system or the practitioner-interventionist and the served or client system. The method(s) used present additional contingencies both common to methods in general as well as method-specific. General contingencies applicable to the use of any method require a level of familiarity with the epistomlogy as well as proficiency in applying it in problem situations to achieve a desired outcome. Many of the factors discussed in connection with contingencies of the interventionist are applicable here since they are independent of method. However, these factors may be mediated based on the manner of use by the practitioner. For example, if the method used is Systems Dynamics, the modeling portion of the intervention is frequently used as a personal tool without involvement of the actors. On the other hand, the elicitation stage of the method generally requires group participation in defining the constants and variables to be used in the simulation models to be generated. For SSM, its less prescriptive Mode 2 version allows for a range of use from a prescriptive seven-stage (Mode 1) version used in participation with the actors of the client system to its use only as a 'personal tool' by the practitioner (Checkland and Scholes, 1990). Here, the mode of use may be governed by the receptivity of the actors (of the client system) to the method. Some methods are designed to be used participatively such as SODA (Eden, 1989), interactive planning (Ackoff, 1981) and strategic choice (Friend and Hickling, 1987). The participative use of a method poses an additional contingency on the system served, namely that of facilitation skills.

A PROPOSED CONTINGENCY MODEL OF INTERVENTION

Intervention, as discussed earlier, may be viewed as a two component system consisting of a served and a serving subsystem in interaction with one another and subject to constraints, open to an environment and producing an outcome as a function of the intervention process. The two subsystems serve as a core of the model. Arguing that contingency theory may be applicable to the intervention process, a set of primary contingency variables pertaining to each subsystem is proposed. These are based on analysis of relevant literature, surveys and interviews of inexperienced and experienced practitioners and personal experience. Figure 1 illustrates the proposed model; the variables have been explicated in the preceding sections.

Its intent is to provide a framework for understanding and accounting for the variables potentially affecting the outcome of an intervention.

REFERENCES

Ackoff, R. (1981). *Creating the Corporate Future*. Wiley, New York.

Argyris, C. (1970). *Intervention Theory and Method.* Addison-Wesley, Reading, MA.

Beckhard, R. (1969). *Organization Development: Strategies and Models*. Addison-Wesley, Reading, MA.

Burns, T. and Stalker, G. (1961). *The Management of Innovation.* Tavistock, London.

Checkland, P. (1981). *Systems Thinking, Systems Practice*. Wiley, Chichester.

Checkland, P. (1999). *SSM: A 30-Year Retrospective.* Wiley, Chichester.

Checkland, P. and Scholes, J. (1990 b). *Soft Systems Methodology in Action.* Wiley, Chichester.

Checkland, P. and Holwell, S. (1998). *Information, Systems and Information Systems.* Wiley, Chichester.

Child, J. (1981). Culture, Contingency, and Capitalism in the Cross National Study of Organizations. In *Research in Organizational Behavior* (Cummings, l., and Staw, B. Eds.) 303-356. JAI Press, Greenwich, CT.

Damanpour, F. (1996). Organizational Complexity and Innovation: Developing and Testing Multiple Contingency Models. *Management Science*, 42(5), 693-716.

Eden, C. (1982). Problem Construction and the Influence of OR. *Interfaces*, 12: 50-60.

Eden, C. (1989). Using Cognitive Mapping for Strategic Options Development and Analysis (SODA). In *Rational Analysis for a Problematic World* (Rosenhead, J. ed.) 21-42. Wiley, Chichester.

Fiedler, F. (1967). *A Theory of Leadership Effectiveness.* McGraw-Hill, New York.

Friend, J., and Hickling, A. (1987). *Planning Under Pressure.* Pergamon Press, Oxford.

Fry, L. and Slocum, J. (1984). Technology, Structure, and Workgroup Effectiveness: A Test of a Contingency Model. *Academy of Management Journal*, 27: 221-246.

Galbraith, J. (1977). *Organizational Design.* Addison-Wesley, Reading, MA.

Greenberg, J. and Baron, R. (1995). *Behavior in Organizations.* 5th ed., Prentice Hall, Englewood Cliffs, NJ.

Harvey, D. and Brown, D. (1982). *An Experiential Approach to Organization Development*, 2nd ed. Prentice Hall, Englewood Cliffs, NJ.

Lawrence, P. and Lorsch, J. (1967). *Organization and Environment.* Harvard University Press, Boston, MA.

Leavitt, H. (1964). Applied Organization Research on Industry. In *New Perspectives in Organizational Research* (Cooper, W., Leavitt, H. and Shelly, M., eds.). Wiley, New York.

Lewis, P. (1994) *Information-Systems Development.* Pitman Publishing, London.

Mingers, J. and Taylor, S. (1992). The Use of Soft Systems Methodology in Practice. *Journal of the Operational Research Society*, 43(4), 321-332.

Mintzberg, H. (1973). *The Nature of Managerial Work.* Harper & Row, New York.

Ormerod, R. (1995). Putting Soft OR Methods to Work – Information Systems Strategy Development at Sainsbury. *Journal of the Operational Research Society*, 46, 227-293.

Perrow, C. (1967). A Framework for the Comparative Analysis of Organizations. *American Sociological Review*, 32, 194-208.

Schein, E. (1969). *Process Consultation: Its Role in Organizational Development.* Addison Wesley, Reading, MA.

Taylor, F. (1947). *Scientific Management.* Harper & Bros., New York.

Thompson, J. (1967). *Organizations in Action.* McGraw-Hill, New York.

Tichy, N. (1974). Agents of Planned Social Change. *Administrative Science Quarterly*, 19, 164-182.

Wilson, B. (1990). *Systems: Concepts, Methodologies and Applications.* Wiley, Chichester.

Woodward, J. (1965). *Industrial Organization: Theory and Practice.* Oxford University Press, Oxford.

SSM-GUIDED RESEARCH TO IMPROVE THE LINKAGES BETWEEN A SCIENCE PROVIDER AND THEIR END-USERS

Janet Reid, Terry Kelly, and Ian Valentine

Institute Natural Resources, College of Sciences,
Massey University, P O Box 11222,
Palmerston North, New Zealand.
J.I.Reid@massey.ac.nz

INTRODUCTION

People make sense of the world differently, some are guided by systems thinking and some are not. This has implications for systemic practitioners and the way they interact with others who do not see the world as they do. This paper describes the evolution of a process of inquiry to date that has been explicitly influenced by systemic researchers. The authors attempt to make sense of their experiences and to highlight some important aspects of the inquiry process to date. Highlighted in particular, are the stages and manner in which the client was involved in the process, and how the way information from the rich-description was conveyed to the client influencing the way in which the problem situation was viewed and action defined by the client.

INQUIRY CONTEXT

The authors of this paper were contracted to undertake research by a New Zealand science provider (the client) with whom they had not previously worked. The client undertakes research and generates knowledge in the area of soil quality management. The primary end-users of the client include Local Government entities (who are delegated the responsibility of managing regional natural resources), agricultural consultants, and Government Ministries.

This inquiry process contributes to a program of research being undertaken by the client that is funded by Government from the Public Good Science Fund. The New Zealand (NZ) structure for funding public good science is based on the premise that the contribution of science to the sustainable management of natural resources is dependent on the successful application of that science by end users. The New Zealand Government, through the Foundation of Research Science and Technology (FoRST)

Edited by Ragsdell *et al.*, Kluwer Academic/Plenum Publishers, New York, 2002.

directs public good science by allocating funds to specific outcome areas in order to encourage the delivery of outcomes by science providers that are of benefit to NZ (FoRST, 2001a). Science providers compete for the opportunity to deliver the specified outcomes and attract funding allocated to this area by FoRST. In addition to directing the area in which research will be undertaken, FoRST also promotes collaborative behaviour as a means of ensuring the findings of research are implemented to benefit NZ.

> "We want to see genuine engagement with potential end-users of research and between research groups and organisations" (FoRST, 2001b).

Therefore, the quality of the linkages between science funders, providers and end-users is considered to be a key factor in attaining success in this funding structure. The research described in this paper contributes to an objective that is part of the research contract between the client (science provider) and FoRST. This objective is directed at improving sustainable soil management by exploring ways that information is generated by the science provider and adopted and utilised by end-users. Although required as part of the contract, there is also a desire on the part of the client to improve the level of implementation of the science by end-users.

SYSTEMIC INTERVENTION

Many documented cases describe the use of Soft Systems Methodology (SSM) by practitioners and researchers motivated by a desire to inform and bring about meaningful change in complex problem situations that involve people (Brocklesby, 1995; Checkland and Scholes, 1990; Reid *et al.*, 1999). This paper explores the implications of researchers adopting a systemic approach of inquiry to undertake research for a client into a problem situation that was not initially defined in a systemic way. Selecting SSM as a framework to guide action in a research project implies an acceptance on the part of the researchers that the concept of a system is a useful intellectual construct (Checkland, 1981). Further, it implies that the researchers possess a desire to intervene in a purposeful way to contribute to meaningful change in that situation and that learning by those involved in the situation will contribute to that change (Checkland, 1985; Midgley, 2000).

When systemic researchers accept the role as change agent they also accept a role in the facilitation of learning. The differentiation Midgley makes between intervention and systemic intervention was found to be useful in terms of describing the focus of the learning that the researchers aimed to facilitate with the client, through the process of inquiry.

> "If intervention is purposeful action by an agent to create change, then *systemic* intervention is purposeful action by an agent to create change *in relation to reflection on boundaries* (Midgley, 2000, p129).

Learning results from interaction that encourages participants to reflect on and redefine the boundary of the problem. In this case, the researchers worked to achieve change by firstly gaining an understanding of the problem situation and by working with the client to redefine the problem boundary and consequently the nature of the inquiry process itself.

PROCESS OF INQUIRY

An initial guide for the researchers was provided by a set of objectives that were negotiated and agreed upon by the research team and the client. These objectives also marked the formalisation of a relationship established on a level of common understanding between the researchers and client. The objectives are output focused, and reflect some assumptions about the nature and boundary of the problem situation to be investigated. Four key objectives were:

1. Determine the specific end-users, and link them to specific types of information provided by the programme;
2. Determine how this information is accessed and utilised;
3. Understand the information flow processes within end-user organisations;
4. Identify ways to improve uptake and use of web-based information and tools by the different end-users.

These objectives directed the researchers to study the end-users and to gain an understanding of the information flows within these organisations, and based on this, identify ways to improve the uptake of information provided by the client. The boundary of the problem situation was initially defined within the objectives as bounding the end users only (Figure 1.).

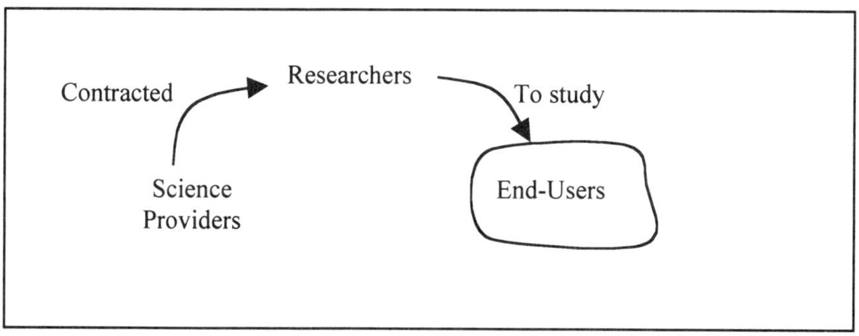

Figure 1. Initial Conceptualisation of the Problem Situation and Inquiry Process

A systemic worldview implicitly underpinned the researchers' approach and conceptualisation of the problem situation under investigation. Consequently, the approach taken was to gather information to enable researchers to develop a rich-picture understanding of the situation, primarily from the perspective of end-users and to a lesser extent from the client. Researchers approached the problem situation with a willingness to accept that the boundary of the problem situation was in fact broader than initially suggested in the research objectives. This was confirmed from preliminary information gathered.

The client identified four end-user organisations, three local Government entities and a Government Ministry. These end-users provided the focus for the researchers' data collection. Semi-structured interviews were undertaken with a cross section of people from each organisation. Individuals were selected on the basis of a snowball sampling technique (Patton, 1990). That is, each interviewee was asked to identify other individuals in the organisation whom they believed would provide a perspective that would enrich the researchers understanding of the problem situation. Specifically, interviewees were asked about their role in the organisation, the information they sought and used and how they accessed that information. Further, those interviewed were asked about their awareness and understanding of the client and the nature of their involvement and relationship with the client.

Researchers developed a rich-description as a way of making sense of the data. The rich-description was developed from a synthesis of the common themes, structures and processes (Checkland, 1981; Wilson and Morren, 1990) that emerged from those interviewed and from the limited discussions that had taken place with the client.

Initial objectives for this research implied that the nature of the problem related to the flow of information between and within end-user organisations. However, the researchers' findings suggested that the factors impacting on the uptake and use of information provided by the client were much broader. Factors impacting on the quality of the linkage between the client and the end-users were identified as a more meaningful way to make sense of the problem situation (Figure 2.).

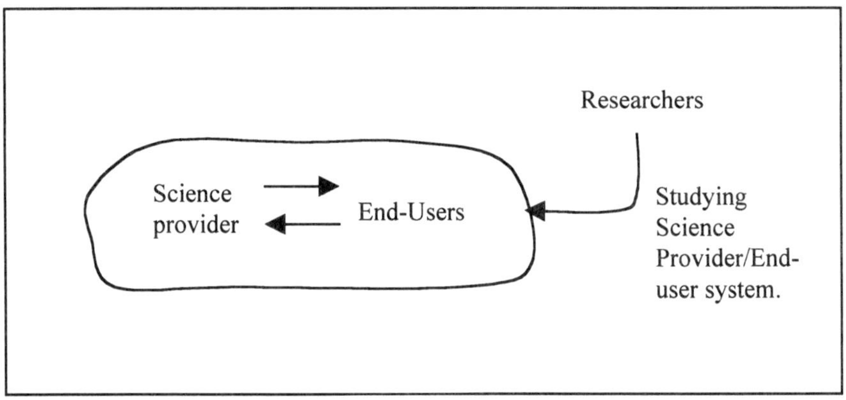

Figure 2. The Problem Situation as the Linkage between the Science Provider and End-users

The researchers identified that change to improve the linkages was more likely to occur if the client redefined the boundary of the problem to include themselves as science providers and their end-users. If a change in thinking by the client was to occur, the client needed to be willing to change and also they needed to be accepting of the role of the researchers in assisting this change. Therefore, it was the intent of the researchers to work with the client to assist them to redefine the boundary of the problem to include not only themselves and their end-users but also the researchers (Figure 3). The researchers were asking the client to renegotiate the initial basis on which the relationship between client and researchers was established, without compromising the achievement of the objectives that underpinned that initial relationship.

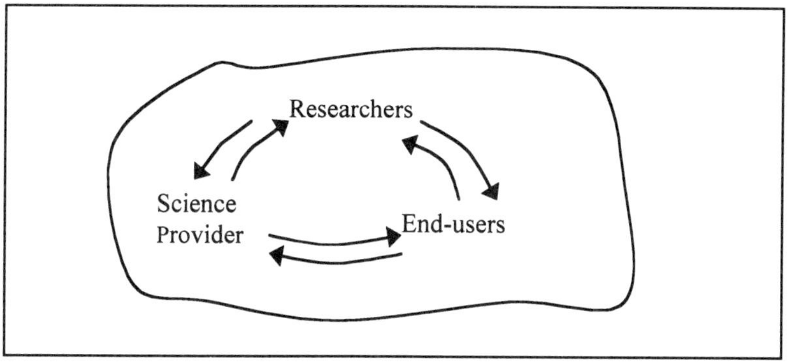

Figure 3. The Researchers, Science Providers and End-users as Part of the Problem Situation

The science provider who is the client for the inquiry process is a group of individuals located at a number of sites in New Zealand. The researchers had principally been communicating with one individual within the client group. At this stage in the inquiry process the researchers sought to involve the larger client group in the process to define action. This intent influenced the way in which the rich-description was expressed for the client and the manner of the interaction with the client by the researchers.

Sharing the Rich-description

Visualisation of the rich-description as a rich-picture proved useful for the researchers. However, as understanding of the situation developed and the decision was made to involve the client group in the process, how to convey the rich-description understanding to the client group became an issue. The development of a logical hierarchy of concepts based on a qualitative data analysis technique (Dey, 1993) was found to be a useful way to analyse and represent an understanding of the rich-description. The hierarchy provided a perspective on the problem situation that was considered likely to encourage the client to redefine the boundary of the problem situation and look to include themselves, the researchers and end-users in the situation.

The quality of the linkage between the client and end-users was a concept that provided the focus for the hierarchy. Drawing from the rich-description the researchers identified key concepts relevant to the quality of the linkage between the client and end-users. The process of analysis enabled the researchers to identify abstract concepts that encompassed factors relevant across all three end-user organisations. The logic for the structure of the hierarchy was based on identifying the key factors that impacted on the concept placed one level higher in the hierarchy. At the highest level of the hierarchy five factors were identified as influencing the quality of the linkages between the client (science provider SP) and their end-users (EU) (Figure 4). These five factors are: the end-users commitment to national or regional responsibilities [1]; the nature of personal relationships between individuals in the science provider group and end-user organisation [2]; the degree to which information from the science provider meets the needs of the end-users [3]; the level of awareness of information the science provider generates [4]; and finally the end-users perception of the science provider group [5].

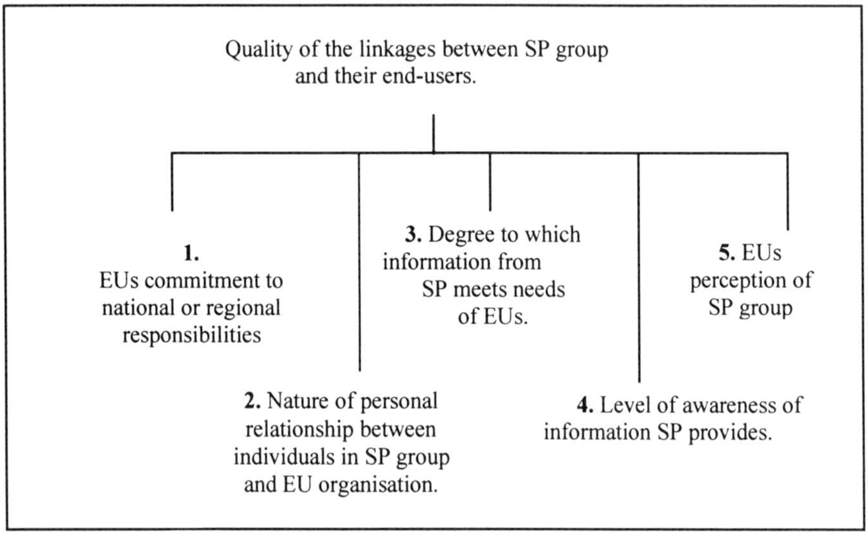

Figure 4. Highest Level Nodes in Logical Hierarchy (SP refers to science provider and EU refers to end users)

The hierarchy was further expanded to include progressively lower levels for each of the five higher levels. To illustrate the expanded nature of the hierarchy a branch of node 2, the nature of personal relationships between individuals in the science provider group and end-user organisations is included below (Figure 5).

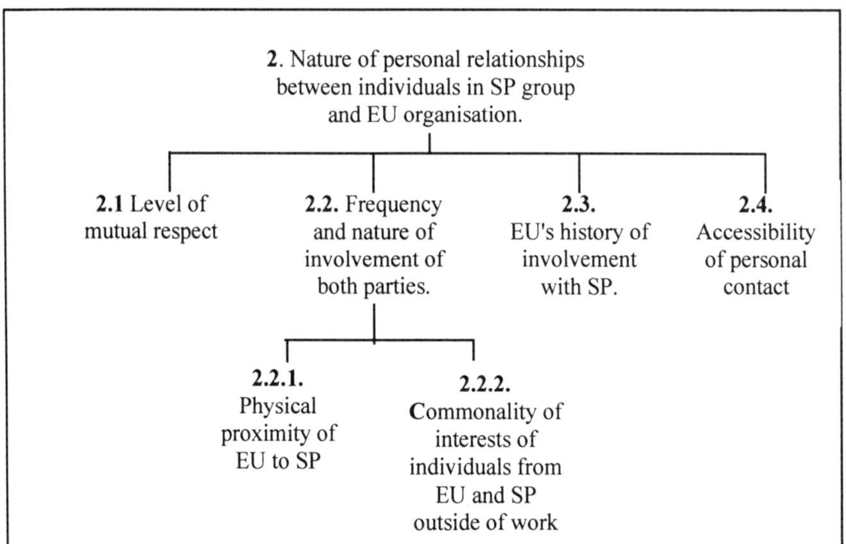

Figure 5. Expanded Hierarchy of Node 2 from the Logical Hierarchy

The nature of the personal relationships between individuals from the science provider and the end-user organisations is influenced by four main factors. These factors are the level of mutual respect that exists between individuals from the end-user and science provider organisations [2.1]; the frequency and nature of the involvement between these individuals [2.2]; the history of that involvement [2.3]; and the accessibility of personal contact between the science provider and the end-users. Both, physical proximity of the science provider relative to their end-users [2.2.1] and the commonality of interests of people from these organisations outside of work [2.2.2] were factors that influenced the frequency and nature of their involvement together [2.2].

Client Workshop: Shifting the Boundary

In order to progress the inquiry process, individuals from the client group agreed to participate in a workshop with the researchers. From the outset the focus of the workshop was broader than the problem as it was defined by the original objectives for the research. The focus of the workshop was on the linkages between the client as a science provider and their end users. The stated aims were threefold:

1. To gain the perspective of individuals in the client group on the flows and utility of soil quality research information between themselves and their end-users;
2. To present preliminary findings on the perspectives of some of the end-users of soil quality research information; and
3. To gain input from the client group as to the next phase in the inquiry process for the researchers.

Researchers facilitated an informal participatory workshop that was designed to be inclusive and encourage the sharing of ideas and learning. In small groups, the science providers firstly were asked to represent visually the flow of information between themselves and their end-users. Secondly, the groups were asked to represent the factors that they believed influenced the linkage between the end-users and themselves as science providers. Once the perspectives of the client group had been shared, the researchers presented the end-users' perspective in the form of the hierarchy of factors influencing the quality of the linkage.

There was general acceptance by the client group that it was useful to view the problem situation as involving both the end-users and themselves as science providers. Their view of the role of the researchers is as yet unclear, although they are willing to continue with the relationship and responded with interest to the findings of the inquiry to date. For some, the hierarchy was viewed as a framework that they could use to guide their future involvement with end-users and to assist them to design their organisational communication strategies.

Within the boundaries laid out in the contract between the client and FoRST, this research needs to achieve certain outputs. However, the opportunity for this research also to add value to the client organisation is accepted. The client expressed a willingness to shift the boundaries and to work with the researchers.

CONCLUDING COMMENTS

The initial contract between the client and the researchers suggested a non-systemic definition of the problem involving the flow of information within and to the end-user organisations. Guided by a systemic worldview implicit within SSM, the researchers deliberately set out to facilitate learning by the client about the problem situation. Through their involvement with the client and the manner in which preliminary research findings were presented to the client, the researchers influenced the client's perspective of the problem boundary and thereby the nature of the problem. This change in perspective by the client also changed the nature of their involvement in the problem situation and with the end-users and the researchers.

The process of inquiry described in this paper is on going. From the researchers' perspective it is hoped that they can continue to work with the client and end-users to improve the quality of the linkage between the parties.

ACKNOWLEDGEMENTS

Financial support for this study was provided by the New Zealand Foundation for Research, Science and Technology as part of contract CO9X0016.

REFERENCES

Brocklesby, J. (1995). "Using Soft Systems Methodology to Identify Competence Requirements in HRM", *International Journal of Manpower,* 16(5/6):70-84.

Checkland, P.B. (1981). *Systems Thinking, Systems Practice,* John Wiley and Sons, Chichester.

Checkland, P.B. (1985). "From Optimising to Learning: A Development of Systems Thinking for the 1990s", *Journal of Operational Research Society,* 36(9):757-767.

Checkland, P.B. and Scholes, J. (1990). *Soft Systems Methodology in Action,* John Wiley and Sons, Chichester.

Dey, I. (1993). *Qualitative Data Analysis.A User-Friendly Guide for Social Scientists,* Routledge, London.

FoRST (Foundation of Research Science and Technology). (2001a). "Public Good Science and Technology", [Online]. Available: http://www.frst.govt.nz/public/pgsf/pgsf2.htm [Retrieved October 8 2001].

FoRST (Foundation of Research Science and Technology). (2001b). "PGS&T and NERF Funding Principles and Process 20001-2004", University Presentation. September 2001.

Midgley, G. (2000). *Systemic Intervention: Philosophy, Methodology and Practice,* Kluwer Academic / Plenum Publishers, New York.

Patton, M.Q. (1990). *Qualitative Evaluation and Research Methods,* (2nd edition), Sage Publications, Newbury Park, USA.

Reid, J.I., Gray, D.I., Kelly, T.C., and Kemp, E.A. (1999). "An Application of SSM in the On-Farm Labour Situation in the New Zealand Dairy Industry", *Systems Research and Behavioral Science,* 16:341-350.

Wilson, K.K. and Morren, Jr. G.E.B. (1990). *Systems Approaches For Improvement in Agriculture and Resource Management,* Macmillan Publishing Company, USA.

A MULTI-METHODOLOGICAL APPROACH TO EMERGENCY CALL HANDLING IN THE METROPOLITAN POLICE SERVICE

Raymond Rowe

Apercu Limited
91 Monkhams Lane
Essex, IG8 0NN
Dr.Rowe@apercu.org.uk

INTRODUCTION

This paper explores a multi-methodological approach to evaluating and improving Metropolitan Police Service (MPS) policy toward emergency telephone call handling from 1999 to 2001. A multi-methodological approach was adopted to identify the problem situation and agree an agenda for change. Soft Systems Methodology (SSM) and a number of other methodological techniques were used to assist in identifying the problem situation. Archive data was gathered throughout the research to inform SSM. Interviews were conducted with emergency telephone call handlers and inform SSM. Questionnaires were designed to investigate policy and procedures for emergency telephone call handlers operating in the MPS and inform SSM. A number of emergency telephone call handlers were observed and identified as case studies to inform SSM.

PROBLEM SITUATION

The Metropolitan Police Service (MPS) operates a sophisticated electronic communication centre, based at New Scotland Yard; the Territorial Policing Central Communications (TPCC) employs 348 staff on a shift basis providing a 24 hour service to London. The most significant functions of TPCC include the following:

1. Answering 999 calls and initiating an appropriate response
2. Providing liaison with the organisations such as the London Ambulance Service (LAS), London Fire and Civil Defence Authority (LFCDA), London Buses, and British Transport Police (BTP)
3. Deploying and managing traffic units across London
4. Initiating an appropriate response to the activation of sensitive alarms (e.g. embassies, palaces)
5. Providing a Central Casualty Bureau (e.g. the World Trade Centre atrocity).

Edited by Ragsdell *et al.*, Kluwer Academic/Plenum Publishers, New York, 2002.

One of the key aspects of call handling is to classify calls according to the speed of response required by the police to the incident. This process is known as call grading and is carried out using a Central Graded Response Manual. The priority for action depends on the urgency and severity of the incident.

TPCC is pledged to meet a performance target of 80% of calls to be answered within 15 seconds of the call being presented under their Performance Charter. Performance against this standard had been falling in 1999 and 2000. Call volume has been increasing year on year for the last 5 years. The total number of calls handled in 1999 was 2,329,073 compared with 2,727,500 in 2000. Although there is a seasonal trend in calls, the increasing number of 999 calls is fairly consistent throughout the year. The problem situation was TPCC's ability to improve their performance target for 2001 using their current staffing levels and introducing skills-based emergency call handling.

WHY A MULTI-METHODOLOGICAL APPROACH WAS USED

A number of distinct but we believe complementary methodologies were used. The author had experience of using a multi-methodological approach during his research into multi-agency decision-making (Rowe, 2000). The links between these are highlighted in Figure 1. Starting with the outside loop archive data was obtained and analysed. The results informed the design of interviews. Information gathered from the interviews informed the design of the questionnaires. Information gathered from the questionnaires identified which emergency call handlers to observe. Information gathered from the observation of emergency call handlers identified case studies. Information gathered from the case studies informed the archive data. The cycle continued again through interviews, questionnaires, observation and case studies. All five methodologies informed Soft System Methodology (SSM) the output of which fed back into the use of those methodologies and assisted in understanding the problem situation.

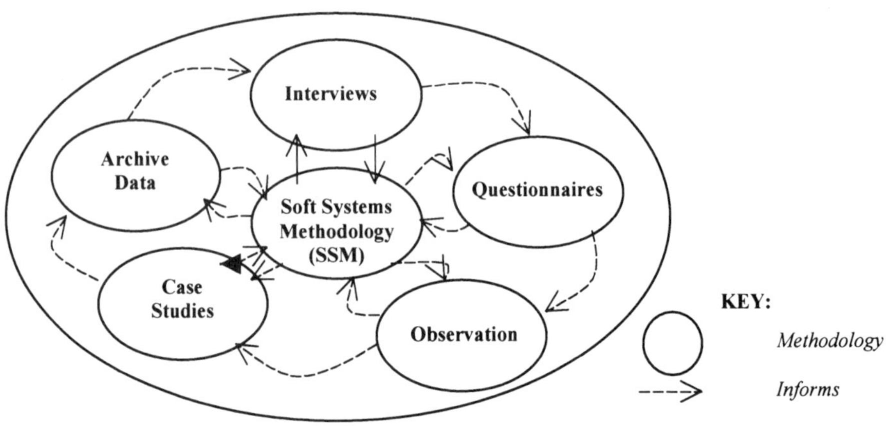

Figure 1. Multi-Methodological Approach: Inter-relations Cycle (Rowe, 2000)

Mingers and Brocklesby (1996) produced a grid, based on Habermas' suggestion that it is useful to distinguish between our relations to, and interactions with, three worlds: the material world, the social world, and the personal world. Mingers and Brocklesby (1996) found it useful to place these three separate domains on a grid that could be used to map characteristics of different methodologies and show how they link together (see Figure 2). By employing Mingers and Brocklesby's (1996) grid the author undertook a detailed study of different methodologies to identify the benefits of linking them for the research. Mingers and Brocklesby (1996) suggest that before adopting a particular methodology or intervention, consideration should be given as to which domain and activity on the grid the methodology or technique can be of assistance. The author found that the completed grid generated a list of possible methodologies and techniques to choose from. SSM was chosen , as indicated in Figure 2, since it offered an appreciation of the 'social' and 'personal' domains that included social practices, power relations, individual beliefs, meanings and emotions. SSM offered an analysis of differing perceptions and weltanschauung and, an opportunity to explore alternative conceptualisations and constructs. Also, SSM offered action to be generated to seek accommodation and consensus. Archive data was gathered to inform the preparation of interviews and questionnaires. The results of these were in turn used to identify emergency call handlers for observation and case studies. Soft Systems Methodology (SSM) was used to generate an agenda for change. Qualitative and quantitative methods were used to evaluate and validate those changes.

	Appreciation of	**Analysis of**	**Exploration of**	**Action to**
Social	social practices power relations **SSM** **Observation Case Study**	distortions conflicts interests **Archive** **Observation Case Study**	ways of altering existing structures **Archive** **Observation Case Study** **Questionnaire**	generate empowerment and enlightenment **Archive** **Case Study** **Interview**
Personal	individual beliefs meanings emotions **SSM** **Interview** **Questionnaire**	differing perceptions and weltanschauung **SSM** **Interview** **Questionnaire**	alternative conceptualisations and constructions **SSM** **Interview** **Questionnaire**	generate accommodation and consensus **SSM**
l	physical circumstances **Archive** **Observation Case Study**	underlying causal structure **Archive** **Interview** **Questionnaire**	alternative physical and structural arrangements **Interview** **Questionnaire**	select and implement best alternatives **SSM** **Interview** **Case study**

<u>**KEY:**</u> Areas where SSM was useful in this research

Figure 2. Mapping Methodology and Techniques to Mingers' and Brocklesby's 1996 Grid: Linking phases and dimensions of an intervention (Mingers and Brocklesby 1996)

Mingers and Brocklesby (1996) support the multi-paradigm approach to research suggesting that:

> 'Adopting a particular paradigm is like viewing the world through a particular instrument such as a telescope, an X-ray machine, or an electron microscope. Each reveals certain aspects but is completely blind to the others. Although they may be pointing at the same place, each instrument produces a totally different, and seemingly incompatible, representation.' (Mingers and Brocklesby, 1996).

The author was aware that terms such as 'paradigm', 'methodology', and 'technique' are commonly used with a variety of meanings. In this paper, therefore, it was decided to use the definitions suggested by Mingers and Brocklesby (1996). Briefly, a *paradigm* is a very general worldview based on a set of fundamental philosophical assumptions that define the nature of possible research and intervention. A *methodology* is a structured set of guidelines or activities to assist an individual in undertaking research or intervention. A methodology will develop, either implicitly or explicitly, within a paradigm and will embody the philosophical assumptions and principles of the paradigm. A *technique* is a specific activity that has a clear and well-defined purpose within the context of a methodology (Mingers and Brocklesby, 1996:104).

USING SSM TO IMPLEMENT SKILLS-BASED EMERGENCY CALL HANDLING

The use of SSM was found to be an appropriate methodology to implement skills-based emergency call handling in the MPS, and there follows a discussion on root definitions, conceptual models and CATWOE or TWOACE (Rowe, 2000) below. A relevant system to implement skills-based emergency call handling was identified; 'A system owned and operated by members of the Metropolitan Police Service to implement skills-based emergency call handling.'

Then a root definition of a system to implement skills-based emergency call handling was identified: 'The system identifies the likely difficulties to be encountered in implementing skills-based emergency call handling aimed at improving the effectiveness of emergency call handling.'

Smyth and Checkland (1976) devised a method for checking whether root definitions were well formulated called the 'CATWOE' criteria. The mnemonic stands for Customers (C), Actors (A), Transformation Process (T), Weltanschauung (W), Owners (O) and Environmental Constraints (E). If all these elements could be readily identified in the root definition then there was a good chance that the RD was viable. The author discovered that when he used the mnemonic CATWOE the importance of the transformation process (T) in testing the RD was not emphasised. Checkland (1981) summarised the work of Smyth and Checkland (1976), stating that:

> 'The core of a root definition of a system will be a *transformation* process (T), the means by which defined inputs are transformed into defined outputs. The transformation will include the direct object of the main activity verbs subsequently required to describe the system.'(Checkland, 1981: 224)

Checkland (1981) stated that *ownership* was the next criterion, followed by *actors, customers, environmental constraints, and weltanschauung.* Checkland suggested that as there was likely to be more than one possible weltanschauung (W), a separate root definition ought to be prepared for each (W) considered relevant. The author developed his own mnemonic, TWOACE (Rowe, 2000) for testing the RD. It was a good aide memoir to remembering the CATWOE criteria but most importantly it highlighted the importance of the transformation process (T) and weltanschauung (W) in testing the RD. TWOACE was considered a more appropriate test for the RD and has contributed to the knowledge of SSM.

Using the Mnemonic TWOACE

1. The transformation process (T) is the conversion of input to output:
   'From an organisation without skills-based emergency call handling to an organisation with skills-based emergency call handling.'
2. The weltanschauung (W) is the worldview that makes T meaningful in context:
   'The use of skills-based emergency call handling is likely to improve the effectiveness of emergency call handling in the MPD.'
3. The owner (O) who could stop T:
   'The Metropolitan Police Service.'
4. The actors (A) are those who would do T, the emergency call handlers:
   'Police and civilian emergency call handlers.'
5. The clients and customers (C) are those who benefit from T or are the victims of T:
   'The police service and the community.'
6. The environmental constraints and elements (E) outside the system which it takes as given:
   'The legal position of handling emergency calls.'

This RD satisfies the TWOACE criteria. The verbs were isolated from this RD and expanded to provide the basic conceptual model (CM) as shown in Figure 3.

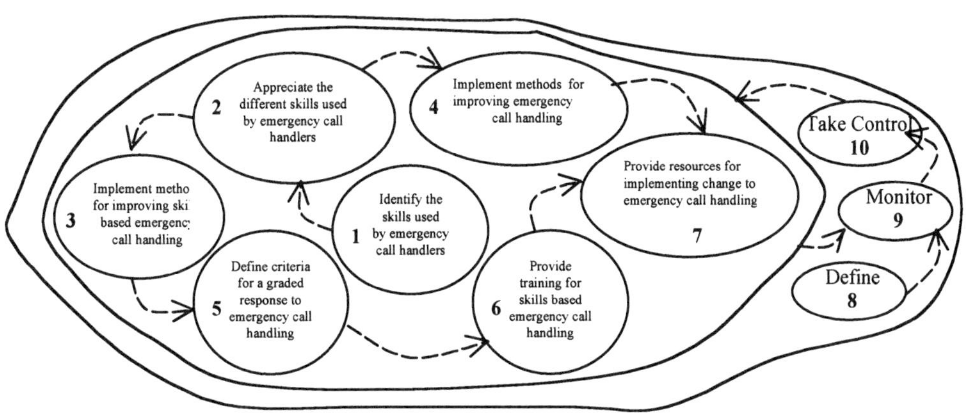

Figure 3. Conceptual Model of a System to Implement Skills-based Emergency Call Handling

The Conceptual Model of a System to Implement Skills-Based Emergency Call Handling

Conceptual model of a system to implement skills-based emergency call handling was identified (see Figure 3):

1. Identify the skills used by emergency call handlers;
2. Appreciate the different skills used by emergency call handlers;
3. Implement methods for improving skills-based emergency call handling;
4. Implement methods for improving emergency call handling;
5. Define criteria for a graded response to emergency call handling;
6. Provide training for skills-based emergency call handling;
7. Provide resources for implementing change to emergency call handling;
8. Define performance criteria 1 – 7 above;
9. Monitor 1 – 7 above;
10. Take controlling action to ensure that the system to implement skills-based emergency call handling works.

Generating an Agenda for Change

Checkland insists that the comparison stage proper should be conducted with the principal actors in the problem situation so that a debate resulting in agreement about 'desirable and culturally feasible' changes will be generated. The owner and actors discussed "**Table 1**. Activity in the conceptual model of a system to implement skills-based emergency call handling" and agreed that it should form an agenda for systemically desirable and culturally feasible change leading to an improved problem situation. The agenda formed the background of recommendations submitted to the MPS TPCC policy makers.

Implementing Recommendations to Improve the Problem Situation

The emergency call handlers at MPS Territorial Policing Central Communications were concerned that the automatic call distribution (ACD) system was rather ad hoc. It did not allocate emergency calls to those call handlers with the necessary skills. This put less experienced call handlers under extra pressure. They were more likely to make mistakes and take longer to resolve the call. This was likely to have an impact on MPS TPCC's ability to improve their performance target. The senior management team (SMT), at MPS TPCC, agreed that the emergency call handling should be skills-based and that changes to the ACD be implemented. They also provided resources for identifying and training skills-based emergency call handling. The SMT adopted the graded response manual, which identified the criteria and appropriate response to emergency calls. The graded response manual was the output of a MPS TPCC working party and was adopted by call handlers elsewhere in the MPS. The senior management team, at MPS TPCC, agreed to monitor the implementation of changes and improvement in the problem situation through their monthly management report.

Table 1. Activities in the Conceptual Model of a System to Implement Skills-based Emergency Call Handling

Activity in Conceptual Model	Exist or not in real situation?	How is it done?	Who is doing it?	Comments: Good/ Bad?	Alternativ Include on Agenda?
1. Identify the skills used by emergency call handlers	No A,C,I,O &Q	Observe and monitor emergency call handlers	SSM analyst / Call Centre Manager	Presently not identified	Yes
2. Appreciate the different skills used by emergency call handlers	No A,C,I,O &Q	Communicate differences through training	SSM analyst / Call Centre Manager	Unable to appreciate different skills	Yes
3. Implement methods for improving skills-based emergency call handling	No A,C,I,O & SSM	TPCC SMT to support agenda for change	SSM analyst / Call Centre Manager	Ad hoc emergency call distribution	Yes
4. Implement methods for improving emergency cal handling	No A,C,I,O & SSM	TPCC SMT to support agenda for change	SSM analyst / Call Centre Manager	Changes to made to ACD	Yes
5. Define criteria for a graded response to emergency call handling	No A,C,I & O	TPCC working party to agree and prepare a graded response manual	TPCC produced Graded Response Manual	Revise current operating procedures	Yes
6. Provide training for skills-based emergency call handling	No A,C,I & O	TPCC SMT to make policy for training	TPCC produced Skills-based training material	Length of training problematic	Yes
7. Provide resources for implementing change to emergency call handling	No A,C,I & O	TPCC SMT to make policy for allocation of resources	Some MAP members	Lack of resources to support change	Yes
8. Define performance criteria 1-7 above	No C & SSM	TPCC SMT monitor monthly report	Call Centre manager to identify data for monthly report	Situation could be improved	Yes
9. Monitoring 1-7 above	No C & SSM	TPCC SMT monitor monthly report	Call Centre manager to identify data for monthly report	Situation could be improved	Yes
10. Take controlling action to ensure that the system to implement skills-based emergency call handling works	No C & SSM	TPCC SMT monitor monthly report	Call Centre manager to identify data for monthly report	Situation could be improved	Yes

Source of information in real situation above:

Key:

A	= Archive Data	TP/CC	= Territorial Policing Central Communications
C	= Case Study		
I	= Interviews	SMT	= Senior Management Team
O	= Observation	ACD	=Automatic Call Distribution
Q	= Questionnaire	SSM	= Soft Systems Methodology

CONCLUSION

The paper set out to explore a multi-methodological approach to evaluating and improving Metropolitan Police Service (MPS) policy toward emergency telephone call handling from 1999 to 2001. The author had experience of using such an approach while researching multi-agency decision-making (Rowe, 2000).

A multi-methodological approach was adopted to improve the problem situation. The MPS Territorial Policing Central Communications (MPS TPCC), based at New Scotland Yard, provides a 24 hour service to London. The most significant functions of the MPS TPCC is answering 999 calls and initiating an appropriate response. The problem situation was MPS TPCC's ability to improve their performance target for 2001 using their current staffing levels and introducing skills-based emergency call handling.

Lessons were learned from the integration of multi-methodological techniques with SSM. The author mapped a variety of multi-methodological approaches on to a grid, based on Mingers and Brocklesby (1996). The continual flow of data and information to inform SSM was gathered from archive data, interviews, questionnaires, observations and case studies. This was used to generate an agenda for change to improve the problem situation. The MPS TPCC implemented the changes and these were evaluated. The emergency telephone call handlers were considered cost effective in dealing with the increasing number of emergency telephone calls. The introduction of a skills-based emergency telephone call handling system has led to greater consistency in police emergency telephone call handling.

The author devised his own mnemonic, TWOACE (Rowe, 2000), for testing the root definition. It was a good aide memoir to remembering the CATWOE criteria but most importantly it highlighted the importance of the transformation process (T) and weltanschauung (W) in testing the root definition. TWOACE was considered a more appropriate test for the root definition and has contributed to the knowledge of SSM.

REFERENCES

Checkland, P.B. (1981). *Systems Thinking, Systems Practice,* John Wiley & Sons, Chichester.

Mingers, J. and Brocklesby, J. (1996). "Multimethodology: Towards a Framework For Critical Pluralism", *Systemist*, 18(3):101-131.

Rowe, R. (2000). *Young Offenders and the Criminal Justice System: A systems science approach to evaluating and improving Metropolitan Police Service policy toward young offenders.* Unpublished PhD Thesis, City University, London.

Smyth, D.S. and Checkland, P.B. (1976). Using a systems approach: the structure of root definitions, *Journal of Applied Systems Analysis*, 5(1):75-83.

USING SSM TO IMPROVE SUPPLY CHAIN EFFECTIVENESS

Ross Smith, David Mackay, Graeme Altmann and Gulender Gencoglu

School of Management Information Systems
Faculty of Business and Law
Deakin University, Geelong, Victoria, Australia 3217
Contact eMail: smithr@deakin.edu.au

INTRODUCTION

Supply chain partners are always striving to better coordinate their activities, to reduce inventories, and improve the responsiveness of the flow of goods and information. Techniques including quick response (QR), efficient consumer response (ECR) and just-in-time (JIT) manufacturing are often used to assist this process. While the mutual exchange of information is an essential component of success, information exchange by itself is insufficient to achieve maximum gains.

In a combined business and government initiative seeking to improve supply chain processes in the Textile, Clothing and Footwear industry, network facilitators have been used to run workshops designed to drive inefficiencies from the supply chain and achieve improved "agility" and "pipeline acceleration". However, despite reporting significant improvements in productivity, participants also comment on a lack of structuring within the workshops. This paper reports on an innovative approach to improve the efficiency of industry supply chains by using Soft Systems Methodology (SSM). SSM has been used in conjunction with existing workshop techniques to provide new insights into Supply Chain Management (SCM) issues.

Initially in this paper, the scene is set with a brief review of some areas of SCM concern, and recent trends. SSM is then briefly introduced, and the objective of the present research program is established. The research approach is then outlined, including a brief background to the workshop participants and their situation. Some of the key results are then outlined, including insights into the way in which SSM might be integrated with an existing SCM workshop technique used for some years now by the Quick Response (QR) Group within Business Victoria. Some of the perceived strengths of the use of select SSM techniques in this context are reported. Finally, some possible future paths for this research are discussed.

Edited by Ragsdell *et al.*, Kluwer Academic/Plenum Publishers, New York, 2002.

SUPPLY CHAIN MANAGEMENT

Interest in SCM has grown steadily since the 1980's, as firms have come to recognise the benefits to be gained from collaborative relationships within and beyond their own organisations. It has been realised that it is not possible for companies to effectively compete in a climate of isolation from their suppliers and other key stakeholders; cooperation brings success for all (Lummus and Vokurka, 1999).

Areas of current SCM concern have been highlighted by Hakanson (1999), and include:

1. Improving customer service - having the right product at the right place at the right time to reduce the number of people purchasing elsewhere;
2. Saving money and reducing costs; and
3. Achieving better cash utilisation - the time it takes to be paid for products delivered.

According to Li (2000), recent trends in supply chain management are now pushing companies to more diverse products, shorter product life cycles, increased competition and ever-increasing globalisation. These trends challenge companies seeking to achieve supply chain efficiencies. Li argues that in order to address such challenges stakeholders in the supply chain must convene regularly to discuss matters of mutual concern.

One might observe that the above concerns and trends are but symptoms of a fundamental SCM principle. Forger (2000) has suggested that what makes one supply chain better than another is collaboration. Collaboration allows companies to predict supply chain events, as well as permitting suppliers and customers to plan around potential glitches while maximizing their return on supply chain opportunities. Further, the impact of collaboration is not limited to improvements in the inter-organisational chain, but can have profound effects within the four walls of each company making up that chain.

In summary, examination of the SCM literature reveals difficulties with the lack of communication between supply chain partners, cultural inconsistencies, and difficulties with the integration of systems between key stakeholders in the supply chain. In short, if one is to take collaboration seriously, to achieve the full potential of a supply chain, one must be able to analyse the social, cultural, and political dimensions of a SCM situation.

SOFT SYSTEMS METHODOLOGY

It is beyond the scope of the present paper to review Soft Systems Methodology (see Checkland and Scholes (1990) and papers therein). Suffice to say that SSM has been widely used as a general systems-based problem-solving framework but little has been written on its application to SCM problems, although Rigby *et al.* (2000) makes oblique reference to some perceived shortcomings of SSM when describing research undertaken in the field of agile supply chains (see also Day, 1999).

Developed in the UK and used around the world, SSM provides a set of procedures and notations for investigating organisational problem situations and taking action to improve those situations. SSM has a well-documented set of processes, supported by a rich set of techniques for finding out about problems and generating systems-based models that can be used to generate systemically desirable and culturally feasible organisational change.

Even though SSM has not been widely applied to addressing problems in supply chain management, the very nature of SSM appears to make it a suitable vehicle to address the social, cultural, and political issues commonly associated with SCM problems. What is evident from a search of the SCM literature is that SCM problems of a human nature are overwhelmingly more common than problems of a technical nature. In this sense, organisations are social systems, as well as business systems - thus the social construction of problematic situations in organisations are complex to understand and manage.

RESEARCH OBJECTIVES

Reflecting the above observations based upon the extant literature, the aim of this study is to address the research question:

> Can SSM be used to complement an existing workshop-based approach for thinking about supply chain management problems by conceptualising and modeling the key activities of suppliers, retailers and other stakeholders?

As such, can SSM techniques such as rich pictures, CATWOE analysis, root definitions and conceptual models be used to reveal issues that were previously unrecognised by those involved in the management of supply chains?

RESEARCH APPROACH

Given the exploratory nature of the present research, a case study approach has been used to address the research question. Yin (1994, cited by Rubin and Babbie (1997)) defines a case study as an empirical inquiry that: *'Investigates a contemporary phenomenon within its real-life context; when the boundaries between phenomenon and context are not clearly evident; and in which multiple sources of evidence are used'.*

Participants in the present case study were a senior supply chain manager, a production planner, a SCM facilitator and two SSM facilitators. Two half-day workshops were used to collect the data. The SCM facilitator was from Business Victoria, a Victorian Government initiative for improving supply chain management processes within the Textile Clothing and Footwear industry (TCF) industry. The facilitator had extensive experience in running SCM workshops but had no knowledge of SSM. The senior supply chain manager and the production planner were from a large Australian company in the TCF industry, specifically a producer of socks and hosiery.

As a producer of socks and hosiery, the organisation manufactures locally as well as importing products that can no longer be cost effectively produced within Australia. The company purchases yarn from local producers and imports yarns. The company sells to the major retail discount stores such as Woolworths, Coles Myer and Big W. They also concentrate on selling their own brands in smaller boutique stores around Australia and are looking at the prospect of exporting niche products, particularly by using Merino wool because it can be marketed as a unique Australian fiber.

The company had been previously involved in meetings initiated and facilitated by a Victorian government SCM facilitator who had spent a considerable amount of time working with TCF companies, forming cluster groups. Those that made up the cluster groups were typically a major retailer, a TCF manufacturer and one or two suppliers to the manufacturer. The aim of convening the cluster groups was to achieve mutual gain by improving their total effectiveness in servicing each other and the end consumer. The format of these earlier meetings was an approach outlined below (see the next section of this paper). It was against the experience of participants in those previous workshops that select SSM concepts and techniques were introduced in the two half-day workshops conducted as part of the present research.

Data collection from the two SSM-based workshops took the form of:

1. Some pre and post workshop semi-structured interviews with some of the workshop participants;
2. Drawings and written responses to questions asked in the workshop situation; and
3. Audio recordings (each workshop was audio recorded with the permission of participants).

Data from the above collected sources were analysed to:

1. Document supply chain management processes currently employed by workshop participants;
2. Identify SCM issues and problems;
3. Identify initial reactions to the SSM techniques introduced in the sessions; and
4. Collect reactions to the experience of using SSM techniques in the workshops.

THE INTEGRATION OF SSM WITH EXISTING SCM APPROACHES

As indicated above, the workshops discussed in this paper built on previous workshops that had been run over several years by the Quick Response (QR) Group within Business Victoria. This QR program has been documented by Perry and Sohal (2000) and is summarised on the left-hand side of Figure 1. Specifically, in these workshops senior managers meet regularly in newly formed cluster groups consisting of one or two representatives of major retailers, manufacturers and suppliers. The processes involved in the QR workshops can be summarised, as shown on the left-hand side of Figure 1, as involving four key features:

1. The generation and circulation of company SCM 'wish lists' within each cluster group;
2. Discussion of the wish lists and the development of group and company action plans;
3. The injection into each group of a consultant selected by the various companies to provide assistance with in-house Quick Response implementation activities; and
4. Follow-up strategic and tactical planning amongst each cluster group.

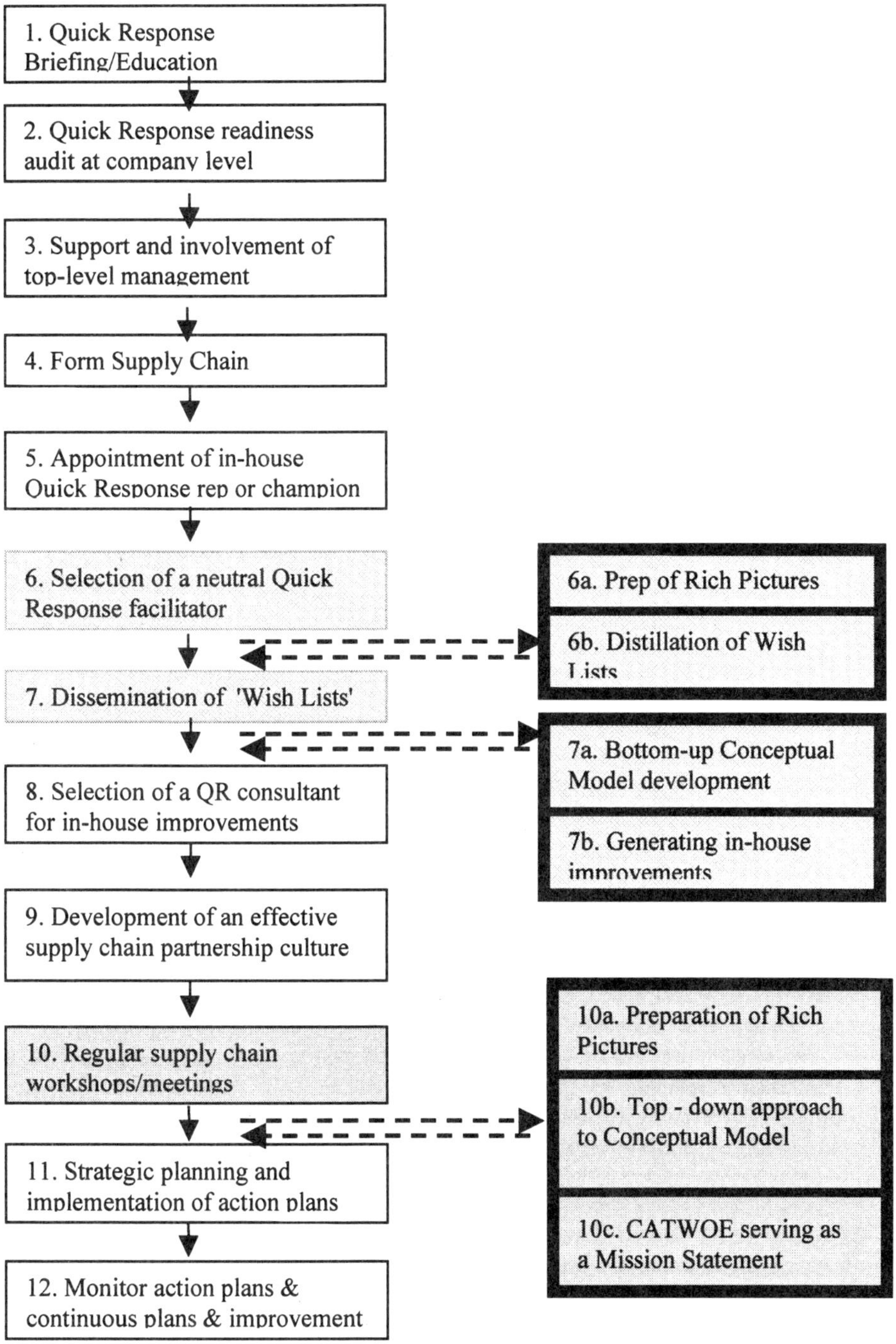

Figure 1. The QR Workshop Process Model of Perry and Sohal (2000) (left side). With additional possible points for the inclusion of SSM techniques (Steps 6a, 6b, 7a, 7b), and with the placement of the additional SSM techniques preferred by workshop participants (Steps 10a, 10b and 10c).

Results from these QR workshops (Perry and Sohal 2000) reveal that improved communication and information flow result, as well as improved performance within the supply chains involved.

Prior to the workshops discussed in this paper the present authors speculated on points at which SSM techniques might be injected into the QR workshop process model of Perry and Sohal (2000). The initial view that was formed is represented by the additional Steps 6a,b and 7a,b shown on the right of Figure 1. Specifically, it was felt that the SSM technique of Rich Picture Building and reflection upon those Rich Pictures might facilitate the preparation of the SCM 'wish list' (i.e. Steps 6a,b). Further it was felt that the use of CATWOE analysis, SSM Root Definitions and Conceptual Modeling of each node in the supply chain (i.e. a "bottom up" development of conceptual models of each node in the supply chain) might provide a useful means of facilitating the identification and specification of in-house improvements (ie. Steps 7a,b).

This possible use of SSM in conjunction with the established QR workshop process model described by Perry and Sohal (2000) was introduced to workshop participants, and reactions were collected. Results and findings from these workshops follow.

RESULTS

The Placement of SSM Techniques within the Workshop

Before looking at the reactions of participants to the individual SSM techniques, it is worth reporting the reactions of those participants to the suggested placement of those techniques within the workshop process. In summary, significant doubt emerged from the participants concerning the suggested placement outlined in Figure 1 (Steps 6a, 6b, 7a and 7b). Instead, a preferred sequencing, outlined also in Figure 1 (Steps 10a, 10b and 10c) emerged.

In summary, the participants argued that:

1. The SCM 'wish list' generation step did not require support from SSM Rich Picture building (see further discussion under 'Workshop Techniques' below). Instead, the wish list approach had the advantage of yielding quickly high value insights that build important enthusiasm for the workshops within participants. Use of graphic SSM Rich Picture building and reflection at this very early stage in the life of the workshop group was seen as unnecessarily delaying the realisation of these quick returns. Instead the workshop participants argued that the Rich Picture techniques would be best introduced subsequently, as part of the cycle of regular supply chain workshops/meetings, where it might be used as a technique to identify some of the less readily apparent SCM issues and concerns (Figure 1, Step 10a).
2. SSM CATWOE analysis and Conceptual Model Building was seen also as best introduced as part of the subsequent cycle of regular supply chain workshops/meetings (Figure 1, Step 10b). Further, it was believed that the value of the conceptual models came from examining the supply chain from a top-down perspective, rather than the bottom-up approach proposed in Step 7a. The CATWOE technique was seen also as a useful means of reviewing/reasoning about the mission of supply chain participants (Figure 1, Step 10c).

Workshop Techniques

As noted above, the workshop participants strongly supported the retention of the use of wish lists, a central component of the previous QR workshops. These were seen as a fast way of focussing workshop participants' thinking on the main supply chain issues and problems, yielding quick returns to the participants. The wish lists typically contain requests for improved working relationships between partners,improved response to customer orders and reduced pipeline waste and delay. In a sense, the wish lists were simply lists of things participants would ideally like to happen in the supply chain in order to improve their own performance and profitability. It was observed that using wish lists seemed to 'break the ice' with participants.

In the course of the workshop, when the SCM facilitator asked the senior supply chain manager to come up with a wish list, as many as 15 'problems' were produced in a manner of minutes. The QR facilitator stated reasons as to the usefulness of the wish list approach. They were:

1. It gives people ownership, therefore it's not just the facilitator coming in and stating that he is going to rework the supply chain;
2. It is the day-to-day things that people are anxious to fix;
3. The senior supply chain managers know intimately all the details, thus the solutions are usually fairly readily achieved; and
4. It leads the people in the workshop to realise that problems can often be solved by a simple alteration to a requirement or a procedure in another cell.

This is not to say that the SSM techniques, when introduced to the workshop participants, were not appreciated. Indeed, the SCM facilitator commented on the fact that the use of SSM substantially added to the wish list approach:

> 'After deriving wish lists you end up with informal results. What you want to do then is to add structure, ...now SSM techniques can be used to add that structure.'

What did emerge was that it was felt that the SSM techniques were better introduced after the quick gains achieved by using the wish list approach had been realised. In this regard it is worth noting that the participants observed that it was not a matter of whether it would be useful to apply SSM to SCM, - rather, where and how SSM could be applied.

The collected observations of participants on their exposure to the SSM Rich Picture Building, CATWOE and Conceptual Modeling techniques are recorded as reasons for using SSM (Table 1), identified strengths (Table 2) and difficulties raised (Table 3). It is beyond the scope of the present paper to discuss each of these in detail. Suffice to say that they provide support for the use of SSM in this context, and some insight into potential difficulties that may need to be addressed in future SSM-enabled SCM workshop sessions.

Table 1. Reasons cited by Participants for using SSM in the SCM Context

Reason
Acknowledges different ideas and perspectives
Promotes understanding of SCM concepts
Prompts questioning, thinking, and critical analysis
Encourages formal consideration of political and social environment
Provides a disciplined, formal way to proceed in an unstructured situation
Iterative nature of the approach

Table 2. Strengths of SSM in the SCM Context cited by Participants

Strength
Conceptual nature of the approach
Adds structure to wish lists
Focuses on the purpose of the chain, systems, needs, outputs …
Practical
Promotes creativity and builds confidence
Provides discipline and a formal structure
Systematic nature of the approach
Able to deal with unstructured situations
Participative and collaborative in nature
Prompts thinking
Allows for questioning of the current supply chain arrangements
Takes account of social, political and cultural issues

Table 3. Difficulties Encountered in the use SSM in the SCM Context cited by Participants

Difficulty
Learning how to begin to draw Rich Pictures
Problems in reaching group consensus
Learning how to represent human activities using SSM conceptual models

DISCUSSION AND CONCLUSIONS

Drawing upon the results reported above, four central themes related specifically to the introduction of SSM techniques and associated concepts have emerged from the research to date:

1. It appears that SSM allowed participants to achieve a better understanding of supply chain management processes because it introduces techniques that allow them to quickly see the views and perspectives of other stakeholders. Participants easily take to the use of rich pictures to visually show supply chain processes. Rich pictures accommodate different peoples' image of the chain, and therefore

eliminate pressures on the manager to come up with verbal descriptions of systems and processes.

2. When an approach involving using CATWOE analysis as a tool for the development of root definitions or SCM 'mission statements' was canvassed, participants felt that by incorporating the elements of customers, actors, transformation, worldview, ownership and environment, they could see and reanalyze the cooperative role of the supply chain.
3. Participants saw that the value of conceptual models in the SCM context comes from examining the supply chain from a top-down perspective. Rather than beginning with each participant in a supply chain developing their own activity model and capturing their perceptions of the current sequence of events, it was argued that it was better to start at the top-level and develop *fresh views* to overcome the danger of locking oneself into current practices.
4. SSM techniques were seen as generally useful in identifying shared and conflicting views. By having the parties come together in a workshop situation they are able to discuss key processes through the use of universal icons and remove those icons that are not appropriate. This allows inhibitors to a smooth functioning supply chain to be identified so that they can be removed.

It should be noted, however, that the experience reported in this paper is limited to two workshop sessions and associated interactions with the workshop participants. It is therefore important that the findings reported be confirmed by the conduct of a future, richer series of such sessions, with participants drawn across all tiers in supply chains, operating in a number of industrial settings. Results to date, however, are promising.

REFERENCES

Checkland, P.B. and Scholes, J. (1990). *Soft Systems Methodology in Action*, Wiley, Chichester.

Day, M. (1999). *Buyer-supplier interation in the UK tableware supply chain.* Unpublished PhD Thesis, University of Keele.

Forger, G. (2000). "Collaboration -The Supply Chain's Defining Factor?", *Supply Chain Management Review.* http://www.manufacturing.net/scm/ (Last Access: 5 Oct. 2001).

Hakanson, B. (1999). "Supply Chain Management: Where Today's Business Competes", *Ascet,* 1. http://www.ascet.com/ (Last Access: 5 October 2001).

Li, Y. (Editor) (2000). *Supply Chain Update - May 2000*, Grainger Center for Supply Chain Management, University of Wisconson-Madison. http://wiscinfo.doit.wisc.edu/grainger/. (Last Access: 5 October 2001).

Lummus, R.R. and Vokurka, R.J. (1999). "Defining supply chain management: a historical perspective and practical guidelines", *Ind. Mgt & Data Systems,* 99(1):11–17.

Perry, M. and Sohal, A.S. (2000). "Quick Response practices and technologies in developing supply chains - A case study", *International Journal of Physical Distribution & Logistics Management* 30(7/8):627-639.

Rigby, C., Day, M., Forrester, P. and Burnett, J. (2000). "Agile supply: rethinking systems thinking, systems practice", *International Journal of Agile Management Systems,* 2/3:178–186.

Rubin, A. and Babbie, E. (1997). *Research Methods for Social Work,* Brooks/Cole Publishing Company, U.S.A.

Yin, R.K. (1994). *Case Study Research, Design and Methods*, Sage Publications, Newbury Park.

IMPROVING SYSTEMS INTERVENTION IN SMEs: REFLECTIONS ON SYSTEMS BOUNDARIES IN PRACTICE

Lorraine Warren[1] and Gillian Ragsdell[2]

[1]University of Lincoln
Lincoln LN6 7TS, UK
lwarren@lincoln.ac.uk

[2]University of Paisley
Paisley PA1 2BE Scotland
gillian.ragsdell@paisley.ac.uk

INTRODUCTION

The role of the systems practitioner is complex with many dimensions to take account of; one dimension is the range of expectations that are to be met – expectations of the organisational stakeholders, of the paying client, of the wider community and personal expectations, for instance. Meeting these expectations while intervening effectively and responsibly in a situation makes numerous demands on the systems practitioner and their skills. Historically, emphasis has been on the expert, technical skills of the practitioner, which in turn has supported a functionalist approach to intervention and left the practitioner outside of the organisational system boundary. There seems to be a shift towards a recognition that it is often the social, human skills of the practitioner that can 'make or break' an intervention. This recognition promotes a more interpretive approach and often brings the practitioner within the organisational system boundary. Inevitably such a shift brings another dimension to the role of the systems practitioner and, with it, more complexity. In this paper, we suggest that an understanding of the position of the practitioner in relation to their system of interest can aid the management of this additional complexity. Analysis of interventions with two small and medium sized enterprises (SMEs) is used as a vehicle to draw out practical illustrations of different boundary positions for an interventionist and to frame discussion of the functionalist – interpretivist spectrum of approaches.

Edited by Ragsdell *et al.*, Kluwer Academic/Plenum Publishers, New York, 2002.

SMEs AND INTERVENTION

Many small firms experience the need for changes in management practice as their companies start to grow and develop. Stage models of growth for small firms, whilst they differ in their detail, highlight the requirement for the implementation of control mechanisms as the firm expands. One of the most well known models is that of Greiner (1972), who identified five stages and their associated managerial problems:

Stage	Impetus for growth	Problem
1	CREATIVITY	Crisis of leadership
2	DIRECTION	Crisis of autonomy
3	DELEGATION	Crisis of control
4	COORDINATION	Crisis of red tape
5	COLLABORATION	Crisis of communication

Greiner's model has been subject to critique for its rather mechanistic, owner-centric nature, but in the small/micro-firm arena, the very nature of growth (however that is defined) inevitably brings with it a shift in the locus of control from the Owner/Manager to others, as the need to delegate and coordinate becomes more pressing.

From a slightly different angle, Steier and Greenwood (2000) argue that successful management of a network is critical to the success of an entrepreneur in establishing and growing a venture. Networks can provide conduits for information, provide legitimacy and provide access to resources without incurring the costs of vertical integration. Yet they are not static:

> "They are constantly evolving, and the major challenge to the entrepreneurial firm becomes the handling of changing criticalities arising from growth: indirect or new links may become more centrally important; other previously critical links become less do. As the patterns of importance change, the entrepreneur faces the need to reconfigure the network in order to improve its efficiency and effectiveness". (Greenwood, 2000 pp 184)

The introduction of an external consultant to a small micro-firm is an example of a reconfiguration point, with the aim of facilitating learning and innovatory practice concerning management systems and to bring legitimacy to the change endeavour. This was so in the cases we have chosen to present here. In systemic terms, reconfiguration can be characterised as a period where boundaries, norms and values are challenged and changed in an overt goal-directed way as part of the formal consultancy process. Furthermore, the very presence of the consultant causes perturbation to the *status quo*, bringing its own boundary challenges.

Case One – Introduction to Aggregate Products Limited (APL)

Aggregate Products Limited is a company that manufactures and supplies a range of building and garden products to civil engineering contractors, large DIY retailers, garden centre companies and direct to the end consumer. It employs around 22 full time production operatives in four separate departments, with an annual turnover in the region of £2 million. The company has been in operation for around 40 years, and has operated as a family business for much of that time. A year ago, APL began a Teaching Company Scheme (TCS) project with the (then) University of Lincolnshire and Humberside. The TCS is a DTI initiative which part-funds joint projects between universities and small companies, with a graduate being employed for two years to carry out a clearly defined project. Two academics, one from the Faculty of Business and Management and one from the Faculty of Engineering, became involved along with the Managing Director and a semi-retired senior member of staff.

The project is essentially a business process re-engineering project designed to build integrated production and management systems to improve strategic and operational performance. Machinery and production techniques will also be upgraded. It is also agreed that the Associate will undertake a Masters level degree, based around the development of a modular Learning Contract developed around the project, drawing on the University's Work-Based Learning skills and ethos. In the past, management and production control systems at APL have been rather *ad hoc*, and the project aims to formalise these systems yet retain flexibility to support the firm as it evolves and production increases. Standard tools of strategic analysis (SWOT; Nolan, 1979; Earl, 1989; McFarlan, 1984) are being used, as suggested by the academics, and concepts of design are taken from current thinking on enterprise integration (Sandoe *et al*, 2001) and the VSM (Espejo *et al*, 1996). The DTI track and monitor each project through a network of regional consultants who arrange meetings at three-monthly intervals. This is in addition to the weekly, minuted meetings that take place between the company representatives and the academic partners. Presently, the project is doing well against its agreed performance indicators.

Case Two – Introduction to Hull Community Radio (HCR)

Reflections on this project have been fully presented elsewhere (Ragsdell and Warren, 1999). The key points pertinent to this discussion will be summarised here.

The authors of this paper were introduced to a small local community radio station, HCR, by a past student of the university who had become impressed by the potential of systems thinking. HCR had a core team of three who wanted to be better able to manage the amount of work they were receiving following some very successful, yet fairly *ad hoc*, ventures they had been involved with in the voluntary sector. They were considering the options of moving to a more formal organisation, such as a co-operative or a public limited company. This had led to considerations of internal structure, hierarchy and control and communication systems. We advised them to speak to financial advisers concerning the establishment of a formal structure - this was not our area of expertise. Our past experience of Beer's (1985) Viable System Model (VSM) led us to suggest its use for the remaining matters of concern.

The VSM was introduced to the group in the 'soft' way, as a diagnostic tool which could be used to see if HCR was promoting the identity, autonomy, cohesion, belonging and performance that can ensure the viability of an organisation. Any discrepancies between the HCR and the model could be addressed and action proposed to bridge any gap. The model was introduced through a series of informal meetings between the authors and the team. It was gratifying to see them readily embrace the model and the theory behind it - also that the possibility of establishing a bureaucratic hierarchy was of concern to all; it was not something they wanted to create.

We were conscious not to railroad proceedings by operating in an expert capacity, but at the same time, our capacity as facilitators was inhibited by the limited time HCR requested; our involvement seemed to be seen as a favour, and we were therefore not to be over-burdened with responsibility. A clearer definition of roles and expectations at the beginning could have avoided this uncertainty. Despite these concerns the HCR team were able to design a new organisational structure for themselves, with our help.

ANALYSIS OF CASE STUDIES

Much of the consultancy literature tends towards a functionalist stance, concentrating on expert-driven interventions into large companies. Such an approach could be characterised as modernist. The soft and critical systems movements have challenged this positioning, preferring a facilitative/interpretive stance, where meaning and consensus emerge through methodologies employing models and schemas as hermeneutic enablers. Soft Systems Methodology (SSM) (Checkland, 1981) was a watershed in the 'interpretive turn' taken by systems thinkers in the 1970s; older 'hard' models such as the Viable System Model, with its roots in Variety Engineering and cybernetics may now be used in 'soft' ways (Espejo *et al,* 1996). In the 1990s, Taket and White (1994) were amongst the first to embrace fully a post-modern perspective, where the events and phenomena of an intervention are regarded as a 'text' that might be subject to narrative analysis. It is claimed that this approach further breaks down the stereotypes of the expert and the professional, reducing the perceived distance between the practitioner and the client, through:

1. recognition of the co-responsible nature of the OR encounter
2. skill transfer in both directions and the empowerment of all involved
3. recognition of difference and working with it

This is a useful set of guidelines, which by and large we would consider that we met, and, in the case of APL, are continuing to be met, although this was not our overt intention. We are not attempting to characterise ourselves as postmodern consultants operating through narrative analysis, nor were we looking to these guidelines at the time of the interventions. Both authors would characterise themselves as practising immanent reflection in the critical tradition throughout the interventions, with what appear so far to be quite successful outcomes. However, they are useful in drawing out some key points of the interventions.

Hull Community Radio

It was important to HCR that the resultant design did not subsume individuality and yet did not focus on the individual, an important aspect of any intervention in the voluntary sector. The VSM presented opportunities to focus on a range of systems without openly formalising and assigning specific responsibilities therein - these could be resolved by discussion outside the contact time with the researchers. This empowering aspect of VSM does not normally draw attention, but it has most welcome potential. Being sensitive about the style of the intervention and sharing the VSM in the way that we did means that the group members are better equipped for managing organisations in the future - the possibility of sustainable improvement for HCR and its members.

Aggregate Products Limited

For the intervention into APL, responsibilities for all parties are clearly mapped out within the DTI schema and the project specification. The onus is on company 'ownership' of both problem and solution; it is the responsibility of the academics to ensure appropriate toolsets and methodologies are used in an ethical manner. Skill transfer also operates the other way, as not only do the academics get the opportunity to 'practise what they preach', but they have the opportunity to refine and develop their understanding of their practise in a small company.

TENSIONS BETWEEN AGENDAS

Having checked our interventions against White and Taket's (1994) guidelines, we now emphasise tensions between underlying philosophical agendas that emerged as the two projects progressed.

Hull Community Radio

It appeared, as individuals and as a group, that the HCR members held what one might consider to be 'postmodern' attitudes to the process of intervention. There was agreement with Takala *et al's*, (2001) characteristics of postmodernist attitudes since there was a clear understanding of:

1. the group (including the researchers) behaving as an open, generative and interactive system
2. the environment being turbulent and unpredictably changing
3. a focus on knowledge creation and sharing, learning process
4. the value of tools of knowledge management
5. faith in human interaction and empathy
6. accreditation of researchers/consultants being gained through trust and continued interaction

Whilst this made the intervention productive from both sides, such a match could be said to have made it relatively easy for us. However, the importance of personal agendas within HCR eventually led to the breakdown of their team, and an end of the interaction prior to what we would have considered the most productive phase, namely the evaluation. On the other hand, a loose project agenda and limited interaction meant that there was little at stake to either party, in the short/medium term, if the interaction were to come to an end. As we remarked at the time, we do wonder if we could have worked harder at maintaining the unitary nature of the HCR situation rather than taking it for granted. With little opportunity to re-explore the situation there was little chance to gauge whether there were any undercurrents of pluralist activity that might have called for other methodologies to be used in parallel with the VSM.

Aggregated Products Limited

The 'postmodern' attitudes listed above are often applied to what are loosely termed 'entrepreneurs' as traits of entrepreneurial success. Clearly, this is a rather universalising theory, given the diversity of motive amongst small/micro firm owners. Owner managers of small family firms may hold rather conservative attitudes to business management. Our impression of the Managing Director of APL is that his attitudes fall somewhere between the two-always seeking new ideas, but enjoying those within a formal structure. The structure of the TCS generally tends to enforce a need for owner-managers to adopt more 'modernist' stakeholder attitudes to the process of intervention, as characterised by Takala *et al* (2001):

1. organisation as a semi-closed adaptive system
2. environment as stable predictable linear change (better forecasting and planning)
3. focus on more formal processes and structures
4. values of accountability and efficiency
5. tools - BPR, TQM, audits, reviews evaluations
6. accreditation through formal standards

Here, the academic consultants are presented through the scheme as sources of external expertise. So not only are there boundaries between the formal roles of different project members, but there might be a requirement in some cases for the owner manager to overcome a philosophical boundary personally in respect of the future operation of the company. The inherently modernist, goal-driven nature of the DTI project specification may appear full of promise to the owner-manager in terms of its promises for the future, yet may be threatening in terms of its requirements. But it may also provide a 'security blanket' for the question of public accountability over funding. As to the boundary between the world of the university and the SME, there is a good team spirit between the academics and the management team; there is some occasional banter stereotyping academic activity, but neither feel threatened by this, so take it in good part - the theoretical side of the project is taken seriously by the company. The academics are sensitive in turning the humour back on the company or the DTI, for fear of causing offence, in that we are 'guests' of the company, and 'employees' of the DTI. Of course, we may be acting over-sensitively! Any tensions concerning the operation of the project tend to be laid at the door of a reified notion of TCS as external - very quickly, rather than undermining the team bonding. The academics also get on well, and there is no tension arising over the necessity for different work commitments to be to the fore at different stages of the project.

CONCLUSION

In conclusion, from our experience, we would argue that there is no 'one best way' to undertake an intervention for a growing small/microfirm. A useful approach for future interventions would be to apply a systemic analysis of stakeholder attitudes (after Takala *et al*, 2001) as we perceive them at the outset, and indeed throughout the project, and how this may impact on tensions at certain boundaries. This may provide some indication of areas that might have to be sensitively managed by the consultant:

1. The owner-manager - presumably the key power player - is s/he tending to the modernist or post modernist in their view of organisations and how the intervention should progress? Are they going to have to accept a new organisational culture not in keeping with their view of the world of work? How 'entrepreneurial' are they/do they seem to be in their attitudes? The shift from *ad hoc*, informal management control systems to more formalised arrangements in periods of increased growth can cause tension for those who have perhaps thrived on 'seat of the pants' approaches in the past. This may cause them to lose interest as the project progresses. The choice of whether to delegate new control functions in favour of more 'entrepreneurial' goals such as generating new contracts is one that can be usefully discussed at the outset, to avoid feelings of impending rigidity in owner managers.
2. Who are other key players in the project, including the interventionists, and how might their underlying attitude impact on the project?
3. Who is the funder of the project and how rigid is the specification in terms of establishing roles and expectations? Does it have a modernist or a postmodernist agenda, and how does this match with the views of those identified in steps 1 and 2? Our HCR experience tells us that even a good match may bring constraints in our perceptions of success. This leads us to 4:
4. What processes for project change procedures are available? The constraints of what can and cannot be changed need to be clarified at the outset - as does the escape route for a researcher who is losing their research outcome due to unforeseen change. Transfer has to be a two-way process.
5. The need to share understanding of theoretical models through the establishment of a shared vocabulary focusing on mutual benefits is essential. The 'death of the expert' should not result in the 'death of mutual respect!'

REFERENCES

Checkland, P. B. (1981). *Systems Thinking, Systems Practice*, Wiley, Chichester.

Beer, S. (1985). *Diagnosing the System for Organizations*, Wiley, Chichester.

Earl, M. (1989). *Managerial Strategies for IT,* Prentice Hall, Hemel Hempstead.

Espejo, R., Schuhmann, W., Schwaninger, M., and Bilello, U. (1996). *Organisational Transformation and Learning*, Wiley, Chichester.

Greiner, L. E. (1972). "Evolution and Revolution as Organisations Grow", *Harvard Business Review*, July/August, pp37.

McFarlan, F.W. (1984). IT changes the way we compete, *Harvard Business Review*, May-June, pp. 98 - 103.

Nolan, R. L. (1979). Managing the crisis in data processing, *Harvard Business Review*, March-April, pp. 115-126.

Ragsdell, G., and Warren, L. (1999). "Learning from Beer: Lessons for Intervention", *OR Insight,* 12(4):16-22.

Sandoe, K., Corbett, G., and Boykin, R. (2001). *Enterprise Integration*, Wiley, Chichester.

Steier, L., and Greenwood, R. (2000). Entrepreneurship and the Evolution of Angel Financial Networks, *Organization Studies*, 21(1):163-192.

Takala, M., Hawk, D., and Rammos, Y. (2001). "On the Opening of Society: Towards a more open and flexible educational system", *Systems Research and Behavioural Science,* 18:290-306.

White, L., and Taket, A. (1994). "The Death of the Expert", *Journal of the Operational Research Society*, 45(7):733-748.

SOFT SYSTEMS METHODOLOGY BASED ON ORGANIZATIONAL KNOWLEDGE-CREATION THEORY

Taketoshi Yoshida

Japan Advanced Institute of Science and Technology
1-1 Asahidai, Tatsunokuchi, Ishikawa 923-1292, Japan
yoshida@jaist.ac.jp

INTRODUCTION

We deal with a problem like the project management of information system development, which is unsuitable for being formulated in any mathematical forms. This sort of problem is usually complex and has multiple objectives, where some of them are often contrary to each other. Our objective in this paper is to discuss a systems approach to such a problem.

As a systems methodology to improve problems latent in problem situations, we may follow Flood and Jackson (1991). For example, if the problem situation is in complex-pluralist problem context, one of the most effective problem-solving system methodologies is the soft systems methodology (SSM) (Checkland, 1981). SSM is based on the action research and systems thinking, and thus gives us a framework to represent the results of learning. And the systems thinking and SSM has been evolved in business and management domains (Jackson, 2000).

In the same domains Senge (1990) proposed the organizational learning theory, which also emphasized the importance of systems thinking as one of the disciplines of the learning organization; the other disciplines are personal mastery, mental models, shared vision, and team learning. The aspects of the organizational learning theory are, of course, one of the most important factors in a goal-seeking problem solving. However, it tends to be passive self-adjustment for adaptation and be trapped in stimulus-response type of learning.

Nonaka and Takeuchi (1995) solved some secrets of Japanese companies which were successful at continuous innovation. In the book they emphasized the importance of knowledge creation in an organization, and further articulated how such a company created new products, new services, and new organizational structures. Although their theory has some affinity with the organizational learning theory, it further articulated the importance of the viewpoint from the dynamic and endless knowledge spiral between tacit and explicit knowledge in epistemological dimension and among individual, group, and organization in ontological dimension. They also mentioned the five enabling conditions that would promote such a knowledge spiral.

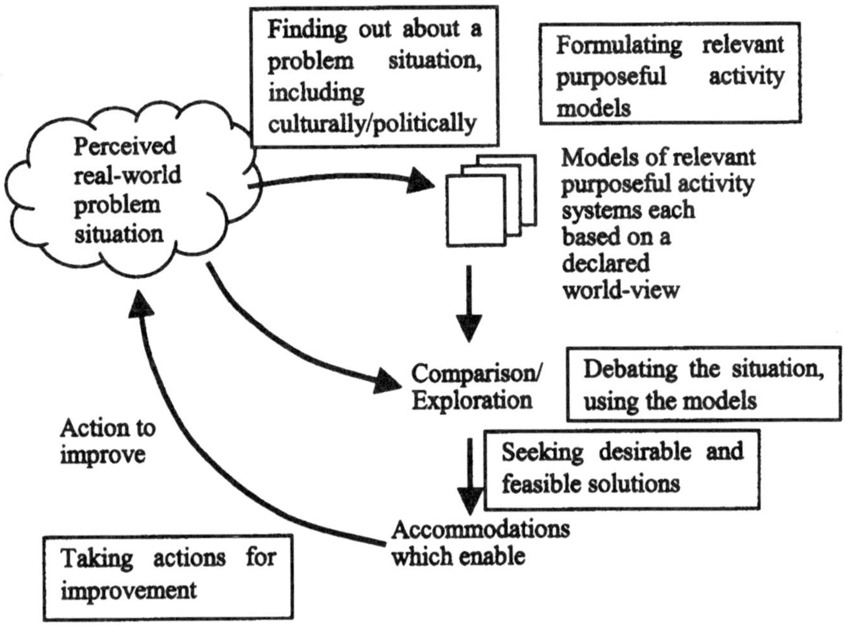

Figure 1. Soft Systems Methodology

For our objective in this paper, we focus on the soft systems methodology and the organizational knowledge-creation theory, and study the complementary relation between them. In the following sessions we briefly explain the soft systems methodology and the organizational knowledge-creation theory for our understanding by using an example of information system development project. Then, we will show that the soft systems methodology can be re-considered from the viewpoint of the organizational knowledge-creation theory. Consequently, we can pay special attention to the knowledge spiral with the creation, acquisition, accumulation, and exploitation of knowledge, which are main parts of intellectual activities of individuals, groups, and organizations.

SOFT SYSTEMS METHODOLOGY

Soft systems methodology (SSM) consists of the following activities (Jackson, 2000), which are based on the exploration and learning processes (Figure 1).

1. Finding out about a problem situation culturally and/or politically (draw rich pictures);
2. Formulating some relevant purposeful activity models (tasks and issues, root definitions, and conceptual models);

3. Debating the situation, using the models, seeking from that debate both
   a. changes which would improve the situation and are regarded as both systematically feasible and culturally desirable, and
   b. the accommodations (a sort of compromise) between conflicting interests which will enable action-to-improve to be taken;
4. Taking action in the situation to bring about improvement;
5. Based on the changed situation, we start the next learning process.

In this way we first draw rich pictures for understanding the problem situation. Second, the relevant tasks and issues are extracted. According to them we define the root definitions and these are examined by CATWOE analysis. Then the conceptual models are depicted as the required transformation in the root definitions. Note that the derivations of root definitions and the related conceptual models are conducted in the world of systems thinking. Finally, we derive the desired and feasible changes from the exploration and comparison between the worlds of reality and systems thinking, where we depend on debate. Furthermore if necessary, the organizational culture and climate are discussed to obtain the accommodations. The corresponding action is taken and then the next cycle of learning begins.

ORGANIZATIONAL KNOWLEDGE-CREATION THEORY

Roughly speaking, organizational knowledge creation is the capability of a company as a whole to create new knowledge, disseminate it throughout the organization, and embody it in products, services, and systems (Nonaka and Takeuchi, 1995, p.3). Focused on explicit and tacit knowledge in epistemological dimension, they mentioned that the key dynamics of organizational knowledge creation is the interaction between these forms of knowledge, and that organizational knowledge creation is the repeated spiral processes of these interactions among individual, group, and organization in ontological dimension. These knowledge interactions are articulated as four modes; socialization, externalization, combination, and internalization, which are hence called SECI model as depicted in Figure 2. Showing our understanding of this model, we employ the following example of an information system development. Main persons or organizations in this example are users of the information system and developers of the information system.

Socialization (from tacit to tacit) is a process of sharing experiences and hence creating tacit knowledge. The developers begin with this socialization by sharing users' tacit knowledge that has been emerged from their daily operations. This process starts with building a field of interaction. Externalization (from tacit to explicit) is a process of articulating tacit knowledge into explicit concepts. This process is triggered by meaningful dialogue or collective reflection. This process articulates the latent ideas and images that have been obtained by achieving the socialization process. Then the members of a project team can share this explicit knowledge. That is, the developers and users jointly make the fundamental concept of the desired information system and share its concept. Combination (from explicit to explicit) is a process of systemizing concepts into an archetype. In this example, an archetype is an information system. Here, the concept is justified and the information system is developed by linking explicit knowledge. Then the developed information system is introduced to the users. Internalization (from explicit to tacit) is a process of embodying explicit knowledge into tacit knowledge. The users utilize the introduced information system in their daily operations and develop tacit knowledge through their experiences. This is the beginning of the second SECI spiral.

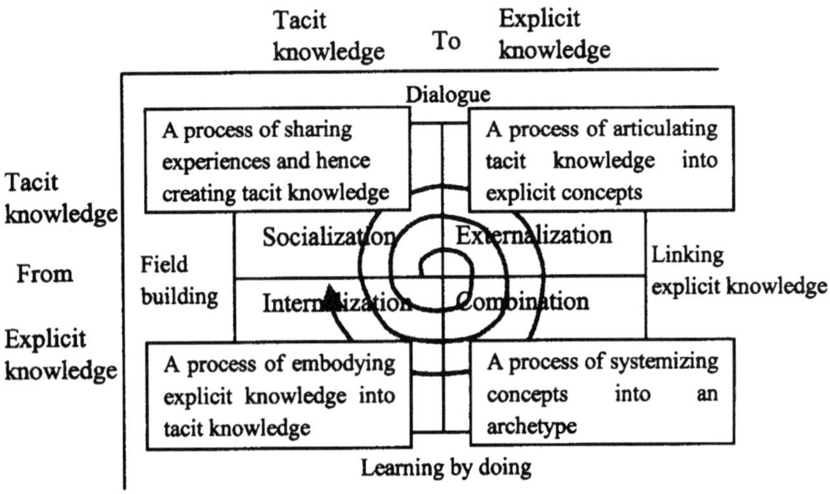

Figure 2. Contents of Knowledge Created by the Four Modes

The interpretation of SECI model totally depends on the problem situation that we focus on. For example, we may apply SECI model to the project of clarifying the problem situation in the users to derive the requirements for the desired information system as an archetype. In this case, we may interpret SECI model in a different way from the above example, where SECI belongs to inside the user company. Note that the essence of the organizational knowledge-creation theory is the management model of knowledge-creating companies and hence can be applied to broad and various types of problems in business and management domains.

In addition to the above, Nonaka and Takeuchi (1995) proposed the importance of the enabling conditions to promote the knowledge spiral. These five enabling conditions for organizational knowledge creation are as follows.

1. Intention
2. Autonomy
3. Fluctuation and creative chaos
4. Redundancy
5. Requisite variety

Owing to these enablers, four modes of SECI model become continuous-time-varying dynamic processes and are promoted in the knowledge spiral.

OBSERVATIONS

Figure 3 shows the relation between the organizational knowledge-creation process and the soft systems methodology on intellectual activities of individual, group, or organization. The developers in the previous example should share, through the socialization process, the experiences that the users are tacitly aware of in their daily operations, because highly subjective insights, intuitions, and hunches are at the root of knowledge creation and innovation (Nonaka and Takeuchi, 1995, p.229). Then the developers externalize some concepts and hence share them among the members by utilizing documents and pictures. The derived concepts must be justified before the combination process. Then the combination process builds an archetype such as a new product, a new service, or a new organizational structure; in the previous example, the information system. The developed archetype is introduced to the users and they proceed to the internalization process again to become aware of the next real-world problem situation. This is one of the cross-leveling of knowledge.

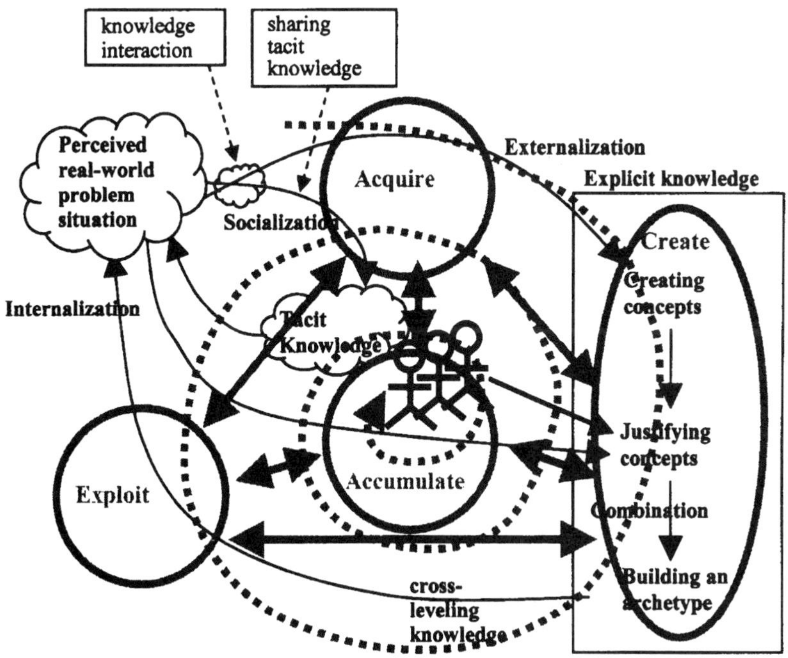

Figure 3. Organizational Knowledge Creation and SSM on Intellectual Activities

As we may see in Figure 3, we introduced the socialization process in addition to the SSM process as depicted in Figure 1, because the dynamic interaction between tacit and explicit knowledge is the key of the success. The externalization is triggered by meaningful dialogue or collective reflection among the members as depicted in Figure 2.

The externalized tasks and issues, rich pictures, conceptual models, and the resultant accommodations from logical and cultural analyses in SSM are the explicit knowledge. The archetype is the output of combination. In order to indicate the importance of shared (or accumulated) tacit knowledge for the justifying-concepts process in Figure 3, we connect it in the centre of the figure to the justifying-concepts activity by an arrow.

Note that a set of three person-type icons in the centre of Figure 3 should be interpreted appropriately in the context. These are individuals, groups, an organization, or organizations in a problem context, where the tacit and explicit knowledge are accumulated.

CONCLUSION AND REMARKS

The essence of the organizational knowledge-creation theory is to develop and promote the organizational capability of the intellectual activities. When we consider SSM from this viewpoint, we may understand the validity of Figure 3. Hence, we can now conclude that the concept of innovation is added to SSM which is based on the learning process. The importance of dialogue and collective reflection is one of our assertions along with SECI process. Furthermore, we assert that tacit knowledge is essential in all intellectual activities and hence in SSM. Finally, we must consider the enabling conditions of organizational knowledge creation. No systems methodology works without them. These are essential to promote the knowledge spiral.

Nonaka and Takeuchi (1995, p.226) argued seven practical implications. One of the implications is that top management should create a knowledge vision and communicate it within the organization. The knowledge vision is a long-term strategy of a company and hence it is vague against the front-line employees whose operations are mostly daily routine. Hence, it is important to show a concrete direction regarding what kind of knowledge they should seek and create. For the purpose we propose following the balanced scorecard approach (Kaplan and Norton, 1996), although we must develop the approach further to being concrete in our context. Applying their approach, we may study and investigate the knowledge vision from four perspectives; financial, customer, internal-business-process, and learning and growth perspectives. Based on these perspectives, we will derive a strategy map (Kaplan and Norton, 2000). And we will further develop the balanced scorecard that clarifies strategic objectives, strategic initiatives, and strategic measures including lag and lead indicators, and their causal relationship. If we have this sort of balanced scorecard, we may expect that top or senior managers, middle managers, and front-line employees share the knowledge vision in their own words. This will be one of the ways to develop and deploy the knowledge enabling conditions.

REFERENCES

Checkland, P. (1981). *Systems Thinking, Systems Practice*, John Wiley and Sons, Chichester.

Jackson, M.C. (ed.) (2000). "Peter Checkland at 70: A Review of Soft Systems Thinking," *Systems Research and Behavioral Science*, **17.**

Flood, R. and Jackson, M.C. (1991). *Creative Problem Solving – Total Systems Intervention*, John Wiley and Sons, Chichester.

Kaplan, R. and Norton, D. (1996). *The Balanced Scorecard*, HBS Press, Boston.

Kaplan, R. and Norton, D. (2000). *The Strategy-Focused Organization*, HBS Press, Boston.

Nonaka, I. and Takeuchi, H. (1995). *The Knowledge-Creating Company*, Oxford University Press, Oxford.

Senge, P.M. (1990). *The Fifth Discipline: The Art and Practice of the Learning Organization*, Currency Doubleday, New York.

Section Three

Knowledge Management

INQUIRING INTO LEARNING AS SYSTEM

Susan Byrne, David Todd, Barbara Simpson, Christine Woods, and Rainer Seidel

The University of Auckland
Private Bag 92019
Auckland
New Zealand
b.simpson@auckland.ac.nz

INTRODUCTION

Possibly the most fundamental skill that is required in the emerging Knowledge Age is the ability to learn. This capability resides not only with individuals, but also at every other systemic level from dyads, groups and teams, to organisations, institutions and society at large. Indeed, learning is a veritable haystack of complex, interacting and inter-related elements that span the levels of the social system. Thus, research into learning that focuses on one or another level of analysis is necessarily limited in its explanatory capacity. It cannot be said that any element or isolated set of elements enables learning; to take this view is to err on the side of naiveté.

This paper has emerged out of a long-term research project that aims to understand, from an organisational learning perspective, the process of innovation in small and medium sized manufacturing enterprises. Since its inception, the project has adopted an inductive, multiple case study approach designed to elucidate the extreme complexity of innovation processes. To date, the research has produced a detailed knowledge typology (Simpson *et al.*, 2000; 2001) that identifies the key elements of technological learning. The typology comprises four knowledge categories, namely Identity, Direction, Capability, and Relationship, that interact with each other in a "generative dance" (Cook and Brown, 1999) of new knowledge creation. Although these four knowledge categories are well grounded in our data, the typology nevertheless fails to capture the dynamic nature of the learning process. In fact the majority of published models of organisational learning are similarly static (e.g. Hitt *et al.*, 2000; Lam, 2000; Spender, 1996). Only very few scholars have endeavoured to extend their theorising into the realms of dynamic process (e.g. Crossan *et al.*, 1999; Nonaka and Takeuchi, 1995).

Reflecting on this parlous state of affairs, it became apparent to us that one of the primary obstacles to researching dynamics is methodology. The majority of research

Edited by Ragsdell et al., Kluwer Academic/Plenum Publishers, New York, 2002.

methods that are available in the social / organisational domain are based on realist assumptions, focussing narrowly on techniques for classification and measurement. But understanding dynamics demands an understanding of time, which by its very nature poses difficulties for this functionalist approach. As Ricoeur explains in his analysis of Augustine's Confessions, time must exist in order to be measured or classified, but "time has no being since the future is not yet, the past is no longer, and the present does not remain" (1984:7). So, to really come to grips with dynamic process, it is necessary to operate within an entirely different ontology, one that permits a relativist perspective on the nature of time and reality.

Faced with this ontological challenge, we began to search for research methods that are not constrained by the usual assumptions of functionalism. Further, we sought an approach that accommodates the systemic interdependencies and high level of complexity inherent in innovation and organisational learning. Our search led us to Soft Systems Methodology (Checkland and Scholes, 1999), which appears to hold the potential to address all of these needs. The aim of this paper then, is to explore the efficacy of Soft Systems Methodology (SSM) as a means of inquiry into dynamic learning processes. We begin with a description of SSM, emphasising the assumptions that underpin this approach, and we then proceed to demonstrate the application of SSM techniques to a specific technological learning incident that is drawn from our case studies. Finally, we conclude with an assessment of the insights that these techniques have provided into the dynamics of a technological learning process.

UNDERSTANDING SSM – FROM ASSUMPTIONS TO TOOLS

The comprehensive justification of any research design involves a statement of not only *how* the research will be done, but also *what* and *why*. Mingers and Brocklesby (1997), for example, suggest that the research process involves several layers of reference comprising a *paradigm* (ontological and epistemological assumptions) that is linked to a related *methodology* (a structured set of activities), which in turn is linked to relevant *techniques* (specific activities) and *tools* (artifacts used to perform particular techniques). As an illustration, a functionalist perspective is informed by paradigmatic assumptions of an external, objectively knowable reality, it employs experimental and quasi-experimental methodologies, it draws on a huge array of techniques that are used for measurement, and there are many instruments and software packages that may be used as tools for data collection and analysis. Similarly, SSM may be described in terms of these layers of reference but before proceeding with this, we must first differentiate between two areas of systemic thought, soft and hard.

The 'soft' systems thinking perspective advocates that the real-world is a jumbled, complex mess that is difficult for stakeholders to understand and therefore take action in. One way of attempting to understand the real world is through the use of 'systems' as a thinking tool, a frame to organise the thinking of those involved. Consequently the notion of a system ultimately becomes an abstraction of reality, or a construct that people use to make sense of the messes that dominate the world (Checkland and Scholes, 1999).

'Hard' systems thinking on the other hand advocates the position that the real world is made up of systems and that through better understanding of these systems, people can take more decisive action. The 'hard systems' approach is based on assumptions of objectivity and rationality, whereas the 'soft systems' approach is more able to accommodate the subjectivity and irrational behaviours that are frequently evident in human decision making. For this reason, the soft stance is more appropriate for our research endeavour.

Checkland and Scholes (1999:A35) articulated the following three rules as constitutive of SSM:

1 *" you must accept and act according to the assumption that social reality is socially constructed, continuously;"*
2 *"you must use explicit intellectual devices consciously to explore, understand and act in the situation in question; and"*
3 *"you must include in the intellectual devices, 'holons' in the form of systems models of purposeful activity built on the basis of declared worldviews".*

The taking of this view within the research is consistent with the constructivist paradigm philosophical underpinnings, hence the tight fit of SSM within the context of the research design. These rules guide us to see SSM from a broader perspective. Firstly they make explicit that SSM is based on the paradigmatic assumptions of constructivism. That is, an ontology that sees realities as multiple, intangible mental constructions that are dependent for their form and content on the individuals or groups who constructed them, and an epistemology that acknowledges the role of the researcher in the construction of knowledge (Guba and Lincoln, 1998). It is these assumptions that provide a philosophical foundation for the methodological framework that SSM represents. Secondly, the constitutive rules guide us to select research techniques and tools (intellectual devices) that are consistent with the methodology. Checkland and Scholes (1999) identified a suite of appropriate techniques for SSM including rich pictures, root definitions, and roles / norms / values analysis, but they also suggested that many other techniques may be equally suitable provided they are applied in a way that is consistent with the principles of SSM and constructivism. For instance, in this paper we suggest that mind maps, stock and flow diagrams, and causal loop diagrams are all appropriate techniques within an SSM analysis.

MULTIPLE TECHNIQUES – MULTIPLE PERSPECTIVES

In this section, we have selected one particular instance of technological learning drawn from our case studies. By applying several different techniques within SSM, we seek to gain multiple different perspectives on the dynamics of the learning process. The data that we present were collected over a period of about 6 months using interviews and discussions with the various actors in the case organisation, as well as researcher observation. The events that we describe took place in a small plastics manufacturing company, which, despite its size, is a major global supplier of creative and colourful playground componentry. The primary manufacturing technology used by the company is rotational moulding, a process that is characterised by relatively poor process control, and hence unpredictable product quality. The learning process that we observed was aimed at overcoming these quality issues (the 'problem of fit') by developing a process control innovation.

Not surprisingly, we discovered a huge amount of complexity in this process. Figure 1 is a rich picture that represents many of the events and interactions that occurred. The general flow is captured in the following narrative, but we hasten to emphasise that this is but one of many narratives that might be drawn from this rich picture.

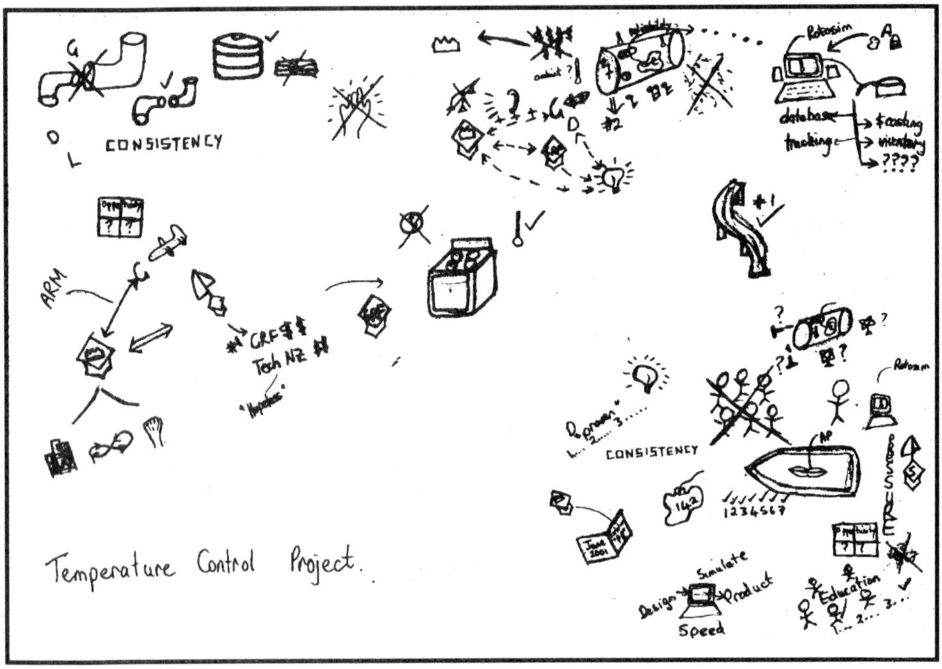

Figure 1. Rich Picture of the Process Control Innovation

Some years before the quality problems began to surface, Herb, the CEO, had met Sid, a Professor of Engineering, at an international conference on rotational moulding. As chance would have it, Sid subsequently took up a position at a University located in the same city as Herb's company. Sid spent the first year of this new appointment "going around small [plastics] companies ... trying to tease out from them what [the University] could do for them".

When the problem of fit-ups surfaced, Herb talked it over with Sid who said "the obvious thing to do is to try to implement some process control". He suggested that a temperature-controlled process would eliminate some of the inconsistencies in the moulded products. Although temperature-monitoring systems were available already, this off-the-shelf technology was too expensive, did not incorporate automated controls, and was incapable of operating in a factory environment. Herb realised this was one way the industry would be moving in the future, and "instead of being left behind" he felt the company "should be able to do it and do it better" than the proprietary systems already available.

Stan, a research student from the University, was brought on board to develop the necessary process control systems. Over a one-year period, he worked closely with the production manager, to find a practical and reliable means of using temperature to

control the rotational moulding process. This innovation is the first practical and reliable rotational moulding temperature control system to be developed and implemented at factory level anywhere in the world.

One consequence of implementing this temperature-controlled process is that the company now has access to a whole raft of new process data. The computer system developed to facilitate the temperature process produces a variety of outputs as a by-product of the "cooking" process. This then raises a number of new opportunities. For instance, the computer-generated records will provide assistance with the inventory system, much welcomed by the accounting department. Exactly how many new opportunities will arise from this innovation has yet to be explored by the company.

Although this narrative has the advantage of being readily understandable and logical, it loses a great deal of the complexity and non-linearity that is represented in the rich picture. But at the other extreme, the rich picture itself is too complex as it stands. It does demonstrate a temporal flow, but more detailed analysis is required to get right down into the micro-dynamics of this learning process.

Mind maps allow the researcher to develop a level of abstraction that is difficult to achieve in rich pictures, which are essentially an unstructured depiction of a story. The technique of mind mapping zooms in on a central topic and imposes a structure around this. In the example that we present here (Figure 2), we have zoomed in on the issue of 'problem of fit', which is represented in the top left corner of the rich picture. The structure that we have imposed on this analysis is derived from the knowledge typology that has been previously reported (Simpson *et al.*, 2001). Specifically, the four primary questions (How are we reflecting?; How are we strategising?; How are we doing what we are doing?; How are we interacting?) relate respectively to the four knowledge categories (Identity, Direction, Capability and Relationship). From there, the mind map allows us to "drill down" into various segments of the innovation and to explore linkages between them. For example, the problem of fit has wasted resources including time, money and plastic, it has resulted in customer complaints, and it has had a negative impact on export potential. Many other linkages that were not obvious in the rich picture may be teased out using the mind map, but this technique is limited in its ability to elucidate causal or sequential linkages.

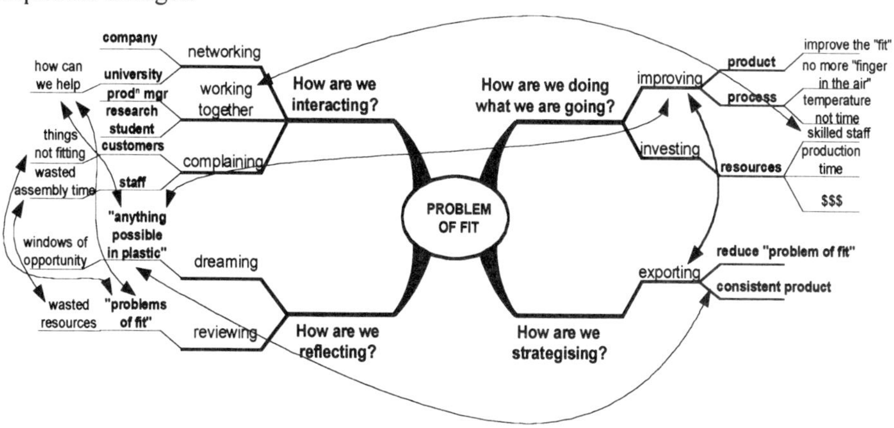

Figure 2. A Mind Map of the Problem of Fit

Figure 3 is the Stock and Flow Diagram (Forrester, 1961) that we used to investigate the problem of fit. It focusses on the decision to reallocate resources that ensued once the problem of fit had been recognised. In making this decision, resources had to be pulled away from day to day activities and redirected towards improving the production process. The improvement that was subsequently implemented not only resulted in less product variation and less defective product, but it also solved a number of other process related problems that had been faced by the company. This in turn lead to new opportunities for innovation.

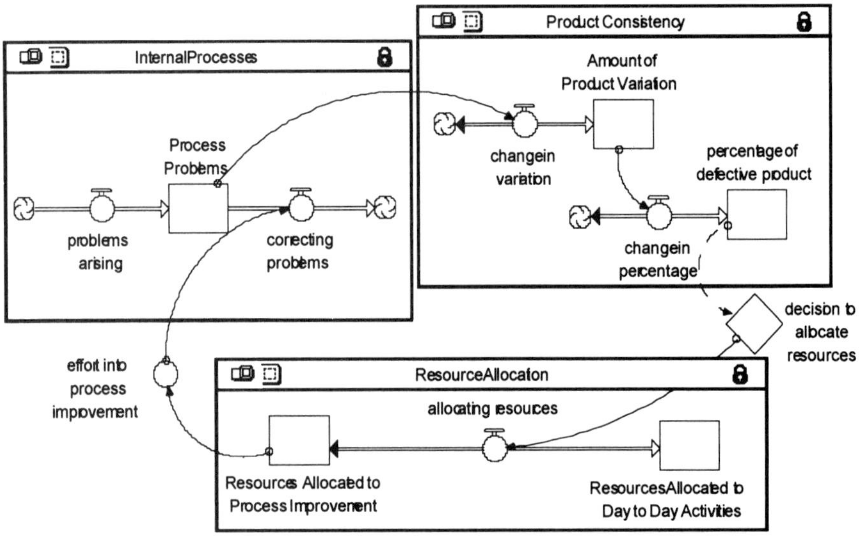

Figure 3. Stocks and Flows Diagram for the Problem of Fit

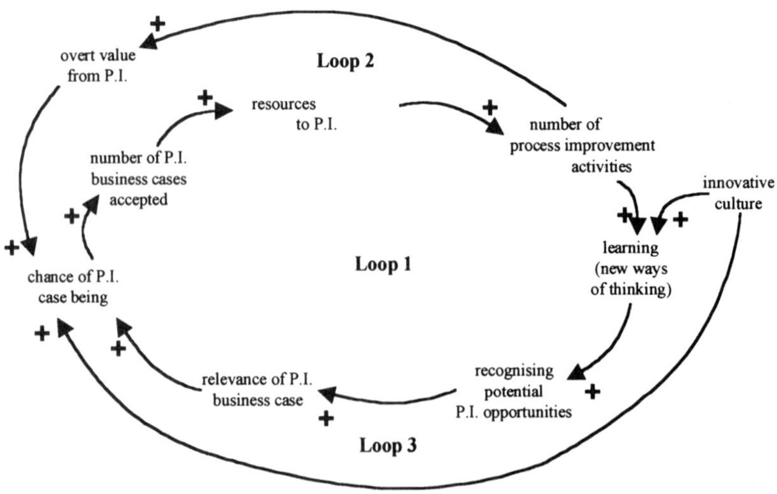

Figure 4. Causal Loop Diagram for Process Improvement (P.I.)

To further investigate the dynamics of new opportunity creation arising out of the original problem of fit, we invoked Causal Loop Diagramming (Sterman, 2000). The diagram (Figure 4) begins with the decision to allocate resources to process improvement (P.I.) and then proceeds to trace the various consequences of this and subsequent decisions. The diagram makes explicit the process of opportunity appraisal that happened in the company, albeit by largely implicit means. This process of resource allocation – learning – appraisal – further resource allocation represents one complete cycle in the diagram (Loop 1). But two other loops are also apparent. Loop 2 highlights the reinforcing influence of successful process improvement upon attitudes towards the value of further improvements, and Loop 3 brings in the influence of an innovative culture. Overall this diagram demonstrates that learning in this instance has been driven by (Direction) the combined impact of the doing (Capability), the perceived value of this doing (Identity), and the culture of the organisation (Relationships). And so we return to the four knowledge categories that underpin our understanding of technological learning.

DISCUSSION AND CONCLUSIONS

Our work on innovation in SMEs has shown us that technological learning is a complex dynamic process covering a diverse range of elements and interactions at various levels of the organisation. The objective of our current research is to develop new insights into this process by using an SSM-based approach. In this paper we use a case study to demonstrate the value of a range of techniques within a consistent paradigm to gain complementary insights into the dynamic nature of technological learning.

Our first technique is the rich picture, which provides a visual 'story' of an innovation process. It captures the complexity of the overall process, and demonstrates the interactions between the various agents involved. However, despite this richness of presentation, the dynamics of the learning process are illustrated at only the most general level, with emphasis on the major steps and interrelationships.

The technique of mind mapping, which is more abstract, structured and rigorous in its application, is used to explore from the general to the specific. The large volume of interactions at this level necessitates the creation of separate mind maps from specific viewpoints (in our case the problem of fit), which is positioned in the centre of the map and used as a starting point for interpreting and sense-making. Our mind map captures all existing linkages and connections within its boundaries, but does not consider the time sequence or any possible causal relationships.

Our third technique is the stock and flow model, which in our view is consistent with the SSM approach as long as the underlying world view is made explicit and the existence of potential multiple models is acknowledged. We use the technique to analyse the situation systemically and to capture the feedback and delays that are inherent in the technological learning process. The discipline of the stock and flow model helps us understand what is 'flowing', what is 'accumulating and depleting' as time passes. The process includes a specific consideration of time, level and rates.

The final technique we have presented in this paper is that of causal loop diagramming. Again, given that we specifically state the worldview from which the diagram is constructed, and acknowledge that multiple interpretations are possible, the use of this technique is acceptable within an overall SSM methodology. We chose to use these diagrams as they are a classic way of capturing interconnections, multiple causalities, and a 'sequence' of events and hence of a 'time' process. From this approach we gain a sense of the dynamic unfolding of the innovation, and of the associated learning process.

Our research has shown us that the application of a range of complementary techniques under one consistent paradigm enables the development of new insights into the nature of technological learning. Each technique provides a new perspective, helps focussing on specific aspects of the process, and opens a new level of understanding without losing the integrity of the overall context. Currently we are investigating the use of further techniques (that cannot be covered in this paper due to space limitations) in order to achieve an even broader understanding and develop refined models of the technological learning process. We are confident that our work will also lead to the development of new insights and improved tools for managers and other practitioners involved in innovation in SMEs.

REFERENCES

Checkland, P., and Scholes, J. (1999). *Soft Systems Thinking: A Thirty Year Retrospective,* John Wiley and Sons, Chichester.

Cook, S.D.N., and Brown, J.S. (1999). "Bridging epistemologies: The generative dance between organizational knowledge and organizational knowing," *Organization Science*, 10(4):381-400.

Crossan, M., Lane, H.W., and White, R.E. (1999). "An organizational learning framework: From intuition to institution," *Academy of Management Review*, 24(3):522-537.

Forrester, J., (1961). *Industrial Dynamics*. Cambridge, Massachusetts: The MIT Press.

Guba, E.G., and Lincoln, Y.S. (1998). Competing paradigms in qualitative research. In *Landscape of Qualitative Research.* (N.K. Denzin and Y.S. Lincoln, eds.) Sage, Thousand Oaks, CA.

Hitt, M.A., Ireland, R.D., and Lee, H. (2000). "Technological learning, knowledge management, firm growth and performance: an introductory essay," *Journal of Engineering and Technology Management*, 17(3-4):231-246.

Lam, A. (2000). "Tacit knowledge, organizational learning and societal institutions: An integrated framework," *Organization Studies*, 21(3):487-513.

Mingers, J., and Brocklesby, J. (1997). "Multimethodology: Towards a Framework for Mixing Methodologies," *Omega: International Journal of Management Science* 25(5):489-509.

Nonaka, I., and Takeuchi, H. (1995). *The knowledge creating company*, Oxford University Press, Oxford.

Ricoeur, P. (1984). *Time and Narrative Volume 1*, University of Chicago Press, Chicago.

Simpson, B., McGregor, J., Seidel, R., Kolb, D., Henley-King, J., and Tweed, D. (2000). "Learning in the Manufacturing Sector". *University of Auckland Business Review*, 2(1):38-50.

Simpson, B., Seidel, R., Byrne, S., and Woods, C. (2001). "Technological learning: Towards an integrated model," *Proceedings of the 4th International Conference on Organizational Learning & Knowledge Management - New Directions*, London Ontario, 1-4 June 2001. Richard Ivey School of Business, University of Western Ontario.

Spender, J. C. (1996). "Making knowledge the basis of dynamic theory of the firm," *Strategic Management Journal*, 17(Winter Special Issue):45-62.

Sterman, J. (2000). *Business Dynamics: Systems Thinking and Modelling for a Complex World*, Irwin McGraw-Hill, Boston, MA.

CREATING CORPORATE VALUE THROUGH KNOWLEDGE MANAGEMENT SYSTEMS

Larry Churchman and Chris Royce

Anglian Water Services, Technology Group
Thorpe Wood House
Peterborough
Cambridgeshire, PE3 6WT

INTRODUCTION

Technology Group is the current business name for the department responsible for technical control of asset provision within Anglian Water Services.

AWS Technology Group is Created

The business vision set by AWG chief executive Chris Mellor is "To become a leader in the creation, development, management and operation of infrastructures in the UK by 2005 and a significant global player by 2010". AWG provides a global "whole life" infrastructure capability.

AWS Technology Group is part of the regulated water utility company Anglian Water Services Ltd. - a wholly owned subsidiary of AWG plc - and emerged from a review of engineering and procurement processes during 1998. The review identified four key underpinning management tenets: relationship, risk, value, and knowledge.

The business purpose for AWS Technology Group is to enable and control the process of asset creation to a consistent functional level, within the parameters set by the regulators of the utility and taking account of environmental and planning needs within a framework of sustainability, whilst driving down the total cost of asset creation. We control the design, construction, and commissioning of treatment works and associated infrastructure.

The formation of AWS Technology Group in 1999 was the trigger to examine existing knowledge systems and to identify knowledge management practices that could be improved. At the same time there was an opportunity to plan a workplace, a management structure, and a style that would enhance knowledge sharing.

Edited by Ragsdell *et al.*, Kluwer Academic/Plenum Publishers, New York, 2002.

DEVELOPING KNOWLEDGE MANAGEMENT

From the start there was a determination to implement a knowledge management strategy internally rather than to hand over the task to consultants. Our feeling was that only we fully understood the relevance of our knowledge.

Building the Concepts

Inspiration for what has been developed within Technology Group has been drawn from the books by Peter Senge, Ikujiro Nonaka and Hirotaka Takeuchi, and Verna Allee. Senge (1990) gives an organic picture of the learning organisation in his "The Fifth Discipline". Nonaka and Takeuchi (1995) tackle some areas of theory built from experience of knowledge working in the classic "The Knowledge-Creating Company". Allee (1997) gives a Total Quality based interpretation of the complexities behind knowledge sharing theory in "The Knowledge Evolution".

Both Dr Ronnie Lessem, of Buckingham University, and Sudhanshu Palsule, currently advisor on organisational development to the United Nations, have been influential in AWG's aspiration to become a learning organisation.

A corporate university was formed to help facilitate the development of a learning organisation, this was able to promote community colleges of best practice, but efforts to implement corporate knowledge management techniques were less successful. The focus for knowledge management should normally be at community of practice level.

Developing Structure

The development of a knowledge management structure within AWS Technology Group has been evolutionary rather than revolutionary. The fact that what had been done in the past had worked reasonably well was an indication that there was functional knowledge management already in place, though not specifically identified as such.

A consistent policy to enhance individual personal development and to strengthen effective team working had been applied over the six preceding years. Most employees had a good working knowledge of Total Quality Management and other business improvement approaches. The key for knowledge management was to demonstrate that individual contributions really could make a difference and that thoughts and ideas were valid business currency.

The key organisational dimension to the AWS Technology Group model is matrix management. In this management matrix, teams of similar disciplines are managed by function heads to form effective communities of practice. For project purposes the team members are reassigned to become project team members under a project manager, but may be members of more than one project team at any one time.

The deliberate selection of the communities of practice management structure has greatly assisted knowledge management. The team manager is also the knowledge manager for that team and it is his/her responsibility to endorse and communicate emergent knowledge to the team and to contribute revision and new sections to our guidance documents as required. The project managers become the knowledge specialists, responsible for ensuring information on projects is maintained. The engineers, scientists and technologists become the knowledge engineers responsible for the creation and documentation of knowledge. Finally, the management team are the knowledge officers responsible for setting standards.

The function heads or middle managers are pivotal in the process of endorsement of emergent knowledge and best practice. In our knowledge realisation mechanism endorsement is seen as key to the uptake and application of new knowledge and techniques as well as a control on the quality of the knowledge base. To establish an understanding of our knowledge needs Figure 1 was drawn to help visualise the interactions that our knowledge management system would enable and facilitate.

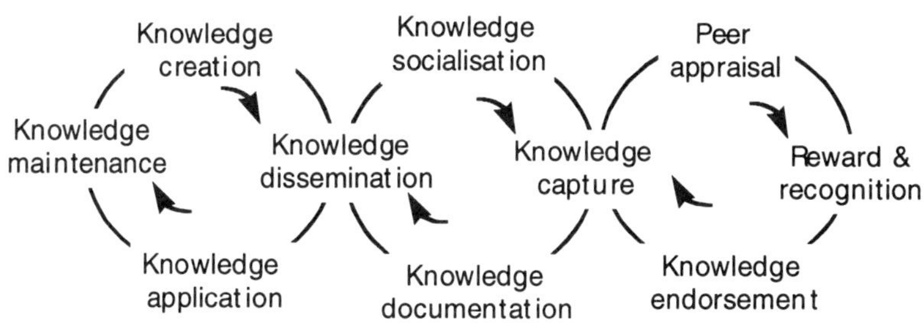

Figure 1. The Technology Group Knowledge Engine

Within AWS Technology Group a significant part of knowledge working entails the assimilation of modified approaches to the application of engineering, process science, etc. There is very little new accretory and unchanging knowledge to be assimilated as the basic tenets of knowledge for each community of practice have been acquired through extensive further and higher education; the majority are professionally and academically qualified.

Knowledge Habitats

An analogy that combines many of the preceding concepts has helped to confirm the AWS Technology Group strategy. If each entity of knowledge is likened to a living organism with the properties of life, then we can consider the building of an accretory knowledge base as like the building of a collection of living organisms - a zoo.

Each organism may have to be captured, in a special, humane trap, before being transported to the zoo. At the zoo numbers of keepers manage the organisms and maintain them with reference to life-span and environmental, dietary and behavioural needs. Dead organisms have to be removed swiftly to prevent cross contamination and there is always plenty of sweeping out to be done!

Zoos therefore present a complex management problem, and remove the creatures they hold from the environment for which they have evolved. Some rare species are bred with the intent to return them to their natural habitat at some time in the future; if the habitat should ever become safe for their return. Commonly when animals are returned to the wild they have lost the ability to survive unassisted.

For us this zoo represents I would suggest a managed computer based knowledge capture system. The lesson seems clear; that living (knowledge) entities can look after themselves when left to evolve in their natural environment. For us the knowledge habitats are the communities of practice. The knowledge management team are, perhaps, the conservators of the environment. The documented knowledge base must respond rapidly to changes in the knowledge environment.

In a normal business environment the significance and relevance of knowledge is constantly changing. Capturing and isolating knowledge from its natural environment can produce unforeseen consequences when it is retrieved for reuse.

KNOWLEDGE REALISATION

The realisation process consists of several steps. First we need to realise (appreciate) that there is worthwhile knowledge in existence. Next is the need to realise (understand) the value and relevance of that knowledge to our business purpose. Finally, by appropriate application the value of that knowledge is realised in a financial sense.

The Business Environment

In AWS Technology Group, most changes that we might wish to make to the process control documents that form our internal, documented knowledge base must be carried out with reference to all the external regulators and the evolving demands of our product end users - the asset managers. In other words the process of examination and endorsement of emergent knowledge is highly significant to us and essential in terms of safety, the health and well-being of our customers, and business sustainability.

Revisions to the knowledge base are made in response to feedback from the application of the documented knowledge by practitioners. Revisions are incorporated at three monthly intervals and remain highlighted for two or three revision cycles. This provides a degree of pacing for the complex stages of appraisal, endorsement, and assimilation. It also provides a period of stability of three months which benefits the relationship with our partnering contractors and suppliers.

There is a parallel and wider knowledge cycle that affects documents generated by national bodies. These include engineering contract and national or European standards, specifications, and codes of practice. We participate in the updating of these knowledge sources where this is appropriate.

The underlying structure for project management in AWS Technology Group is based on PRINCE2 methodology. Our asset creation management systems are the result of an interweaving of PRINCE2, value management, and corporate investment control procedures.

Six Stages of Value Management

The principal tool for the application of our team based knowledge and for the identification and application of emergent knowledge is the Value Management process.

There are currently six value management stages attached to any project. VM workshops 1 to 3 are used to establish a need for investment and to select the best option in terms of value for money and appropriateness of solution. VM4 is a series of structured meetings to fine tune the design of the selected option against a cost/ performance measure - value engineering.

The VM5 and VM6 project review workshops have been added recently; VM5 to be applied at handover of a construction and VM6 to be applied after a period of operation of that project product.

Most value management workshops generate a report that indicates the strategic approach applied at the workshop and the outcomes from the workshop, including cost and value assessments.

VM Output

The value management workshops help us to identify entities of knowledge that might prove useful to the asset creation process, but they will generally require a process of evaluation and endorsement before they become ready for incorporation in the documented knowledge base.

This evaluation may be made by the project team during the course of the project and endorsement be effected by project success. On the other hand an output or issue from a value management workshop would be owned by the relevant function head and converted into one of several actions ranging from team (community of practice) briefing to the rewriting of part of our key control documents.

The process of communicating knowledge entities to other associated business units is not fully resolved, but is partly assured by their agreed participation in the value management workshops relating to schemes in which they have involvement.

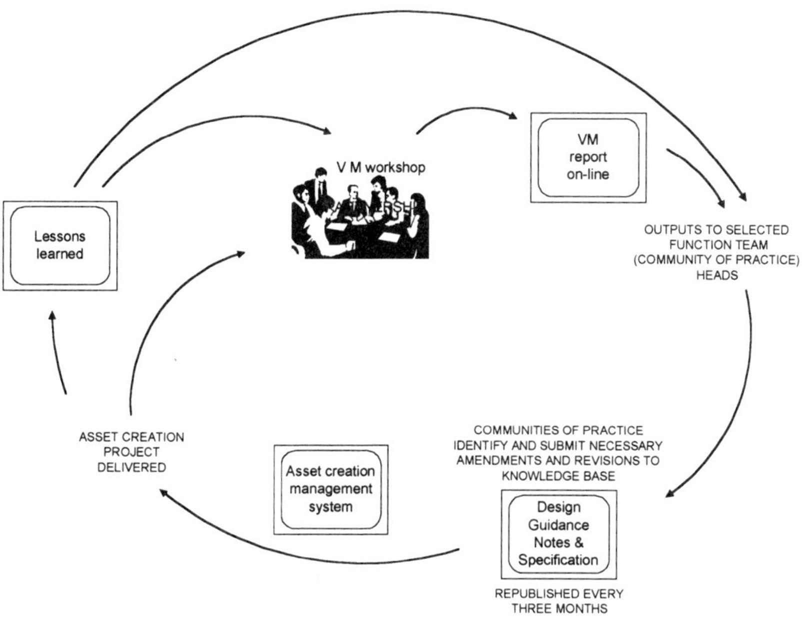

Figure 2. Typical Knowledge Realisation Cycle

Figure 2 represents a knowledge realisation cycle related to asset creation. From one project under normal controls there emerge a number of lessons that are logged during the course of solution selection, design, construction, and commissioning of the new asset. Lessons linked to safety or other key issues are propelled directly to the relevant function head for analysis and decision. The remainder are presented to key players in the project team at the "value management 5", post construction review, workshop and become embodied in the workshop report. It is the duty of the development engineer to convert this report into discrete actions and to allocate owners to these actions: the usual owners will once again be function heads who may delegate tasks to expert members of their team, or community of practice.

Direct Knowledge Working

As part of the process of continuing professional development, staff are invited to attend lunch time sessions in which approved packages of instruction (lectures) are delivered on contemporary issues.

Approximately monthly sessions on selected projects provide an opportunity for project teams to share the experience of a recent project with a self invited group of technologists. Discussions focus on construction techniques, partnering, treatment processes, etc.

Some of those attending will be able to feed such information back into VM2-4 workshops - keeping the knowledge alive in its natural habitat. Some part of this knowledge will eventually be embodied in the Design Guidance Notes and Engineering Specification as it becomes further tried and tested.

Supporting Asset Creation

To achieve a consistent functional level of water treatment and transport there is reliance on our design guidance notes, the contract specification and various guides to risk and value management and partnering relationship management.

Our approach to sustainability is being formalised by the introduction of typical environmental footprint studies. This represents an expansion of an approach that has always encompassed environmental impact assessments.

Cost is being more tightly controlled through the partnering alliances, through supply chain management, through the selective use of cost consultants, and through application of value management.

An early commitment to reduce paper usage is reflected in the asset creation guidance and control documentation being communicated via an intranet within the company and via an extranet to partnering contractors and suppliers.

Partnering contractors receive the contract specification detail and the content of the internal design guidance knowledge base via the extranet. This is part of the foundation of the partnering relationship where shared knowledge contributes to a mutual understanding of the design approach applied to AWS asset creation. In the long term this strategy is designed to bring about:

1. Evolution towards an optimum approach to asset creation.
2. A greater degree of consistency of approach by different partnering contractors.
3. A greater degree of consistency of approach across the region of East Anglia.
4. A more modular design approach and consequent reduced design, build and operational costs with reducing environmental impact.

SUMMARY

The Knowledge Realisation Model illustrated in Table 1 identifies some of the components of AWS Technology Group knowledge management. This highlights the complexity and the integration that is required to drive forward a value creating knowledge management system.

Table 1. Knowledge Realisation Model

Knowledge Realisation Model		
Corporate Systems	**Team Systems**	**Individual Systems**
Knowledge Chain (primary)		
Knowledge creation Knowledge dissemination Knowledge application Knowledge maintenance	Knowledge socialisation Knowledge capture Knowledge documentation	Peer appraisal Reward and recognition Knowledge endorsement
Supporting System (secondary)		
Quality Assurance Value Management Value Engineering PRINCE2 Benefits realisation Knowledge realisation Balanced Scorecard	Form of the organisation Specialist communities of practice	Personal Development Investors in People (gaining, sharing, using) Bright ideas scheme
Created Value		
Corporate awareness New realities Global Knowledge Supply Chain benefits Innovation	Technical Management Asset Management Project Management Product Management	Individual development Breaking of paradigms Personal reward

Much of what is achieved is taken on trust, since measuring the effectiveness of the components of such a system is a complex and self defeating task in cost terms. Our performance is measured in terms of successes and failures of water treatment and transport with the assets that we have helped to create. Our financial performance is directly measured.

Because AWS Technology Group knowledge management was developed through a mostly empirical approach there has been no identifiable system development cost. Information technology support was used where it was already available. No new IT infrastructure was developed specifically for knowledge management. We use a mixture of Microsoft tools including MS Office, MS Exchange and MS FrontPage for the Intranet/Extranet.

We are confident that through better knowledge management we can provide ever more effective and sustainable solutions and steadily reduce the cost of the provision of these in real terms. This will contribute to steadily improving customer satisfaction.

REFERENCES

Allee, V. (1997). *The Knowledge Evolution - Expanding Organizational Intelligence*, Butterworth-Heinemann, Boston.

Nonaka, I. and Takeuchi, H. (1995). *The Knowledge-Creating Company*, Oxford University Press, Oxford.

Senge, P. (1990). *The Fifth Discipline*, Doubleday, New York.

INFORMATION, KNOWLEDGE AND TEAMS

Malcolm Crowe

University of Paisley
Paisley PA1 2BE, UK

INTRODUCTION

This paper will examine the shifts in perspective that have taken place in the Information Systems community in the last decade, as reflected in the published literature. Specifically, a trend is detectable where authors increasingly treat information almost as if it were data, rather than being inside the heads of the participants. Currently, it would seem that when authors write about knowledge, they do intend something inside people's heads, and consequently considerable critical attention is being paid to writings on Organisational Learning or Knowledge Management. We discuss some of these controversies briefly in this paper.

The literature in this discipline is large, and this paper cannot attempt to give a complete bibliography. Nevertheless it draws together some strands from the recent literature that appear to indicate a surprisingly coherent framework for ongoing discussion, and seeks to draw some tentative conclusions on the future of information systems analysis and software support for increasingly informal and collaborative business processes.

EARLY INFORMATION SYSTEMS RESEARCH

Many of the debates discussed in this paper have their origins in the contributions of systems theory and the analyses of the results of systems intervention experiments of the 1960s (Checkland 1981, Susman and Evered 1978). Research on Information Systems in the 1980s (Friedman 1989, Galliers 1992) necessarily drew on experiences with the computer-based systems of the previous decade, which from today's perspective used comparatively simple data processing procedures.

A feature of Information Systems Research has always been to focus on phenomenology rather than ontology, on process rather than goals, on ethnography rather than prescription (Burrell and Morgan, 1979). Despite this emphasis on phenomenology, a review of the literature on systems and information (Checkland and Holwell, 1998) identified numerous competing definitions of what an organisation *is*. More recently, it

Edited by Ragsdell *et al.*, Kluwer Academic/Plenum Publishers, New York, 2002.

would appear that fewer authors have felt it worthwhile to invent further definitions of what is meant by "information, systems and information systems", and on the whole this is to be welcomed.

In addition, mathematical methods seem to be no longer in fashion, so that few authors now appear to believe that a good idea needs to be dressed up in borrowed mathematical clothes, as appeared to be necessary in the 1970s. The systems movement has distanced itself from its origins in General Systems Theory and the predictive differential equations of dynamical systems (Bertalanffy, 1956). But the intellective notion of system, as in Checkland's work, has also, regrettably, been somewhat eclipsed by the software solutions designer's notion of system, with a consequential impoverishment of the notion of information system.

It is no longer fashionable, either, to spend time debating the roles of postmodernism and interpretivism in information systems. Possibly this may be viewed as a success, since the culture of information systems research has certainly moved on (Crowe et al, 1994). Where once the focus of concern was on how physical artefacts such as marks on paper and data can be construed as information and meaning through individual understanding and interpretation (hermeneutics) (Checkland, 1981), the emphasis is now on learning organisations and the formation of knowledge as information made useful by being shared.

INFORMATION AND DATA

In business and industry, using computers once meant data processing. Today, with the rise of software for handling and presenting text and multimedia, computers are used much more immediately by many sorts of employee: not just data entry clerks and accountants. It is common to say that computing has become pervasive in leisure and home activities too, with the success of the internet and personal computing. Thus computers, and specifically the Internet, are now generally seen as sources of "information" (loosely defined) rather than mere data.

This change in the context of information systems research has been reflected in the literature, which once contained serious warnings about the difference between data and information (Checkland, 1981). It would appear that this difference is seen as less of a major issue today than it once was.

In the literature we read formulations such as "'information' will be taken to mean data to which some meaning has been added" (Martin, 2001). In the past, some authors (e.g. Checkland, 1989, Stowell and West, 1994) created similar formulations in order to imply a fundamental distinction between data and information. Now, however, looking at the context of such statements in recent writing, it would seem that much less of an issue is being made of the difference.

Formerly, the notion of meaning being added implied of necessity the intervention of individual understanding and sense-making: Although some reminders or clues as to the meaning that the individual or group assigned to the data might well be added (as further data), mere addition of more data would not turn data into information. Interpretation of the "meaning" would still depend on the understanding of individuals of the extra symbols that had been added to convey it. The idea that such understanding might be shared among groups also seemed problematic (e.g. Ciborra, 1993).

Points of this kind were useful during the era when some thoughtless claims to objectivity by systems analysts were in conflict with insights from information systems research (Doyle and Wood, 1991). But sharing of language and the association of meaning are everyday facts of life, and computer data, especially when in the form of ordinary text, can be considered as no harder to interpret than written or spoken language (Wittgenstein, 1953). It may therefore now be reasonable to regard the philosophical difficulties associated with the use of a shared *language* as a matter for specialist study (Chomsky, 1988) or the museum of ideas, rather than the current work of information systems practitioners. Indeed the attempt to associate any universal meaning with any particular utterance seems today rather quaint (Rorty, 1989). Thus for practical purposes in the information systems field, the problems of representing meaning seem little different, in principle, from the problems of representing anything else: in all cases there need to be some shared symbols and concepts, and the aspiration to a shared understanding.

The trend we have noted above is not without sharp critics. Mingers (2001) complains that "information systems is currently precisely epitomized by [the approach of 'disembodied cartesianism'] since what it sets out to do is to capture that, *and only that*, which can be represented by an abstract symbolism suitable for electronic storage and processing." One might reply that any written language is such an abstract symbolism. It would seem however that Mingers' real target was Knowledge Management, which we return to later in this paper.

There is nevertheless a consensus that attention needs to be paid to the context of the information system. Important aspects of the context are the notions of the value and usability of streams of data, which are currently seen as operating at a level where individual interpretation are still considered to be important. At present in the literature, it would appear that this level is called "knowledge". We consider this level next.

INFORMATION AND KNOWLEDGE

There is a clear progression from the kinds of data that might be recorded by a mechanical sensor, or contained in a purchase order, via the addition of contextual hints and the formation of aggregates of such data, to matters of judgement such as might be contained in legislation, corporate policies or international treaties. Progressively such information becomes more difficult to treat automatically, and interpretation becomes more of an issue. If information is taken to be data to which a meaning has been attached, then the next step is currently called knowledge.

The traditional view of knowledge as expressed by Suchman (1987), was that

> shared knowledge is a potentially innumerable body of empirical assumptions and presumptions that stands behind every explicit action or utterance, and from which participants in interaction selectively draw in understanding each other actions.

This passages places knowledge inside the heads of the participants, when they start to interpret the actions and utterances of others. As ideas have evolved, information and knowledge are seen, at least, as more closely interlinked: knowledge grows during learning and is distilled from various sorts of information including observation of the consequences of situated actions.

If anything has changed in current writings, it is a more explicit recognition of the ascent from data through information and meaning to knowledge, as in Mingers (2001)

> everyday knowledge is an ever-present personal background *through which* we experience the world; .. it is inherently *social* rather than individual, *assuming* an already-established sharing of meaning..

More explicitly, Martin (2001) speaks of "'knowledge' as information placed in context and given added value through the addition of judgement and/or action. ... whereas information is all around us, knowledge emerges from creative connections drawn by recipients".

The information-theorist's notion of the quantity of information contained in some data is related to the changes in expected probabilities of events (i.e. of different sorts of further data). In a sense, this is a notion of the value of the information: in this sense, perhaps, one might even hope to construct a sort of Shannon's Theorem for knowledge.

For others, the creative connections are more organisational. Hayes (2001) views knowledge as residing in an evolving, continuously renewed set of relations of persons, their actions, and the context they are situated within – ten years ago such a statement would have been made of information, or of the emergent notion of the business organisation, whose relationship with knowledge we will consider next.

KNOWLEDGE AND ORGANISATION

There is a sense in which knowledge and organisation are tightly bound together. The organisation, as providing the motivating context for the activity of a group, also provides the context for considering the value and usability of a data stream, and both of these aspects combine in the idea of knowledge. We speak of knowledge (or learning) when we are considering the impact of information on the activities or capabilities of individuals, organisations, or academic disciplines.

Knowledge is therefore indispensable: there is knowledge required *a priori* to any undertaking, including the establishment of an organisation or a team. We return later to how the activities of the organisation result in the growth of some sorts of knowledge. Knowledge, however, appears as a precursor to the organisation and hence to management. One task of management, from this viewpoint, is to protect and enhance the core capabilities of the organisation.

One could speak of this responsibility as one of Knowledge Management. But great care is the required, since while the composition and activities of teams and their infrastructure can be managed, their generation of knowledge can be merely facilitated.

As Mingers (2001) remarks, the pitfalls in speaking of knowledge management include

> (i) *objectification* – the idea that knowledge is a tangible object that can be processed, stored and manipulated, (ii) *commodification* – that knowledge has a specified value, and (iii) *codification* – that it can be codified and represented, and especially that knowledge can be made explicit..

Marshall and Sapsed (2000) provide an extended critique along these lines. In this context, the notion of organisational culture seems more helpful, since no-one seems yet to have proposed "culture management" though many managers aspire to change the culture of their organisation.

CULTURE AND PERSPECTIVES

Both knowledge and organisation contribute to, possibly even determine and are determined by, the culture of shared meanings and pragmatics (Crowe, 1999; Cronin, 2001). Culture is about the appropriateness of language or actions to particular situations, "what would count as an answer" to "what would be understood as a question".

Hayes (2001) notes that Boland and Tenkasi (1995) saw organisation as being

> characterised by a process of distributed cognition in which multiple communities of specialised knowledge workers, each dealing with a part of an overall organisational problem, interact to create the patterns of sense making and behaviour displayed by the organisation as a whole.

In considering the use made of computer based assistance ("groupware") by skilled participants in such processes to manage perspectives and interactions, he notes an evolution from Boland and Tenkasi's view of groupware as merely providing repositories to encode and share data. He even states this as a requirement for designers of future groupware systems to consider "the creation of boundary objects, entities which can, in some sense be shared across different social worlds, yet have a variable significance between them".

Such entities are a commonplace in all hierarchical forms of management, where each person acts as a boundary object between the activities below and the management above. It is an everyday occurrence for people to take account of the differences in perspective between management and the grass roots as they apply to any transaction. This is not to say that such differences in perspective do not cause problems: an unskilled operator of the management game will stand accused of being two-faces, a poor communicator, etc. We are encouraged by the formulation of Hayes' paper, however, to consider other kinds of, possibly more computationally tractable, boundary objects.

In such less familiar circumstances, however, the difference in perspective being applied on the two sides could pose severe problems. If such a boundary object formed part of an automatic mechanism, it might go unnoticed for some time that data prepared on one side of it was being misunderstood when considered from the other side, and there might, as Ciborra and Lanzara (1994) noted, develop a reluctance to generate any data that might be viewed through the boundary.

TEAMS AND LEARNING

Teams, almost by definition, are groups of people charged with a task. In carrying out their task, however routine and repetitive it may seem to the outsider, they gain experience and therefore learning. Such learning takes place simultaneously at several levels. First, to the extent that the task is successfully carried out, in addition to the learning associated with mechanistic actions, there is the process of taking account that this is can be successfully done for this situation and these circumstances; then that it could also be successfully done with some variation in conditions and circumstances, and more widely as a generic ability to undertake this general sort of task. Secondly, there can be a focus on the roles in the activity, taking into account the individual contribution of team members, the team as a group, and the organisational culture which notes that here is a type of activity we can do in this way. To the extent that the team is carrying out the task as part of the activities of the organisation, the organisation can be said to learn.

Possibly the most interesting sorts of tasks are those that bring about noticeable changes in abilities and culture of the organisation. Generally we expect these tasks to be where the experience of working together, the nature of the task, and the information being considered are all relatively novel to the team. In these circumstances are found the riskiest and the most rewarding projects, since the nature of the new knowledge that may be gained, and hence its effect on the team's environment (the organisation) is not known in advance.

The literature on Organisational Learning (OL) seems to present a rather confused picture. There are many models, which "seem to come out of the blue, as most, if not all, the writers have approached the subject from scratch" (Lähteenmäki, 2001), in a similar way to the work on information systems and organisations discussed earlier. Considering this critique, it does seem that the focus in OL literature has been on high-profile management initiatives which promise to transform the business, rather than on facilitating a learning culture through patient attention to infrastructure and practice.

INFORMATION SYSTEMS ANALYSIS

Another important aspect of the context of Information Systems research has been the slow rejection of mechanistic and hierarchical approaches to information systems provision. In the literature many writers were responding to this trend in the early 1990s. For example, Ciborra (1993) gave several reasons why the construction of a system was not analogous to a manufacturing process: first, the non-deterministic world where contingencies are known only to a limited extent; second, the nature of the task – the product to be manufactured is complex, new and volatile in its specifications; and

> Third, the *nature of the product* makes a difference. Constructing a system is not simply building a car.. The final outcome of a system project is in fact a real *organisational change*. People's roles, skills, decision-making power, career patterns, work organisation, communication etc., may be affected by the new system… If the actual use of any routine, the practical application of rules, or the integration of any data-processing procedure are the outcome of a daily negotiated order within the organisation.., then systems development is no exception.

On this basis, the introduction of computer-based processing into work practices should be seen as gradual, consensual, and enabling. The literature is full of histories of failures of grandiose mechanistic systems, and these failures often divert attention from the successes of information systems: word processing, spreadsheets, databases, communications, internet searching. Software applications that are easily tailored to the changing needs of a business have become the norm, and there seems to be increasing acceptance that the expertise inside employees' heads cannot be reduced to the kind of surface, functional information captured in a short interview with a systems analyst. Software to enable people to do what they do rather than to replace them has become more widespread and more widely welcomed.

Similarly, in considering the possible infrastructural support for the processes considered above, including software, we find that the most effective changes have been incremental (Ciborra 1996, Crowe and Kydd 1998). Groupware needs to start from the established basis of electronic mail.

In this context, information systems analysts should seek to build in reflective aids into software, so that what has been learned in any exercise is briefly considered from the variety of viewpoints outlined above. For example, if electronic documents are used such as e-mail or templates, simple devices could be included to capture participants' views of the value in organisational learning of a particular scenario, and whether a reference to it should be included in a digest of activity (such as Frequently Asked Questions, or training materials for new staff). Subsequent comments on the usefulness of such digests could also be valuable, if similarly collected.

Similarly, attention should be paid to leveraging accepted software and supported practices rather than risking change resistance with a novel system. For example, electronic messaging is popular and widely used: simple additions could assist in classifying the messages as associated with different business processes, so that the messages and replies were automatically handled differently. Thus some types of messages could automatically be placed in a group folder, have the company advertisement or a disclaimer appended, be digitally signed etc. Slightly more subtle additions could automatically propose a classification which the user could refine.

The difficulty with such modest proposals is that while similar strategies have proved successful, they lack the glamour of the quick fix or the large sale. It is to be hoped that the benefits of careful attention to organisational knowledge, from the viewpoint advocated above, will encourage management not to seek instant grandiose solutions, but to adopt incremental strategies that value the learning and experience embodied in the members of their organisation.

REFERENCES

Bertalanffy, L. von and Rapoport, A. (eds.) (1956). *General Systems Yearbook of the Society for the Advancement of General Systems Theory*, vol.1, Braun-Brunfield, Ann Arbor, MI.

Boland F. and Tenkasi, R. V. (1995). "Perspective making and perspective taking in communities of knowing," *Organisation Science*, 6:350-372.

Burrell, G. and Morgan, G. (1979). *Sociological Paradigms and Organisational Analysis*, Gower, Aldershot.

Checkland, P. B. (1981). *Systems Thinking, Systems Practice*. Wiley, Chichester.

Checkland, P. B. (1989). "Soft Systems Methodology," in *Rational Analysis for a Problematic World*, (J. Rosenhead, ed.), Wiley, Chichester.

Checkland, P. B. and Holwell, S. (1998). *Information, Systems and Information Systems - making sense of the field*, Wiley, Chichester.

Ciborra, C. U. (1993). *Teams, Markets and Systems: Business Innovation and Information Technology*, Cambridge University Press, Cambridge.

Ciborra, C. U. and Lanzara, G. F. (1994). "Formative contexts and information technology: understanding the dynamics of innovation in organizations," *Accounting, Management and Information Technology*, 4:61-86.

Ciborra, C. U. ed. (1996). *Groupware & Teamwork*, Wiley, Chichester.

Chomsky, N. (1988). *Language and Problems of Knowledge*, MIT Press, Cambridge MA.

Cronin, B. (2001). "Knowledge Management, organizational culture and Anglo-American higher education," *J. Information Science*, 27:129-137.

Crowe, M. K., Beeby, R. B., and Gammack, J. G. (1994). *Constructing Systems and Information: a process view*, McGraw-Hill, Maidenhead (available as a full-text electronic book at http://library.paisley.ac.uk/).

Crowe, M. K. and Kydd, R. A. (1998). "Delegation and Interference: The Personal Workstation and the Corporate Network," *Proceedings*, EuropIA'98 7th International Conference on the Applications/Implications of Computer Networking in Architecture, Construction, Design and Urban Planning pp.195-208.

Crowe, M.K. and Kydd, R.A. (1999). "Culture, ideas, idiosyncrasy," in *Synergy Matters: working with systems in the 21st century*, (A.M. Castell,, A.J. Gregory, G. Hindle, M. James and G. Ragsdell, eds), Kluwer Academic/Plenum Publishers, New York.

Doyle, K. and Wood, R. (1991), "Systems Thinking, Systems Practice: Dangerous Liaisons", *Systemist* **13**, p. 28-30.

Friedman, A. L., with Cornford, D. S. (1989). *Computer Systems Development: history, organisation and implementation*, Wiley, Chichester.

Galliers, R. D. (ed.) (1992). *Information Systems Research: issues, methods and practical guidelines*, McGraw-Hill, Maidenhead.

Hayes, N. (2001). "Boundless and bounded interactions in the knowledge work process: the role of groupware technologies," *Information and Organization*, 11:79-101.

Lähteenmäki, S., Toivonen, J., and Mattila, M. (2001). "Critical aspects of Organizational Learning Research and Proposals for its Measurement," *British Journal of Management*, 12:113-129.

Marshall, N. and Sapsed, J. (2000). "The Limits of Disembodied Knowledge: Challenges of Inter-project Learning in the Production of Complex Products and Systems," in *Knowledge Management: concepts and controversies*, University of Warwick, 10-11 February 2000 (http://bprc.warwick.ac.uk/papers.html).

Martin, W. J. (2001). "The role of knowledge content in e-commerce," *J. Information Science*, 27:180-185.

Mingers, J. (2001). "Embodying information systems: the contribution of phenomenology," *Information and Organization*, 11:103-128.

Rorty, R. (1989). *Contingency, irony, and solidarity*, Cambridge University Press, Cambridge.

Stowell, F. A. and West, D. (1994). *Client-Led Design: a systemic approach to information systems definition*, McGraw-Hill, Maidenhead.

Suchman, L. A. (1987). *Plans and situated actions*, Cambridge University Press, Cambridge.

Susman, G. and Evered, R. D. (1978). An assessment of the scientific merits of action research, *Administrative Science Quarterly*, 23:582-603.

Wittgenstein, L. (1953) *Philosophical Investigations*, Blackwell, Oxford.

A HUMAN VALUE SYSTEM BASED METHODOLOGY FOR AN ORGANISATIONAL DEVELOPMENT CONSULTANCY

Misha Hebel and Bjørn Mathiesen

Brignition Ltd,
Newlyn, Haw Lane
Olveston, Bristol BS35 4EQ (UK)
Telephone: +44 (0) 7799 061547
E-mail: mishah@brignition.com

INTRODUCTION

The successful management of change is often asserted to be dependent on the alignment of values (Hall, 1994: Hofstede, 1991 and 1994). The delivery and facilitation of this process by identifying values and integrating the combining cultures is frequently outsourced as it requires skills, an outside perspective and resources rarely kept in-house, even in larger companies. This paper describes the design and development of a holistic methodology providing a flexible but structured toolkit to facilitate learning and manage business processes after global mergers and acquisitions (M&As). The methodology was developed during the course of research undertaken for an Organisational Development (OD) consultancy.

BACKGROUND

The Brignition consultancy was founded on 30th September 1999 and started trading in January 2000. It was born as a reaction to the experience of its founders [Mathiesen and Cren] who felt limited by large consultancies inflexible use of methodology and production of formularised solutions. Mathiesen and Cren shared a view that building relationships on a shared set of values and a commitment to value-adding was the core of useful consulting and working in effective teams. Whilst wishing to combine ethical and commercially viable aims they were also well aware from their ex-colleague, Roald Nomme, that some claim only 15% of the consultants' contribution is profitable for the client (Freedman and Zackrison, 2000).

The name is a contraction of 'Bridging' and 'Ignition', the two overarching aims of the companies approach. The purpose is to *bridge* the gaps that exists in an organisation

Edited by Ragsdell *et al.*, Kluwer Academic/Plenum Publishers, New York, 2002.

after a merger or other major upheaval where a future state is sought. The response to which can be likened to a grieving process as familiar structures and relationships disappear, there is upheaval and fear but life has to go on. This results in new opportunities emerging and gaining momentum. The *ignition* refers to starting a process of change, kindling the organisational learning and leadership development, sparking life into opportunities present in difficult situations full of emotions and conflict (Goleman, 1996).

This basic philosophy has been developed by acquiring new people and skills and ensuring assumptions are articulated at every opportunity. Core to the development has been the building of relationships and constant testing of usefulness and reflection on learning. This paper documents the building of a practitioner methodology with the aim of evaluating it further in the heat of conference discussion and debate.

METHODOLOGY AND CONTEXT

Avison and Fitzgerald (1997) argue that a methodology is more than a collection of procedures, techniques, tools and documentation aids - it has a philosophical view and assumptions that guide its emphasis. In other words methodology refers to the overall approach to a research, design and delivery process, from the theoretical underpinning to the collection and analysis of the data to active solutions. "Like theories, methodologies cannot be true or false only more or less useful" (Silverman, 1994, p.2). The underpinning of the methodology outlined in this paper is a need to be useful, indeed helpful, by being flexible to client needs rather than dictating right or wrong.

There are many different types of methodology, however the key methodologies in many disciplines range between two main research paradigms, positivistic and phenomenological, which can be interpreted as the extremes of a continuum (Hussey and Hussey, 1997). The positivistic approach lends itself to quantitative analysis whereas the phenomenological approach lends itself to qualitative analysis. Each methodology is moved some way along the continuum according to the practitioners' assumptions.

Qualitative methodologies are concerned with whether patterns, concepts and theories in a particular environment can be applied to other environments. To do this requires a comprehensive understanding of the activities and behaviour being studied (Vogt, 1993). Research and methodologies are often validated by triangulation which has vital strengths in encouraging productive research, enhancing qualitative methods and allowing the complementary use of quantitative methods which are more empirical and number based (Jick, 1979).

Systemic methodology has taken two main routes since the 1950's. This paper tends towards more holistic and practical problem solving than computer design methodologies - encouraging the practitioner to deconstruct reality and step outside their instinctive worldviews. For example the early use of Checkland's Soft Systems Methodology (SSM) encouraged a non text based summary of the situation (rich pictures) followed by conceptual analysis then a comparison of the two states (Checkland and Scholes, 1990; Patching, 1990; Open University, 2000).

Core to this methodology and others such as the Viable Systems Model (VSM) (Open University, 2000) was the interconnectedness of any situation, an acknowledgement that where human activities take place there will always be consequences to actions sometimes surprising and far beyond usual expectations thanks to emergence and synergy. Values also exhibit strong systemic qualities and it can be argued that their combination in both individuals and groups also prompts emergent properties which in turn can result in the unexpected.

Applied methodology cannot be separated from the context in which it is used. In order to be successful new organisations need to consolidate and develop their collective knowledge but barriers inevitably exist (Santosus, 2001). Unfortunately the pursuit of knowledge can become lost in emotive games of control and the perceived regaining of personal dignity often lost in a merger - or job change of any kind - when the transition period throws up doubts and uncertainties (Goleman, 1996).

This is where a need for outside help, not necessarily unemotional but capable of a broader perspective becomes most useful. Senior management teams in organisations need help to step outside their existing perspectives because otherwise they end up trapped doing the same things they've always done. This results in a dominant management logic which is the way managers conceptualise the business and make critical resource allocation decisions about technologies, product development, distribution, advertising, or human resource management. This prevails and all the while a company is successful this is reinforced and valued. Unfortunately it can also result in cognitive dissonance and a paralysis when changes occur.

Sveiby (1998) argues that humans are very good at doing things tacitly, so good that they can express only a fraction of it in words. It is argued that a much better way is to show how things are done as information is a very poor vehicle for transferring knowledge. This is because over the years innumerable patterns are built up in our brains that serve as unconscious rules of procedure to cope with every conceivable situation. These rules save us a great deal of energy, enabling us to act quickly and effectively without having to think about what we are doing the whole time. These rules of procedure are also an essential part of acquiring and improving skills. A variation on this model has been developed by Hebel (2000) as the transformation process of building a value system. It demonstrates the constant reinforcement of patterns and subsequently values over time with some examples of how they are influenced.

But these patterns or rules are also a limitation. Since they are always there, they affect new knowledge like a filter or a pair of spectacles. This is supported by the observation that the values developed as a result of these patterns may also act as a filter where each experiences is sub-consciously and very rapidly compared to the values model or dominant logic in our minds (Hebel 1998a and b). In this way all new knowledge is always coloured by our previous knowledge - and this is why it can be argued that there exists no "objective" knowledge and so no one way of implementing a solution or methodology.

What this points to is the emotionally turbulent situation that managers in M&As as well as many consultants find themselves in - on the surface it is about finding the right processes, cutting costs, streamlining etc. However, sense needs to be made of the changes, feelings about the change needs to be handled sensitively and a constructive way forward not only has to be identified but implemented in a very practical and values-concordant way.

It should be noted that when a group attempts to superimpose a new set of values on subordinates who are discontent, the result is often that they act out the surface signals of the new culture. They do this cynically and without internalisation so any methodology needs to help identify how to reconcile the fit between the approach and the benefits to the workforce, that would not just give a commitment to all but also rebuild trust, subsequently developing morale and motivation. This requires an ethical stance and a constant reference to the assumptions embedded in the consultants value system.

So the acceptance and success of a methodology in practice depends not only on environmental factors but also depends heavily on the human values of those involved - for example practitioner, client, theoretician, benefactors - being matched sufficiently for them to communicate and act with some understanding.

Of course it may also be argued that the obstacles are so great that they may prohibit usefulness and that methodology may inhibit creativity and limit options. However the structure and coherence it offers in its communication are thought to outweigh the disadvantages in a practical situation.

From this summarised theoretical basis values are reviewed and applied as something more than positive ideals. Prior work has indicated that people were prone to giving organisationally acceptable or normative answers to questions directly related to values (Hofstede, 1994; Hebel, 1998a). Consequently when consultancy is conducted these are constantly born in mind.

The following section builds on the theory by briefly describing the practical tools currently being developed and which it is anticipated, will be much refined by the time of the conference.

EVOLUTION OF A METHODOLOGY

The methodology described in this paper has evolved from tacit and explicit use of holistic systems concepts and real world practice of the company members (Brignition, 2001). Brignition operate in a multi-cultural, multi-lingual environment with technology an integral but not dominant part of their delivery. Collectively they draw on six different languages and cultures; two genders; between 1 and 30 years employment experience in business and public sector in many European countries; education to at least Masters level in business, theology, systems theory and organisational psychology; plus a wide range of personal interests. In many senses they are a true interdisciplinary systems group.

They are very aware of the real commercial aspects of consultancy and the sometimes negative perception held of that activity. These include a view that consultants are greedy and hold organisations to ransom. Ultimately killing confidence and sometimes doing great financial damage. This environment although unwelcoming presents challenges and opportunities welcomed by the members of Brignition. It is the lack of acknowledgement of this area that prompts a desire to articulate a methodology to include diversity and the opportunity to contest conventional wisdom.

The methodology is primarily qualitative in that patterns are sought and applied to different environments. The antecedents of Brignition members and their value systems are recognised as an important and unavoidable basis for how they perceive the present and the future. From a background of leadership development and knowledge management Nomme describes practical holistic thinking as "to think about the connections between employees, the company, and society" (Fishman 1999). From experience of performance measurement and value systems, Hebel (2000) notes that human values and values systems form the basis of individual personalities and consequently they affect all activity involving people. Cren (1998) argues that there is a clear advantage in speaking a common language. Høyem draws on experience in counselling and has reflected on the real application of a values measurement tool (Dramstad and Høyem, 2001).

Although in practice it is primarily a business process (BP) and OD methodology, the development also draws strongly on the concepts of systems theories and Information System (IS) methodologies. The successful communication of information between consultant and an organisation's management and employees is essential if it is to be successful. After a merger many companies find themselves in possession of much data or unstructured facts but little information or explicit knowledge (Avision and Fitzgerald, 1997). Different sections of the organisation often operate on assumptions about how others work like they do or are so totally alienated from one another that an integrated solution cannot even be perceived. Checkland and Holwell (1998) add an intermediate stage between data and information called 'Capta' which specifies the cognitive or appreciative aspect that must be established before data can go on to something meaningful.

Methodology is seen as containing the following elements and stages; Aims, Philosophy, Values, Methods (application and processes), Tools and Evaluation. The length of this paper restricts elaboration of these stages to that which is most unique and hopefully interesting.

Core to this methodology's philosophy is establishing a sound, values compatible way of working. This started by surfacing values and discussing their synergy. So far those identified are flexibility, help, trust, sincerity, clarity, dialogue, relationships, creativity and learning. This discussion is ongoing and is translated back into working environments. It is considered a necessary platform for delivery which given the multicultural nature of the working environment must be scaleable (concept/product/integration) and open to standardisation.

Values in Practice

Clarity, trust and dialogue are supported by deciding the company's aims, determining individual capabilities and using good interview techniques i.e. no prejudgement of importance, observing, hearing not paraphrasing, and analysis techniques i.e. gathering impressions and emerging interpretations, thus making the qualitative data more reliable (Hatch, 1999). Reliability is further improved by being systematic and rigorous in approach. This means being explicit in the use of methodology, data collection methods, techniques and procedures, and theories and frameworks both with clients and between the consultants.

Flexibility and learning means constantly reviewing the framework of good practice: constraints must be considered such as time, existing experience, bought-in experience, client needs, available personnel, knowledge of personnel skills, ability to handle conceptual stuff.

Help, rather than increasing dependency on the consultant, is a core value and it is hoped to convey this through sincerity, communication and sensitivity. As a group, strengths and weaknesses can be shared without fear of censure and reflection adds to learning.

Relationships are cultivated both within and outside the company acknowledging that no relationship can be objective and intervention, however brief, always has some consequence. This is supported by offering a creative entrepreneurs attitude to larger consultancy. These are embodied by the personal values of Initiative, Drive and Energy.

Many of the techniques used to facilitate the delivery of such values are well known and included surveys, rich pictures, mind mapping, counselling, brainstorming, other methodologies, facilitated discussions, values measurement, business model comparison, and business process analysis. However, a number of these are increasingly being transferred to an web-enabled environment (eRoom, 2001) that can cross time and physical barriers. The following section introduces a couple of succinct ways in which the way the methodology can be described beyond a familiar set of tools and it is these areas that are currently being evaluated.

Using a Mnemonic

A familiar way to summarise the stages of a process or methodology the mnemonic or use of initials has been a favourite of systems and business thinkers alike. This particular one evolved from the sharing of humour - a core activity within the group - and the need to find a word that could be occasionally substituted when the shared language of English failed.

Systemic - We are holistic in our approach, aware of interconnections, complexity and consequences of our actions

Tutor-based - Facilitation, counselling, mentoring are key to development in demanding transitional phases

Usable - We want to leave organisations with something they can continue to develop with or without our help

Friendly - We are interested in building mutually beneficial relationships and mutuality, respect and humour must be a part of this

Forthright - We don't want to play games or pursue avenues that are destructive or misleading to any partner - trust and honesty are important in all areas including key relationships or in commercial arrangements

In effect this summarises the philosophy of the company and provides the core of the methodology which has been much discussed in order to present a coherent approach. The worldview is that if that if transparency is not possible then how will the distrust of consultants be overcome.

Using Metaphors

To try to find another way of conveying that it is understood how complex a clients working environment is and how we can add value to their situation a number of metaphors have also been developed. Due to the space constraint they only outlined here.

1) Fruit bowl - Our skills and services can be likened to a selection of fruits that can be chosen according to palette and situation. There is flexibility in choice and solutions can be varied.

2) Picture frames - The perspectives we deal with are a consequence of each situation being framed by the people involved with it. The existing situation might be likened to an old, ornate, gilded frame. We try to help people to reframe their images to something more contemporary.

3) Racing track - Appreciation of environment and opportunities - Business is like formula one racing, fast and furious, international with potentially huge risks. For the sport to exist various things must be in place, investment, participants and spectators, safety regulations, technical know-how, support teams and regular checks. We act as pit-stop staff allowing businesses to pause, think, recharge and refit.

The advantage of such metaphors is that consultant and client can elaborate on them; used as the basis of a creative discussion and can help others to develop their own metaphors. On their own the metaphors are only a beginning: nevertheless very often that is what is needed in a merger situation, bridging the gaps between parts of an organisation and igniting employees enthusiasm for change. This is supported by Davis (2001) who argues that research and consultancy should be coaching people in organisations to enable them to become self-organised learners.

Cautionary Note

It is obvious that this methodology is in its early stages and the development has had its conflicts. It was felt quite keenly that it should not be linear or prohibitive but flexible and creative. This has resulted in an ever moving structure and risks looking ill-thought through and haphazard. However it is also reflective of a changing business environment and an evolutionary process. Further application and continual reflection and learning are the only way of maintaining the process.

Because of the experience of using other methodologies it is anticipated that there may be difficulty in practical implementation. In addition, the consultant context means follow through and evaluation is not always possible. Champions move on, further restructures occur and costs start to become prohibitive because people don't connect to the value any longer.

CONCLUSION

This paper describes both the birth and initial development of a methodology. There is no doubt that it is values based and flexible with a strong systemic element of holism and synergy drawn from experiences with previous employers. Specific systems methodologies were not explicit at the outset however the core concepts have much in common with them. It is growing and developing and the process of defining a methodology has proved invaluable for internal coherence of the company. At this moment in time the company defines methodology as a structured way of approaching client situations and problems that includes an underlying philosophy of helping others to help themselves. The company aims to help clients bridge the gap between a state of upheaval and one of organisational growth - so they can continue to learn and grow in their own way.

REFERENCES

Avison, D.E. and Fitzgerald, G. (1997). *Information Systems Development: Methodologies, Techniques and Tools* (2nd ed.), McGraw-Hill, Maidenhead.

Brignition Ltd. (2001). http://www.brignition.com/ (1st October 2001).

Checkland, P.B. and Scholes, J. (1990). *Soft Systems Methodology in Action*, John Wiley and Sons, Chichester.

Cren, S. (1998). Minutes of Executive Board of Feltham First held on 29 April 1998. http://www.feltham-first.org.uk/general/minutes/MinutesApril29th1998.htm (6 September 2001).

Checkland, P.B. and Holwell, S. (1998). *Information, Systems and Information Systems*, John Wiley and Sons, Chichester.

Davis, C. (2001). Learning to Benefit from Information Systems, *PhD thesis UWE Bristol.*

Dramstad, H. P. and Høyem, D. (2001). Values are OK, but there is more to it; it has to have effect! *Internal publication for Wallenius Wilhelmsen Lines.*

eRoom (2001). *The Digital Workplace*, http://www.eroom.com/ (1st October 2001).

Fishman, C. (1999). The Way to Enough, Fast Company Magazine Issue 26:160 July/August 1999 http://www.fastcompany.com/online/26/norskhydro.html (28 August 2001).

Freedman, A.M., and Zackrison, R.E. (2000). *An Executive Guide to Employing Consultants*, Gower.

Goleman, D. (1996). *Emotional Intelligence*, Bloomsbury, London.

Hall, B.P. (1994). *Values Shift - A guide to Personal and Organisational Transformation*, Twin Lights Publishers, Rockport.

Hatch, M. (1999). Listening to the Experts, B826 The Management Research Project, The Open University Business School, Audio 2, Side 1.

Hebel, M. (1998a). Values Systems - A way to greater understanding, *Systems Practice and Action Research, **11**(4):381-402.*

Hebel, M. (1998b). Human Values in MIS - Fit or MISFfit?, *Systemist, **20**:140-152.*

Hebel, M. (2000). Human Values and the Management of Technological Change, *Cognition, Technology and Work, 2:106-115.*

Hofstede, G. (1991). *Cultures and Organisations, Intercultural Cooperation and its importance for survival*, Harper Collins, London.

Hofstede, G. (1994). *Uncommon Sense About Organizations*, Sage, London.

Hussey, J. and Hussey, R. (1997). *Business Research: a practical guide for undergraduate and postgraduate students*, MacMillan Press Ltd.

Jick, T, D. (1979). Mixing Qualitative and Quantitative Methods: Triangulation in Action, *Administrative Science Quarterly*, December, 24, pp. 602-11.

Open University (2000). *T306 Managing Complexity: A Systems Approach*, The Open University, Milton Keynes.

Patching, D. (1990). *Practical Soft Systems Analysis*, Pitman Publishing, London.

Santosus, M. (2001). Thanks for the Memories: Northrup Grumman case study, *CIO Magazine*, http://www.cio.com/archive/090101/thanks.html *(1st September 2001).*

Silverman, D. (1994). *Interpreting Qualitative Data*, Sage, London.

Sveiby, K (1998). http://www.sveiby.com.au/TacitTest.htm (25 Sept 2001).

Vogt, W. P. (1993). *Dictionary of Statistics and Methodology*, Sage, Newbury Park.

SYSTEMIC LEARNING WITH A SEARCHLIGHT APPROACH

Stig C Holmberg

Division of Informatics
Department of Information Technology and Media
Mid Sweden University, SE-831 25 ÖSTERSUND, Sweden
Shbg@ieee.org

INTRODUCTION

Knowledge handling and knowledge enhancement seems to become an issue of increasing importance and actualisation in contemporary society. In business as well as in private life, new skills and competencies appear to be required in an increasingly faster rate. Further, with current technology at hand, it seems reasonable trying to support this knowledge endeavour with computer assistance. However, as discussed by Winograd and Flores (1986), such a support may be designed for either replacing the man by the computer or having the computer to support the man. On this point, Winograd and Flores (1986) have already demonstrated that the computer has its strong properties where the human being is restricted and vice versa. Hence, the support strategy will in this paper be taken as the preferred one.

In this context, Holmberg (2001) has further identified Anticipatory Modelling and Computing (AMC) as the essential difference between Popper's (1979) Bucket Theory of Science (BTS) and his Searchlight Theory of Science (STS). A third input, at last, comes from Simon (1995), who has elucidated severe restrictions in human intellectual capacity, or with his own words, our 'bounded rationality.'

Hence, the purpose of the research reported in this paper is to find means for increasing the human knowledge handling capacity. In approaching that objective, the following main questions will be discussed:

1. Is Popper's (1979) searchlight theory of science (STS) a suitable base for human knowledge handling and acquisition?
2. Is a computer support, i.e. an Anticipatory Searchlight Engine (ASE), of such STS-based knowledge enhancing feasible and possible?
3. With what functions can in that case ASE most suitably, support us?
4. What does a reasonable robust and effective first tentative design of an ASE look like?

Edited by Ragsdell *et al.*, Kluwer Academic/Plenum Publishers, New York, 2002.

LIMITS TO HUMAN KNOWLEDGE

The total knowledge space (K), i.e. all there is to know, may be a space with many dimensions (R^n) and a complex topology. For our purpose here, however, it may be divided into four two-dimensional fields according to Figure 1. Those are, (1) what you do not know that you know, i.e. unconscious knowledge, (2) what you know that you know, (3) what you know that you do not know, and (4) what you do not know that you do not know. Here it has to be observed that knowledge is infinite, i.e. the fourth field, K4 in figure 1, extends all the way into infinity, i.e. K has no outer boundary.

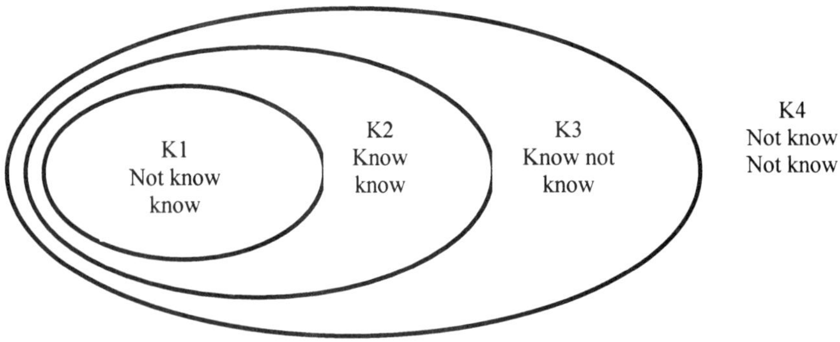

Figure 1. The Four Different Regions K1 to K4 of the Total Knowledge Space K

Figure 1 can in this context also be used as an illustration of Simon's (1995) analysis of our bounded rationality. He points out that our rationality, i.e. our knowledge, is critically bounded in at least three different ways. First, we know just an infinitesimal part of all that there is to know. Hence, the small fraction K2 we know put in relation to the total knowledge space **K** can be expressed as in equation 1. With other words, independently of how much we as individuals happen to know, in relation to the total knowledge space our individual knowledge quotient will always be zero.

The second restriction concerns our bounded capacity for computing. This restriction means that we arrive to compute just a few of all the innumerable implications and combinations of the things we, after all, happens to know. Even this is a severe restriction as the very heart of creation or design is the finding or invention of new and creative combinations of things we already know.

$$K2 / (K1 + K2 + K3 + K4) = K2 / K = K2 / \infty = 0 \quad (1)$$

K1, unconscious knowledge
K2, conscious knowledge
K3, conscious unawareness
K4, unconscious unawareness
K1 + K2 + K3 + K4 = K
K, total knowledge space

Finally, the third restriction is the very low space of our attention. "The Magical Number Seven" is often used as a metaphor for how many knowledge "chunks" we can keep in short-term memory, i.e. our attention. Here, it is also important to observe that a knowledge element can only be transferred from long to short term memory if it is recognised and if it is indexed. This implies that a memory may not be seen as a container, as in BST, but as a intricate mechanism for pattern recognition, indexing, access, and retrieval. This later interpretation is being more in line with STS.

Further, even if it is not part of Simon's discussion we obviously also are restricted by a rather narrow learning bandwidth. This means that we are able to increase K2 with just a finite and limited number of knowledge units, or bits, per time unit. Because of the Bremermann limit of 10^{93} (Bremermann, 1962) this dire fact will become equally true whatever computer support we may apply. In fact, this transcomputational problem is a neat and very striking demonstration that a BTS approach will never work!

Those restrictions, which an appropriate computer support may help us to overcome or diminish, are summarised in Table 1.

Table 1. Restrictions to Human Rationality according to Simon (1995)

1. Restricted memory content and restricted memory access mechanism
2. Restricted computational (associative) capacity
3. Restricted space of attention (magic number seven)
4. Restricted learning bandwidth.

PROPOSED MEASURES

As has already been said, the computer has its strong properties where the human being is restricted and vice versa. Hence, the natural approach appears to be to design insightfully a Man-Machine system for helping us in both better organising our current knowledge and effectively guiding us in learning the right new things at the appropriate moment. Further, the assumption made here is that such a system may favourably be based on Popper's (1979) Searchlight Theory and the emerging research results on Anticipatory Computing (Nadin, 2000; Dubois, 2001).

Towards a Searchlight Theory Guided by Anticipatory Computing

As was said in the introduction, Karl Popper (1979) has made a clear distinction between what he calls, '*the bucket theory of science' (BTS),* and the '*searchlight theory of science' (STS).* The starting point of BTS being the assumption that before we can know or say anything about the world, we must have gathered perceptions or sense experiences. Hence, our experience or knowledge consists either of accumulated perceptions or otherwise of accumulated, sorted, and classified sense expressions. Due to this view, our mind works like a container - or a bucket – in which knowledge elements randomly accumulate during our lifetime. Another significant property is that BTS perceive knowledge as consisting of discrete and unrelated knowledge elements or facts.

According to STS, on the other hand, our hypotheses, expectations and theories precede the observations. With help of completed observations the hypothesis at hand may be tested, and, if necessary, modified. In other words, those hypotheses may be seen as searchlights, which successively guide us further and further into the knowledge space. Learning becomes an infinite activity, but a new searchlight is guided and oriented by not only the preceding one, but even by its context and current expectations and objectives. This means that important connections with anticipatory theory and anticipatory computing here emerge as obvious. In contrast to BTS, in STS knowledge is understood as a continuum with strong relations between the parts.

Further, in STS observations are planned and selective perceptions, which are guided by an anticipatory system. Consequently, the better our searchlight, the more useful and relevant our knowledge about the world will be. Here it is obvious that active anticipation may be seen as the essential difference between a system learning according to BTS and STS respectively.

Rosen (1985) gave the original but tentative definition of an anticipatory system. According to this, an anticipatory system "contains a predictive model of itself and / or of its environment that allows it to change state at an instant in accord with the model's predictions pertaining to a latter instant". Elaborated and refined definitions have been proposed by later research (Dubois, 2001). Anyhow, an anticipatory system can still be seen as a system, which acts and takes measures in advance in order to be in a better position at a later instant. Put in relation to STS, anticipation could help in guiding the searchlight in the right direction. This means a situation where we learn the right things just in time.

Hence, the computer assistance we are looking for may be conceptualised as an Anticipatory Searchlight Engine (ASE). This device will be designed according to anticipatory principles and a STS understanding of learning. The principle objective of AES is the guiding of our learning searchlight in the desirable direction at the right moment and the reinforcement of our bounded rationality. Such a development would be fully in line with Stock's (1993) vision of a "Metaman", i.e. man and machine in perfect symbiosis and synergy. However, further achievements within anticipatory research (Dubois, 2001) are necessary before an ASE will be fully realisable.

Desirable Properties of the Anticipatory Searchlight Engine

In order to overcome the weak points imposed upon us by our bounded rationality (Simon, 1995), it seems reasonable that the supporting Anticipatory Searchlight Engine (ASE) as an initial approach ought to be given the five main functional capabilities proposed in table 2.

Table 2. Main Functions of AES

1. Assist in creating hypotheses
2. Assist in widening and directing focus of attention
3. Assist in calculating implications
4. Assist in indexing, recognition and information retrieval
5. Assist in decreasing our thrownness

At first, ASE ought to help us creating new hypotheses by making new combinations of the knowledge we already dispose of. This is the first and most crucial step in the knowledge process set out both in STS (Popper, 1979) and design science (Simon, 1995). Here it is interesting to see that both Nadin (2000) and van de Vijver (1998) have proposed an enlarged definition of anticipation. This means, an anticipation not only along the time dimension, but more seen as an infinitely enfolding co-relation, or an anticipation with a flattening of time. Clearly this enlarged type of anticipation fits very well into Popper's creation of new hypotheses. Hence, to the degree that an anticipatory process may guide the combination of ideas into new hypotheses, as discussed for example by Nadin (2000), instead of being entirely random, the approach will gain in effectiveness.

Secondly, ASE may also help us widen our narrow focus of attention. In this way, more things may be taken into account at the same time and more relations and connections between items may be identified. This may require some sort of improved visualisation tool as discussed for example by Simon (1995). Further, in his "la Méthode", Morin (1977) argues that learning is not a question of memorising "everything" in an accumulative or encyclopaedic meaning. Contrary, it is an issue of identifying strategic links between apparently disjoint fields and of finding the crucial relations in disorganised areas. That learning in Morin's sense will increase our repertoire for actions according to Ashby's (1958) fundamental law of requisite variety. Satisfactory knowledge is to dispose of an acceptable action for each possible challenge.

A third aid ASE can give us concerns computability. Hence, with ASE we may be able to compute more implications of alternative or hypothetical decisions and courses of action. This improved ability will sharpen our hypothesis testing skills in a desirable manner.

Further, ASE has to help us in retrieving the necessary information for a given task. This has an interesting connection with Popper's (1979) world one, two, and three. The first one being the outer real world, the two others being the knowledge held in people's collective minds and knowledge stored in libraries and databases of different forms respectively. The net and other modern technology has drastically increased our access to the information stored in libraries and databases but it has done nothing to the living knowledge in our minds. However, only knowledge in our minds can be used in creating new ideas and hypothesis or for trigging actions. Hence, we need help to know what information we need to obtain. Once the information is identified, it may easily be retrieved from any source, both internal and external. This function will consequently include an indexing system but also, and more importantly, pattern recognition and identification of necessary data and facts.

At last, the inevitable thrownness situation (Winograd and Flores, 1986) may be discharged if an anticipatory system can tell us in advance what situations to expect. In short, this means that we can not avoid from acting in any situation, but all our actions will be based on unstable and tentative representations. Hence, it will be unavoidable to make mistakes but with better anticipation and better hypotheses we may hope that those mistakes will be less grave and that they will provide better learning. Van de Vijver (1998) designates this as "*failing better*". With such a support, it would also be possible to prepare oneself to a certain degree, for example by testing different courses of action in advance.

TENTATIVE DESIGN OF AN ANTICIPATORY SEARCHLIGHT ENGINE

The Anticipatory Searchlight Engine (ASE), i.e. the computer part of a forthcoming synergetic man-machine knowledge booster, will be implemented in an ordinary personal computer environment. The preliminary or tentative design of ASE, which is given here, is the pragmatic conclusion of the analysis and synthesis, which has already been discussed earlier in this paper. The design is open ended and living, i.e., it will be modified and refined as new experiences and lessons are gained from implmentation efforts and prototyping experiments.

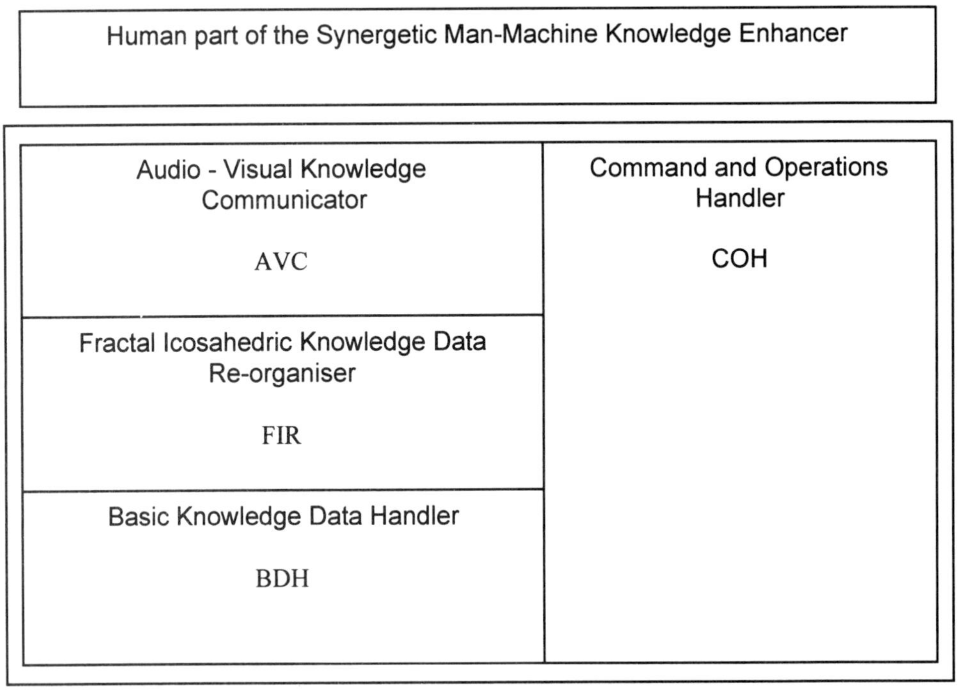

Figure 2. Basic Modules in the Tentative ASE Design

Beer's (1994) icosahedrically organised infoset, originally developed by Fuller (1992), is tentatively taken as the basic knowledge organiser in ASE. Hence, each of the twelve vertices of the regular icosahedron (RI) represents a knowledge entity, or a knowledge cluster, and each of the thirty edges represents a relation, or relation vector, between two such knowledge entities. Due to the RI:s organisation it will be no special ordering or priority between the elements in the structure. Further, the RI has fractal and recursive organisation as each vertex may contain a whole RI on the next lower level. In the other direction, the RI in focus is contained in a vertex of the RI on the next higher level. An individual vertex in the total knowledge structure will hence be fully identified by three co-ordinates denoting, vertex of RI in focus, level of parent RI, and vertex in parent RI.

Given the knowledge structure based on the RI, the next dominant feature is that the whole structure is dynamic. This means that it continuously re-organises itself and knowledge entities are moved up, down, or sidewise to new vertices. In this way new and interesting combinations or knowledge patterns may emerge. A third main feature is the display and interaction function. Here it will be possible to see the RI in perspective but also as a plane figure with preserved topology. By using colours in a similar way to Beer (1994) it will be possible to highlight certain features and properties. With the interaction function it is possible to turn around the RI or to move between vertices or to shift vertex (subject) or edge (relation) in focus. It is, of course, also possible to move up or down between different levels in the fractal knowledge structure (le Moigne, 1994).

The first tentative AES design contains four main modules according to Figure 2. The user, i.e. the human part in the synergetic man-machine knowledge enhancer, operates the system via the Commands and Operations Handler (COH). Via COH different commands and necessary data are entered and system messages and prompts are displayed. This module is implemented according to currently accepted rules for graphical user interfaces and computer – user interaction.

The Basic Knowledge Data Handler (BKH) supports a conventional rational data base with basic knowledge elements. A basic knowledge element can in this context be nearly anything that the user decides to have accessible. It can be stored in the local base but it can also be a reference to an external source, for example on the web. There are no predefined concepts or categories in this database but the user may define any relations between stored entities. All such relations, however, are fuzzy relations. With each stored item there are also, a counter and a time stamp. Hence, it is possible to see when and how often an item has been accessed.

In the Fractal Icosahedric Knowledge Data Re-organiser (FIR) the items in BKH are allotted positions in the fractal RI structure. This process is partly random and partly guided by the items' access rates. However, the user may at any moment change the exact rules for this allocation process. This process is also going on all the time. This means that the structure is dynamic and no item may be locked to a certain position in the RI structure for very long.

The last module is the Audio-Visual Knowledge Data Communicator (AVC). This unit is responsible for presenting the current configuration of the knowledge structure. The basic presentation unit is the regular icosahedron, in both plane and three-dimensional form. This, however, can be complemented with several other data formats.

CONCLUSIONS

Closing back on the initial research questions raised in this paper the following conclusions may be drawn. First, Popper's (1979) Searchlight Theory of Science (STS) is underpinned by striking and convincing research findings of several eminent researchers. Hence, it seems reasonable to take STS as a base for any prosperous knowledge creation activity. Further, as the human rationality obviously is severely bounded in areas, there the computer has its most superior properties; it appears straightforward trying to combine the human and the computer into a synergetic man-machine system.

Hence such a system, an Anticipatory Searchlight Engine (ASE), at first has to enlarge our limited focus of attention. Further, ASE will also help us compute new combinations by continuously forming new relations between items in our knowledge bank. Anticipatory theory, in its enlarged sense, in combination with Beer's icosahedric infoset (1994) has proved itself being a promising base in the design of such a device.

Further, the anticipatory character of ASE will emerge fully first if it is applied in a real case. Given, for example, that we want to investigate a specific subject area. That goal will then work as a searchlight for the ASE according to the principles of Exploratory or Evolutionary Anticipation (Holmberg, 2000).

The tentative design of ASE, which is given here, will be further tested and refined during ongoing experimental prototyping.

REFERENCES

Ashby, W.R. (1958). "Requisite Variety and its Implications for the Control of Complex Systems," *Cybernetica*, I:83-99.

Beer, S. (1994). *Beyond Dispute, The Invention of Team Syntegrity,* Wiley, Chichester.

Bremermann, H.J. (1962). "Optimization Through Evolution and Recombination", in *Self-Organizing Systems,* (M.C. Yovits, G.T. Jacobi, G.D. Goldstein, eds), Spartan Books, Washington, D.D.

Dubois, D. (ed), (2001). *Computing Anticipatory Systems, CASYS – Fourth International Conference,* AIP Conference Proceedings 573, American Institute of Physics, Melville.

Fuller, R.B. (1992), *Cosmography, (Posthumous*), MacMillan Publishing Company, New York.

Holmberg, S.C. (2000). "Designing and Prototyping Towards Anticipatory Applications", in *Computing Anticipatory Systems, CASYS – Third International Conference.* (D. Dubois, ed.), AIP Conference Proceedings, American Institute of Physics, Melville.

Holmberg, S.C. (2001). "An Anticipatory Searchlight Approach", *Int. J. of Computing Anticipatory Systems,* 9:70-83.

le Moigne, J-L. (1994). *Le Constructivisme,* Tome 1 – 2. ESF éditeur, Paris.

Morin, E. (1977). *La méthode, la nature de la nature,* Éditions du Seuil, Paris.

Nadin, M. (2000). "Anticipation: a Spooky Computation", *Int. J. of Computing Anticipatory Systems*, 6:3-47.

Popper, K.R. (1979). *Objective Knowledge, An Evolutionary Approach*, Revised Edition, Oxford University Press, Oxford.

Rosen, R. (1985). *Anticipatory Systems,* Pergamon Press, Oxford.

Simon, H.A. (1995). "Problem Forming, Problem Finding, and Problem Solving in Design", in *Design and Systems: General Applications of Methodology. Praxiology: The International Annual of Practical Philosophy and Methodology*, (A. Collen and W. Gasparski, eds.), Vol. 3, Transaction Publishers, London.

Stock, G. (1993). *Metaman – the merging of human and machines into a global superorganinis,* Simon and Schuster, New York.

van de Vijver, G. (1998). "*Anticipatory Systems: A Short Philosophical Note*", in *Computing Anticipatory Systems, CASYS – First International Conference*, AIP Conference Proceedings, (D. Dubois, ed.), American Institute of Physics, Melville, N Y.

Winograd, T. and Flores, F. (1986). *Understanding Computers and Cognition, a New Foundation for Design,* Addison Wesley, Reading.

THE KNOWLEDGE WORKER AND THE KNOWLEDGE AGE

James G. Howell

Division of Computing and Information Systems
University of Paisley
Paisley
Scotland, UK.
e-mail: Howe-ci0@wpmail.paisley.ac.uk

INTRODUCTION

Historically, new forms of journeyman emerge to suit new forms of work. The form of work is often defined by the materials and tools of the journeyman's trade and the environment in which they ply their trade (Howell, 1999; Howell and Gammack, 1997). The Knowledge Worker, a term first coined by Drucker (1959), now generally refers to a worker associated with a form of work which we might meaningfully categorise as "knowledge work" (Drucker, 1991; Collins, 1998, Nomikos, 1989). The importance of the social context of problem solving in knowledge work has been identified (Hagglund, 1989; Ngwenyama and Klein, 1994; Suchman 1983, 1987; West 1990, 1992). If recognition of the social context is important at the individual level, then this is also true at the group level. At the group level of the business, businesses require to realize that knowledge is the new competitive resource to which value is added to enable the successful business to peddle in the free marketplace, and value can be translated into helping the organisation or individual to achieve some form of competitive advantage (Despres and Hiltrop, 1995). The value of most products and services will depend on how knowledge-based intangibles can be developed (Quinn, 1992). The Knowledge Worker is often associated with a technological skill set and access to technology that allows him/her to access a "data repository" (Kappes and Thomas, 1993). Generally, the knowledge worker is associated with information rich business environments and the networked organization (Coulson-Thomas, 1991).

The knowledge worker is seen as being able to "turn" his/her raw material – data and/or information – into knowledge. If we are to understand this new phenomenon of the knowledge worker it becomes imperative that we find a more general understanding of knowledge, the "turning" process and the knowledge worker's skills or as Ngenyama and Klein (1994) suggest a Universe of Discourse (UOD). This would help provide a basis for a more coherent discourse about the knowledge worker.

Edited by Ragsdell *et al.*, Kluwer Academic/Plenum Publishers, New York, 2002.

This paper will try to enrich this discourse by adopting different approaches to understanding the knowledge worker.

THE NEW BUSINESS: FROM DATA TO KNOWLEDGE

During the fifties and sixties the new computer age spoke of data, as in data processing. In the seventies and eighties, when it was recognized that users had a constituency, data became transformed into information. Traditionally, information has been seen as "processed data" (Mingers, 1995). In the nineties, another transformation took place: from information to knowledge. There has been a shift from data intensive to knowledge intensive organizations (Ngenyama and Klein, 1994). At the individual level, it is suggested that this new artisan - the knowledge worker - can "turn" data into knowledge. This "turning" or transformational process from data to knowledge is not easily understood. The basic concepts of "data", "information" and "knowledge" are confused, as well as the relation between them (Checkland and Holwell, 1997). It is oft wrongly inferred that some simple scalar relationship exists as suggested by Despres and Hiltrop (1995),

> "It is possible .... to define knowledge work as systematic activity that traffics in data, manipulates information and develops knowledge."

In this definition of knowledge work, an association between knowledge work and systematic activity is made plain, perhaps without significant intention. Wiig et al (1997) describe a scalar spectrum of data, information, knowledge and wisdom.

Knowledge Management and Intellectual Capital are hot issues with today's business. "Intellectual Capital is the knowledge of everyone who works for it (the business)". Knowledge Management and Intellectual Capital are being seen as a new fad and an antidote to Business Process Reengineering (Royston, 2001). Generally, a performative[1] view of knowledge is being adopted in relation to the modern business in the Knowledge Age. Knowledge does not have an ontological status. It may be simply a state of mind, but the fact is that the expression 'I know' commonly has a performative rather than a descriptive use (Ayers, 1974). Knowledge in action, or performative knowledge, where the action is characterized in decision-making is often the desired product of the knowledge worker (Roy et al., 2000; Cushman et al., 1999; Harris, 1999).

It is expected that "a true knowledge worker has to design information to contribute to action that produces value" (Dunham, 2001). There is a common thread, encapsulated, in the author's view, in the concept of the knowledge worker being able to process information to create value for the business.

---

[1] Performative as identified in the classification of Ayers (1974) of performative and representational. Performative also in the sense used by Ngenyama and Klein (1994) as action constitutive knowledge. A fuller typology of knowledge is discussed in Agnew et al., 1986; Wielenga and Breuker, 1986; Johnson et al., 1987; Slatter, 1990; King and McAulay, 1991.

WHO IS THE KNOWLEDGE WORKER?

Adopting a realist ontological stance, enriching our understanding of the Knowledge Worker may be achieved by taking different approaches, such as:

1. seeking to understand the emergent business environment in which the knowledge worker is commonly associated with,
2. understanding societal changes that may have given rise to this new phenomenon,
3. better understanding the new "journeyman's" associated set of skills, knowledge and competencies,
4. seeking to better understand the materials they "turn" and the "turning" process.

Taking the latter first, potentially offering an insight and understanding that has been historically useful from the times of the Middle Ages (Howell and Gammack, 1997). This approach provides a very limited approach to understanding the knowledge worker because the transformational process from data to knowledge is not easily understood. The basic concepts of "data", "information" and "knowledge" are confused – as well as the relation between them (Checkland and Holwell, 1997). The "Age of Information" has given way to the "Knowledge Age". There is little agreement about the fundamental concept of information (Mingers, 1995; Goguen 1997) and, in turn, little agreement about the fundamental concept of knowledge.

Although there is a deep problem with knowledge (Ayers, 1974) and in understanding the elemental or source material of data, research continues to study the transformational process. In gaining knowledge from individuals or groups, a key step of Knowledge Acquisition has been recognized. This key step of the process is often conducted by a Knowledge Engineer working with a domain expert. In the process of gaining or acquiring knowledge, Ngenyama and Klein (1994) recognize six fundamental problems that need to be examined and understood if we are to be able to constitute a general theory of knowledge acquisition.

In 1993, the US Army undertook a project to examine how knowledge workers in their ranks use information during decision-making processes. The study was premised on the view that "A large number of Army personnel can be classified as *knowledge workers* - people who produce not tangible products, but some form of processed or enhanced information" (Kappes and Thomas, 1993). The project adopts the view that knowledge is created through access to information by knowledge workers. The design phase of the project sought to create objects encapsulating appropriate information. There seemed little attempt to capture tacit or implicit knowledge. This project sought to research a specific set of skills and knowledge associated with the new journeyman - the knowledge worker.

More and more businesses are moving towards competency-based recruitment and appraisal (Howell, 1999). Inspection of the general competencies often reveals a mix of these intellective and behavioural "skills" (Table 1).

Table 1. The Competence Framework

Competencies	Behaviour/Intellective Skills
Getting the job done	Handles conflicting priorities without compromising effectiveness.
Teamwork and working with others	Builds and uses an effective network of contacts to the benefit of projects and colleagues..
Communicating	Conveys ideas and facts clearly, concisely and with persuasion.
Thinking and Analysing	Quickly grasps intricacies of new work, projects or instructions.
Managing people	Gains commitment from, and motivates staff to perform well. Understands and implements policies and Equal Opportunities objectives.
Managing financial and other resources	Manages change to achieve results with minimum disruption.

It is helpful to be able to find a framework for understanding the mix of competencies, individually and as a whole. Kolb (1984) provides one approach for mapping the terrain of managerial competence based on what is called the 'competency circle'. The two dimensional map (shown below) shows the specialised adaptive competencies of managerial knowledge around the experiential learning cycle based on their association with the four basic modes of the experiential learning process,

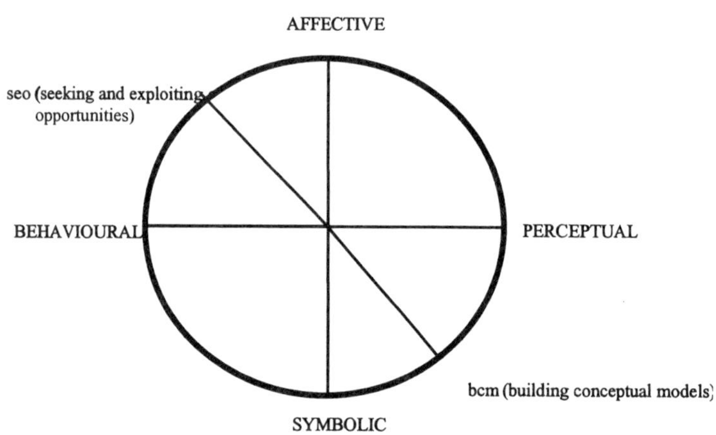

Figure 1. The Competency Circle

affective competencies (e.g. being sensitive to people's feelings)
symbolic competencies (e.g. building conceptual models)
perceptual competencies (e.g. gathering information)
behavioural competencies (e.g. making decisions)

The experiential learning cycle can eventually be expected to lead to a formation of competencies that might be described as integrative. This does not simply imply a particular set of abilities that can be directly trained and added to the other four sets of competencies, but confers an ability to move fluidly among particular modes (or arcs of the competency circle) as required in a given situation. This fluidity may be engendered by the reflective practice of systems thinking. Although the competency circle is attractively holistic as a tool, it is not easily adaptable to display integration of competencies. Profiles of competencies can be measured and plotted for different professions at different levels, resulting in spatial or 'star maps' of managerial competence. Integrative competence can be seen as the highest level of development, achievable through time and able to be shown on any candidate model. This tool can be usefully applied to helping understand the range of competencies recognized in examples of the knowledge worker or competencies expressly desired, if we can view the knowledge worker as a transdisciplinary individual.

The importance of the social context in understanding knowledge work has been identified earlier in this paper. At the societal level, Bell (1980; 1987) argues that society is entering a new phase of evolution to a post-industrial society, marked by a predominance of professional, scientific and technician groups. The growing information handling requirements of business, the development of information products and the primacy of scientific research will bring forth a new and powerful class of (knowledge) workers. Post-industrial society will be based on professional skill rather than the ownership of private property. Kumar (1981) remains unconvinced of these (Bell's) arguments supporting this post-industrial form of society. He concludes that post-industrial forms of analysis are unconvincing, naïve and intellectually conservative.

> "Defining and analyzing knowledge work which, in spite of the muddle and confusion which surrounds it, tends to emerge as the privilege of a minority elite" (Collins, 1998).

Collins adopts an industrial relations perspective for a post-industrial society in his paper reviewing the seminal work of Bell. Inconclusively, he suggests that the knowledge worker may be everyone and no-one.

In 1959 Drucker coined the term Knowledge Worker, and provided a general definition that has predated the "Information Age", and, in turn, the "Knowledge Age. In 2001 more specific definitions of the Knowledge Worker, in relation to the Knowledge Age can be identified. For example, UK-based consultancy Kudos (Dunham, 2001) suggests that:

"...there will be a new breed of employee, whom we'll call the knowledge worker. The knowledge worker will meet a number of these broad characteristics. He or she will:

work in a team - either local or virtual

work with information which provides a competitive or beneficial advantage (I say beneficial because there are still a few organizations left who do not work for real or notional profit)

typically use IT for day-to-day activities

work in an area which has a direct impact on the bottom line (whatever that bottom line may be)".

> "The extraction of information, its selection (what to leave out!) and its structure will become increasingly important. These knowledge workers, the managers, the professionals and the specialists whose expertise is hard to replicate and takes years to develop, are becoming the most coveted assets an organization has in a world of chronic skills shortages." (Dunham, ibid.)

Chronic skill shortages are commonly suggested. It is not clear that what kinds of skills these are in relation to the knowledge worker. It has become all too easy to maintain a focus on technological developments and to focus on the latest set of technological "skills" without consideration of the context within which skills can be developed.

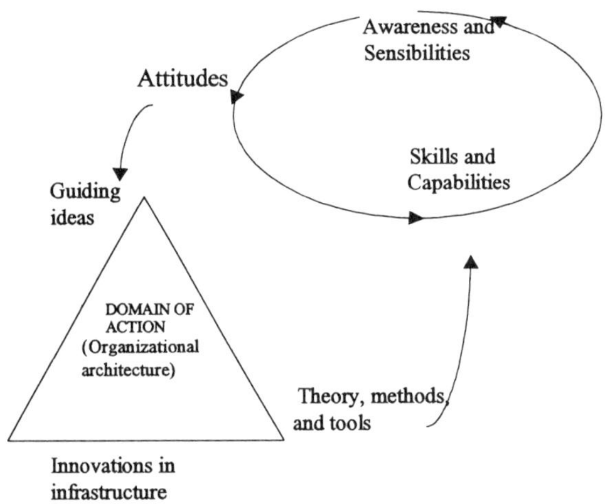

Figure 2. The Learning Organisation

As applications become more accessible and usable, and technologies become more transparent, many more people will become engaged in forms of work which do not require detailed knowledge of the technologies. Instead of requiring detailed knowledge, classic human and business themes are likely to dominate specific forms of activity. One model of this presented by Senge et al (1995) shows the ideas and fads of the day which shape mundane activity and technologies as transient and ephemeral compared to the domain of deep change where meaningful work and development is realised.

As new skills and awareness increase, reflective practitioners can engage in purposeful activity without having to be concerned more than minimally with the technological media of the day. Whilst recognising the undeniable potential for technology to make a difference at this level of organisational activity, to understand more deeply the impact of the mechanistic variations, we require to provide a basis for the reflective practice of systems thinking. The merely mechanistic is then contextualised through Systems Thinking.

CONCLUSIONS

The phenomena of this new type of worker and the type of work that they engage in are poorly understood (Despres and Hiltrop, 1995). Fundamental to understanding the knowledge worker is understanding the transformational process by which they "turn" information" into knowledge of value to the business at that level. Value is created and peddled in the free marketplace but the transformational process is difficult to understand and/ or elicit. Underlying research continues to point towards mechanistic models of transformation that do little to contribute to understanding the complexity and richness of the process, such as the US army example noted in this paper. Little consideration is given to, for example, the reflective practice of the knowledge worker, perhaps due to the "terra incognito" of implicit knowledge (Ngwenyama and Klein, 1994). Information systems can offer a worthwhile contribution to this work of understanding the knowledge worker, particularly in understanding a transformational process that may not benefit by being looked on as systematic or through a lens whose wearer seeks to apply mechanistic models to improve understanding.

REFERENCES

Agnew, N. M, Brown, J. L. and Lynch J. G. (1986). "Extending the reach of knowledge engineering", *Future Computing Systems*, 1:115-141.

Ayers, A. J. (1974). *The Problem of Knowledge*, Penguin, Suffolk, UK.

Bell, D. (1980). *Sociological Journeys: Essays 1960-1980*, Heinemann Education, New York.

Bell, D. (1987). "The Post-Industrial Society: A Conceptual Schema" in *Evolution of an Information Society*, Cawkell A E (ed), AdLib, London.

Checkland, P. B. and Holwell, S. (1997). *Information, Systems and Information Systems: Making Sense of the Field*, Wiley, Chichester.

Collins D. (1998). "Knowledge Work or Working Knowledge?: Ambiguity and Confusion in the Analysis of the 'Knowledge Age'", *Journal of Systemic Knowledge Management*, March.

Coulson-Thomas, C. (1991). "IT and New Forms of Organisation for Knowledge Workers: Opportunity and Implementation", *Employee Relations*, 13(4), MCB University Press.

Cushman, A., Fleming, M., Rosser, B., Hunter, R., and Harris, K. (1999). "The Knowledge management scenario: trends and directions for 1998-2003", http://gartner6.gartnerweb.com.ggbin/ggtoc.

Despres, C. and Hiltrop, J. M. (1995). "Human Resource Management in the Information Age: Current Practice and Perspectives on the Future", *Employee Relations*, 17(1):9-23.

Drucker, P. (1959). *Landmarks of Tomorrow*, Heinemann, London.

Drucker, P. (1991). "The new productivity challenge", *Harvard Business Review*, Nov.-Dec., pp 69-79.

Dunham, R. (2001). "Information Worker to Knowledge Worker", *Enterprise Design*, www.khi.com, 4th Sept.

Goguen, J. A. (1997). "Towards a Social, Ethical Theory of Information" in *Social Science Research, Technical Systems and Cooperative Work: Beyond the Great Divide*, (G. Rowker, ed.), Erlbaum.

Hagglund, S. (1989). "Emerging systems for computer based knowledge processing in office work", Office Technology and People, 4:135-145.

Harris, K. (1999). "So, you're a knowledge worker: what do you do all day", http://gartner6.gartnerweb.com.ggbin/ggtoc.

Howell, J. G. (1999). "Educating the Knowledge Worker" in *Synergy Matters, Proceedings of the 5th UKSS Conference,* (Castell et al., eds.), Plenum, New York.

Howell, J. G. and Gammack, J. G. (1997). "Systems, Crafts and Sustainability" in *Systems for Sustainability, Proceedings of the 4th UKSS conference*, (Stowell et al., eds.), Plenum, New York.

Johnson, P. E., Zualkernan, I. and Garber, S. (1987). "Specification of expertise", *International Journal of Man-Machine Studies*, 26:161-181.

Kappes, S. and Thomas, B. (1993). "A Model for Knowledge Worker Information Support", *USACERL Technical Report FF-93/10*, p 1.

King, M. and McAulay, L. (1991). Barriers to adopting management expert systems, *Expert Systems*, 8:139-147.

Kolb, D. A. (1984). *Experiential Learning: Experience as the Source of Learning and Development*, Prentice-Hall, Englewood Cliffs, NJ.

Kumar, K. (1981). *Prophecy and Progress: The Sociology of Industrial and Post-Industrial Society*, Pelican, London.

Mingers, J. M. (1995). "Information and meaning: foundations for an intersubjective account", J. *Information Systems*, 5:285-306.

Ngwenyama, O. K. and Klein, H. K. (1994). "An exploration of expertise of knowledge workers: towards a definition of the universe of discourse for knowledge acquisition", *J. Information Systems*, 4:129-140.

Nomikos, G. E. (1989). "Managing Knowledge Workers for Productivity", *National Productivity Review*, 8(2).

Quinn, J. B. (1992). *Intelligent Enterprise: A Knowledge and Service based Paradigm for Industry*, The Free Press, New York.

Roy, M., Falardeau, J., and Pelletier, C. (2000). "Support Systems for Knowledge Workers: The Need for New Development Approaches", *Journal of Knowledge Management Practice*, Oct.

Royston, S. (2001). "Enter the Knowledge Worker, Kudos-UK, www.KUDOS-UK.com, 4th Sept.

Senge, P., Ross, R., Roberts, C., Smith, B. and Kleiner, A. (1995). *The Fifth Discipline Fieldbook*, Nicholas Brierley Publishing Ltd., London, UK.

Slatter, P. E. (1990). "Models of Expertise in Knowledge Engineering" in *Knowledge Engineering*, (A. Hojjat, ed.), North Holland, Amsterdam, pp. 130-156.

Suchman, L. A. (1983). "Office Procedures as Practical Action: Models of Work and Systems Design", *ACM Transactions on Office Systems*, 1:320-327.

Suchman, L. A. (1987). *Plans and Situated Action: The Problem of Human-Machine Communication*, Cambridge University Press, New York.

West, D. (1990). "'Appreciation', 'expertise', and knowledge elicitation: the relevance of Vicker's ideas to the design of expert systems", *Journal of Applied Systems Analysis*, 17:71-78.

West, D. (1992). "Knowledge elicitation as an enquiring system", *J. Information Systems*, 2:31-44.

Wielinga, B. J. and Breuker, J. A. (1986). "Models of expertise", in *Proceedings of ECAI '86*, 1:306-319.

Wiig, K. et al. (1997). *Leveraging Knowledge for Business Performance*, Witts Business School, South Africa.

DESIGN OF A KNOWLEDGE MANAGEMENT SYSTEM FOR QUALITY IMPROVEMENT IN A PUBLIC EDUCATION SYSTEM IN MÉXICO

Cuauhtémoc Olmedo

Industrial and Systems Engineering Department,
ITESM Toluca. Eduardo Monroy Cárdenas # 2000,
Toluca, 50110, México

INTRODUCTION

This paper describes a proposed design for a knowledge management system (KMS) that would support quality improvement efforts for a public education organization, where strong sociodemographic and budgetary differences exist between operating units.

There are approximately 7000 schools, half of them located in urban communities and half in rural settings These units vary in number of teachers, from one teacher-principal to 25 teachers. Teachers seek urban positions, thus creating a flow of people from the rural locations, where many begin their practice, to urban areas, depriving the rural areas from experienced teachers.

The National Quality Award Model (NQA) for México (*Modelo de Dirección por Calidad...*, 2000), is derived from the Malcolm Baldridge Award, and includes a knowledge management section, point 4.0 "Organizational Knowledge", which subdivides in 4.1 "Communication" and 4.2 "Intellectual Capital". The project reported here was divided in two streams, both interrelated through the National Quality Award guidelines: an *Improvement Projects* stream, on which this report is based, and a *Quality Model Development* stream.

Three major flows of knowledge were identified in the organization: *formal training*, dealing mainly with pedagogical issues; *on-the-job training*, dealing mainly with quality and practices issues; and *everyday knowledge* dealing with the structuring of the other two streams to specific situations. While these were understood as separate flows, each being applied without explicitly considering the others, we propose that they must be interrelated and thus included knowledge stocks in our model that feed from all flows; these stock are labeled *Everyday* knowledge stock, *Quality* knowledge stock, and *Technical* knowledge stock.

The organization is in part structured as follows: the schools are managed by a principal, who reports to a supervisor. Each supervisor controls an average of 10 schools and reports to a region coordinator, who usually controls 10 supervisors. The original purpose of the supervisor´s job was in part academic and in part administrative; however, most of the work they currently perform is adminstrative. For academic support, Teacher Centers were created, reporting to the region coordinator and giving advice in pedagogical matters to teachers, who can freely access these centers.

WHAT SHOULD WE UNDERSTAND AS KNOWLEDGE IN THIS SITUATION?

For the NQA (*Modelo de Dirección por Calidad...*, 2000, 76) knowledge management is the set of practices used to identify, develop and apply in an integral and systematic fashion the knowledge an organization such as its technology, policies, procedures, databases and documents, as well as the experience and skills not explicitly mentioned by the employees.

Our involvement in the situation led us to explore further for elements of the knowledge stock and related processes that could help in the design of a KMS for this situation. We believe that a major source of conflict in previous improvement strategies was a tendency of the experts to overlook the knowledge users had and consequently provided only *supply fixes* (Pacey, 1994). This led to the inclusion in our model of an activity called "knowledge fusion".

But what is the knowledge that users had? Our inquiry led us to the theories of everyday life by Heller (1998) and Schutz and Luckman (1977). While teachers and principals had *technical* and *quality* knowledge, they lacked awareness of the importance of *everyday knowledge,* implicit in culture, which allowed actors to structure a situation and evaluate the impact of improvement projects: this everyday knowledge is integrated of diverse entities, such as interaction protocols, meaning structures, previous experiences, and in general a background against which practices can be evaluated.

STREAM 1: FORMAL TRAINING... AND ITS TURBULENT CHANGES

Starting some 7 years ago an important restructuring of public education practices was decided, including a shift in the pedagogical paradigm (Kuhn, 1970): from a memoristic, "conductist" approach to a constructivist approach, where meaningful learning is central. This created confusion and resistance, since the older teachers were trained *exclusively* in the conductist approach, while the younger teachers were trained mainly in the constructivist approach but were also trained in the conductist approach. Many of the supervisors were trained and had worked inside the previous paradigm; while most of them took additional training, they felt unable to provide guidelines for practice but were unwilling to relinquish their power positions.

STREAM 2: ON THE JOB TRAINING... AND ITS VARIETY

Teachers and principals and supervisors undergo periodic training in pedagogical and quality issues; however, this training created inconsistencies within the knowledge stocks of people and their roles: some of the knowledge in quality that people had previously received turned out to be inappropriate for the situation and new knowledge had to be diffused; several of the quality tools, such as statistical process control charts, are inadequate, since the assumptions behind these tools are not valid in the educational context. This created some conflict, even within the intervention group, since some of the knowledge, which had proved its value in other contexts, was simply useless here.

Another face of the situation is that the new pedagogical paradigm modifies the balance of power between teachers and students in the classroom: now the teacher is not the *source* of knowledge but rather a *facilitator* who must help students integrate their experiences and learning into wider systems of knowledge and meaning. However, many teachers, principals and supervisors are not trained for this new role, which requires not only different skills, but also different attitudes.

STREAM 3: EVERYDAY KNOWLEDGE: THE GLUE

Due to the diversity of social settings, teaching must adapt general procedures and skills to local conditions; many of these conditions are not pedagogical in strict sense, but may limit the success of the teaching effort. For instance, in rural communities it is frequent that children skip school because they are required for harvesting. In other contexts, due to local or family traditions, education is not viewed as an important activity for children and the schools must begin by persuading parents in order to get their support for the teaching activities.

All this leads us to recognize the importance of the *meaning structure* within which knowledge is structured by the individual. As Heller (1970) and Schutz & Luckmann (1977) explain, knowledge is valued within this meaning structure, according to which, often unconsciously, the actor determines what experiences are important and how should they be integrated with the knowledge already in stock. Teachers then must find ways to help students integrate knowledge within this meaning structure, which they, as their students, often ignore and are ill equipped to uncover. Out of the classroom, social situations usually arise where solutions must take into account this meaning structure. As a consequence, improvement projects in the different areas of a school must consider this meaning structure if they are to be of any use.

RESTRICTIONS FOR KNOWLEDGE MANAGEMENT IMPLEMENTATION

Promotion within this organization is achieved by accumulating points obtained by satisfactorily performing certain tasks. However, the point scale in use was derived under the previous paradigm, where innovation was not considered important. This represents an obstacle for the diffusion of different forms of knowledge since employees do not see the link between these and their advancement.

In order to get a measure of knowledge we must observe the consequences of knowing, since knowledge is observed in competent action. So, performance measures must be developed around the *application* of knowledge to some purpose, in our case, to quality improvement efforts, but the problem of changing the monitoring system still exists: some people are able to obtaining points without really improving their practice.

A CONCEPTUAL DESIGN FOR A KNOWLEDGE MANAGEMENT SYSTEM

The purpose of the proposed KMS is to provide useful knowledge to people in the operating units so that they can design, implement and learn from improvement projects, considering budgetary restrictions, previous training, which conflicted with new practices and created resistance to change, and the variety of social settings within which activity takes place. It is important to note that there are unintentional processes which must be considered in the design of intentional activity.

The design methodology used, based in Checkland & Schöles (1990) and Wilson (1984) and not included for space reasons, takes as an input a *skeleton,* which is a list of elements and processes related to knowledge derived from the literature; these activities must be modified and integrated by the designers to fit the situation. A partial skeleton is presented in "Table 1. Knowledge elements" and "Table 3. Knowledge processes table." The model is presented in figure 1 and a mapping to the skeleton appears in "Table 2. Mapping of processes and elements to roles and mechanisms."

BRIEF DESCRIPTION OF RELEVANT ACTIVITES

Activity 2: Develop and Use *Learning* Spaces [...]

This a very unstructured activity since each of the knowledge processes may be performed in several ways and the presence of people is essential, as opposed to other activities that may be performed asynchronously. The presence of an experienced facilitator is imperative. This activity could be coordinated by the Teacher Centers, since these are currently accepted as places to meet and discuss issues and their coordinators have skills in group dynamics. It is also an opportunity for higher management to express direction and align efforts.

Activity 6: Knowledge Fusion

People in this situation value the knowledge they hold. When new knowledge is diffused, they compare it to their previously held knowledge and accept it or reject it according to the usefulness they perceive in it. However, necessary new knowledge may conflict with previous knowledge, so that this activity is directed toward making explicit the assumptions behind earlier and newer elements of knowledge, so that the rationale for "discarding" old knowledge is clear and accepted. This fusion may be undertaken with the use of models and dialogue sessions, where people are allowed to express their reasons for resistance. This should be responsibility of Teacher Center coordinators, for they understand theoretical issues and have the practical tools to guide understanding.

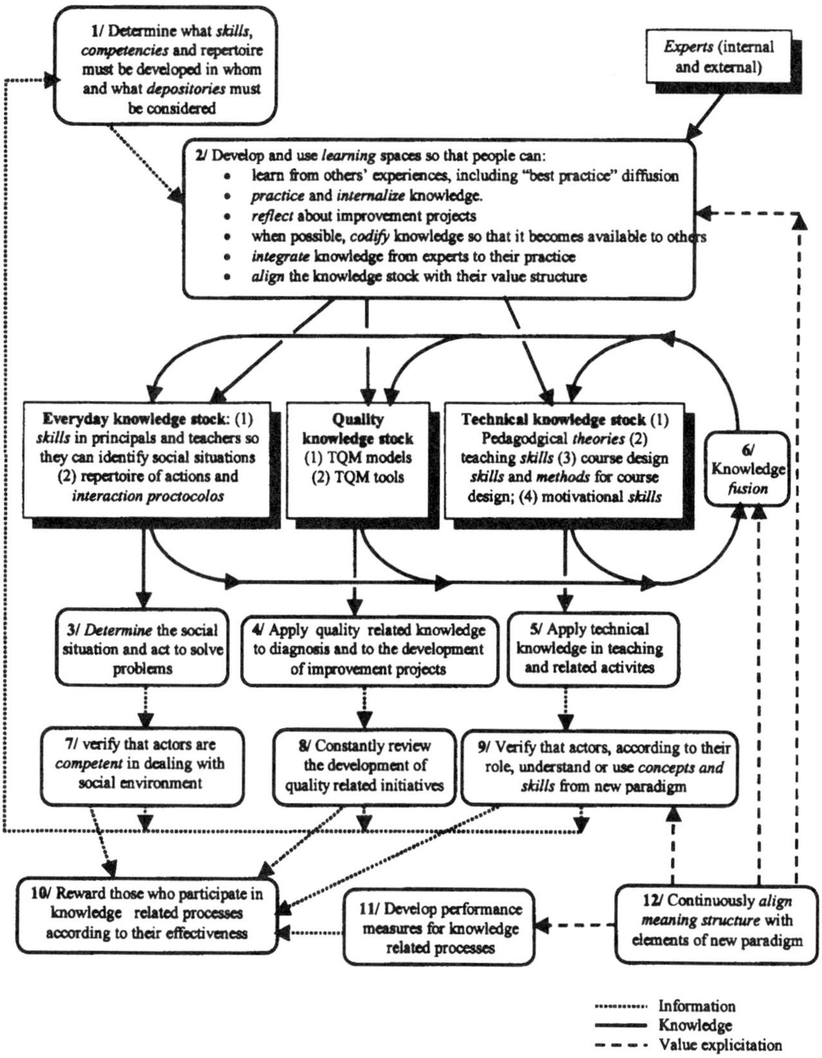

Figure 1. A Proposed Knowledge Management System

Activity 10: Reward those who Participate in Knowledge-Related Processes According to their Effectiveness

The current reward structure does not recognize the efforts people make in improving practices, and may promote secrecy. This may be problematic in those parts of the organization where evaluation is coercive rather than improvement oriented.

Activity 12: Continuously Align Meaning Structure with Elements of New Paradigm

The meaning structure, at the organizational and local levels, must be made explicit, so that people responsible for projects and other initiatives can discover the restrictions and opportunities that culture tacitly offers. This is a responsibility of principals, at the local level, and of coordinators at the organizational level.

CONCLUSIONS

From this reflection, some methodological observations to be drawn: (i) case study research provides relevant input for the first cycle of action research, so that proposed solutions have a firmer rooting; (ii) "old" knowledge *must* be considered as an input to intervention: knowledge, understood as the structuring of experiences (Schutz & Luckmann, 1977; Heller, 1998), cannot be thrown away and thus people construct new knowledge on sediments of old one; (iv) the multilayered and paradoxical essence of knowledge was recognized as the intervention progressed: at any moment, knowledge was required to design and perform activities, which produced (a) new knowledge, (b) the need for more knowledge, and (c) the realization that initial knowledge was insufficient or inadequate thus leading to reflection about and critique of results obtained.

Table 1. Knowledge Elements Table (partial)

Element [code]	***Description***	***Type***	***Authors***
Practices **[PR]**	Sets of activities involving specific skills and procedures that can be identified as leading to a desired result	Tacit/ Explicit	
Competencies**[CO]**	Capacity of applying a skill in a superior form	Tacit	
Skills **[SK]**	Capacity to do something in a proper way because of practice or training.	Tacit	
Situations and types **[SI]**	Interrelations of roles, objects, subsets of language, purposes, rules and other elements that are distinguished as particular by members of a culture, and which restrict behavior and action repertoire.	Tacit	Heller, 1998; Schutz and Luckmann, 1977; Hall, 1989.
Interaction protocols **[IP]**	Sequences of socially accepted acts that are performed in specific situations in order to exchange or communicate with others.	Tacit	Heller, 1998; Hall, 1989; Searle, 1994.
Theories and models **[TM]**	Systemic arrangement of concepts; in addition, theories have undergone some kind of empirical testing. Includes TQM models.	Explicit	
Technology **[TE]**	Knowledge applied in a systematic way for a purpose; usually has "hard" and "soft" components	Explicit	Rogers, 1983 /Pacey, 1994
Procedures and methods **[PM]**	Sequences of activities to be performed to obtain a specific result. Includes TQM tools.	Explicit	
Meaning structure**[MS]**	System of values and other elements of culture that give meaning to experiences	Tacit	Schütz and Luckmann, 1977

Table 2. Mapping of Processes and Elements to Roles and Mechanisms

Activity	*Related elements*	*Related processes*	*Main Related Roles*	*Mechanisms*
1/ Determine what *skills*, *competencies* and repertoire must be developed in whom and what *depositories* must be considered	PR, CO, SK	CTL	Principal Supervisor	Peer evaluation Testing
2/ Develop and use *learning* spaces…	IP, PR, SK, PM, TE, MS	PRI, ALI, EXT, SOC, SED, DIF	Teacher Center	Meetings Workshops Congresses Teacher Center facilities
3/ *Determine* the social situation and act to solve problems	SI, MS	EXP	Teacher Principal	Dialogue with social actors
4/ Apply quality related knowledge to diagnosis and to the development of improvement projects	TM, PM, PR	PRI	Principal Teacher	Work Groups
5/ Apply technical knowledge in teaching and related activities	TM, SK, PR, TE, MS	PRI	Teacher	Work Groups
6/ Knowledge *fusion*	SI, TM, MS	OCS, SED, ALI	Teacher Center Teacher	Work Groups
7/ Verify that actors are *competent* in dealing with social environment	CO, SK, SI, IP, MS	CTL	Supervisor	Peer evaluation Testing Dialogue with social actors
8/ Constantly review the development of quality related initiatives	TM, PM, PR	CTL	Supervisor	Peer evaluation Testing
9/ Verify that actors, according to their role, understand or use *concepts and skills* from new paradigm	TM, SK, PR, MS	CTL, ALI	Teacher Center Principal	Peer evaluation Testing
10/ Reward those who participate in knowledge related processes according to their effectiveness	PR, MS	CTL	Supervisor Principal	Promotion Bonus
11/ Develop performance measures for knowledge related processes	TM, SK, PR, IP, MS	CTL	(all)	Work Groups
12/ Continuously *align meaning structure* with elements of new paradigm	CO, SK, PR, TM, MS	EXP, OCS, ALI	Supervisor Principal	Meetings Workshops

Table 3. Knowledge Processes Table (partial)

Process	*Description*	*Authors*
Ordering / combination/ structuring **[OCS]**	New experiences are integrated with previous experiences according to their significance. Leads to systemic knowledge: it is a process of ordering by means of which a knowledge system is generated, by combining diverse bodies of explicit knowledge.	Schutz and Lukmann, 1977 / Nonaka andTakeuchi, 1995
Sedimentation **[SED]**	New experiences are constantly integrated into the knowledge stock of the actor; thus old experiences become part of a series of layers upon which new experiences are integrated.	Schütz and Luckmann, 1977
Diffusion **[DIF]**	Innovations are adopted by actors within a social systems at a rate that depends on several factors, such as previous practices and values, relative advantage and observability of the innovation.	Rogers, 1983
Explicitation/ Determination **[EXP]**	When a situation is encountered, two cases may arise: the actor has sufficient knowledge as to identify the situation with some type or he/she must obtain new knowledge, understood as the structuration of experiences, until the situation can be classified.	Schutz and Lukmann, 1977
Socialization **[SOC]**	Leads to harmonized knowledge: consists in sharing experiences, making the use of language unnecessary.	Nonaka and Takeuchi, 1995
Externalization Codification **[EXT]**	Dialogue or collective reflection, by means of metaphors, analogies, models or propositions, lead to conceptual knowledge: tacit knowledge is captured in concepts by means of language.	Nonaka and Takeuchi, 1995
Practice/ Internalization **[PRI]**	Leads to operational knowledge: it is a process of conversion of explicit knowledge into tacit knowledge within the person; related to *learning by doing* and to *knowledge in action.*	Nonaka and Takeuchi, 1995/ Schön, 1987
Measurement / control **[CTL]**	Knowledge is intangible, so that performance measures must be developed around the *application* of knowledge.	Miller, 1996
Alignment **[ALI]**	In order to be of value, knowledge must be aligned with the culture of the social system where it thrives.	Carrillo, 1998

REFERENCES

Carrillo, J. (1998). "Managing Knowledge-Based Value Systems", *Journal of Knowledge Management,* 1(4): 280-286.

Checkland, P., and Scholes, J. (1990). *Soft systems Methodology in Action,* Wiley, Chichester.

Hall, E., (1989). *Beyond Culture,* Anchor, New York.

Heller, A. (1998). *Sociología de la vida cotidiana,* Península, Barcelona.

Kuhn, T. (1970). *The structure of scientific revolutions,* Chicago Univ. Press, Chicago.

Nonaka, I. and Takeguchi, H. (1995). *The knowledge creating company,* Oxford Univ. Press, N.Y.

Miller, R. (1996). *Measuring what people know,* OECD, Paris.

Modelo de Dirección por Calidad: una herramienta efectiva y práctica para incrementar la rentabilidad de su empresa (2000). SECOFI, México

Pacey, A. (1994). *The culture of technology,* MIT Press, Cambridge.

Rogers, E. (1983). *Diffusion of innovations,* The Free Press, New York.
Schön, D. (1987). *Educating the reflective practitioner,* Jossey Bass, San Francisco.
Schutz, A. and Luckman, T. (1977). *Las estructuras del mundo de la vida,* Amorrortu, Bs. Aires.
Searle, J. (1994). *Actos de habla,* Cátedra, Madrid.
Senge, P. (1990). *The Fifth Discipline,* Doubleday Currency, New York.
Wilson, B. (1984). *Systems: concepts, methodologies and applications,* Wiley, Chichester.

KNOWLEDGE IN SELF-ORGANIZATION PROCESS: HOPE (FINITE) OF SYSTEM THINKING

Antonín Rosicky

Department of System Analysis
University of Economics, Prague
W. Churchill Square 4, 130 67 Praha 3, CR
rosicky@vse.cz

INTRODUCTION: SOCIETY OF HIGH MODERNITY

An apposite appellation of dramatically forced changes in contemporary society encounters dilemmas resulting from their newly emergent properties. Moreover aspiration to entitle nature of contemporary environment briefly brings reductions of its growing complexity. Stressing some particular attributes original labels have qualified essential transformation in economic, political and cultural life by comparing with (seemingly) periods as post-industrial, post-capitalist or post-modern. The name 'information society' is based on recognized importance of information within society however it is inherited from evident features of modern information technology. Inconsistency of such associates with a foggy concept of information now replaced by the term 'knowledge society'.

Nevertheless the notion of 'global world' is more significant not only for its broad enlargement in recent times, but namely for considerably modified views. Overcoming an anthropocentric position it considers rather the whole world than society and stresses human participation on the process of (cultural) evolution. This conception adverts to many newly arising problems and points rather at necessity of change than at change itself. Such opinion is not inhered only to humanists or ecological activists, but also many gurus from business area. For example Charles Handy (1996:28) argues: "*Now the enemy is us and our own societies, because what we're fighting against is our own sense of values, our own principles*" as he supplies a few pages later: "*You can't stumble backwards into the future*".

In the face of difficulties to define global 'world' many authors outline some important features of 'globalizations' that is a natural process of cultural evolution. Giddens (1990) impeaches peremptory character of running changes and notes that the evolution epitomizes rather intensification of industrial (modern) style of thinking and adequate institutions. On that account he labels our period as *'society of high modernity'*

Edited by Ragsdell *et al.*, Kluwer Academic/Plenum Publishers, New York, 2002.

that is connecting with an everywhere present existence of general expertise. Giddens defines such expertise in the sense of 'systems of technical and knowledge (expertise) that organizes many domains of the material and social environments and our workaday life'. Concept of such expertise identifies new expressive feature of our global environment supplying traditional notion of its (increasing) turbulence. From Giddens' apprehension truly comports with the concept of '*second industrial revolution*', which is maintained by many 'technological boosters'. The second estimation seems me to be more inspiring by its frank coupling with the concept of knowledge (information) and way of its sharing within modern society. This topic gains new and very important dimension if we accept the perception of Giddens' work-fellow Urich Beck (1992). He points at a paradox of our trust and dependence on such trust that often disagrees - sometimes in a fundamental way - with other expertise concerning the same situation.

FROM HARD TO SOFT KNOWLEDGE

Outlined problems bespeak dilemmas appertain to credibility of used knowledge based on Cartesian rationality and belief in truth of 'Scientific' recognition. Probably many have greater or lesser doubt about its infallibility and own system thinking or accept its importance at least. Nevertheless, most human activities result from deep but narrow knowledge and its mechanistic application in name of Human Progress.

There is no doubt that just advent and further expansion of the science as a source of knowledge was a motive force of (cultural) evolution of last centuries culminating in industrial or modern era going with an unprecedented transformation of our environment. There are human's intentions and consecutive activities, reinforced by great amount of energy from the past (out of immediate circle like coal, oil) and based on 'objective knowledge' that actually participate on the evolutionary process. Although many humanists censure rationalism of the Science, both implacable communities that don't share common language (Snow, 1962) are based on the same anthropic notion. It approves oneself as a belief in a truth and human reason (ratio) on the one side and humanity anthropocentrically oriented on human welfare on the other side.

Juncture of both tendencies result to the problematic acceptance of cognition of modern science usually connected with relativity theory and quantum physic. However notion of complexity as well as dealing with it seems to be a stumbling block of traditional science. Analytical approach can enable understanding fractional problems or single parts of intricate or complicated (rather than complex) wholes. Appropriate methods solve rather mental difficulties than problems of actual (intrinsic) system's complexity and enthrone new problem how to understand a catch of the whole. Mechanistic conception infers properties of the whole from exact relations of changeless components what is pertinent for deterministic systems and optionally for stochastically determined systems (Ashby, 1956). Accordant arrangement of system also corresponds to notion of order in that such described state of affairs of which knowledge together with an acquaintance of particular enables correct expectations concerning the rest or future (see von Hayek, 1973).

Ordering of all designed systems is grounded on such arrangement knowledge: In the case of technical artifacts (machines) resulting systems have deterministic character while softer systems such as human (rational) decision making or systems of human activities (Checkland, 1981) strive for dealing with various form of uncertainty and use stochastic models. In both cases artifacts cover knowledge of their designers: First case represents the traditional practices of systems engineering (and first order cybernetics) and applied knowledge based on (objective) physical laws and exact models are embedded into real systems as their internal information. The second case comprises modeler's knowledge that represents his conception of much more complex reality including notion problems and optionally also way of their solving. Such models then involve not only author's knowledge of facts and laws but also his worldview, intentions, values, readiness to carry risk etc. Using of model that is built up by some others, increasing namely by repetitive use in computer's software, we adopt also such personal features. In this way we lower information only on (quantitative) denomination of selected attributes (variables) usually incorrectly labeled as data.

Described practices represent probably most typical case of shared and expanded expertise that is seemingly shielded by rationality, while actually arises from knowledge many, often anonymous experts, designers (namely designers of IS) and knowledge providers (including various consultants, but also journalists, teachers and others middlemen). Far from actual context emerges within social systems an anonymous power (Rosicky, 1999) that influences our actions (and feelings), increases systems' totalitarianism on the one side and decreases its adaptability and consequently also cultural aspect of evolution. With regards to similar consequences of similar 'hard methods' and distinguished consequences interpretative approaches have been developed and used. Probably Checkland's soft systems methodology (Checkland, 1981) exemplifies not only most famous way but also actual a turning point of systems thinking. Above all three aspects seem to have an essential importance in this paper:

- Acceptance of more (diverse) point of views and system's concepts,
- Orientation towards (dynamical) processes in place of (static) entities and finally
- Use of various (not only exact) models, including narrative models based on natural language.

Adverted fundamentals in many respects resemble some from principles recently used in popular ideas of knowledge, knowledge management and learning organization. Or better they correspond with opinions of those authors that assert human nature knowledge and distinguish it from knowledge in the sense used in artificial intelligence domain. For what remains 'paradigm shift' resulting from concept of soft system does not lie in the space between determinism and uncertainty (and solved by 'soft computing') but rather in the nature of human and complexity of social systems.

OBSERVER WITHIN SELF-ORGANIZING SYSTEMS

Nevertheless interpretative inspiration, so close to idea of democracy, refers to problems of optimal (sub-optimal) or correct knowledge and solution and ways of its retrieval. In spite of good thoughts like idea of democracy that is not domination of the most but view to opinion of the minor community (Simon, 1983) face to face to reality and grave problems diverges. In better cases actual solving results from mechanistic methods based on poll, however various rules can partly prevent some iniquity. Nevertheless the question of best or proper knowledge is more complicated and any 'most fair-minded' plebiscite and not even expert's discussion cannot assure truth knowledge. From consistently holistic position only correct (best) knowledge doesn't exist and aspiration to find it is rather affair of finding of credible knowledge in discourse in Foucault's sense (Foucault, 1971) of word. In this sense discourse constitutes rather credibility of the knowledge as relatively consistent set of more or less formalized (told) rules considered to be valid for observed field (entity, system) of the human interest. Appropriate procedures that should attest credibility of the knowledge are based on shared paradigms, present theories, shared values, forcible norms, i.e. on knowledge of involved people as well as on their intentions and also on relations of power.

Such enunciation raises many new issues appertain to understanding nature of purposely (pragmatically) based human cognition including abstract concepts of such phenomena as system, information and knowledge on the one side and their real merits (substance) on the other side. In systems theory and practice they are modified to many others concerning particular topics as follows:

- Concept of emergence away from 'mind body duality' optionally material nature of the Universe and human abstraction resulting from its cognition;
- Constitution of systems as complex autonomy wholes of those properties results from actual interactions of system's components as well as system's interaction with environment (with systems within environment);
- Nature of information and its role in the process of system's organization and participation evolution of the Universe contemplating its various forms and ;
- Disposition of knowledge and its coupling with information and its (their) peculiar role within social systems an their evolution alternatively notion human and information within cultural stage of evolution.

Mentioned arguments epitomize other paradigm shift in system theory and cybernetics that is at the very least comparable with moving from hard to soft thinking. They cohere with radical changes and new ways of contemplation in established disciplines such as phenomenology in philosophy, conception of thermodynamics systems far from equilibrium in physic or contemporary evolutionary biology. By the same mail new disciplines and theories frequently based on system's ideas come out and enrich conceptions of system conversely. Attention deals mainly with theory of dissipative systems (structures), synergetic and chaos theory and recently with theory of complexity. Consideration deserves concept of autopoiesis that bridges space between modern systems theory and second order cybernetics and closes constitution of the new worldview in this way. In spite of the more or less communal topics each from mentioned disciplines evolves relative independently and focus on some aspects. This fact brings problems of mutual cohesion and together with non-uniform terminology causes goodly difficulties to establish corporate framework supporting radical understanding.

Its fundamental novelty lies in the explication, understanding and acceptance of many significant features and concepts such as:

- A difference between 'the world how is perceived by (external) observer' on the one hand and 'the world how actually exists' on the other. While observer looks at systems from the outside, insight into the real world necessitates 'view from the inside'. Existing difference between reality and its subjective images is emphasized by objective reality on the one side and many of subjective realities on the other.
- Internal spontaneous processes of complex systems appear from the outside as a system's self-organization. Circular relationships play a basic role in the involvement of actual (former) structures in their everlasting reforming viewed as

fluctuations. Outwardly such systems maintain state(s) of dynamical equilibrium and keep identity that can be irreversibly changed in an unpredictable way.

Quite new insight strong exceeding familiar and native perception makes an understanding and assumption of new ideas very challenging. Appropriate problems are closely tied up with natural language and many concepts convert common meaning that needs not to be of importance at first appearance. Let us remind only a few examples: Accordingly to Prigogine's idea '*from being to becoming*' we can describe this distinction more fairly: While an external observer apprehends the world as a set of static and purposefully discriminated entities the real world actually consists from complex interaction of many various wholes (systems). Observer's intention and ability to differentiate decides if such whole will be considered (diversified) as (autonomous) 'systems', 'system's component' or 'system's environment' (higher system). Bertalanffy's notion of systems as '*a set of interrelated elements*' appears to be insufficient and complex self-organized systems need to think upon connections of part(s) and wholes. In other words part or individual participate on the complexion of the whole that generates its environment (see Heisenberg's 'Part and Whole' (1979)).

Also many fundamental concepts are not learned suitably to their generally constitutive meaning and they are distinctly interpreted in various domains or are not used in some disciplines at all although their use could complete better understanding in a useful way. Considering the topic of this paper I would remember a few of them: Above mentioned discussion enlarges traditional discrimination of system's quality in the sense of teleology and teleonomy for new categories of *'purposiveness'* and *'purposefulness'* of systems (Ulrich, 1983). The concept of 'average value' about which system fluctuates is usually false – while each from fluctuations is actual (materialized) structure, average value is not a real state, but rather an abstract (virtual, ideal) pattern. And finally to concept of attractor that is next to entirely presented as continuous curve (or pixel) and its notion is focused on 'regular or strange' behavior. However such 'trace of singular states materialized in past' defaces (bedims) its character in future as '*a region of a dynamical system's state space that the system can enter but not leave*'.

Accepting adumbrated concepts we can stress the reference to the changed worldview: In defiance of various proclamations hard system thinking confesses positivistic (objective) cognition, while interpretative (soft) approaches inhere constructivist worldview that accept diversity of individual system's notions.

The observer' position is seen within whole environment from that he subconsciously receives information(s) while his conscious is focused on perceived situation. Also circularity of his interaction with the environment plays an important role - circular process embodies reality, receiving information, observer's knowledge and activities. This circle points to dynamical wholeness of information (received by observer from environment) and his implicit knowledge as well to integrity of physical and cognitive activities involving utterance (newly emerging information). In this way also two concepts of information distinguished by Kampis (1991) (information as knowledge and information as an action) appears as suitable abstraction while nature of both (knowledge and information) is holistic. Finally, the autopoetic notion considers (Maturana, Varela, 1998) organizational closure of whole nervous system (not only the brain) and the biological nature of cognition of structural 'coupling' and 'languaging'.

Such seemingly idly intricate explication of internal cognitive processes frames new background for better understanding of the nature of information. Consequently close association of information and knowledge outlines Banathy's opinion concerning '*tight interrelation of evolution and information*' (Banathy, 1996) optionally role of human and his knowledge in (cultural) evolution.

INFORMATION AS A GENERAL CONCEPT

Accepting inherent system dynamic as actual interaction of its components materializing on wide nets of structures we can propose new concept of information. Such general notion outlines information in its prime sense of the Latin term *'informatio'* that denotes 'to shape' or 'to actuate form'. Resulting expression/emergence of autonomous system that is usually labeled as receiver of information is just a pattern of fluctuating internal structures. For an external observer it appears to be the appearance (state, utterance) or behavior or (re)action.

For better understanding we must distinguish information itself from its meaning. From described view information comes in sight as a diversity of some particular structure. Away from quantitative notion of variety (as a set of elements) and static conception diversity is rather understood as versatility in the sense of 'changeable whole of differences' (as a distinctions that can be made between various objects and concepts). Here it seems to me to remind Ashby's underestimated idea (Ashby, 1956:2/1) determining the difference as the most cardinal concept of cybernetics. Also famous Bateson's sentence (1972: 452) sounds similarly: "*What we mean by information – elementary unit of information – is a difference which makes a difference*".

The concept does not identify information with matter or energy however it lays stress on the fact that it is closely tied up with particular matter and/or energy. In other words: Information objectively exists as diversity of matter or energy but its meaning emerges only from interaction with other structure. Mutual interaction depends on physical properties of both structures and the same information tied up with particular matter/energy can interact only with some physical structures and gains 'meaning' in this way, while other structures within environment rest inert. With genetic information its biological 'meaning' is actuated only in that cases, when its signs (triplets) are connected with four chemical elements (structures) named as *adenin* (A), *cytosin* (C), *thymin* (T) and *guanin* (G). Nature of such meaning lies in participation of formatting proteins and materializes only in particular biological environment (structures). Both possible source of defects (or deviations) e.g. information itself or interacting (structures of) environment, can as whole and very complex process (see Kampis, 1991) disable or damage (change) resulting structures.

Such information (structure) can approve oneself through (in)formation of other structures (systems). It is a case of record information that forms (rather than gains) meaning only in interaction in suitable system's structures that are named as receptors or scanner. In the case we insist on the idea of meaningful information we can speak about potential information. However notion of the meaning outreaches common anthropomorphic views and depends on the nature of receiving system and its complexity. Actual meaning of information is generally viewed as internal processes of interacting structures and depended not only on information itself but also on complexity (and properties) of interacting structures/systems. From external positions it appears as system's reaction on received information and commonly we talk about information that initiates response of receiver (Rapoport, 1986). Many from such responses need not to be evident outwardly and consist only from re-creation (re-formation) of intrinsic structures. Most internal changes maintain the pattern of external information but some of them can bring on its modification. Similarly as intrinsic system's structures fluctuate also intrinsic information (variations of their matter/energy change (fluctuate).

NATURE OF (HUMAN) KNOWLEDGE

Accepting mentioned concept of information and its meaning dependent on the complexity of interacting systems we can appreciate autonomy as an essential feature of living organisms. They interact with environment as individuals distinguishing diversity as information from systems and its meaning (re)create internal structures and partly approve oneself as receiver's behavior. Increasing complexity comes with an emergence of nervous systems and with a various level (type) of its emergence aligned with variously labeled activities, properties and states from perceiving, cognizance, and awareness to self-consciousness, and capability to learn (from experience) and (re)create individual knowledge.

Just concepts of learning and knowledge gain a wide publicity however their nature remains fogged as well as notion of information (and consciousness at last). Many problems understanding them arise from Cartesian duality of the body and mind. According to Pribram (1999) we can reckon many activities and phenomena (inc. mind, consciousness, thinking and also knowledge) for an emergence of the brain, or better of the whole nervous system. Mentioned shift from brain to nervous systems is important with regards to holistic conception of nervous systems, its enormous complexity and self-organizing character. Maturana and Varela (1998) have elucidated the biological nature of cognition and organizational closure of nervous systems. Important is the concept of 'structural coupling', characterized by authors Maturana and Varela, (1998:75): *We speak of structural coupling whenever there is a history of recurrent interaction leading to the structural congruence between two (or more) systems.*

Knowledge is rather an ongoing process within an extremely complex nervous system that maintains and re-creates peculiar structures than static representation or commodity (entity). Such knowledge can evolve in immense numbers of different 'attractors' with various 'saddles' that associate with more stable stages linked with varying awaked states.

Accepting the idea of information as diversity of structure we can consider implicit knowledge to be part of intrinsic information of individual that has character of phenotype resulting from an individual's interaction with environment. However richness of information processing resulting (and producing) complexity of nervous systems covers not only information received from but also present knowledge (structures, states) and information about 'personal states'. In this sense we can see Miller's types of information(s) (Miller, 1978) compared with commonly used terminology (Table 1).

Table 1. Types of Information Participating on Processes of Nervous Systems (Miller, 1978)

Miller's types of information	familiar sense (name)
Information about self and own parts	Feeling and emotions
Information from the past	Actual (former) implicit knowledge
Information from the world outside	Common concept of information expressed through language as data (chosen discrete signals) but also analogous signals.

Complicity of all three types of information characterizes wholeness of individual knowledge that is not state but uninterrupted dynamical process of the complex nervous systems. While such concept of knowledge is convenient to the notion of mind or memory, realized (up-dated) knowledge to resembles rather attractor or thinking and feeling (evolved ideas and emotions). Some sensible states comport with atrractor saddles - those more explicit are associated with 'concepts' (and symbols) and those less discernible are labeled as 'endocepts' that are not tied up signs and therefore cannot be communicated. None-the-less endocepts present 'primitive organization of former experiences' and Arieti (1976) characterizes them as prior-to-conscious" cognition that are *'quite unknown and rather felt like global atmosphere*'. Embracing an notion of 'memes' as '*an information pattern, held in an individual's memory, which is capable of being copied...*' (Heylighen, 1995) that are basic element of cultural evolution an idea of their connections with described concepts and endocepts. Just feelings and emotions (information about self) influence nervous structures and enable discrimination in loosely characterized (only felt) domain polarized by terms "felicity – tribulation" or "good – wrong". Such ability is essential for formation of individual (intrinsic, physiological) needs and values and participate on an emergence of human intentionality. Searle's concept (Searle,1986) indicates human possibility to orient himself within the world and to go whistle (to act intentionally) and gives to knowledge competency to embody indifferent (inert) information and pragmatic human agency.

An egregious feature of human consciousness is an ability of utterance resulting from an expression of some internal states (structures, saddles) that are coupled with symbols (signs, marks, tags) and concepts. They are building stones of (human) language that have character of 'component system' (Kampis, 1991) and enables to externalize internal knowledge fluently. Maturana and Varela (1998) explain consensual character of language emerging from interaction of human individuals within (various) societies. This fact leads to human ability to mutual share of some internal knowledge through language. Such externalized part of internal knowledge closely tied up with language is commonly mused and anthropocentrically reduced notion information. To distinguish this type of information from others we will use term conceptual information that is qualified by an existence of shared (meaningful) signs and principles (rules) of their linking into higher linguistic wholes bearing information that can initiate (re-creation of) individual knowledge forming its meaning and/or human activities.

METAMORPHOSIS OF INFORMATION AND KNOWLEDGE

Advance understanding of information and knowledge does not exist without new concept of data, that is closely associated with formal aspects of language. We can understand language as component system or (from semiotics) systems that consist from sets of signs and rules for their linking into wider wholes (Kampis (1991). While basic building stones are (mutual) distinguishable letters (or phonemes in spoken language) meaningful signs (structures) emerging from human communication are only words (and higher wholes - sentences). Also (physical) structures of elemental signs (letters, words) are representing consensual forms labeled data that together with consensual rules shape syntactic level of conceptual information. Many formulas of data as 'representation of facts, concepts, or instruction in a formalized manner suitable for communication, interpretation, or processing by humans or by automatic means' (e.g. British Standards BS 3527, Part 1, 1976) emphasize formalized aspects. However concept of the representation itself is misguided similarly as frequently quoted linear chain DATA → INFORMATION → KNOWLEDGE that resign on question information's origin. Insight information in the sense of suggested dynamical concept, points to its complexity resulting from circular character of human information processing.

Let us highlight two linked loops of cognitive processes: Primary is based on incessant perception through all human sensors that receive many different information(s) from the environment directly and mostly also subconsciously. It has an empirical character that produces immediate subjective experience and initiates appropriate actions including re-forming of implicit knowledge (structures), physical actions (motion) but also utterances (circulates, externalizes) conceptual information. The second cognitive loop is based on sharing of indifferent syntactic (level of) information and its meaning is tied up with implicit knowledge that was formed within consensual 'linguistic domain'. Information shared in this cognitive loop is mostly received knowingly and initiates limited number of sensors (usually eye and/or ear) and describes particular states (of the world) and mediates experience of some other. It is an abstract level of cognition enabling inert consideration of virtual world (systems) out of reach of immediate experiences. Such information itself is value-free however it initiates complex implicit knowledge, including values, feeling and emotions, and consequently also the same (similar) actions like primary loops. However abstraction and use of symbols make possible complex human communication and also design of large and complicated (rather than complex) systems that can be later realized as artifacts. Participation of both cognitive loops that associate perceived situation with subjective and culture shared experiences on continual constitution (re-form) of implicit knowledge specifies its individual character that is far from objectivity but it is neither pure subjective.

The suggested concept outlines the character of conceptual (cultural) information that always results from actual, versatile and knowledge of nonesuch singular person. Maturana and Varela (1998:28) state: '*Everything said is said by someone*'. Or is it the question 'Whether (conceptual) information rather than information or knowledge?' is superfluous or too academic. The same being under the sway of dressy trends and abstractedly from used words *'treat information and knowledge as if they where something could be possessed'* (in. Gibson, 1996:136). Or is it only inculpable ignorance (wrong knowledge) when we regard information as a source, currently lower information systems on the level of determined 'input-output systems' and really speak about 'knowledge of an organization'?

Explicit discrimination between information and knowledge, as any discrimination, constitutes information and (in circular relation) supports understanding an the one side and appropriate activities on the other. Therefore we shall restrict the term '*knowledge*' to internal (tacit) and intentional wholeness of the mind while the term *'explicit knowledge'* we should replace by original term information. However single difference is not competent and we need to distinguish (and to know and/or to understand in this way) between conceptual information and the general concept of information. Assumption of outlined diversity highlights two essential facts:

- Treatment with (conceptual) information epitomizes various ways in which we share individual knowledge(s) in substance.
- Grand conception of information that reduces uncertainty must be related to a whole; accepting recent concept of becoming (rather than being) this idea turn abstraction depending on the observer's knowledge.

Many similar feelings are accepted in business practice and argue for an effort towards a new approach to complexity, information and human role within systems (business, society). Unfortunately they often have a rather intuitive character, remain on the surface of complex problems and without wider coherence and understanding of actual system's complexity they are usually applied inconsistently and inwardly to the traditional thinking and narrowly defined purposes.

CONCLUSION: FINITE HOPE OF PRAGMATIC SOCIETY

Davenport, without an explanation of fundamentals, beholds *'continuous conversion of knowledge into information and information into knowledge as key element'* of successful dealing with knowledge. Later he alleges (Marchand et al, 2000:166). *'Knowledge is information within people's mind, without a knowing, self-awareness person there is no knowledge'.* Similarly, Nonaka (Nonaka et al., 2000) defend the implicit character of knowledge: '*It is our strong conviction that knowledge cannot be managed, only enabled...*'. Opinions of gurus from business in Gibson's book (1996) converge in a similar way, stressing modern system thinking. Kelly states (in. Gibson, 1996:252):

> 'The old models don't work any more, because of the increasing complexity of our human-made artifacts. And by artifacts I mean the things that we produce, like appliances and gadgets as well as institutions and processes... And at that point, they take on the kind of complexity that is basically ungovernable by a mechanical mindset. It's just too complex to control in a linear way.'

Changing framework of modern science mentioned above as well as growing senses of problems and looking for their practical solutions in new ways embodies many Kuhn' s attributes going before radical paradigm shift. Contemporary situation commemorates the Middle Ages to modern period: While ecclesiastical dogmas were at that time replaced by modern science, today character of objective knowledge produced in the 'enlightenment' spirit cannot encompass contemporary complexity. Snow points to the anthropocentric oriented duality of rationality and humanities mentioned above, however pronounces expectation to genesis of new type of science. In similar sense Klír (1991), who holds a constructivist view, lays his hopes on system science and associates it with computer simulations. However advocated positions refer to problem of veracity of knowledge that forms nature of used models (expertise) as well as a fundamental of unpredictable evolution of complex systems. From this position seems Senge's conviction that *'benefits of (information) technology were not in the technology alone – but in the interplay of technologies and new conceptual understandings of complex systems'* (in. Gibson, 1996:125) to be corresponding with suggested thought.

Probably some few doubt about circular influence of systems thinking and modern science in the way, described above. However I would emphasize the scientific recognition in the sense of the mentioned 'view from inside' that involves human itself inc. and his limited cognition and intentional knowledge. However such perceiving of the world that transcends familiar (empiric) perceiving is extremely challenging a we can truly ask after distinctions of appropriate mental abilities and consequences for social systems. Advancement and growing dimension of (shared) human knowledge is associated with discerned complexity and increasing abstraction that is integral part of the commonly shared knowledge (including technology that is based on it). I would like to reject strongly an idea of technological determinism - technology itself forms the environment, offers possibilities, but only reinforces and accelerates 'enlightenment trends of industrial (modern) society'. Information technology involves such process in the sense that includes (meta)knowledge that deals with conceptual information (actually individual knowledge). But it is only intentional human knowledge, which tells way of technology's use as well as manipulated information and cannot be divided from essential epistemological topics.

The concept of knowledge resulting from limited cognition, creating information and initiating real activities as well as its credibility seems to be crucial issue not only of contemporary society but also cultural evolution at all. Gidden's trust to abstract systems and shared expertise stands for our growing (often verbatim vitally) dependency on them and at the same time we better or worse care for their security. However such systems based on general knowledge necessarily have problems to be cope with some unique local and temporal conditions. To guarantee their wider adaptability and endurance we must use knowledge rising from cognition of an immediate situation and support human abilities in this way. Therefore we should focus primarily on such institutions as pragmatically influenced science and education that depend on grants that spring from business that is based on traditional rationality. On the contrary mass media as other important institution that support common shared knowledge dreams about objectivity and indifference of information that provide. Advocating liberal thought in the spirit of liberal humanity they advocate *liberty of press* on the one side and are liable to commercial pressures on the other. In both cases they stay on the position of objective information or one message (information) than can be compensated by other message (information). Emergence of such 'artificial information institutions wholes' is a natural and irreversible process and we must accept it for the future. However we can influence individual competency with critical systems thinking and with knowledge of our cognizance and responsibility towards wider systems and radical paradigm shift to change conversion of devotional belief into the 'enlightenment credence'.

However radical change of the prior paradigm has been forced by business (Queen Isabelle granted Columbus due to her interest on commerce with spice). Contemporary paradigm shift stays rather against the traditional business and economy, it does not offer growth of welfare, rather survival. Understanding this results from shared knowledge. Can we hope to change it? What must come about to accept it? What can we do?

REFERENCES

Arieti, S. (1976). *Creativity: The Magic Synthesis*, Basic Books, New York.

Ashby R. (1956). *An Introduction to Cybernetics*, Chapman Hall, London.

Banathy, B. (1996). Information-based Design of Social Systems, *Behavioural Science*, Vol. 41, pp.104-123.

Bateson, G. (1972). *Steps to Ecology of Mind*, Northvale, Jason Aronson Inc.

Beck, U. (1992). *Risk Society: Towards a New Modernity*, Sage Publications, London.

Checkland, P. (1981). *Systems Thinking, Systems Practice*; Wiley, Chichester.

Foucault, M. (1971). Orders of discourse, *Social Science Information*, Vol. 10, No. 2, pp. 7-30.

Gibson, R. (ed.) (1996). *Rethinking the Future*, Nikolas Breadley Pub., London.

Heisenberg, W. (1979). *Der Teil und das Ganze – Gespräche im Umkreis der Atomphisic*, Deutscher Taschenbucehr Verlag GmbH & Co.KG, München.

von Hayek, F.A. (1973). *Law, Legislation and Liberty*, Routledge, London.

Heylighen, F. (1995). *Memetics, catchword*, http://www.cict.demon.co.uk/ (10. 8. 2001).

Giddens, A. (1990). *The Consequences of Modernity*, Polity Press, Cambridge.

Kampis, G. (1991). *Self-modifying Systems in Biology and Cognitive Science*, Pergamon Press, Oxford, NY.

Klír, G. (1991). *Facets of Systems Science*, Pergamon Press, Oxford.

von Krogh, G., Ichijo, K., and Nonaka, I. (2000). *Enabling Knowledge Creation,* Oxford Univ. Press, London.

Kuhn, T.S. (1962). *The Structure of Scientific Revolution*, The University of Chicago Press.

Marchand D., Davenport, T., and Dickson T. (2000). *Mastering Information Management*, Prentice Hall, London.

Maturana, H., and Varela, F. (1998). *The Tree of Knowledge*, Shamhala, Boston.

Mingers, J. (1995). *Self-Producing Systems - Implication and Applications of Autopoiesis*, Plenum, New York.

Miller, J. (1978). *Living Systems*, McGraw-Hill, New York.

Pribram, K.H. (1999). *Mozek a mysl / Brain and Mind* (selected papers), Galery, Praha.

Rapoport, A. (1986). *General System Theory*, Abacus Press, Turnbridge Wells, UK.

Rosicky, A. (1998). The Danger of an Anonymous Power, in: *Proceedings from IDIMT '98 Conference*, Univesitätverlag R. Trauner, Linz.

Rosicky, A. (1999). Information within the (Human) System: Considerations of Freedom, Power and Order, in: *Synergy Matters*, Plenum, New York, Proceedings from 6th UKSS International Conference in Lincoln, UK.

Searle, J. (1986). Mind, Brains and Science, Harvard University Press.

Simon, H. (1983). What Is Industrial Democracy?, in: *Economic Impact*, Vol. 34, No. 3, pp. 76, 83.

Snow, C.P. (1962). *Science and Gouvernment, The Godkin Luctures at Harvard University*, New York: New American Library.

Ulrich, W. (1994). *Critical Heuristics of Social Planning*, Wiley, Chichester.

THE VALUE OF KNOWLEDGE WORKERS – WHAT AFFECTS IT AND WHO DECIDES?

Aidan Ward and Steve Armstrong

Ward, Antelope Partnership, 19, Fawkham Road, Longfield, Kent DA3 7QP, tel: +44 1474 702259, fax +44 1474 704671
aidan_ward@antelopes.com
Armstrong, Computing Department, Faculty of Maths and Computing, The Open University, Milton Keynes, England MK7 6AA, tel: +44 1908 654056, fax: +44 1908 652140,
s.armstrong@open.ac.uk

INTRODUCTION

One crucial side effect of the development of the knowledge economy and its army of knowledge workers is that the value of knowledge gets conflated with the job market. Knowledge finds its place in a market for the contributions made by knowledge workers. In effect there is a market in knowledge. It remains implicit and its mechanisms are not studied.

This paper looks at the uses and value of knowledge in an attempt to understand how the demand side of this market might operate. Companies who need knowledge attempt to acquire it through hiring people who possess the looked-for knowledge: this constitutes the demand side. The paper also speculates about some supply side dynamics. The supply side is the acquisition of knowledge as a part of a proposition to the job market.

We start by addressing the sort of decisions that confront managers of companies wanting to acquire knowledge. What do these managers think that knowledge is? We then look at decision making as a situated practice and look at the role of power in inhibiting effective decision-making.

MANAGEMENT DECISIONS

An operational definition of knowledge as relevant to business organisations has to concern itself with management decisions. We presume that there is a close connection between knowledge and the ability to make an informed decision. Both knowledge of

Edited by Ragsdell *et al.*, Kluwer Academic/Plenum Publishers, New York, 2002.

context and know-how of actioning the decision are immediately relevant. Information about the content of the decision is useless without this background.

What is a good decision and what is a poor decision? What contribution does "knowledge" as conceived of by those taking decisions play? We can analyse a management decision after it has been taken both for its effectiveness *per se*, and for any deficiencies in knowledge or content information that contributed to a less than effective decision. We still need a value system in which we can distinguish more desirable from less desirable outcomes from those decisions, and of course that value system will largely determine what useful knowledge consists of. As an example of this effect consider Henry Kissinger's question to Zhou Enlai "What is your view of the French Revolution" to which Zhou's suitably inscrutable answer was "It's too early to tell" (e.g. Suyin, 1995).

A more extended example will focus the need for an operational definition of knowledge. Benchmarking is a widespread practice within industry and there is a Government sponsored version called Best Value for public services. The idea is that by comparing output measures of various organisations involved in similar activities, the more effective ones can export their practices to the less effective. There are a number of prominent critics of Best Value that appears, in practice, to promote a copycat mediocrity and extreme lack of innovation (e.g. Seddon, 2001). Can we compare knowledge of what everybody else is doing with knowledge of what might be possible? Clearly it matters greatly what we think of as knowledge or as appropriate knowledge.

Our operational definition of knowledge at this point is "that information, context and know-how which allows managers to make sound and effective business decisions". This is how managers faced with business decisions think about their potential lack of knowledge and their possible need to acquire it.

The two fundamental measures for business are cost and risk. Cost has been prioritised traditionally, largely because it is more readily measurable. Risk is arguably more fundamental as a determining factor in business outcomes. In an economist's view of making decisions, the appropriate comparison is opportunity cost: if I choose this option and in doing so I close down this other option, then what have I gained and what have I lost? The same structure applies (only more so) to opportunity risk: we might speculate that seemingly rational and safe decisions to follow "best practice" actually carry huge opportunity risk because they preclude much more advantageous decisions.

At the project to build Heathrow Terminal 5, these opportunity risks are being dealt with explicitly. The main contractors are engaged on the basis that their combined target costings for the agreed programme of work amount to more than the economic worth of the completed terminal (given huge uncertainty in ten year projections). They therefore work jointly to find ways of completing the work at a lower cost on the understanding that they can share the savings with the client. The crucial knowledge is the as yet undefined ability to negotiate and implement cross boundary work practices that deliver real savings when implemented. Unless they succeed in this phase the work will not proceed (Thomas, 1999).

POWER AND KNOWLEDGE

In practice, these descriptions of the role of knowledge in decision making are still too formal and rational. Real decisions are made in a more pressured and situated way, with a large component of skill and know-how that are not available for explicit analysis (Klein, 1999). Knowledgeable people often make intuitive decisions ahead of the facts, decisions that are astounding in their accuracy. So knowledge can be seen as an adjunct of a person, the knowledge we are interested in is knowledge that in practice supports effective action. If you want effective decisions that require knowledge, you cannot necessarily abstract the knowledge from the person making the decision.

We know quite a lot about what inhibits effective action. Fear and doubt certainly cause delay. A perception of powerlessness and lack of options can be a complete block. We also know that without a possibility of action we are blind to information in the first place, so that knowledge cannot be gained without the power to act. This is another interpretation of benchmarking and best practice, that the people concerned are only authorised to make essentially defensive adjustments to practice. All they know and all they can know is how they rank amongst providers of this particular service and how to move within those rankings.

So opportunity risk largely arises in a power context. If in practice I am constrained in what I can do, of course the opportunities that are available beyond that constraint are not attainable and will not even be seen. Any way of seeing things is also a way of not seeing things. Any knowledge for action is also an unknowing about other actions. The knowledge we see as valuable is the direct root of our opportunity risk. Zhou Enlai says in effect that he values his awareness more than his potential opinions.

For an extended example of this effect, look at the fashion, particularly in the USA, of ranking employees on their contribution to the company. Critics have focused on the massive demotivation this produces in those ranked low. We know that of the order of 95% of performance is determined by the context of work and is not under the immediate control of the employee. So a scale is introduced (with serious consequences in times of lay-offs) which asserts that employees have power over their destiny when serious knowledge would have to conclude that they did not. When we concentrate upon our knowledge about how organisations work, other sorts of knowledge and other sorts of opportunity become unavailable. Process improvement has to contend with the politics of changing the ranking of employees.

So when we look at knowledge in situated decision-making we find that it is contingent on the power to take the decisions in question. We can't have an oracle that gives the right answer to a decision question, unless that oracle actually faces the decisions and their consequences. The idea that a consultant advisor knows the "right" decision to take where the responsible manager does not is a widespread but highly misleading myth. The countervailing piece of folk wisdom is "anyone can make a mistake but it takes a consultant to really mess up".

KNOWLEDGE AND UNCERTAINTY

Knowledge is a form of security, protection from the turbulence of the business environment. If we know how to carry out this action, and because we know how to do it we have a value in the world, we are respected for our knowledge. Knowledge appears to be a bulwark against uncertainty in the job market. It is this property of knowledge that anchors it in the political. There are other decisions that can be made in the face of uncertainty that privilege other knowledge and may invalidate our present knowledge altogether. Technology choices have this character, for instance.

For example, if I choose a particular database technology I need these sort of experts with a particular sort of knowledge and if I choose that sort of expert system I need some quite different knowledge. Knowledge workers must support the technology they derive their value from in a partisan way, in order to protect the value of their knowledge in the market. They must prioritise their skill context above the problem context.

So if a company manager wants to treat knowledge as an asset, it is an asset contingent on his opportunity risk. There is in principle no limit to the asset value we can lose by paying attention to opportunities other than the ones we have traditionally pursued. This is the conservatism of best practice: we know how we struggle to do even what we know, let's not even think about trying to do something we don't know.

The root of uncertainty is in relationships between people (Marris, 1996). My primary uncertainty is whether you will respect the agreements and the reality we share or whether your other attachments will cause your attachment to me to change. If all relationships in a culture are contingent and repressive in nature, then the knowledge that is valued is only that of how to prevail. Group knowledge becomes impossible because it implies joint action. So culture directly affects what sorts of things are known and what sorts of things can be known. Like the apocryphal production line worker forced out by his colleagues for working too fast, we cannot graft knowledge onto a culture that rejects it.

In a chaotic situation order will emerge if it is allowed to (Wheatley, 1999). The important factor here is time and our ability to live with and to study the chaos for its patterns, its behaviour, and its potential. The common assumption that chaos is dangerous and must be suppressed by imposing order, even if that order is arbitrary, only contributes to chaos in the larger system. Thus a whole order of vital knowledge is predicated on being able to live or survive with an apparent lack of knowledge, lack of control and lack of order. Sometimes knowledge is the ability not to act, to refrain from making decisions that by their nature will be premature.

An example here would be the almost total lack of research into the effect of agricultural practices on micro-organisms in the soil (Humphrys, 2001). We claim to know how to farm efficiently despite mounting evidence that such confidence is completely misplaced. We fail to fund vital research because it can only destroy current agricultural knowledge and the social and political structures it supports.

Not only must the possessor of knowledge have the power to act on that knowledge, but the nature of that knowledge and that action are contingent on the context of relationships and organisational structure. The key problem for organisations looking to acquire knowledge is that effective use of that knowledge will require organisational change that is also not understood.

TIMESCALES FOR THE DEVELOPMENT OF KNOWLEDGE

If knowledge is to support a particular set of management decisions and actions, it must be available at the time those decisions and actions need to be taken. Getting effective know-how in place is subject to the constraints of how quickly the organisational change implied can be achieved.

In most cases, knowledge can neither be developed nor transmitted in timescales that suit the perceived need. In the agriculture example above, the necessary technical research could be done within months, but the gaining of critical mass for significant action will take many years. Staying with agriculture the same observation could be made of the progress of the BSE epidemic.

The two important factors in determining timescales are the speed with which the relevant body of people can change their action stance, and the timescale over which individuals can develop the maturity to contribute to the effective use of knowledge. Maturity is dealt with in a section below. In Zhou's quote above he is also saying that important changes take centuries to come about. From our Western viewpoint, we associate the Chinese with an ability to make small changes that they think will pay off in tens of years' time.

If we turn and look at the propagation of what people call knowledge in business environments, we are seeing something much more ephemeral. People are gaining, using, and abandoning something that certainly amounts to skill and technical expertise, the questions would have to be about whether this "knowledge" contributed to effective management decision making. If we looked, for instance, at the fashion for ERP systems that swept the corporate world and led to huge demand for people who understood how the technology worked, we would have to place alongside that the lack of benefit from such systems and the major business failures that occurred. From the perspective of this paper, the crucial knowledge was not in place and could not be acquired, leading to poor management decisions.

In the world of software development, large proportions of staff are contractors. They say they are contractors because they do not want to be tied into any particular employer who they generally regard as not understanding the needs of developers and development. Their normal experience is of being pushed to do work that they think is not important or not being done the right way. They need to be able to leave to find a better place to work. Viewed from the other side, businesses cannot find enough committed IT staff and are forced to rely on contractors. They specify narrow sets of skills and roles to agencies who find people to fit in the pigeonholes (Armstrong and Ward, 1999). There is simultaneously a pursuit of new technical knowledge in the small and a complete abdication of dealing with the business and organisational knowledge needed to improve decisions. We would claim this effect is driven by a mismatch in timescales.

In this scenario, technology change appears to drive the timescales. Knowledge workers are politically attached to technologies as we have described so technology decisions in organisations have wide ramifications. The perceived "best" people follow the perceived "hot" technologies so there is immense pressure to track the leading edge "knowledge". All the situated aspects of knowledge outlined above militate against any organisation being able to use this leading edge knowledge effectively. The supply side dynamics completely overwhelm the demand side constraints.

MANAGEMENT MATURITY

What does it take to "know" about the knowledge market effects described here? Is there a way for managers to escape a cycle that drives them from a poor understanding of how to acquire knowledge to taking on people who further distort the shape of the available decisions?

Clearly if we want knowledge to be transferable, manageable for the general good, we must take account of people's ability to use it. We cannot assume that knowledge is an asset that can be unlocked by whoever it is given to.

If managers need knowledge to make effective decisions, they also need the maturity to play an effective role in the necessary actions that follow. There is good research showing the distribution of managers across a scale of maturity levels, showing how they think about their role and how able they are to get effective contributions from others in achieving goals (Fisher and Torbert, 2000). We speculate that the sort of knowledge looked for by managers at different maturity levels varies with their level:

1. Diplomats want to please and therefore need knowledge of the social order and how to make suitably deft moves that keep them friends with all the power players. All knowledge is relative to personal power.
2. Technicians have a single management tool and need knowledge about how to use that tool effectively in a wide range of settings. All knowledge comes from a single perspective.
3. Achievers know how to co-opt others into their agenda and need knowledge of how to bend people to their will and to subvert the agendas of others. Knowledge has demonstrably to drive the bottom line.
4. Strategists no longer need to be at the centre, so they need knowledge of the wider world in which they make their plays. Knowledge can be reconfigured and reconnected, made useful in varied contexts.

So only people who can stand aside from the rush to achieve are capable of acquiring new knowledge in a way that benefits them and their organisation. As ever to those who have, more shall be given and to those that have not, even what little they have shall be taken away.

Pluralist Organisations

Finally, there is a steady move towards pluralist views of the world and of organisations that recognise many valid stories about the meaning of what is going on and the values that can be seen in the actions taking place. It is possible to manage businesses (as a strategist!) by managing the parallel development of stories that give the business access to insights into its work.

From a post-modern and pluralist perspective, there are many valid knowledge domains that may well be mutually inconsistent and even incompatible. A strategist will feel no compulsion to rationalise or harmonise these in order to make "progress". We could say that this is the debate that underpins the European Union (EU) and what the practical meaning of subsidiarity might be. We still have business organisations that cling to a central planning view of the world that puts consistency above flexibility, although such a belief is on the wane (Mulgan, 1997).

OPERATION OF THE KNOWLEDGE MARKET

As with any new market, the bad tends to drive out the good - Gresham's Law (Bannock et al., 1998). In the early development of stock markets, customers in general got a seriously raw deal, and that raw deal was closely connected to their lack of understanding of investment as skill (Schwed, 1995). The work of recruitment agencies as one mechanism of the knowledge market seems to echo the same lack of maturity in understanding what is being sold, what is being bought and where value is added in the process. Margins are huge, volume is huge and understanding of the commodities themselves is generally lacking. The customers rarely get the benefit they are looking for and are promised. Despite the fact that the mechanism does not work to the benefit of its customers, they continue to use it because it is the only game in town, because they cannot get to a perspective where they can see its failings, and because there are persistent myths of dramatic success.

The great advantage of a market is to free up commercial relationships by playing down existing business relationships. This ought to allow high value knowledge offerings to succeed more quickly. The disadvantage of a market is that the darlings of a market seem to be successful and valuable when it may only be a collective fantasy or bubble. We can see that the shakeout of the e-commerce sector has pricked many of these bubbles that seemed to contribute so much vital knowledge to existing companies.

We can see how distorted the market is by looking at education and training provision in the light of the factors above. Very little educational practice teaches the factors that will affect the viability of what is being taught. It is assumed in educational quarters that the bubbles in the knowledge market represent real gains and real value and that they should be tracked by educational practice. The knowledge that is required would instead accelerate the development of management maturity and an ability to see the nature of bubbles.

The factors in determining what is effective knowledge outlined above are designed to inform both buyers and sellers in the knowledge market where long term value might lie. The problem, as hinted at in the subtitle ("who decides") is that the market makers have little or no interest in the fundamentals. Recruitment consultants in general advise neither their clients about how to improve the value they get from their choice of recruits, nor the recruits about how to better market crucial skills.

Put yourself in the shoes of a company director who knows that the IT underpinnings of his company are collapsing, and who sort of understands that the lack of integration between business adventures and supporting investment in infrastructure is coming home to roost. People are queuing to sell skilled rescuers, replacement systems, interim management, outsourcing and all sorts of business bandwagon. None of these injections of "knowledge" affect the knowledge gap, which is about how to trade-off adventures and necessary redevelopment of infrastructure. Each business bandwagon recognises the desire for a quick fix at the expense of forming the conditions for needing another quick fix next year. Each one brings about another revolution at the expense of evolution (e.g. McWhinney, 1992).

REFERENCES

Armstrong, S. and Ward, A. (1999). "What is the point of working in an industry where there is a better chance than evens of your work being shelved?" in *Synergy Matters – Working with Systems in the 21st Century*, (A. M. Castell, A. J. Gregory, G. Hindle, M. James and G. Ragsdell, eds.), Kluwer Academic / Plenum Publishers, New York.

Bannock, G., Baxter, R.E. and Davis, E. (1998). *The Penguin Dictionary of Economics*, Penguin Books, West Drayton, Middlesex.

Fisher, D. and Torbert, W.R. (2000). *Personal and Organisational Transformations*, Edge\Work Press, Boston.

Humphrys, J. (2001). *The Great Food Gamble*, Hodder and Stoughton, London.

Klein, G. (1999). *Sources of Power: How People Make Decisions*, MIT Press, Cambridge, Massachusetts.

Marris, P. (1996). *The Politics of Uncertainty*, Routledge, London.

McWhinney, W. (1992). *Paths of Change*, Sage Publications, Newbury Park

Mulgan, G. (1997). *Connexity*, Chatto and Windus, London.

Schwed, F. (1995). *Where are all the customers' yachts?* Wiley, NewYork.

Seddon, J. (2001). *The Better Way to Best Value*, Vanguard Education, Buckingham.

Suyin, H. (1995). *Eldest Son : Zhou Enlai and the Making of Modern China*, Kodansha America, New York.

Thomas, M. (1999). *STORM*, BAA Internal Publication.

Wheatley, M. (1999). *Leadership and the New Science*, Barret-Koehler Publishers, San Francisco.

INTRODUCTION TO KNOWLEDGE PROFILING

M.Yolles

Liverpool John Moores University
Byrom Street, Liverpool L3 3AF
England, UK

THE CYBERNETIC THEORY OF VIABLE SYSTEMS

Viable system theory is a cybernetic approach that is concerned with autonomous, active, purposeful, and adaptive social organisation that can operate in complex situations and survive. Since complex situations entail variety differentiation, in surviving a viable system responds to changing situations by generating sufficient variety through self-organisation to deal with the situational variety it encounters (called requisite variety). A viable organisation is able to support adaptability and change while maintaining stability in its behaviour. In particular an organisation is viable if it can maintain stable states of behaviour as it adapts to perturbations from its environment.

Viable organisations seek ways of improving their ability to survive in complex situations. This is often coupled with the idea that they have fluid patterns of knowledge, and organisational survival hinges upon an ability to create and manage knowledge. Knowledge creation/recognition is therefore of prime importance to organisations. Knowledge creation is associated with worldview (Yolles, 1999), defined in terms of a belief system (beliefs attitudes and values), language, and a cognitive space of related concepts that link together to form a pattern of knowledge.

To enable us to discuss cognitive, organising and behavioural features of a social, we have differentiated it into three domains that are ontologically related. The worldview of the social interacts with that of its membership in the cognitive domain, and this drives purposeful adaptiveness that occurs in an organising domain. The social is able to behave because of its physical manifestation, sometimes seen as a system, which has form and thus structure, and process, and these are assigned to an energetic behavioural domain. It should be clear that behaviour is not possible without a structural form, and if the structure changes, then so will the possibilities for behaviour. Indeed, structure operates as an enabler of behavioural possibilities, as well as a bounding constraint. We have already indicated that the behavioural domain usually houses what many refer to as a systems representation of the social organisation. What we call the cognitive domain houses the metasystem from which organisational decisions normally emanate. The

Edited by Ragsdell *et al.*, Kluwer Academic/Plenum Publishers, New York, 2002.

organising domain would therefore have been seen as a gap that connects the system and metasystem, and that may be subject to surprises. It is strategic and political in nature, and operates through information. The three cognitive, organising, and behavioural domains are analytically and empirically independent.

There are properties associated with each of these domains (Table 1) that are a development of Habermas' (1974) theory of knowledge constitutive interest. We refer to this as the Viable Systems Domain Model (VSDM), and it has commodities associated with each domain. The property of the cognitive domain is cognitive influence, connected with the commodity knowledge. The property associated with the organisation domain is cognitive purpose, and connected to information. The behavioural domain is associated with cognitive interest, and operates through the commodity of data. The three domains are ontologically related, but are analytically independent.

KNOWLEDGE PROFILING AND TYPE

The nature of viable systems theory is that it connects people through their mental processes to social groups (e.g., worldviews and patterns of knowledge). In a newly formed social group normative cognitive elements (concepts and processes) develop through the communicative interactions that occur amongst its membership. This provides the basis for group coherence, and the capability of developing a cognitive domain that can direct the organisation and promote self-organisation. From these beginning propositions develops a theory of the organisation that also becomes a theory of the social group.

The idea of knowledge creation has been replaced by local knowledge creation, and from this has arisen the notion of knowledge profile that can replace Kolb's (1974) idea of individuals having learning style. Thus, the nature of a person's knowledge profile will determine the way that they will capture and respond to explicit knowledge. In principle the ideas of knowledge profile can also be applied to the social group, so that coherent groups can have a sociality that determine their abilities to effectively perform their operations.

Marshall (1995) was interested in knowledge type within the context of complex decision-making. Her model is a knowledge schema that consists of four component in the decision making process:- *identification knowledge*: mental organisation of individual's knowledge and experience; *elaboration knowledge*: access to a generic framework that contains the essential elements of all these similar experiences; *planning knowledge*: use of this framework to plan solutions; *execution knowledge*: use of skills and procedures to execute the solution. The model was initially for use in problem solving, and later (Marshall *et al.*, 2000) for decision-making. Marshall's schema, which is empirically supported, can be conceptually supported by relating it to Schutz and Luckmann's (1974) definition of the schema relevancies that are connected to narrative. We can easily argue that the context Marshall's decision-making is expressible as a narrative. To illustrate this we note that a narrative is a reflective experienced coherent story.

Table 1. VSDM Relationship between Human Cognitive Interests, Purpose, and Influences

COGNITIVE INTERESTS OF THE BEHAVIOURAL DOMAIN (DATA)		
Technical	**Practical**	**Critical Deconstraining**
Work. This enables people to achieve goals and generate material well-being. It involves technical ability to undertake action in the environment, and the ability to make prediction and establish control.	Interaction. This requires that people as individuals and groups in a social system gain and develop the possibilities of an understanding of each others subjective views. It is consistent with a practical interest in mutual understanding that can address disagreements, which can be a threat to the social form of life.	Degree of emancipation. For organisational viability, the realising of individual potential is most effective when people: (i) liberate themselves from the constraints imposed by power structures (ii) learn through precipitation in social and political processes to control their own destinies.
COGNITIVE PURPOSES WITHIN THE ORGANISING DOMAIN (INFORMATION)		
Cybernetical	**Rational**	**Ideological**
Intention. This is through the creation and strategic pursuit of goals and aims that may change over time, enables people through control and comunications processes to redirect their futures.	Logico-relational. Enables missions, goals, and aims to be defined and approached through planning. It involves logical, relational, and rational abilities to organise thought and action and thus to define sets of possible systemic and behaviour possibilities.	Manner of thinking. An intellectual framework through which policy makers observe and interpret reality. This has a politically correct ethical and moral orientation. It provides an image of the future that enables action through politically correct strategic policy. It gives a politically correct view of stages of historical development, in respect of interaction with the external environment.
COGNITIVE INFLUENCES WITHIN THE COGNITIVE DOMAIN (KNOWLEDGE)		
Social	**Cultural**	**Political**
Formation. Enables individuals/groups to be influenced by knowledges that relate to our social environment. This has a consequence for our social structures and processes that define our social forms that are related to our intentions and behaviours.	Belief. Influences occur from knowledges that derive from the cognitive organisation (the set of beliefs, attitudes, values) of other worldviews. It ultimately determines how we interact and defines our logico-relational understandings.	Freedom. Influences occur from knowledges that affect our polity determined, in part, by how we think about the constraints on group and individual freedoms, and in connection with this to organise and behave. It ultimately has impact on our ideology and our degree of emancipation.

Thus decision-making processes are fundamentally embedded within a continuing narrative. This means that we should be able to assign Marshall's schema to the Schutz and Luckmann classification. This results in Marshall's four-stage schema being reducable to three schema relevancies: identification knowledge, elaboration (with planning) knowledge, and execution knowledge (Table 2).

KNOWLEDGE MIGRATION

It is possible to create a knowledge cycle based on the model embedded in table 1. Local knowledge creation occurs through a process of knowledge migration that is essentially an identification knowledge process. It relates to the other phases of knowledge process in a knowledge creation cycle given in Figure 1. This links to the three domains of Table 1, and depicts the three phases of the knowledge process: migration, appreciation, and action. Migration is associated with the cognitive domain, appreciation with the organising domain, and action with the behavioural domain. The way that migration occurs is conditioned by cognitive influence, appreciation though cognitive purpose, and action through cognitive intention. Each phase process has an input and an output. A feedback loop is able to condition each phase process directly, or through its input. The nature of the feedback has implication for the way in which the social group responds to situations and appreciates the need of semantic communications that make them broadly meaningful (Yolles, 2000a). The model would appear to be consistent with that of Marshall, since the control loops in Figure 1 enable: unequal frequency of phase use; non-sequential ordering of phases; and the possibility of rapid oscillation between phases so that phase progress can in effect occur in parallel.

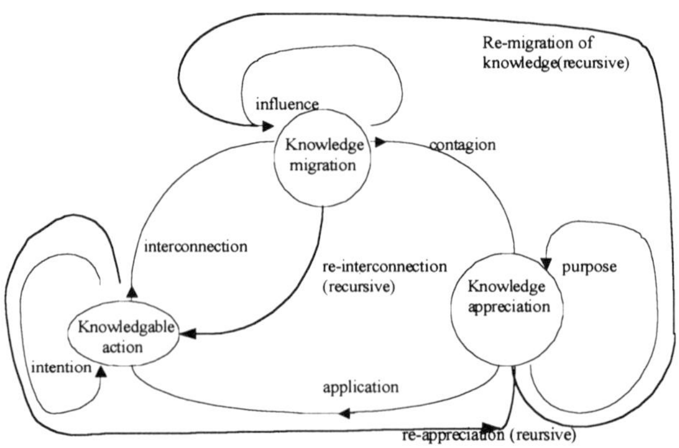

Figure 1. The Knowledge Cycle

As part of knowledge migration, new knowledge is locally generated for viewholder actors. While this may be seen as part of a socialisation process, it may also be seen as a spontaneous local viewer thing when the process of knowledge migration operates as a knowledge creation trigger.

Newly migrated knowledge may be shared and re-shared within the suprasystem, because the new knowledge created by one worldview holder will have a local definition that will be different for others. As a result, the originally migrated knowledge will have to be re-migrated in a feedback loop. This is fundamentally consistent with the notion of paradigm incommensurability, since every worldview will have its own distinct pattern of meaning that will be different from every other one. This does not stop the knowledge from being "contagious" to relevant others within a given suprasystem through the continuous semantic communication process that they participate in, that involves recursive migration (that is re-migration and re-remigration) of knowledge. Each recursive knowledge migration has the potential of new local knowledge creation for each worldview holder in the lifeworld. As knowledge is migrated, it is likely to pass through a morphogenic process, and sometimes a metamorphic one that makes it new to the group.

The process of knowledge appreciation can follow knowledge migration. An appreciation of how migrated knowledge can be of use to a relevant other is essential if they are to be able to harness it within a behavioural world. Knowledge appreciation by relevant others is dependent upon knowledge contagion to these others. However, this is filtered through patterns of knowledge that relate to ideology and ethics. This filtering process also enables recognition that information has been transmitted. In addition, the evaluation reference criteria derive from knowledge about intention and logico-relational cognitive purposes. Interestingly, this connects with the Marshall (1995) idea of planning knowledge - the knowledge of which pathways to select in order to achieve a solution.

A consequence of the process of knowledge appreciation is its intelligent application. We say intelligent because its obverse, rote application, may not require knowledge appreciation or even migration. Knowledge application can occur behaviourally within a superstructure and a substructure. Superstructure identifies the institutionalised political and cultural aspects of a situation, and is also issue related. Substructure is task orientated, and relates to the mode and means of production (e.g. technology) and the social relations (e.g. roles and relationships) that accompany them.

The process of knowledgeable action is dependent upon the application of knowledge. Knowledgeable action is action that occurs with awareness of what is being done within a behavioural world. Knowledgeable action in a situation is dependent upon knowledge application to the tasks that are perceived to require to be addressed within the situation. This is filtered through knowledge that activates weltanschauung derived emancipative capabilities, that enable knowledgeable action to occur. The evaluation reference criteria derive from knowledge about actor interests through work and interaction. It relates to the Marshall (1995) idea of execution knowledge, that is seen as the computational skills and procedures required to execute a behaviour.

A consequence of the process of knowledgeable action that derives from knowledge migration is the creation of new definition of relationships between identifiable actors. It gives meaning to work related activities, and particularly with respect to those that are involved in interaction.

KNOWLEDGE PROFILING

We can offer two propositions, developed originally in discussion with Paul Iles, that conceptualise the relationship between people and knowledge type, one of which is an adaptation of the other. They are not intended to be conflictual, but should be seen as profiling dimensions Table 2).

The first proposition is that any individual may have a penchant for one type of knowledge, and therefore is particularly good at working in a given type of knowledge. Thus, those who are particularly good technology translators (Iles and Yolles, 2000) or creators of knowledge have a penchant for knowledge identification. In unitary penchants an individual will have a bent towards only one of the three types of knowledge knowledge: identifiers, elaborators, or executors. If they are plural they will have more than one of the three penchants to some degree of simultaneously. A knowledge penchant may also be associated with the local generation of knowledge, which is simply a learning process. If it is possible to evaluate the learning capabilities of a person in a type of knowledge, then it can be referred to as a learning index.

Table 2. Nature of distinct Knowledge Types with Polar Extreme Values

Knowledge type	Polar Attribute of Knowledge Type	
Executors Supports the ability to carry out or perform activities.	**Fundamentalist** Behaviour conforms to some fundamental prescription independent of circumstance. It is useful where conformity is essential.	**Pragmatist** Behaviour reflects the demands of circumstance
Elaborators Supports both elaborators (who understand how to deal with the relationships between cultural attributes) and planners (who through their understanding of cultural attributes and its patterns of knowledge are able to determine possible trajectories for action).	**Insulated** Able to maintain performance in spite of surrounding distractions. Unlikely to be able to respond to constructive criticism. Likely to be strategically bound.	**Responsive** Able to respond to the needs or situations of others. May limit ability to centre on tasks by being over-responsive. Likely to be strategically open.
Identifiers Supports the creation or translation of ideas and concepts; its members are appreciative of the nature of knowledge.	**Ideationalist** Centres on the knowledge. Good at acquiring or creating concepts or ideas. No know-how to develop them for material implementation.	**Sensatist** To do with the senses. Able to develop or engineer existing ideas for material implementation. Good concept translator. Cannot generate new ideas or concepts.

The second proposition is not interested in the relative knowledge penchant of an individual, but rather to the nature of the knowledge penchant itself that we refer to as orientation. To explore this, we define each knowledge type in terms of two extreme polar positions. We have created three pairs of polar extremes for each knowledge type. Let us consider each in turn. After Sorokin (1937), we can classify two polar extremes for identifiers as *sensate* and *ideational*. Sensate culture is to do with the senses, and can be seen to be utilitarian and materialistic. Ideational culture relates to ideas, and an example might be the adherence to say the idea in essence.

Consider now the appreciation phase of knowledge creation. This has associated with it those who might be called elaborators. It is possible to classify two polar types of elaborators, those who are either responsive to or isolated from others and the new knowledge that they offer. Finally, closely associated with the phase of knowledgeåble action are executors. Two types of executors may be posited. Fundamentalists adhere to notions very strictly, whilst pragmatists provide for some degree of leeway in the way that they adhere to notions. It is not necessary to be either fundamentalist or pragmatist. There may be phases in between them, in the same way, for example, there may be between insulated and responsive elaborators, or sensate and ideational identifiers. Thus for instance, an identifier may be able to mix sensate and ideational perspectives, in a condition referred to as idealistic.

These polar extremes are seen to exist on a continuous axes, and an individual will have knowledge attributes that exist somewhere between the two polar extreme positions (we shall see this more clearly in a moment). This construction provides us with the possibility of postulating knowledge types that are richer than Kolb's "learning styles" that links directly to the table 1 domains (Yolles, 2000b; Iles *et al.*, 2000). A number of learning indexes have been proposed that relate to knowledge creation (Figure 2 Table 4). The indices have been named so as to demonstrate how they tie into the learning theory discussed earlier. Having postulated different knowledge profile types, an individual of a particular knowledge type may also have a local penchant within that type.

The indices are measure of knowledge workers. The most interesting for our purposes, is that of knowledge creation "style", since it implies learning style. Associated with each phase of knowledge creation are types of knowledge workers. Those who are particularly good at migrating knowledge are seen as knowledge identifiers. Following the notions of Sorokin (Yolles, 1999), we can classify two cultural classes of identifiers, sensate and ideational. Sensate culture is to do with the senses, and can be seen to be utilitarian and materialistic. Ideational culture relates to ideas, and an example might be the adherence to say spirituality or some perhaps unachievable goal. The appreciation phase of knowledge creation has associated with it those who might be called elaborators. It is possible to classify two polar types of elaborators, those who are responsive to new knowledge, and those who are not. Finally, closely associated with the phase of knowledgeable action are executors. Two types of executors are proposed to exist. Fundamentalists obey to the letter a set of laws or fundamental principles that serve as the foundation of a given worldview, and to which they adhere strictly. Pragmatists (according to the Webster College Dictionary 1990 edition) operate, not through theories, but through practical approaches. The dictionary further adds that pragmatists can be utilitarian, un-idealistic, materialistic and unsentimental. However, we would contend that these really require testing since the categories encroach on our other knowledge profile types. Pragmatists, in the end, provide for some degree of leeway in the way that they adhere to notions. It should be pointed out at this juncture, that the way that in considering the typology of knowledge profiles, it is not necessary to be fundamentalist. That is for example, an identifier may not be only a sensate or an ideational orientation. They may have an orientation that enables them to mix sensate and ideational perspectives, in a condition that Sorokin refers to as idealistic. The same idea can apply to executors and identifiers.

CONCLUSION

A theory of the organisation has developed around Viable Systems Theory. It is attuned to a critical perspective, and one of its consequences within the context of knowledge management has been the arrival of the idea of knowledge migration. This highlights the lack of certainty that one has about the quality of explicit knowledge transfer between two or more individuals. Knowledge migration affects the way in which we see situations and evaluate their nature and consequence. The creation of a knowledge profile and its measurement provides an interesting possibility for the examination of the knowledge attributes of individuals. However, there is always a difficulty with the creation of questionnaires based evaluations that the questions that one thinks that one is asking are not the questions that are being asked. This has never been more apparent than in the development of marketing techniques, when respondents to questions would respond to their perception of the image that they would portray by responding to questions in a certain way. Consequently, psychometric validation is still required to ensure that the questions that we think that we are asking are in fact the questions that we are asking. The development of knowledge profiles has been developed so far within a pilot study that has indicated its feasibility as a way of developing evaluations about peoples knowledge capabilities. It has implications not only for the perceived needs of job specification, but also for the nature of team composition as discussed by Belbin (1996) for instance, a topic that will be addressed in the near future.

REFERENCES

Belbin, M. (1996). *The Coming Shape of Organization,* Butterworth-Heinemann, London.

Habermas, J. (1974). *Theory and Practice,* Heinamann, London.

Iles, P. and Yolles, M. (2001). "HRM and Knowledge Management: responding to the challenge", *Research and Practice in Human Resource Management*, 9(1):3-33.

Iles, P.A. and Yolles, M. (2000). "Across the Great Divide: HRD, technology translation and knowledge migration in bridging the knowledge gap between SMEs and Universities", *Human Resource Development International*, 4(1) 1-35, ISSN13678868

Iles, P.A., Yolles, M. and Altman, Y., 2000, HRM and Knowledge Management: responding to the challenge. *Journal of Research and Practice in HRM* (Special Issue on Knowledge Management - invited paper) 8(2)1-31

Kolb, D.A., (1974). *Organisational Psychology: An Experiential Approach.* Prentice-Hall, Englewood Cliffs, New Jersey.

Marshall, S.P. (1995). *Schemes in Problem Solving.* Cambridge University Press, Cambridge,

Marshall, S.P., Wilson, D., M., Knust, S.K., Smith, M.E., Garcia, R.J., 2000, Shared Decision-Making Knowledge in Tactical Situations, Sept. C4ISR Coop. Res. Prog. Conf. available at: http//www.dodccrp.org/Proceedings/DOCS/wcd00000/wcd000e5.htm

Sorokin, P.A., 1937, *Social and Cultural Dynamics*. Amer. Book. Co. N.Y.

Schutz, A. and Luckmann, T. (1974). *The Structures of the Lifeworld*, Heinemann, London. http://sites.netscape.net/mcyrhul/intelligent_organisations.html

Yolles, M.I. (1999). *Management Systems: A Viable Approach*, Financial Times Pitman, London.

Yolles, M.I. (2000a). "The Theory of Viable Joint Ventures", *Cybernetics and Sys*, 31(4):371-396.

Yolles, M.I. (2000b). "The Viable Theory of Knowledge Management", *Research Memorandum of the Janus Centre for Res. in Mgt Sys and Cybernetics*, 3(1), Info. Management Centre, Liverpool Business School.

Section Four

Environmental Management

PARTICIPATIVE ACTION RESEARCH (PAR) FOR RURAL COMMUNITY DEVELOPMENT IN SOUTH SULAWESI, INDONESIA

Hasnawaty Habibie, N. Sriskandarajah, and R.G. Packham

Centre for Systemic Development, University of Western Sydney, Locked Bag 1797, Penrith South DC, NSW 1797, Australia.
E-mail: R.Packham@uws.edu.au

INTRODUCTION

In Indonesia, community development is one of the major strategies for improving the conditions of smallholder farmers in rural areas. Such programs are usually delivered to farmers through extension services using a Transfer of Technology (TOT) approach - a technical orientation primarily concerned with agricultural production issues, rather than being concerned with livelihood issues such as social equity or ecological sustainability (Sriskandarajah *et al.*, 1993; Cornwall *et al.*, 1994). This approach has been challenged as being non-participatory and top-down in its implementation (King, 2000).

Participative approaches can be developed with a reliance on local farming knowledge and techniques, and with the capacity to adjust the management of diverse on-farm resources and inputs to different local conditions, as well to incorporate contemporary scientific understanding. In this study Participative Action Research (PAR) was adopted as an alternative approach to Transfer of Technology, to first identify the problems and needs of the farmers and then to develop strategies jointly with them to meet these needs. The focus was on using both individual and group discussion as opportunities for farmers to learn to become responsible for their own development.

PAR: A RESEARCH METHODOLOGY

PAR in community development requires the practitioner to play the role of a catalyst in the research process. It means that practitioners should view all individuals with whom they are working both as subjects and as active participants in the research process, rather than as objects to be studied (Selener, 1997). In PAR some of the people of the community under study participate actively by forming a research team with the professional researchers throughout the research process, from the initial design, through data gathering and analysis to the final presentation of the results and discussion of their action implementations (Whyte, 1991). Further, the people who engaged in PAR act as co-learners actively engaged in the quest for information and ideas to guide their future actions - dialogue between them is an important feature and source of validity. Through this process, they become the producers and owners of their own information. PAR enables people to solve their problems and to be self-reliant and self-critical (Selener, 1997). Dialectical relations between theory and practice are essential to the validity of the practice of this approach.

The PAR that underpinned this study was based on the work of Bawden 1991; McTaggart, 1991; Greenwood and Levin, 1998 and Kemmis and McTaggart 2000. The research methodology also utilised techniques from the tradition of participatory rural appraisal (PRA) as the process of inquiry (Theis and Grady, 1991). These included brainstorming, livelihood analysis, transect mapping, direct observation, open-ended interview, semi-structured interviews and workshops. The following were key principles that were used in this context: Involving people as subjects; appraising for local knowledge and skill; sharing ideas and information; applying dialogue and negotiating strategies; assuring that people have influence over development decisions; and maintaining a concern for facilitating learning process as much as outcomes

This project commenced with the formation of a small action research team, and developed through a spiral of cycles of planning, acting, observing, reflecting and then re-planning (McTaggart, 1991). The action research team entered the cycle by exploring the situation with the stakeholders, collecting data from many perspectives. The methodology adopted had as its aim the engagement of stakeholders in the generation of knowledge about their lives as farmers and livestock extension officers

THE PAR CONTEXT

Tombolo village was the context for investigating the use of PAR as an overall strategy to support stakeholders (livestock extension officers and smallholder livestock farmers) learning how to learn to improve their own situation. It was evident from the initial inquiry process of this study that farmers saw most of their problems as related to a lack of finance, and that direct government intervention was the only solution. In contrast, the livestock extension officers viewed the problems as being technological. These views were consistent with the top-down TOT development approach. Many authors (such as Chambers 1993; Scoones and Thomson, 1994; Bawden 1995; Pretty 1995) have criticised this TOT approach, questioning the relevance and the effectiveness of top down development models in complex and dynamic human activity systems such as the context of this study. These authors suggest a need to engage stakeholders more participatively in investigating what the causes of problems are, and what possible solution might be. In such processes, farmers need access to new information so as to be able to participate in deciding how to improve their situations. The focus needs to be on the process of learning, with the actions of stakeholders being informed by their own personal constructs of the reality of their situation.

LEARNING-PARTNERSHIPS

As the learning process required more transparent dialogue and negotiation in decision making than was usual, farmers needed to participate in generating this knowledge instead of just listening to how the livestock extension officers were going to try to solve their problems for them. Similarly, the livestock extension people had to accept the questioning of their effectiveness by farmers, and had to reduce any subtle use of their power over the knowledge delivery process. This dialogue was achieved through the formation of a number of learning partnerships, starting with the action research team but including the formation of a leaning group and a community discussion group. Figure 1 illustrates the arena and aim of these learning partnerships.

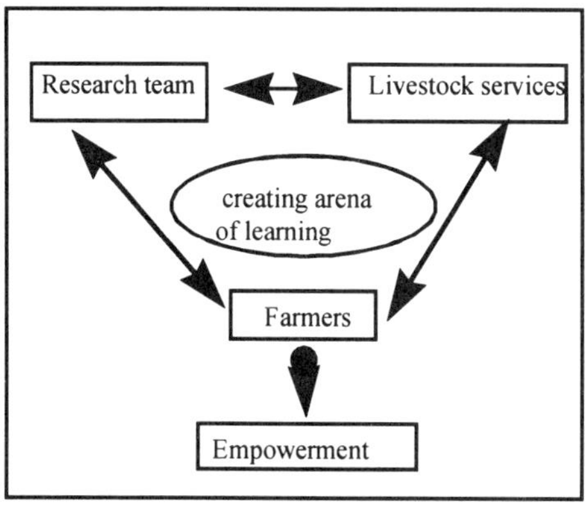

Figure 1. The Learning-partnerships in Tombolo Village

The Action Research Team

When starting to work participatively with a community, it is important to have an inter-disciplinary team membership, which includes local people. This contributes to the validity of the findings, as it encourages the expression of diverse views and the ongoing questioning of individual assumptions via critical dialogue. The research team for this project consisted of four people, and included the senior author, who had a background as a government agricultural extension officer in the area, but who also had master's level education in systemic approaches and action research. The other three were: a female agricultural extension officer who had previously worked in the village; a male socio-economic specialist who held a position with the provincial government, but who had previously worked for a NGO in this village; and the elected village leader (male).

The first task the team undertook was to discuss their own attitudes to community development. The question addressed was: what does community development mean? The research team came to an agreed position that it was: a process that enables people to (a) realise their potential; (b) develop self confidence; and (c) develop better knowledge and skills. The question that then arose was how was this going to be achieved? The research team decided that their focus would be on helping farmers within their own situation, where contextual problems could be recognised and appropriate solution sought from any available source that was relevant to the local conditions.

The Learning Group

Volunteers willing to represent the community and be committed to this research project formed the learning group. The group consisted of seven farmers. The members of the group were recognised as leaders amongst the farmers, and all agreed to allocate time to meet regularly. This group acted as a forum for the whole community, reflecting community thinking to the research team, as well as discussing appropriate issues with the wider community. This group actively participated in the research process from the data collection through to implementing action. Regular, weekly meeting were held between this group and the action research team for the first six months of the project, and then every two weeks after that.

The principles used by this group were to share information, to respect each other, and to be communicative so as to enhance mutual learning. It was important to overcome the usual perspective of these farmers that others would do everything for them. In this context the research facilitator plays an important role in encouraging participants to achieve a deeper awareness of themselves, creating a dialogue throughout the evolving partnership that addresses such questions as: Who are we? Why are we here? What is our purpose? What do we believe? How can we work together? Emancipation (gaining freedom) means rejecting damaging images of one's own culture, replacing these with pride and acquiring abilities for self-reliance in order to functions autonomously (Freire, 1972). The methods used by the learning group were: Informal interviews, diaries for self-collecting data, diagrams, field visits, regular meetings, informal conversations, and action-reflection.

Community Discussion Group

The function of this group was to test ideas with, and to generate knowledge from, the wider community of stakeholders before action was implemented. Six meetings were held during the course of the research, open to the whole community. Following the input of ideas from the action research team, which had already been discussed with the learning group, this broader group generated the possible and probable interventions that could be implemented to help improve the situation. The focus was the identification of needed changes, and the direction that such changes might take. Participants (men and women farmers, livestock extension officers, and local government staff) were encouraged to impart their ideas and concerns about the topics introduced to the meetings. Participants were also asked to respect all views shared by others. Methods used in the facilitation of the Group discussions included: Brainstorming, livelihood analysis, needs assessment and priorities, problem trees, and diagrams.

THE PAR PROCESS

Weekly meetings were held in the village between the action research team and the learning group. The learning group members were initially encouraged to describe relevant experience in relation to their livestock farming, such as production aspects, feeding, diseases, breeding and marketing. They were also asked to reflect on the impact of the socio-economic factors that affected them, and how they coped with these. They were encouraged to discuss their expectations of livestock extension staff.

The discussion topics focused on more specific themes as problematic issues emerged, exploring in more detail the relationships of such issues to other technical and social factors. For example, the learning group commented that a scarcity of fodder for animal feeding was an important issue that needed to be resolved. The research team then conducted a series of focus group meetings to collect pertinent information, and to identify how to integrate forage into the farming systems. The learning group then used this information to identify the availability of fodder species in their area or in neighbouring villages. They also identified fodder management practices and other factors that related to this issue, such as rainfall, soil, irrigation systems and socio-economic factors. The group thus moved through each of the AR steps of planning, acting, observing and reflecting. Outcomes were fed back to the whole community, and from one of these meetings it was decided to run a workshop on developing forage technology.

At the workshop, farmers created their own model of planting forage that was appropriate to their local condition, rejecting the adoption of a technology called *Three Strata Forage Systems (TSFS)* (Nitis, 1989) that had been introduced to them at the start of the workshop. The community clearly understood the long-term benefits of TSFS, and they were told about how it had been used successfully in other parts of Indonesia. It was evident from discussions that most of the farmers were aware of the advantages of planting forage for animal feeding, but they could not accept the *TSFS*. Farmers said: "*We are aware of the advantages of TSFS in the long-term period, but we could not apply it in our farms, as we do not have enough land to do that, we do not want to replace our crops. We think that we still can grow forage in our fields based on our land condition*".

The farmers also saw a limitation to the availability of supplies of the necessary forage species in this area. They indicated a preference for growing forage based on their particular land conditions such as the planting of fodder trees (*Gliricidia*) as fences around the boundaries of homesteads and fields, and Napier grass (*Pennisetum purpurium*) in waste ground adjacent to existing crops, in unused land within the homestead area, and under existing kapuk trees.

Two action steps then eventuated. Firstly, each learning group farmer developed their own model for forage planting, depending on their current land use and condition. Secondly, some of them developed demonstration plots on their own farms, so they could gain experience and show others what to do. These particular learning group farmers collaborated with the research team and livestock extension officers to set up these plots. The farmer-participants recorded their activities, and learning group meetings discussed the problems and issues arising in these demonstration plots. This developed ownership of the exercise, and confidence in decision-making. For example, farmers developed confidence in their ability to understand and use technical and scientific knowledge, such as with the cutting of *Napier grass* for animal feeding generally every 6-7 weeks. Farmers had found that palatability was higher at this time, even though the technical recommendation stated that this grass could be cut between 6-10 weeks. Farmers valued the scientific information for providing general concepts and to understand why certain practices were recommended. As one farmer said: "*We know that cattle like young leaves, but I've never known all the reasons why...now I know that*". The demonstration plots also found that the *Napier grass* can grow well under existing *kapuk tree*, even though the land was categorised as stony land and not useful for cultivation.

This illustrates how the PAR process created the space for stakeholders to become partners in the research. Learning became an active process, combining *finding out* about the problem with *taking action* to improve it (Sriskandarajah *et. al.* 1993; Brown and Packham 1999). Livestock extension officers were also being indirectly re-oriented towards participatory processes by seeing how these worked in practice.

THE OUTCOMES

The following outcomes emerged.

In Relation to Improving Farm Practices

1. Planting forage improved the productivity of cattle. The growth rates of offspring improved from 0.3kg/day to 0.5kg/day.
2. The farmers moved from a system of tethering animals in fields to eat weeds, to a cut-and-carry stall feeding system, utilising more nutritious fodder sources.
3. The farmers found that this tethering system also enabled more manure to be collected for use as fertiliser.
4. Farmers observed that forages such as *Gliricidia* and *Napier grass* produced a lot of leaves and that they had the potential to improve the soil structure, especially when supplied with fertiliser.

In Relation to the Learning Group

1. The group acted as a forum for the wider community.
2. The group also represented the community in relating to others organisations.
3. They also helped to establish cow-fattening groups, with the aim of sharing relevant problems and experiences.
4. Being a part of such a group enabled farmers to understand some of the broader economic and socio-political issues affecting their farm, such as measuring animal weight by buyers, and access to capital, helping to ensure a fair price for farmers.
5. This act brought awareness about the need for questioning the *status quo*.

Improving Extension Strategies

Livestock extension staff found that the way they worked with farmers changed. Previously they had the view that only scientific knowledge could solve farmers' production problems. This change was expressed by one of the livestock extension officers: "*For ten years I have worked with farmers, but I never think that the farmer is a knowledge creator. Formerly, I worked to transfer technology to them without reflecting on what these technologies might mean to the farmers. Now, I realise that farmers have their own local knowledge... and now I need to use my service to improve farmers' understanding of the various choices open to them that are appropriate to their local condition. Learning with farmers not only enables farmers to improve their own practice; it also improves the way I act with them*".

Improving Relationship within the Village

Communication improved amongst farmers, and between farmers and livestock extension staff. One farmer expressed his new experiences in relation to the livestock extension staff: "*This study created space for me to meet with other farmers and livestock extension workers to discuss any issues related to our farming. It also improved my confidence to impart my opinion in meetings, particularly meetings with outsiders such as livestock extension workers or development workers. Formerly, I just tried to be a good listener to them and felt they knew everything.... that created a gap between us. Through this learning process, we have become co-learners... they also learned from us*".

A livestock extension officer similarly described his new experience of working with farmers: "I improved my ability to work with farmers, as this study brought us together to solve the problems, not to sell the technology to the farmer. For example, this participative approach does not put me as the extension officer, as having to do all the giving and the farmers at the receiving end; we learned as partners in finding solutions to farmers' problems".

Improving Understanding of Farmers

Farmers found that this study helped them to understand their problems in relation to other contexts. For example, they became aware that low animal production was not caused by their inability to get a financial loan, but to a relationship with many other factors, including animal feeding, rainfall, and relative price of farm inputs. This point was highlighted by one farmer as follows: "*Now, I am able to see what the real problems are for me, I understand my problems in relation to the other factors such as rainfall, government policy and input price*".

CONCLUSION

This study found that there was considerable potential for the application of PAR to rural community development in Indonesia. A key recognised need was for facilitated processes to help stakeholders make the transformation from an expectation of being told what to do, to becoming co-learners and developers of their own situations. This was achieved in this project in a participative way, where the power of decision-making was shared amongst all stakeholders: Learning partnerships appropriate to this specific situation were the means used to achieve this. An important lesson learnt from this study was that a high adoption of technology will occur if the technology fulfils the needs of farmers, and has been adapted to local circumstances. We believe that Participative Action Research has great potential in agricultural extension, and can enhance the current Transfer of Technology approach to produce clear and definable beneficial outcomes.

REFERENCES

Bawden, R. (1991). "Towards action researching systems", Chapter 2 in: *Action research for change and development.* (O. Zubber-Skerritt, ed.), Griffith University Press, Brisbane.

Bawden, R. (1995). "On the systems dimension in farming systems research", *Journal of farming systems research and extension.* 5(2):1-18.

Brown, M. and Packham, R. (1999). *Organisational learning, critical systems thinking and systemic learning.* Research memorandum, Centre for Systems Studies, University of Hull. UK.

Chambers, R. (1993). *Challenging the professions: Frontiers for rural development.* Intermediate technology publications. UK.

Cornwall, A., Guijt, I. and Welbourn, A. (1994). "Extending the horizon of agricultural research and extension: Methodological challenges", *Agriculture and Human Values.* Spring-Summer 11 (2 &3):38-52.

Freire, P. (1972). *Pedagogy of the oppressed.* Penguin books Ltd. England.

Greenwood, D.J and Levin, M. (1998). *Introduction to action research: Social research for change.* Sage, London.

Kemmis, S. and McTaggart, R. (2000). "Participatory action research", in. *Handbook for qualitative research,* (2nd edition) (N.K Denzin and Y.S. Lincoln, eds.), Sage, London.

King, C.A. (2000). *Systemic Processes for facilitating social learning: Challenging the legacy.* PhD Thesis. Swedish University of Agricultural Sciences.

McTaggart, R. (1991). "Principles for participatory action research", *Adult Education Quarterly.* 41(3):168-187.

Nitis, I.M. (1989). Three Strata Forage Systems. *Asian Livestock* XIV, 4: 37-40.

Pretty, J.N. (1995). Participatory learning for sustainable agriculture, *World Development,* 23(8):1247-1263.

Selener, D. (1997). *Participatory action research and social change.* Cornell University, USA.

Scoones, I. and Thomson, J. (1994). *Beyond farmers first: Rural people's knowledge, agricultural research and extension practice.* Intermediated technology publication, UK.

Sriskandarajah, N., Packham. R.G. and Fisher, R.J. (1993). "Learning Systems: An alternative route to need assessment, service delivery and enhancement of smallholder welfare", in *Livestock Services for Smallholders - A Critical Evaluation.* (P.W. Daniels, S. Holden, E. Lewin and S. Dadi, eds.), Director General of Livestock Services, Indonesia.

Whyte, W.F. (1991). *Participatory Action Research* Sage, Newbury Park, UK.

Theis, J. and Grady, H.M. (1991). *Participatory Rapid Appraisal For Community Development,* International Institute For Environment and Development (IIED), London, UK.

TOWARDS A NEW AGENDA FOR OPERATIONAL RESEARCH AND ENVIRONMENTAL MANAGEMENT

Gerald Midgley[1] and Martin Reynolds[2]

[1]Centre for Systems Studies, Business School,
University of Hull, Hull, HU6 7RX, UK.
G.R.Midgley@hull.ac.uk
[2]Centre for Complexity and Change, Open University,
Walton Hall, Milton Keynes, MK7 6AA, UK.
M.D.Reynolds@open.ac.uk

INTRODUCTION

In 1999, the Operational Research Society granted charitable funding to a systemic intervention project, based in the Centre for Systems Studies at the University of Hull, designed to create an agenda for the future role of operational research (OR) in environmental planning and management. Our final report has recently been published in the form of a book entitled *Operational Research and Environmental Management: A New Agenda* (Midgley and Reynolds, 2001). This short paper summarises our findings.

CONTEXT

Although OR techniques have been widely used in environmental planning and management, the label 'OR' appears to have a low profile in the discourse about appropriate methodologies and methods (Bloemhof-Ruwaard *et al*, 1995; Daniel et al, 1997). Our own literature search reveals that, for every paper on environmental planning and management that is explicit about using OR methods, there are at least five making claims to methodological innovation that are using the same or similar methods without any reference to OR (Midgley and Reynolds, 2001).

Nevertheless, in our view, the methodological focus of OR is of great value, primarily because it is so broad: it embraces a technical focus (e.g., on the assimilation and control of relevant information); it provides ideas about participation and communication between stakeholders; *and* it can invite reflection on values. Therefore, it is our belief that OR can help make environmental management a broad-based, dynamic, applied practice of central relevance to government, industry and the voluntary sector.

Edited by Ragsdell *et al.*, Kluwer Academic/Plenum Publishers, New York, 2002.

This is why we decided to initiate the research reported here. There is a need for an agenda for the use of OR in environmental planning and management that makes the actual and potential contribution of OR more visible, and which sets out the developments needed in OR if its potential is to be realised.

THE PURPOSES OF THE PROJECT

In October 1999, work started on a one-year systemic intervention project to support OR practitioners in creating just such an agenda. The project had three primary objectives:

1. To make existing good OR practice in environmental planning and management more visible;
2. To explore the further potential of using OR techniques for environmental planning; and
3. To ask how OR would have to be further developed if it is to make an increased and sustained contribution to expert support for environmental management.

Points two and three represent an agenda for development and change. Two action-orientated, subsidiary aims also informed the study:

4. To engender commitment from OR practitioners to the agenda through a process by which they could participate in its generation.
5. To produce a development plan for improving the institutional infrastructure to enhance the ability of interested OR practitioners to undertake the work set out in the agenda.

METHODOLOGY

Critical Systems Thinking (CST), as represented in the work of Midgley (1996, 2000), provided the guiding methodological framework for the intervention. Midgley lists the key CST principles as:

1. *Improvement*—defined temporarily and locally, but in a widely informed manner, taking issues of power (which may affect the definition) into account;
2. *Boundary critique*—regularly questioning and exploring value and boundary judgements, both with respect to the methodological approach adopted and the substantive subject matter being investigated; and
3. *Methodological pluralism*—learning from other methodologies and drawing in methods from those methodologies.

Participants in this study did not only define improvement in terms of environmental protection, but also the more proactive improvement of approaches to environmental and associated social development. Boundary critique proved crucial, as what counts as an environmental issue was a thorny and recurring question addressed in locally meaningful ways throughout the project. Also, participants generated many insights into the ways in which OR methods can either marginalise or empower stakeholders in environmental management projects. Finally, the practice of methodological pluralism enabled us to ensure that our methods remained flexible and responsive to the great variety of situations we faced. The methods we used were drawn from Qualitative Applied Social Science (e.g., Silverman, 2000), Interactive Planning (Ackoff, 1981), Soft Systems Methodology (Checkland, 1981; Checkland and Scholes, 1990) and Critical Systems Heuristics (Ulrich, 1983).

In this article we have chosen not to provide any further details of the enactment of the CST principles. For more information about CST, see Flood and Romm (1996), Jackson (2000) and Midgley (2000).

RESEARCH PROCESS

The study ran through four phases, as follows:

Phase 1: Groundwork and Stakeholder Analysis

A multi-agency steering group was established, and a review of the literature on OR and environmental planning was undertaken. A stakeholder analysis was then initiated, drawing upon both the literature and the views of our steering group. Four stakeholder groups were identified: *professional experts* (operational researchers) associated with environmental planning; and users of professional expertise including agencies of *government*, *business*, and the *voluntary sector* (each operating at local, national, and international levels of planning).

Phase 2: Interviews

Two cycles of semi-structured interviews were undertaken: one with stakeholders identified in phase 1, and a second with significant others suggested by those interviewed in the first cycle. 50 respondents agreed to be interviewed in 46 interview sessions. A sectoral breakdown of agencies reveals 11 government, 13 business, 13 voluntary sector, and 11 academic. Phase 2 culminated in the production of an interim report offering feedback to respondents and providing a stimulus to launch phase 3.

Phase 3: Workshops and Mini-conference

Two one-day workshops (in London and Sheffield) took place. Interested parties were invited to explore how better expert support could be provided, based on the outputs of Phase 2. The workshops were designed to establish ideal 'mission statements' associated with possible future agendas, and to explore the parameters in which such statements might be realised. We used a synergy of Idealised Design (from Ackoff's, 1981, methodology of Interactive Planning) and Critical Systems Heuristics (Ulrich, 1983) to structure the two regional workshops (see our report for full details). These provided source material for a two-day mini-conference in Hull (employing adaptations of some of Checkland's, 1981, methods from Soft Systems Methodology) at which OR practitioners and academics developed a fuller agenda for future collaboration.

Phase 4: Reporting

Three working documents were produced during the course of the study: an interim report, a workshops report and a mini-conference report. These were designed to elicit feedback from interviewees and other participants to support the learning process. Presentations (eliciting feedback) were also made to the Manchester Chamber of Commerce and Industry Environmental Committee; a meeting of the Operational Research Society Environmental Study Group; OR42 (the Operational Research Society Annual Conference); and the Industrial Ecology Conference in Berkeley, USA. The main written output is our final report (Midgley and Reynolds, 2001), and this will be supported by a series of articles for both practitioner and academic audiences.

FINDINGS

The traditions of OR and environmental planning share some common concerns. First, both have wide boundaries in terms of clientele, the range of methodological approaches used, and attention to multiple (and often conflicting) values. Second, both traditions have an interest in fostering purposeful interdisciplinarity. Third, both OR and environmental planning are concerned with the *implementation* of, as well as the *design* of, planning strategies.

Three generic issues were found to recur in both the environmental management literature and the interview data generated in our study:

1. *Complexity and uncertainty* (regarding the unpredictability of natural and social phenomena);
2. *Multiple and often conflicting values* (of those involved in environmental planning); and
3. *Political effects* (on those not involved in planning processes, including non-human nature).

An examination of how these generic issues are perceived in the different sectors (public, business and voluntary) revealed clear patterns, summarised in the following four points:

First, each sector can be shown to have concerns relating to each of the three issue categories (complexity and uncertainty; multiple, conflicting values; and political effects).

Second, for each sector there is considerable conflict between interpretations of how each issue category should be addressed. For example, in dealing with issues of complexity and uncertainty, some businesses are seeking to adopt and promote a 'learning culture', taking heed of wider economic, social and environmental affairs in long-term planning. However, others still reduce the idea of 'sustainability' to short-term economic prosperity.

Third, issues of complexity and uncertainty dominate the public sector, with attention primarily focused on developing appropriate 'indicators'. Competing values are the main concern of business organisations, with attention being paid to minimising risks by improving stakeholder interaction. Political issues dominate the voluntary sector, with concerns about representing marginalised interests and widening the net of meaningful participation in planning processes. These might be termed the *primary* issue categories associated with each sector.

Fourth, for each sector the two *secondary* issue categories tend to cluster around the primary issue category. For example, in the public sector, conflicting values and issues of social exclusion tend to be dealt with *in relation* to the formation of indicators to deal with complex and uncertain realities.

The issues discussed above are generic and therefore arguably quite abstract. Substantive issues like transport, green belt policy, pollution, energy, waste, genetically modified organisms—and even wider concerns relating to sustainable development, global warming, world trade, population growth, the elimination of poverty, etc.—can be more specifically examined using the same parameters. That is, any environmental issue being addressed could potentially involve each of the three user groups (from the public, business and voluntary sectors), as well as some form of 'expert' function. Likewise, any substantive issue might be analysed in terms of all three generic issue types as discussed above.

Clearly, in the increasingly complex, interdisciplinary and politicised world of environmental planning, if we want to enhance expert support using OR, it will be vital to do more than just deal with the technical difficulties associated with modelling the natural world. This is not to say that the technical issues are trivial or unimportant (far from it), but it will also be necessary to address the more messy social worlds of values and ethics in which both OR support and environmental issues are embedded. A major challenge for OR practitioners will be to develop methodologies and methods that are capable of dealing with *all three* of the generic themes identified in this research (complexity and uncertainty, multiple values and political effects).

DEVELOPING THE AGENDA FOR OPERATIONAL RESEARCH

Through the workshops and mini-conference, three distinct (though strongly interrelated) agendas took shape:

1. *Develop OR* (with a focus on methodological issues);
2. *Promote Interaction* (with a focus on issues of interdisciplinarity, intersectoral co-operation, etc.); and
3. *Promote Public Participation* (with a focus on issues of accountability and social inclusion).

Each agenda was subject to a process of analysis and 'conceptual modelling' (Checkland, 1981; Checkland and Scholes, 1990) at the mini-conference. Participants asked themselves, what is the transformation being sought? Who are the intended beneficiaries? Who or what might be made a victim (and should something be done about this)? Who should act to implement the agenda? What worldview underlies the agenda? Who should those implementing the agenda be accountable to? And what environmental constraints will have to be taken as given? The answers to these questions led the group to define key activities needed to realise the stated purposes of the agendas.

For agenda 1 (*Develop OR*), the activities centred on establishing an on-going research project to relate methods with problem situations relevant to environmental management. The need for extensive testing of OR methods in case studies was stressed, as was the need to communicate the results of these tests to enhance the OR knowledge base for environmental management. Importantly, however, the idea of relating methods to problem contexts was not conceived as the production of a mechanical rule book for OR practice. Rather, it was seen as involving the reconceptualisation of OR as a *reflective practice.* Amongst other things, this will involve questioning purposes (rather than taking them as given); focusing on the big picture; multi-sectoral thinking; including multiple agents in defining problems; drawing upon and mixing multiple methods; and embracing environmental issues alongside social ones (rather than taking either environmental or social issues as prime).

In agenda 2 (*Promote Interaction*), the activities centred on developing 'skills', 'knowledge' and 'communication channels'. Whilst interaction was mainly focused on important issues of interdisciplinarity, the agenda was also concerned with promoting intersectoral relationships. People said that OR needed to move from being a primarily 'backroom', problem-solving form of expertise to being a more pro-active discipline where raising awareness of issues amongst stakeholders and problem structuring are key activities. Also, it will require OR practitioners to be more outward looking and facilitative than is currently the norm.

Agenda 3 (*Promote Public Participation*) recognised the difficulties of having a catch-all public participation remit: it is not realistic to try to engage 'the public' in improving OR in general. Rather, the emphasis needs to be on local participation in projects, taking care to differentiate between general public expressions of concern and special interest group involvements.

The three agendas can usefully be regarded as nested systems: agenda 3 nesting in agenda 2, which in turn nests in agenda 1. Therefore, ensuring local public participation in projects is one aspect of keeping OR interactive and outward looking, and should have an impact on how interdisciplinary and intersectoral communications are conducted. Similarly, both of these agendas have important implications for developing the methodology of OR.

RECOMMENDATIONS

The action plans for each of the agendas are arguably the primary outputs of this research, together with the bond formed amongst the participating OR practitioners. However, the group was intent on keeping its feet on the ground, and realised that enthusiasm in a mini-conference will not automatically translate into actual change unless some preparatory actions are taken. These are necessary primarily because there is currently an insufficient critical mass of activists in the OR community with an interest in environmental management to make all the plans a reality. However, concrete steps for changing this situation, with (in our view) a high probability of success, were identified.

These steps were translated in our final report into a series of recommendations to both the Operational Research Society and OR practitioners more generally (see Midgley and Reynolds, 2001, for details).

CONCLUSION

In conclusion, it seems to us that there are two scenarios for the future of engagements between OR and environmental planning and management. First, there is a 'business as usual' scenario in which OR will continue to have a relatively low profile. However, an alternative scenario is for OR to become more purposeful and visible as a discipline: celebrating past achievements and heralding future prospects—though crucially always remaining alert to possible weaknesses, and striving to overcome them. In writing our final report, we unequivocally advocated the second scenario as being in the interests of both OR and environmental planning. It is one that remains true to Ackoff's (1974) plea for OR to do more than merely attempt to *predict* the future. Rather, its practitioners should be engaged in its *design*.

REFERENCES

Ackoff, R. (1974). *Redesigning the Future: A Systems Approach to Societal Problems*, Wiley, Chichester.

Ackoff, R. (1981). *Creating the Corporate Future*, Wiley, Chichester.

Bloemhof-Ruwaard, J.M., VanBeek, P., Hordijk, L., and Van Wassenhove, L.N. (1995). "Interactions between Operational Research and Environmental Management," *European Journal of Operational Research*, 85: 229-243.

Checkland, P. (1981). *Systems Thinking, Systems Practice*, Wiley, Chichester.

Checkland, P. and Scholes, J. (1990). *Soft Systems Methodology in Action*, Wiley, Chichester.

Daniel, S.E., Diakoulaki, D.C. and Pappis, C.P. (1997). "Operations Research and Environment Planning," *European Journal of Operational Research*, 102: 248-263.

Flood, R.L. and Romm, N.R.A. (eds.) (1996). *Critical Systems Thinking: Current Research and Practice*, Plenum, New York.

Jackson, M.C. (2000). *Systems Approaches to Management*, Kluwer/Plenum, New York.

Midgley, G. (1996). "What is this Thing called Critical Systems Thinking?" in *Critical Systems Thinking: Current Research and Practice*, (R.L. Flood and N.R.A. Romm, eds.), Plenum, New York.

Midgley, G. (2000). *Systemic Intervention: Philosophy, Methodology, and Practice*, Kluwer/Plenum, New York.

Midgley, G. and Reynolds, M. (2001). *Operational Research and Environmental Management: A New Agenda*, Operational Research Society, Birmingham.

Silverman, D. (2000). *Doing Qualitative Research: A Practical Handbook*, Sage, London.

Ulrich, W. (1983). *Critical Heuristics of Social Planning: A New Approach to Practical Philosophy*, Haupt, Berne.

RHYTHM OF THE RIVERS: AN ECOSYSTEM APPROACH TO HUMAN HEALTH ON THE AMAZON FRONTIER

Tamsyn P. Murray[1, 2] and Roger Packham[1]

[1] Centre for Systemic Development, University of Western Sydney Hawkesbury Campus, Locked mail Bag 1797, Penrith South DC, NSW 1797, Australia.
[2] Centro Internacional de Agriculture Tropical (CIAT) Eco-Regional Centre, Federico Basadre A.P. 558, Pucallpa Peru.
Email: tamsynmurray@optushome.com.au

INTRODUCTION

Since 1996, a team of Canadian and Peruvian researchers has been developing an ecosystem approach to human health. The ecosystem approach provides an interdisciplinary, holistic guide to how to investigate complex socio-ecological problems, drawing on, and bringing together, a wide variety of methods, actors and scales of investigation (Murray *et al.*, 2001). The development of the approach began with a conceptual framework that brought together the most recent understanding of ecosystems as complex systems with secondary data and exploratory fieldwork. This was followed by its application in understanding and improving a problematic situation in the Peruvian Amazon. Equipped with the ecosystem approach, the Ucayali region of Peru was re-visited in an attempt to provide new insights into the driving forces of human health and to investigate the key determinants of, and linkages between, ecosystem and human health. Ultimately the aim of the project was to synchronize health and land use policies and improve the health of local people through better management of their natural resources.

THE RESEARCH SETTING

The research site comprises the frontier region of Ucayali, the Eastern-most province of the Peru. This area and the surrounding upper rainforests are of great biological significance, containing 23% and 44% of known tropical plant and bird diversity in the Neotropics (IUCN, 1996). Despite the natural diversity and fecundity of the setting, rural communities face a range of nutritional and health problems. Moreover, access to health care is restricted because most medical professionals are based in the urban areas, and inadequate infrastructure and seasonal flooding isolates frontier communities. As a result, the population's nutritional status and health is largely dependent on each family's ability to ensure sufficient nutrient intake through different food production activities, their hygiene practices, and their knowledge of how to treat tropical diseases with wild and domesticated medicinal plants.

In the Amazon lowlands, the annual flooding and extensive inundation of large areas of the floodplain challenges every aspect of human settlement. The hydrological cycle and the resulting 8-15 metre change in river levels dictate migratory patterns of fish and wild animals, agricultural production cycles, and the seasonal availability of forest foods. In addition, the resulting habitat changes affect vector prevalence, disease outbreak and its periodicity (Murray *et al.*, 2000). Understanding how the adaptive livelihood strategies of frontier communities relate to and are synchronized with the constantly changing floodplain provides insight into the determinants of food security, nutritional status, and human health.

Since the early 1970s international and national research centres in Ucayali have focused their efforts on combating deforestation. Based on their analysis of the system, deforestation is the result of declining land productivity and soil degradation that forces farmers to expand into forested areas. Research was directed towards technologies that would increase farm productivity and extend the economic life of the land. Scientists confined their studies to the farm level and included only agricultural researchers. They therefore failed to take into account important socio-cultural and political issues at regional and national levels. Solutions based on such an understanding tended to be technological, emphasizing economic incentives, production strategies and access to improved varieties and/or species. Cattle production with improved pastures was selected as the most efficient system to adopt. After 20 years only a handful of farmers had adopted the technologies. Moreover lack of access to credit meant that few farmers were able to purchase cattle and the resulting over-abundance of *bracchiaria*, the improved pasture grass promoted by the researchers, has now created a serious fire hazard during the dry season.

In 1996 Canadian and Peruvian researchers gathered all existing data on Ucayali and developed a rich and detailed history of the region, including ecological, social and political dimensions at the farm, community, regional and national levels. Based on this initial systemic review it was determined that the Ucayali region presented a very dynamic and complex set of interacting issues and problems, and that research and management practices designed to solve specific economic and environmental problems in a linear fashion, oblivious of context, either failed or created more problems than they solved. Relationships between different sectors, key regional and national constraints and information gaps were all identified. From this it was decided that human health was the key presenting issue.

THE CONCEPTUAL BASIS OF THE ECOSYSTEM APPROACH TO HUMAN HEALTH

There are three guiding principles that form the foundation of the ecosystem approach; (i) methodological pluralism, (ii) multi-level investigation and (iii) local participation and action research.

Methodological Pluralism: Multiple methods, multiple disciplines

No complex socio-ecological system can be captured using a single model or method, as no single perception is able to provide a comprehensive or adequate view of reality (Checkland, 1981; Puccia and Levins, 1985; Funtowicz and Ravetz, 1994). Therefore, research on ecosystems necessitates a variety of forms of inquiry, multiple sources of evidence and dialogue with persons representing different worldviews. Different methods are needed to address different forms of complexity and to answer different kinds of questions (Checkland and Scholes, 1990; Midgley, 2000; Holling, 1995).

The research team assessed which methods would be most useful, appropriate and effective at answering particular questions. For data on health and nutritional status, medical diagnostic tests, household surveys, anthropometry, and precise food recall were complemented with ethnographic and participatory methods exploring local diagnoses and understanding of health. For data on ecosystem dynamics and natural resource use, landscape level geographic information systems (GIS), spatial mapping and soil tests were complemented with in-depth household surveys and community-level participatory methods detailing the livelihood systems of families. In this way we were able to compare the results from the methods designed and driven by researchers with those that were led by the community themselves.

Methods were complementary in their ability to verify results. The sequence of methods used allowed us to compensate for weaknesses in individual methods, and to build knowledge systematically. For example qualitative ethnographic techniques provided insight into local priorities and needs, yet extensive household surveys determined the extent to which these issues were common among all community members and which groups were most at risk. In addition, different methods were used to include different actors. For example, surveys involved the Ministry of Health, while mapping, timelines and other participatory research techniques involved different members of the community.

Multi-level Investigation: Multiple issues, multiple constraints

Ecosystems can be seen to exist within nested hierarchies or holarchies (Allen and Hoekstra, 1992; Checkland and Scholes, 1990). They are comprised of smaller systems while at the same time being part of a larger whole. A household is therefore part of a community, while similarly being made up of different individuals. The different levels (household, community, region and nation) evolve within a variety of ecological and socio-economic contexts and constraints (Allen *et al.*, 1993; Conway, 1987).

Recognizing that often the determinants of individual human health may occur at levels higher within the ecosystem hierarchy, variables at four spatial scales were investigated: the individual, family, community and region/landscape. Issues are linked to and have consequence for others higher and lower within the nested hierarchy. The ecosystem approach focuses investigation on the cross-scale interactions of key variables that explain the complexity and multi-dimensionality of health. For example, landscape spatial mapping determined the extent to which families depended on an area larger than their farm or community for food and income. Data on seasonal flooding levels was correlated with water quality and parasitic infections to investigate links between the hydrological cycle and disease periodicity. Comparing the nutritional status of the family with each individual determined the impact of intra-household food allocation and existence of gender and age inequities.

Local Participation and Action Research: Multiple actors, multiple goals

Sustainable ecosystem management and human health cannot be described by a simple package to be imposed. Ways to improve health, or sustain natural resources are more accurately described as a "process for learning" (Pretty, 1998). As there are always multiple system goals, the trade-offs need to be negotiated among stakeholders. Research to improve these issues therefore requires the legitimate involvement of the local community, be they government agencies, women groups, non-governmental organizations (NGOs), or the community members themselves. Together, professional researchers and local stakeholders define the problems to be examined, generate knowledge, analyze the findings and take action. Participatory action research recognizes its inevitable intervention in social settings. Methods are designed to explicitly investigate a situation in order to change it. Researchers are agents of change, recognizing and acknowledging their role as catalysts in the community (Greenwood and Levin, 1998; Wadsworth, 1998).

The ecosystem approach ultimately intends to provide long-term solutions to health. The ability to sustain introduced changes depends on the community's genuine involvement in and commitment to the intervention strategies. They are both the actors and the owners of the process. Long-term change depends on changes in the governance structures as well as the relationship between the community and the relevant decision-makers. Both these factors affect the adaptive capacity of communities to respond to future problems, when outsiders are no longer present. Issues related to governance are especially pertinent on the frontier where there exists a legal and institutional vacuum. Absence of the state in these remote regions where unregulated common-pool resources are the basis of the community's livelihood, necessitates the building of community-based management structures that can ensure the fair and sustainable use of these natural resources.

The team represented the different actors and organizations in the region and therefore included stakeholders from women's groups, native groups, community groups, research organizations, Ministries of Health, Agriculture and Fisheries and the local university. The involvement of these key actors ensured that efforts maintained their relevancy to local needs and ultimately that the information gathered was owned and used by local organizations. The team was responsible for identifying the different issues relevant to human health and how they were related to one another. They developed the selection criteria for the study communities, decided on the timing of the three field visits and identified ways of involving community members in the project.

Data and Process Approaches

When reliable information on morbidity and mortality failed to be captured by the individual surveys, the use of focus groups allowed for much greater engagement of the participants. The recognition that others suffered similar illnesses was a catalyst that prompted further exploration of treatments and prevention measures. From these discussions women asked if they could have more meetings where specific illnesses that they identified, could be discussed in more detail. They were eager to analyze their problems. This example highlights the distinction between a data approach and a process approach (Gijt and Berdegue, 1999). At the outset, the researchers sought data, however, in reality situation improvement demanded a process through which understanding could be generated. What was lacking, besides exact figures on morbidity, was the creation of a space or process through which the women could better understand their problems and develop tools to do something about it.

The experience with parasite testing similarly reinforces the importance of a process approach. Despite the fact that the analysis of stool samples possesses the qualities of hard, data-oriented science, when implemented in the field, its ability to initiate a process more powerful than the actual information collected was unexpected. During the field trips, the parasitologist set up his table and microscope outside the schools in each community. As the mothers arrived with their children's samples he analyzed them in their presence. Once he found a parasite with the microscope, the mother and in many cases father was able to see them for themselves. Observing the parasites first hand had a dramatic effect. The importance of purifying water became real, as they could understand why poor water quality and hygiene were detrimental to their children's health. Subsequently the team was inundated with a series of questions regarding water treatment and other means by which they could reduce the risks of parasitic infection. In response to this interest the team developed large colourful posters of the main parasites, illustrating their cycles, sources and simple ways to avoid infection. The mothers asked to have more meetings where they could discuss how to improve basic hygiene and reduce contamination.

APPLICATION OF THE ECOSYSTEM APPROACH: SEASONALITY AND HEALTH

One of the key findings that emerged with the application of the ecosystem approach was the inextricable link between seasonality and health. This critical issue was overlooked by previous research that used survey instruments delivered only once and often during the dry season when communities are most accessible. However, the involvement of local stakeholders representing many different groups and areas in the region, made it clear that the impact of flooding and the accompanying environmental changes was the most critical variable in understanding the stresses faced by all rural communities in Ucayali. Based on these suggestions three time periods were identified which best reflected the changes that occur over the year. Field visits were therefore timed with the height of the dry season (June-August), height of the wet season (February-April) and the beginning of the rains (October-December). To ensure that the region's heterogeneity was reflected in the study, eight *colonist* and *native* communities were selected in both the floodplain and upland forests. In addition to differences in ethnicity and ecosystem type, communities varied in their degree of access to and involvement in the market economy, the time of settlement, and dominant land use strategy (slash and burn agriculture, fishing, cattle ranching, and oil palm plantations). In total 345 families participated. The action research team then selected an array of different methods to capture the dynamic nature of the constantly changing floodplain and its effects on human health. The following variables were chosen to reflect the interaction of human health and environmental change.

Table 1: Variables Linking Human Health and Ecosystem Dynamics

Health	Natural Resource Use	Environment
Nutritional status	Income levels and diversity	Water quality
Diarrhoeal and respiratory infections	Production levels and diversity	Flooding
Iron status	Diversity of food sources	Hygiene
Parasitic infections	Labour demands	
Infant and maternal mortality	Migration	

Traditional measurement methods such as surveys, medical tests and food recall, were complemented with more participatory techniques, such as mapping, timelines, seasonal calendars, ethnographic sorting, and focus groups. The ultimate goal for the research team and each community was to determine those critical periods when poor health, scarce food resources, and declining water quality and environmental health combine to create an especially desperate situation when intervention is most necessary. This would provide communities as well as the Ministry of Health with the ability to target their programs to those most in need when they were most in need.

In this paper, findings are confined to some of those collected by the community, as the survey data is currently being processed. Figure 1 demonstrates the different cycles of disease, food availability and income level as perceived by members of the five floodplain communities. The numbers indicate relative values; one is low, three is high and two is medium. For example in the floodplain, disease (in this case diarrhoeal infections) is most prevalent in the dry months (June-August) when the water quality is at its worst. In the uplands, there are similar problems in the summer, though they also face malarial outbreaks in the wet season. In the floodplain, fish provide the most important source of protein rich foods during the summer months. However from January through to April, little food is available as the flooded lands inhibit farming, hunting and gathering and fish disperse into the flooded forests. This situation is exacerbated by the rising incidence of respiratory infections that result from the flooded conditions. In the uplands, food shortages occur during the rainy season when fish prices are high and the agricultural harvest is still months away. Lastly, income in the floodplain is linked first to the sale of timber, usually sold in April once it has all been harvested and carried along the flooded rivers, and second, to the sale of the agricultural harvest in September. In contrast, in the uplands, income is highest in January and February when crops reach maturity and are sold in the Pucallpa market. These communities have few alternative sources of income.

In summary, the critical time period for floodplain communities is during the height of the wet season. At this time, respiratory infections are prevalent, fish and wild forest foods are scarce, and agricultural land is flooded. However, in the upland forests, the incidence of parasitic infections and diarrhoeal episodes is greatest during the dry season due largely to the poor water quality as wells and other sources of water for drinking and bathing dry up. In addition, nutrient intake is low as the harvest is months away.

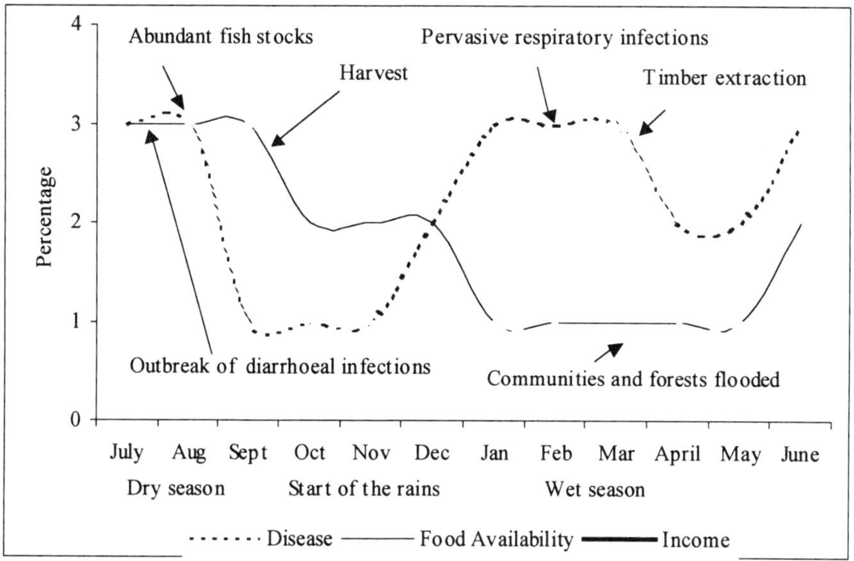

Figure 1: Seasonal Cycles of Disease, Food Availability and Income in the Ucayali Floodplain (adapted from Murray *et al.*, 2001)

CONCLUSION

This research, using the ecosystem approach with its three guiding principles, has demonstrated that firstly, human health can only be understood in light of the constantly changing ecology of the floodplain and upland terraces. Such ecosystem dynamics dictate the patterns of resource use as well as nutrition, food security and disease. To promote more sustainable resource use that is better able to meet the nutrient and health requirements of the population, interventions should therefore focus on ways to take advantage of the annual floods rather than being drowned economically by them. Recognizing that there are annual and predictable health risks and challenges to food security, governments can target those critical periods when communities are most in need. The Ministry of Health can intensify their food aid efforts, provide accessible water purification measures and increase the vigilance of local health workers during these identified times, and the Ministry of Agriculture can work with farmers selecting crops that combat food insecurity as well as incorporating methods of storage for the food scarce periods. Moreover differences between the cycles of diseases and food insecurity in the floodplain and the uplands demand the adaptation of government policies and programs to these different ecosystems.

Secondly, human health can only be understood at the landscape level. As food security and nutrition is determined primarily by the family's ability to migrate across the region, producing and gathering food from a diversity of resources, understanding when and why there is low nutrient intake, must take into account landscape level variables. Research and development activities confined to the farm or even the community, fail to capture the complete picture of the livelihood strategies in Ucayali and the link between activities at different scales. In addition, the dependence of people on a diverse range of ecological resources means that natural resources management strategies should be designed and applied at a landscape level. Community-based management may offer a legal and institutional framework through which an area, absent of government and private land ownership, can find a mechanism to monitor and control resource use. The diversity of resource users and their great differences in power, demand a means of addressing conflicts over common-pool resources and the need for collaborative arrangements. This will be one of the greatest challenges facing researchers, governments and residents of the Amazon.

REFERENCES

Allen, T.F.H. and Hoekstra, T.W. (1992). *Toward a Unified Ecology*. Columbia University Press, New York.

Allen, T.F., King, A.W., Milne, B., Johnson, A. and Turner, S. (1993). "The problem of scaling", *Evolutionary Trends in Plants,* 7(1):3-8.

Checkland, P. (1981). *Systems Thinking, Systems Practice,* Wiley, Chichester.

Checkland, P. and Scholes, J. (1990). *Soft Systems Methodology in Action,* Wiley, Chichester.

Conway, G.R. (1987). "The properties of agroecosystems", *Agricultural Systems,* 24: 95-117.

Funtowicz, S. and Ravetz, J. (1994). "Emergent complex systems", *Futures,* 26:568-582.

Gijt, I. and Berdegue, J. (1999). Personal communication.

Greenwood, D. and Levin, M. (1998). *Introduction to Action Research: Social Research for Social Change,* Sage Publications, Thousand Oaks.

Holling, C.S. (1995). "What barriers? What bridges?", In *Barriers and bridges to the renewal of ecosystems and institutions,* (L.H. Gunderson, C.S. Holling and S. Light, eds), Columbia University Press, New York.

IUCN (1996). *The Conservation Atlas of Tropical Forests: The Americas*. C.S. Harcourt & J.A. Sayer (eds) Simon and Schuster, New York, 354pp.

Midgley, G. (2000). *Systemic Intervention: Philosophy, Methodology and Practice.* Kluwer Academic/ Plenum Publishers, New York.

Murray, T., Sanchez, J. and Waltner-Toews, D. (2000). "An Ecosystem Approach to Human Health on the Amazon Frontier: Methodological Insights", *Cadernos de Saude Publica* Public Health Reports, FioCruz, Rio de Janeiro, Brazil. 17(1) Supplement:180-193.

Murray, T., Kay, J., Waltner-Toews, D., and Raez-Luna, E. (2001). "Linking Human and Ecosystem Health on the Amazon Frontier: An Adaptive Ecosystem Approach", in *Conservation Medicine.* (G. Tabor, M. Pearl, M. Reed, R. Ostfeld, A. Aguirre, J. Patz, and C. House, eds), Oxford University Press, Oxford (in press).

Pretty, J.N. (1998). "Supporting policies and practice for scaling up sustainable agriculture", in *Facilitating Sustainable Agriculture,* (N.G. Roling, and M.A. Wagemakers, eds.), Cambridge University Press, Cambridge.

Puccia, J.C. and Levins, R. (1985). *Qualitative Modeling of Complex Systems*. Harvard University Press, Cambridge, Massachusetts.

Wadsworth, Y. (1998). "What is Participatory Action Research*?", Action Research International*, Paper 2. http://www.scu.edu.au/schools/sawd/ari/ariwadsworth.html.

Section Five

Critical Systems Thinking

TURNAROUND SYSTEMS ALIGNMENT: A METHODOLOGY FOR DEALING WITH COMPLEX-COERCIVE PROBLEMS

Tony Groom

Blythe Farm
Mill Street
Gamlingay
Cambridgeshire SG19 3JW

INTRODUCTION

This paper considers those situations where an organisation is facing a financial crisis as a result of causal factors rooted in its people systems. The causes might be structural (command and control), cultural, social, or political but fundamentally they involve people and how they relate to the organisation and how they behave towards each other. This area of research is subject to considerable argument about the ethics and morality, especially as the objectives of any intervention are normally financially motivated. In recognition of these dilemmas, this paper looks at a continuum of influence, 'empowerment' at one end and 'coercive persuasion' at the other with the objective of finding an acceptable model for introducing rapid change to avoid bankruptcy.

The background to the paper is the lack of consensus among turnaround specialists as to how to achieve transformational change when trying to save a company from bankruptcy. The frequent use of downsizing in a financial crisis flies in the face of research by Lämsä (1999) that found this approach to be flawed as it has such a negative impact on those who remain.

This paper considers how an approach to transforming might fit within a theoretical framework based on Total Systems Intervention (Flood and Jackson, 1991). After considering its relevance, the paper looks at systems based intervention as a theoretical framework. Cultural bankruptcy is developed as a metaphor to describe organisations that are likely to become financially bankrupt as a result of their socio-behavioural-political systems. A 'system of systems methodologies' framework (Jackson and Keys, 1984) is used to locate cultural bankruptcy as a complex-coercive problem-situation. Approaches to similar situations are discussed before a new approach, Turnaround Systems Alignment (TSA) is proposed as offering a suitable methodology.

Edited by Ragsdell *et al.*, Kluwer Academic/Plenum Publishers, New York, 2002.

The conclusions are used to review the concept of cultural bankruptcy and consider the limitations of TSA.

Relevance

This paper is relevant as it looks at organisations that have failed to adapt. The consequence of this failure to change often manifests itself in a financial crisis. It is therefore of interest to all employees, managers, investors, lenders and other stakeholders who need a suitable framework for addressing the problem. Although identifying symptoms of the problem is often the result of a review prompted by a financial crisis, there are few models for dealing with the causal factors.

One of the benefits of developing the cultural-bankruptcy metaphor is that it can help predict whether or not a company is heading for financial bankruptcy. This is particularly important as early identification of the symptoms can help avoid the drastic measures of a TSA approach by using other systems-based intervention methodologies.

Having considered relevance, it is necessary to look at the proposed theoretical framework for implementing change.

TOTAL SYSTEMS INTERVENTION

The theoretical framework behind this paper is based on Total Systems Intervention (TSI). TSI was developed as an alternative view to mechanistic thinking (Flood and Jackson, 1991) and has been documented by Checkland (1999) in his thirty year retrospective.

The concept of TSI was introduced as a meta-methodology by Flood and Jackson (1991). Much later this was used to develop a hierarchy of interdependent systems tools that might be used during an intervention (Ragsdell, 2000). The building blocks for TSI are metaphors which were advocated by Morgan (1986) as a way of viewing and understanding situations. Flood and Jackson (1991) used five dominant metaphors as their starting point when looking at situations within the TSI framework.

Metaphors

The five dominant metaphors of Flood and Jackson (1991):

1. *Machine metaphor* - a closed system view where routine and bureaucracy provide a stable but constrained environment for a tight knit team.
2. *Organic metaphor* - an open system view of a complex but homeostatic internal environment that adapts to its external environment.
3. *Neuro-cybernetic (brain) metaphor* - a viable system view where learning, communication and control dominate, and the focus is on information processing and viability.
4. *Culture metaphor* - a social system view of familiar ways of doing things based on accepted behaviour that is established over time.
5. *Political metaphor* - a conflict system view where relationships between individuals and groups are competitive and based on the pursuit of power.

These dominant metaphors are not mutually exclusive as several might be used to describe aspects of the same system or problem. These and other metaphors (*psychic prisons, flux and transformation,* and *instruments of domination,*) are discussed in detail by Morgan (1986).

System of Systems Methodologies

Jackson and Keys (1984) used a grouping of problem contexts that classified situations in a two dimensional grid to match problem complexity with participant relationships. Problems were defined as being either simple or complex, where simple systems were predictable in that outputs could be pre-determined and were governed by well-defined laws. Complex systems on the other hand were generally socially constructed and involved interactions where the outcomes could not be accurately predicted.

The participants dimension was taken from an industrial classification that defines three categories of relationships: unitary, pluralist, and coercive. Unitary relationships are those where participants (individuals or groups) share values, interests and purpose, and where everyone takes part in decision-making. Pluralist relationships involve everyone in the process but agreement among participants is only reached through compromise and accommodation. Coercive relationships, on the other hand, involve conflict and disagreement where acceptance, submission or compliance is achieved through a degree of coercion. The two dimensions form a six-celled matrix of problem-situations.

Flood and Jackson (1991) used Jackson and Keys' (1984) matrix to develop a 'system of systems methodologies' that classifies systems-based intervention tools by problem-situation.

The basis of this paper was the absence of an appropriate intervention tool for dealing with complex-coercive situations. The lack of a suitable tool may have been related to Flood and Jackson's (1991) metaphor for such a situation where they superimposed five dominant metaphors onto the grouping of problem contexts. This identified the 'prison' metaphor with both simple and complex situations.

It was felt that if a suitable metaphor could be defined, then this would help identify an appropriate systems-based intervention methodology for dealing with complex-coercive situations.

CULTURAL BANKRUPTCY

Cultural bankruptcy - is proposed as a metaphor for describing organisations that are facing financial bankruptcy as a result of their socio-behavioural-political systems. It is inferred that an organisation's cultural systems can precipitate financial problems.

Cultural bankruptcy may be the result of an organisation failing to adapt or the existence of fundamental flaws in any change that is adopted. Failure to adapt tends to be linked to a 'resistance to change' that implies obstruction although all too often it is linked to poor management (Eccles, 1994). The metaphor is emotive as it seeks to focus attention on the need for change to the socio-behavioural-political systems of the subject organisation.

The inference of this definition is that the economic performance of the organisation can be turned round if appropriate change is introduced. The problem is that "such change [of corporate cultures] is complex, takes time, and requires leadership" (Kotter and Heskett, 1992, p12). However, it should be noted that turnarounds are characterised by a lack of time and money (Bibeault, 1999). The implications of this are discussed later in this paper.

Flood and Jackson (1991) used the *political* metaphor to describe complex-coercive problem-situations since they involve relative power. *Cultural bankruptcy* is presented as a more appropriate metaphor since it uses a broader set of components to describe a problem-situation that relates to people.

COMPLEX-COERCIVE PROBLEM SOLVING

Before considering whether or not TSA is a suitable intervention model, it is valuable to look at two very different approaches to implementing change.

Empowerment

Empowering employees has become increasingly popular as a way of involving them in the management of change. While it can be based on an action learning approach (Argyris, 1985), it also conforms to a systems-based philosophy for intervention (Flood and Jackson 1991). Empowerment implies self-determination by employees that requires a level of trust along with devolution of power by management. However, in times of crisis, managers tend to reign in authority and override decisions based on empowerment that may have taken place (Eccles, 1994).

Soft system methodology (SSM) (Checkland, 1981) is an example of a TSI model that is based on empowerment. This was used by Ragsdell (2000) to "engineer[ing] a paradigm shift" where participants were encouraged to get involved by drawing rich pictures to define their current situation. Ragsdell then used metaphors as a development tool to describe the ideal (desired) environment. By linking current with ideal situations, Ragsdell got the participants to take ownership of the actions needed to deliver the paradigm shift.

The TSI framework classifies SSM as a tool for dealing with complex-pluralist problems rather than complex-coercive ones. Furthermore, it is doubted whether any empowerment based tool is suitable when Eccles (1994, p3) argued that it was "of little value in times of rapid, radical, frame-breaking change; it [empowerment] works better in more stable circumstances".

Coercive Persuasion

At the other extreme, Schein (1999) looked at coercion as a way of changing behaviour. Schein developed the idea of a continuum that was based on influence where learning was individually (self) driven while coercive persuasion was organisationally (group) driven.

Schein cites an extreme example of coercion by Communist China who attempted to reframe civilian prisoners during the Korean conflict of the 1950s. Prisoners were subjected to an enforced regime of physical, social and psychological influences as part of a process of disorientation, confession and then learning as part of a process of reframing their world. The confession was a means of getting prisoners to face up to the situation, whether 'perceived' or 'real'. Once the problem was acknowledged by prisoners, they could begin the process of cognitive redefinition. The tactics used to reframe prisoners' viewpoint, although extraordinarily effective they should be considered deplorable and certainly do not conform to a TSI philosophy.

Schein (1999) concluded that the position occupied by an individual on the learning-coercion continuum was determined by their perceived level of freedom to leave the situation. He made the distinction of cognitive redefinition as being either the result of self-driven learning or coercion. The implications of this are considered later as part of a discussion about whether a complex-coercive methodology is compatible with a TSI philosophy.

TURNAROUND SYSTEMS ALIGNMENT

Having argued that cultural bankruptcy is a complex-coercive problem, it is appropriate to consider a suitable methodology for dealing with such situations that also conforms to a TSI philosophy. If this can be demonstrated then such a tool should be incorporated in the grouping of systems methodologies.

Turnaround Systems Alignment (TSA) is proposed as a tool that is specifically aimed at aligning the interests of people within an organisation when the situation is classified as complex-coercive. This involves confronting the political, social and behavioural issues that may be preventing such alignment. As an intervention process, it is based on a cyclical process of improving the situation by quantum leaps. It is similar to action research (Argyris, *et al.*, 1985) in that it incorporates research and action but is differentiated by not being an iterative process since change takes place in quantum leaps.

The process involves a simultaneous top-down communication process and bottom-up research process. The top-down component sets out clearly what is expected while the bottom-up component is looking to identify misalignment within the various systems and people of the business.

The initial top-down component is likely to communicate the purpose of the organisation through statements of intent such as *vision, mission, objectives, and/or values*. At the same time, the initial bottom-up research engages people at different levels and seeks to identify the current situation from a wide variety of perspectives within the organisation. The approach is suitably flexible to allow a variety of research methods to be used, including but not limited to interviews, focus groups and surveys.

Top-down Communication Process

Vision, mission, objectives, and values – Establish purpose by setting statements of intent for the business as part of a coherent strategy.

Bottom-up research process – considers the business as a hierarchical set of systems and sub-systems where the main systems are:

Marketing & operations
People, culture & organisation
Systems & resources

Having gathered initial information, the areas and degree of misalignment between individuals, groups and systems are identified. This is then used to develop the next stage in the cycle where top management sets out their ideal (desired) situation by way of objectives to align the broad system sets with the original statements of purpose.

These objectives are then communicated by a top-down process to set expectations that changes behaviour in pursuit of the ideal situation. At the same time further research is conducted to look at the reaction from individuals, groups and systems as sub-systems and identify misaligned.

It can be seen that each cycle of the process refines the level of detail that is investigated while at the same time identifying misalignment. It can also be seen that the process involves people however it is not about empowerment. Instead it is aimed at identifying those groups and individuals who are resisting change. Where resistance is identified, these people are isolated and if necessary removed. This later sanction may be regarded as coercive although it only relates to divisive individuals.

The Process is Based on a Number of Turnaround Principles

- *Set a clear statement of purpose* - vision, objectives, values and expectations
- *Initiate change* - it is imperative for survival
- *Cash is king* - it buys time to implement change
- *Communicate* - internally and externally with all stakeholders
- *Lead by example* – respect others, take responsibility and act with integrity
- *Act decisively* – act swiftly and accept mistakes will be made.

The TSA process based on the Turnaround Principles is presented as a systems-based intervention tool because it represents a structured approach that does not pre-determine the outcome. It starts with the premise that change is necessary and that everyone is involved. Sine people are involved in the process, they retain an element of free-will that does not locate the level of influence at the coercive end of Schein's continuum. However the proposed system does allow for the removal of those who resist change which act might be regarded as coercive.

In dealing with *cultural bankruptcy*, TSA may be viewed as a "quick fix" tool aimed at survival. A concern is the lack of long-term commitment when rapid management change is introduced, however a change in culture is often needed if a business is to learn to adapt in the future. Once survival is assured, alternative TSI methodologies may be more appropriate to take over from TSA.

Is TSA a Suitable Model for TSI?

There are considerable grounds for arguing that TSA does not conform to "critical systems thinking" as the philosophy that underpins TSI. None the least of these is TSA's proposition that coercive behaviour is sometimes necessary to deal with divisive influence of employees. Flood and Jackson (1991) use three criteria that must be met, if an approach is to conform with critical systems thinking: "complementarism", "sociological awareness" and the promotion of "human well-being and emancipation". TSA is tested for suitability against these criteria.

The "complementarism" criteria requires TSA to be applicable and specific for addressing the problem-situations for which it is most suitable. These are defined by the *cultural bankruptcy* metaphor. As a methodology TSA is not intended to be purely pragmatist or isolationist, although it does have many elements of both. For those who would argue that TSA is a pragmatic approach, it offers a tool-kit rather than a pick and mix strategy; however others might see it as being isolationist where it supports some coercive behaviour that could also be regarded as imperialist. This apparent contradiction is needed to deal with complexity where TSA integrates both hard and soft approaches to intervention.

While the soft elements conform to a philosophy of social awareness within critical systems thinking, the application of the hard components does not. However it is argued that the approach to dealing with divisive behaviour by isolating those involved and where necessary removing them is not about the application. Instead it is a method of stabilising the social group of those who stay. The behaviour of management may be seen as coercive (hard) but the process is normative as it seeks to align the agendas of people in an organisation. While execution of any decision is generally made by managers, they are influenced by the information gathering process through the bottom-up research process.

CONCLUSIONS

The background to the paper was the lack of consensus among turnaround specialists as to how to deal with crisis-situations. There has been very little research into managing organisational change as part of a corporate turnaround. Most of the changes introduced by turnaround practitioners are based on personal experience rather than theory. The limited research carried out has adopted a grounded theoretical approach to producing generic strategies. In spite of these, downsizing in a financial crisis remains the preferred approach and unfortunately those who remain generally suffer considerably from the experience (Lämsä, 1999).

This paper set out to look at corporate turnarounds within a critical systems thinking philosophy by considering them as coercive-complex problem-situations. A metaphor, *cultural bankruptcy* was defined as being specific to organisations where the people systems were flawed such that if they were not addressed the result would be financial bankruptcy. A methodology, *Turnaround Systems Alignment* was then proposed as being one that was both suitable for dealing with such situations and one that conformed with the TSI philosophy.

Turnaround Systems Alignment (TSA) can be summarised as an intervention model that simultaneously operates top-down communication by management to clarify roles and set expectations while at the same time conducting bottom-up research to find out about how people view their role and expectations. The model specifically relates to organisations facing *cultural bankruptcy* where it relies on six fundamental Turnaround Principles.

Limitations of this Paper

The proposed *cultural bankruptcy* metaphor and TSA methodology were aimed at filling a gap in the Total Systems Intervention (TSI) framework (Flood and Jackson, 1991). TSI had identified the need for a suitable methodology to deal with coercive-complex problems. While the limitations of TSA have been discussed, it can be argued that it does not conform to aspects of the TSI philosophy. Critical systems thinking as the philosophical foundation for TSI, requires the promotion of "human well-being and emancipation". While TSA incorporates an open and honest approach to people and encourages their involvement in much of the decision-making, it ultimately employs an authoritarian approach to power and control. Where necessary reorganisation may have to be imposed on a business to create a more mechanistic structure as a temporary measure adopted in the early stages of a turnaround. While leadership is one of the Turnaround Principles that supports TSA, such a mechanistic approach however temporary, might not be considered emancipatory. In spite of this, a successful turnaround should be regarded as providing 'the greatest good for the greatest number', where coercion as a tool available within TSA, that may be used when appropriate.

REFERENCES

Argyris, C., Putnam, R. and McLain Smith, D. (1985). *Action Science: Concepts, Methods and Skills for Research and Intervention.* Jossey-Bass, San Francisco.

Bibeault, D.B. (1999). *Corporate Turnaround: How Managers Turn Losers Into Winners!.* Beard Books, Washington.

Checkland, P.B. (1981). *Systems Thinking, Systems Practice.* Wiley, Chichester.

Checkland, P.B. (1999). *Systems Thinking, Systems Practice - SSM: A 30-Year Retrospective.* Wiley, Chichester.

Eccles, T. (1994). *Succeeding with change: Implementing action-driven strategies.* McGraw-Hill, Maidenhead, England.

Flood, R.L. and Jackson, M.C. (1991). *Creative Problem Solving: Total Systems Intervention.* Wiley, Chichester.

Jackson, M.C. and Keys, P. (1984). "Towards a system of systems methodologies". *Journal of Operational Research Society.* 35:473-86.

Kotter, J.P. and Heskett, J.L. (1992). *Corporate Culture and Performance.* Free Press, New York.

Lämsä, A. (1999). "Organizational downsizing - an ethical versus managerial viewpoint". *Leadership & Organization Development Journal.* 20(7):345-352.

Morgan, G. (1986). *Images of Organization.* Sage, Beverley Hills.

Ragsdell, G. (2000). "Engineering a paradigm shift? An holistic approach to organisational change management". *Journal of Organizational Change Management.* 13(2):104-120.

Schein, E.H. (1999). "Empowerment, coercive persuasion and organizational learning: do they connect?" *The Learning Organization.* 6(4):163-172.

Slatter, S. and Lovett, D. (1999). *Corporate Turnaround: Managing Companies in Distress.* Penguin Books, London.

ON AN APPLICATION OF CRITICAL SYSTEMS HEURISTICS IN A FRAMEWORK FOR EVALUATION OF SOFTWARE DEVELOPMENT PRODUCTIVITY

O Petkova[1] and D Petkov[2],

[1]Central Connecticut State University, petkovao@ccsu.edu
[2]University of Natal, petkov@nu.ac.za

INTRODUCTION

The multiple dimensions of the factors affecting software development productivity represent a challenge for an interdisciplinary research to their nature. Because of the significant role played by people, software engineering is already one of the computer science disciplines that is closest to the social sciences (Kemerer, 1998:66). Lederer and Prasad (1998:138) report that past published research in that area includes experiments on cost estimation using algorithmic techniques, without formal techniques but only using expert judgment, using analogy or case based reasoning, expert systems, simulation and in a few cases action research and case studies. The majority of these approaches are of a technical nature. Pfleeger (2001) concludes that the statistics for most models are disappointing, indicating that no model appears to have captured the essential characteristics and their relationships for all types of development. Similar concerns about the accuracy of normative models for software development effort estimation have been expressed by Subramanian and Breslawski (1993) and others.

One of the starting points for this paper is the understanding that improvements in software development productivity may follow only after the development of a better understanding of the relationships between the factors affecting it within a given project. Hence the need for a holistic modelling approach to the complex relationships between those factors that are relevant for a particular MIS environment. The goal of this paper is to provide a discussion of the role of Critical Systems Heuristics in a larger framework for evaluation of the factors affecting software development productivity. It is assumed that it is this framework that is transferable as a process from the conditions of one organisation to another, and not the actual form and strength of the relationships between the factors concerned. The latter will inevitably vary from one environment to another as a result of the unique features of any social system. The results of this work are aimed at gaining a better insight into the importance of particular factors affecting software development productivity. As a result management can concentrate on them. Thus, implicitly preconditions are created for the improvement of software development.

It can be stated that the underlying philosophical assumptions of this research are based on critical systems thinking in the sense that the purpose of the evaluation of factors affecting software development productivity is seen to be the creation of the preconditions for improvement of productivity. Following Midgley (1996) and Midgley (2000), the notion of improvement plays an important role as an emancipatory idea in critical systems thinking. In line with Jackson (1991), "emancipatory" here is seen as synonymous with releasing the full potential of the abilities of software developers. In order to make this idea operational, it includes not only a mixture of elements from Soft Systems Methodology (see Checkland and Holwell, 1998) and Multicriteria Decision Analysis (MCDA), but also elements of the recently simplified form of Critical Systems Heuristics (Ulrich, 1998). The latter component is in the form of a checklist of questions adapted to the problem area on the basis of Ulrich's critically-heuristic boundary questions for systematic boundary critique. The treatment of software development productivity within Critical Systems Thinking has not been reported before according to the best knowledge of the authors, hence this paper's potential contribution to the generally scarce literature on Critical Systems Thinking applied to Information Systems as a whole.

CRITICAL SOCIAL THEORY AND CRITICAL SYSTEMS HEURISTICS AND THEIR RELEVANCE TO THE PROBLEM OF EVALUATION OF SOFTWARE DEVELOPMENT PRODUCTIVITY FACTORS

Ngwenyama (1991) summarises the main assumptions of Critical Social Theory as: People have the power to change the world; Knowledge of the social world is value laden; Reason and critique are inseparable; Theory and practice must be interconnected; Reason and critique must be reflective in practice.

According to Habermas (1984), communicative action is the original mode of language use, and is morally privileged. The emphasis of this type of action is creating shared meanings through sense making. This is very close to some theories about the essence of Information Systems. Acknowledging the role of Critical Social Theory, it must be noted, however, that the concern here is predominantly about the strand of the contemporary systems movement called Critical Systems Thinking (CST). There have been two strands in critical systems thinking. The first one was presented at first in Ulrich (1983) and is called Critical Systems Heuristics (CSH). The second one is called by Jackson (1997) Critical Systems Thinking UK type, and its practical version is known as Total Systems Intervention (TSI), (Flood and Jackson, 1991). For reasons of space, the following discussion will concentrate on the first one.

The account on CSH, presented here, will be based on Ulrich (1996) and Ulrich (1998). The aim of CSH, which draws on ideas by Churchman, is to give people a voice in matters that are important to them. That is based on a definition of social planning as the art of promoting improvement. The latter is associated with both the knowledge and ethics of planning. Improvement rarely means the same thing for everyone concerned. Promoting it inescapably implies preference; that is to say, it implies choice between the needs and values of different groups of people. An important notion in Ulrich's theory is boundary judgments. These are the facts and values to be considered that define the boundaries of the planning effort. Another one is justification break-offs, which indicate what is to be left out because they define the point at which justification ends. In terms of systems thinking these mean what should be considered as part of the systems and what should not. It has to be noted that there is an interdependence of facts and values.

According to Ulrich (1996:17), boundary judgments are a core concept of systems thinking, the fundamental critical kernel of systems thinking. Another aspect that is considered important is the problem of holism: a sufficient justification of improvement would require us to consider the whole system of all the facts and values that could lead to alternative conceivable options for understanding improvement. Critical Heuristics uses a conceptual framework that consists, among other things, of a set of twelve basic types of boundary judgments which can be put together as a checklist of twelve boundary questions (Ulrich, 1996; 1998). They are grouped in four categories by Ulrich (1998): the plan's sources of motivation, the plan's sources of control, the plan's sources of knowledge, the plan's sources of legitimisation.

Each question can be answered from different viewpoints, not only from the standpoint of those involved, but also from the perspective of those concerned and potentially affected. That is the basis of another key concept of critical heuristics, that is the process of unfolding the normative and empirical selectivity of plans (Ulrich, 1996:35).

In terms of practical application, CSH is considered by some as over-theoretical and lacking in usefulness with respect to "real-world" problem situations (Ellis, 1995:211). This has probably been taken into account by Ulrich, who embarked in 1996 on a new project to develop a new CST approach that is accessible not only to well trained decision makers, but to the majority of citizens (Ulrich, 1998). Flood and Jackson (1991) have allocated a relatively narrow role to CSH as an emancipatory method for addressing "coercive" situations, without recognising its role as a design - oriented approach to planning tasks in general as is the claim in Ulrich, (1996:35). It seems that the two critical elements: the creative element of discovery or unfolding issues and ideas (heuristics), and providing solutions to the problem ("critique"), indicate an advantage of CSH when compared with other approaches, e.g. Soft Systems Methodology.

In spite of the effort of Jackson (1992), there have been relatively few subsequent publications on the practical application of Critical Systems Thinking in Information Systems Research. Most of them are surveyed by Ngwenyama and Lee (1997). However, there is no evidence for the operationalization of Multimethodology (Mingers and Gill, 1997) to practical problems in the field of Information Systems, and this is another justification of this research. On the basis of the preceding discussion, it may be concluded that the results from the evolution of a pluralist agenda within Systems Thinking (see Jackson, 2000) seem to be encouraging for the formulation of a pluralist approach for the evaluation of factors affecting software development productivity.

A PLURALIST MULTIMETHODOLOGY FRAMEWORK FOR EVALUATION OF THE FACTORS AFFECTING SOFTWARE DEVELOPMENT PRODUCTIVITY AND HOW IT WAS APPLIED

The framework for evaluation of factors affecting software development productivity is summarized in Petkova and Roode (1999) and hence will not be discussed here in detail. The framework deals with systemic issues affecting software development productivity and their prioritisation. It involves three stages forming a cycle: *Critique* (it uses rich pictures, brainstorming, CATWOE analysis, stakeholder analysis), *Judgment* (includes MCDA, boundary judgment questions) and *Action* (includes MCDA, boundary judgment questions). The cycle idea is based on (Midgley, 1997), while the mix of the techniques involved has not been reported before. The findings of the brainstorming sessions and stakeholder identification, as well as the rich pictures and CATWOE analysis, help in the problem formulation step of Multicriteria Decision Analysis. It has to be pointed that MCDA is used here as a descriptive approach to decision making along the interpretivist paradigm. Finally the boundary judgments questions of CSH introduce the necessary elements of the Neo-humanist paradigm. The fact that critical systems thinking is used for the formulation of the nature of the evaluation framework does not exclude, however, the need to combine both interpretivist and functionalist approaches in it. For the purposes of this intervention it was assumed that the different paradigms represented by the various techniques involved are not managed from above the paradigms, but rather through the exchange of insights from the use of the different techniques as they are applied in the process of evaluation of the factors affecting software development productivity (Petkova and Roode, 1999).

The framework for evaluation of factors affecting software development productivity was experimentally validated on a complex software project involving relatively a large number of stakeholders. The user company in question is a metal processing plant. It is organized in two major sub-businesses: extrusion and roll production. It is a leader in its area in the whole of Africa. In 1997 it initiated a large expansion project. A major component of it was allocated for massive renovation of the hardware and software base of the Information Systems division of the company. The project required a lot of expertise that was not available readily within its own ranks. It went through the route of outsourcing.

In this case two companies provided outsourcing solutions. One was engaged in the provision of software support for the integration of the automation of technological processes with the Shop Floor Execution System modules of an integrated Management Information System. The second outsourcer was engaged in the provision of support for the Planning and Scheduling component of the integrated system.

The implementation of the framework was conducted in sessions with two teams separately. The first team consisted of four members of the user company project team. The second one was of four members from the outsourcing provider for the Shop Floor Execution System. The first session with each of the teams produced sufficient material for the Critique and Judgment stages of the process. It was followed by a two month period for reflection on the results from the first session. Later there was a second session, in which possible actions were explored with respect to the improvement of software development productivity. It involved a review of the results from the first session, and the formulation of answers to a questionnaire based largely upon Critical

Systems Heuristics by Ulrich (1998). The purpose was to provide another iteration on the Judgment phase and possible material for the Action phase of the framework. However, the latter stage is included only for illustrative purposes in order to provide a complete picture for the shape of the intervention and its outcomes. It is not directly within the scope of this research, which rather concentrates only on the evaluation process. The following section will present the findings for this paper based on the second session with each of the evaluation teams while further details on the intervention can be found in Petkova and Roode (1999).

CRITICAL HEURISTIC SELF-APPRAISAL OF THE FRAMEWORK FOR EVALUATION OF FACTORS AFFECTING SOFTWARE DEVELOPMENT PRODUCTIVITY

The purpose of the second session with the groups (representing the users and the outsourcing providers) was to provide critically heuristic self-reflection on the process of evaluation of productivity factors within the IT project under concern. That activity should be simple enough to be used by software developers on their own without the need for consultants, assisting in the evaluation process. The second session took the form of a structured in-depth individual interviews with the participants mentioned at the end of the preceding section. The outcome of these interviews reflects further on the beliefs of each of the evaluation groups. The analysis of these outcomes sought potential common grounds in those beliefs as a basis for actions on the way forward with this project.

The last four questions deal with issues of experimental validation of the framework for evaluation of factors affecting software development productivity, which are not discussed here for space reasons. The first eight self appraisal questions represented an operationalization of the relevant issues in the checklist of critically heuristic boundary questions defined by Ulrich (1998:11). They are related to four important aspects of the systemic enquiry: motivation, power, knowledge and legitimization. The following subsections present the summary of findings within the responses to these questions.

Identifying Sources of Motivation within the Software Project of Concern

The first question dealt with finding out what would drive ***developers' motivation***:

- *What ought to be the measure of improvement in software development productivity on this project? In what way would the consequences taken together, constitute an improvement?* The analysis of the statements by the participating team members shows a lack of common vision about what ought to be a measure of improvement in productivity. A suitable step forward would be the reprioritization of the project goals and improved coordination between the user company and the outsourcing providers.

Ulrich (1998) has formulated two more questions on this aspect, dealing with the client and the purpose of the intervention, which were not considered here. The reason is that the combination of the elements of SAST, brainstorming and SSM, included in the framework, provides a deeper insight into these two categories.

Identifying Sources of Power within the Project

Two questions were considered here, with respect to the categories ***resources*** **and** ***decision environment***:

- *What resources and other conditions of success are to be controlled by the management of this project?* There is a common opinion that more qualified staff is needed. More realistic time schedules would contribute to the smoother running of the project. It is disturbing that the outsourcing provider indicates a problem with the provision of adequate hardware for development. The latter is a technical, internal issue to that provider, and should have been resolved by its management at earlier stages of the project.
- *What conditions should be left to the environment of the project and should not be controlled by the project management for ensuring better results?* All parties agree to the need for certain autonomy to be given to the staff involved in the development process. This will leave them room to make decisions about presence on site and involvement in operational issues. It would be better to avoid everyday problems with the existing system like enhancement requests, Year 2000 problems, etc.

Identifying Sources of Knowledge

Following Ulrich (1998), the questions from this group focused on such categories as ***expert, expertise and guarantor for implementation***.

- *Is there a need for a separate expert to deal with issues of software development productivity, or should it be part of normal activities of the management of the team?* There is agreement on the need for such additional expertise on issues related to software development productivity. However, it is acknowledged, at least in one response, that ideally the management should be trained with the necessary skills to deal with this issue as it has the power to effect any desired changes.
- *Is there any missing relevant knowledge after these sessions that the management of the team should acquire in order to provide occasional evaluations of factors affecting software development productivity when necessary?* The answers indicate that no such knowledge is missing. This is a proof that the framework is relatively easy to understand and use in the practice of software development companies.
- *Who can be assumed to be the guarantor that any findings related to improvement in software development productivity can be implemented and secure improvement?* In this case there is a division in opinion, as the user company representatives think that such a guarantor can be the management of the company. On the other hand, the outsourcing provider representative suggests that it should be someone on the outsourcing side. If the problem of improvement of productivity is to be evaluated holistically, then the view that management of the user company should be a guarantor for securing improvement becomes more relevant.

Identifying Sources of Legitimisation of the Outcomes of the Software Project

Two categories were of concern in formulating the questions from this section of the interview: the ***witness to the interests of those affected***, and the ***worldview*** associated with the idea of improvement in software development productivity.

- *Who should be the witness of the interests of those stakeholders that are not involved in the inquiry process?* It is assumed by all participants that this role might best be played by the project management. This is problematic as it reflects on the inertia of people who are afraid of raising their voices and silently accept decisions on their future made by management alone.
- *What world view ought to be determining? That is, what should be the vision of improvement of both those involved in such an inquiry and those affected, and should we do with differing visions, if any?* There is a shared vision between the parties involved in the project: to achieve world class production status for the user company, as was indicated in the initial brainstorming session. This can be seen as a positive prerequisite for the implementation of measures leading to improved software development productivity.

CONCLUSION

It can be noted that the results from the first and the second sessions on the evaluation of factors affecting software development productivity for the IT project discussed revealed a significant real problems that were recognized within this project by the company management. Thus it can be concluded that the framework helped to focus on the most vital issues that could affect the future of the software project under concern.

The process of evaluation of factors affecting software development productivity has not been explored so far in a way involving a combination of MCDA and Critical Systems Thinking. The journey towards improvement of software development productivity was influenced also by the reflective power of the critically heuristic boundary judgment questions. Their application demonstrated their practicality to four aspects of the systemic intervention to improve software development productivity within the discussed project: motivation, power, knowledge and legitimisation.

The framework is transportable to any software development environment. It was found that the chosen combination of techniques in the evaluation framework assists in the learning process of gaining better appreciation for the interrelationships between factors affecting software development productivity.

REFERENCES

Checkland, P. and Holwell, S. (1998). *Information, Systems and Information Systems*, Wiley, Chichester.

Ellis, K. (1995). Critical Considerations in the Development of Systems Thinking and Practice, *Systems Practice*, 8(2):199-214.

Flood, R.L. and Jackson, M.C. (1991). *Creative Problem Solving: Total Systems Intervention*, Wiley, Chichester.

Habermas, J. (1984). *The Theory of Communicative Action Vol.1: Reason and Rationalisation of Society*, Heinemann, London.

Jackson, M.C. (1991). *Systems Methodologies for the Management Sciences*, Plenum, New York.

Jackson, M.C. (1992). An Integrated Programme for Critical Thinking in Information Systems Research, *Journal. of Information Systems,* 2:83-94.

Jackson, M.C. (1997). Critical Systems Thinking and Information Systems Research, in. *Information Systems: An Emerging Discipline*, (J. Mingers and F.A. Stowell, eds.), McGraw-Hill, Maidenhead.

Jackson, M.C. (2000). *Systems Approaches to Management,* Kluwer/Plenum, Boston.

Kemerer, C.F. (1998). Progress, Obstacles, and Opportunities in Software Engineering Economics, *Communications of ACM*, 41(8):63-66.

Lederer, A.L. and Prasad, J. (1998). A Causal Model for Software Cost Estimating Error, *IEEE Transactions on Software Engineering*, 24(2):137-148.

Midgley, G. (1996). Systemic Intervention: A Critical Systems Perspective, in, *Action Research and Critical Systems Thinking,* (R. Flood and N. Romm, eds), Plenum Press, New York.

Midgley, G. (1997). Mixing Methods: Developing Systemic Intervention, in *Multimethodology: the Theory and Practice of Integrating OR and Systems Methodologies*, (J. Mingers and A. Gill, eds), Wiley, Chichester.

Midgley, G. (2000). *Systemic Intervention: Philosophy, Methodology and Practice,* Kluwer/Plenum, Boston.

Mingers, J. and Gill, A. (eds). (1997). *Multimethodology: the Theory and Practice of Integrating OR and Systems Methodologies*, Wiley, Chichester.

Ngwenyama, O.K. (1991). The Critical Social Theory Approach to Information Systems: Problems and Challenges, in *Information Systems Research: Contemporary Approaches and Emergent Traditions*, (H-E. Nissen, H.K. Klein and R. Hirschheim, eds), Elsevier Science Publ., New York.

Ngwenyama, O. K. and Lee A.S. (1997). Communication Richness in Electronic Mail: Critical Social Theory and the Contextuality of Meaning, *MIS Quarterly*, 21(2):145-167.

Petkova, O. and Roode, J.D. (1999), An Application of a Framework for Evaluation of the Factors Affecting Software Development Productivity in the Context of Particular Organizational Environment, *South African Computing Journal,* 24:26-32.

Pfleeger, S.L. (2001). *Software Engineering: Theory and Practice*, (2nd edition), Prentice-Hall Inc., New Jersey.

Subramanian, G.H. and Breslawski S. (1993). Dimensionality Reduction in Software Development Effort Estimation, *Journal of Systems and Software*, 21:187-196.

Ulrich, W. (1983). *Critical Heuristics of Social Planning: A New Approach to Practical Philosophy*. Haupt, Bern.

Ulrich, W. (1996). *A Primer to Critical Systems Heuristics for Action Researchers, in* Forum One, Centre for Systems Studies, University of Hull, Hull.

Ulrich, W. (1998). *Systems Thinking as if People Mattered: Critical Systems Thinking for Citizens and Managers,* Working Paper No.23, Lincoln School of Management.

A TRUSTING CONSTRUCTIVIST VIEW OF SYSTEMS THINKING IN THE KNOWLEDGE AGE

Norma Romm

University of Hull Business School
Centre for Systems Studies
Cottingham Road
Hull HU6 7RX
United Kingdom
Email: N.R.Romm@hull.ac.uk

INTRODUCTION

This paper explores the contribution that systems thinkers could be seen as making in the (co-called) "knowledge age". In the knowledge age, people's skills in finding ways of developing knowledge are often highly prized. My suggestion is that it is important that acts of "knowing" developed by those who define themselves as systems thinkers become recognized as themselves being potential points of intervention in the systems being studied. (See also Romm, 1990, 1995, 1998; Jackson, 1993, 2000; Midgley, 1996, 2000; Jervis, 1997; Keys, 1997; and Banathy, 1999.) Intervention effects should be considered as a matter of concern already at the moment of "comprehension" (and not only at the moment of "application"). Already at the point of aiming to develop knowledge, our possible complicity in creating realities should be regarded as a relevant concern. This affects the way in which constructions offered are treated and assessed. In line with what I call a "trusting constructivist" approach (Romm, 2001), I suggest that instead of expecting that authors try to defend themselves on the basis of the likelihood that their constructions provide us with (more) informed understanding of some posited (externally existing) realities, we can award trust on different grounds. We can make judgments in regard to people's accountability by considering the quality of their discursive engagements with a variety of visions and concerns that might be raised by others. This has implications for the way in which we understand the status of people's knowing endeavors.

Edited by Ragsdell *et al.*, Kluwer Academic/Plenum Publishers, New York, 2002.

ACCOUNTING FOR OUR WAY OF APPROACHING "THE WORLD"

Hacking indicates that "profound and ancient philosophical disputes" surround the question of how our knowing endeavors ought to be conceived in society. He suggests that one can locate "irreconcilable differences" between realists and constructionists on this matter, and that it is quite possible that "honorable thinkers [on the matter] may eternally disagree" (1999, p. 68). In this paper I do not presume to resolve the impasse in argumentation between realist-oriented thinkers – who believe that, as Hacking notes, "the whole point of inquiry is to find out about the world" (1999, p. 83) – and constructivist-oriented thinkers, who emphasize that we can only ever have access to "the world" as presented to us. But I do attempt to cast additional light on the realist-constructivist dispute through the way in which I develop constructivist argumentation in the direction of what I call "trusting constructivism" (Romm, 2001).

Constructivist thinking broadly understood is associated with the view that, as Hacking notes, whatever constructions are developed in processes of inquiry, cannot be checked against realities posited to exist outside these process. This kind of thinking has been developed to act as a counterpoint to a more realist-oriented understanding of the process of both natural and social science. As far as the *natural* sciences are concerned, Davis (1997) supports constructivist-oriented thinking when he suggests that what is taken as the "natural world" can be argued to manifest itself in a variety of ways, moving from potentiality to actuality through human involvement in it. According to his account, we cannot presume (as does realism) that there is one world "out there" to be uncovered through the efforts of those committed to finding out about "reality". Alternative realities may become shaped through the very way in which science is undertaken. And we can never have proof about the shape of "the world" outside our observations. Meanwhile, in the doing of *social* science, many authors have made similar commentary. For instance, Hacking points out, in relation to social surveys that are undertaken, that we should not ignore "the possibility that a survey interview [itself] may become a process of persuasion" (1999, p. 161). That is, in the process of administering survey interviews, the setting of categories in terms of which people may consider questions asked, already might affect people's ways of thinking about the issues and hence may bring forth the "opinions" supposedly reported upon. And when results of our surveys are published, again this may well serve to frame people's understandings around the issues. This applies not only to survey research but indeed to any style of social research that may be undertaken in efforts to "find out" about the social world (cf. Romm, 2001, pp. 281-296).

If we decide to concentrate on our complicity as inquirers in potentially bringing forth "the worlds" on which we supposedly report, how might we take responsibility for the way in which we develop constructions? My suggestion in this regard is that the choices involved in our developing our constructions need to be accounted for through an orientation of what I call discursive accountability (Romm, 1996a,b). This entails operating in terms of an orientation to engage sensitively with, and take into account, competing visions and concerns as might be raised by others in relation to the issues under consideration. It does not imply our trying to show – as far as possible – that visions developed are best in keeping with "the facts" provided by the evidence of reality to date. On the contrary it implies that we recognize that ways of defining our relationship with "the evidence" are choices that themselves need to be accounted for.

But now the question arises as to how our accountability might be assessed (within constructivist thinking).

ASSESSING ACCOUNTABILITY IN TERMS OF A TRUSTING CONSTRUCTIVIST VIEW OF KNOWLEDGE-CONSTRUCTION

In working with what I call a trusting constructivist (as opposed to realist) view of the status of knowledge-constructions as choices, I suggest that we can hold others to account for their ways of seeing (and defining attendant possibilities for acting) in terms of our assessments of their quality of discursive engagement with different visions and concerns. People can earn trust in their processes of "knowing" to the extent that they are able to show that they have indeed engaged seriously with values and concerns introduced by others in relation to their own. As far as I see it, this manner of enacting trust relationships is not a matter of people expecting the other party to think and act in ways that might have seemed predictable (or even probable) in terms of instrumental calculations of outcomes. It is rather a matter of people expecting the trusted party to think and act by showing respect for others as persons in the sense of their being prepared to take seriously into account ideas and concerns that may (potentially or actually) be introduced into the arena of discussion. (See also Tyler and Degoey, 1996, p. 349.)

Interestingly, Solomon and Flores note that "the great advantage of a trust culture ... is the ability – the need – to appreciate other people's circumstances and points of view" (2001, p. 28). They argue that we should not "overlook the good that comes from people expressing their values and passions, each person from his or her perspective, which leads to fruitful discussion, corrective criticism, and the emergence of a shared willingness to explore new possibilities" (2001, p. 28). My argument in this context is that in order for trust to become built in terms that make provision for people to appreciate others' (experienced) circumstances and points of view, we need to review current cultural pressures on us to define knowledge as representation of posited realities supposedly pre-structured in a definite way independent of human mediation. (See Hacking, 1999, p. 60; and McIntyre-Mills, 2000, p. 7.) By opening a discussion on how knowledge should be defined in the first place, people would not need to feel compelled to defend their reasoning and attendant ways of seeing "realistic" options for action as rooted in "the evidence" provided by reality. But they are still required to account for their approach by showing that it represents an engagement with other people's visions and concerns brought forward into the arena of discussion. It is on this basis that their discursive accountability may be expressed (and assessed).

Gregory highlights that in processes of discursive encounter people may be called upon to recognize that what they themselves hold dear, are choices that require (re)consideration in relation to others. She suggests that in exploring (and participating in) processes of dialogue, we need to "take account of emotional, intuitive and other psychological features of the actors participating in the dialogue" (Gregory, 2000, p.496). She indicates that, as she sees it, the motivation of people to be prepared to reflect on their emotional attachment to specific concerns, can be derived from their appreciation that others too are required likewise to undertake such self-reflection as part of the process of dialogue (2000, p. 500). She hints that it is on these terms that trust building can ensue.

As Solomon and Flores note in relation to (their) inquiries into the building of trust, "thinking and talking about trust will not only influence our beliefs but also change our behavior in the world and with one another" (2001, p. 14). I believe that (as with other proffered constructions) the narratives that we use to think about the development of our trust relationships should be regarded as just that – they may be conceived as narratives that, as presented, invoke and evoke ways of seeing as well as possibilities for action (in this case, in relation to ways of earning and awarding trust). Constructions around trust (as with other constructions) may have self-promoting properties. Or, as Hacking puts it, they can be considered as (potentially at least) having "looping effects" (1999, p. 104).

Laszlo points out that in constructing his vision of social evolution, he presents a story about, *inter alia*, how people might learn to create "authentic communities", which he defines as characterized by "open and trustful interactions, [where] the tendency is toward differences-holding since a better understanding of similarities and differences enhances ability for further empathy and understanding" (2001, p. 311). Laszlo suggests that trustful interactions based on an appreciation of difference as well as similarity (of outlook) offers a new way of learning and of becoming – by allowing people to reach deep levels of understanding, respect and collaboration. Laszlo's vision of the status of the cognitive construct of "evolution", concurs with Gill's argument that the constructs developed by (systemic) thinkers need to be acknowledged as being part of "the system". As she notes, "systemic change efforts occur within constructions and reconstructions of meaning" (Gill, 2001, p. 343). In developing what I call a trusting constructivist position, I concentrate on the possibility of reconstructing the meaning that we give to the term "knowledge" when we speak about the "knowledge age" – and I forward this possibility for consideration and instantiation. I suggest that at least we need to consider some implications of our definitions of "knowledge" in terms of our way of developing trust relationships when people hold different starting positions and different conceptions of realistic/feasible possibilities for action.

If we expect people to operate (in order to earn trust) by trying to convince others of the likelihood that their ways of seeing and acting are informed as far as can be known in relation to "the realities", this might mitigate against the possibility of their trying to earn trust on other bases. As I have indicated, another basis of trust earning may be by expressing a propensity to bear in mind a variety of options for seeing and acting as the basis for decision-making. In the latter context of trust earning and trust awarding, the actual outcome of people's discursive engagements with alternatives is a secondary matter as far as the development of the trust relationship is concerned. As Flores and Solomon note when they consider trust relationships (though they do not specifically concentrate on definitions of knowing as I have done in this paper), "in a sense, trust is not so much about evidence and outcomes at all, except as secondary concerns. Authentic trust is ultimately about relationships, and what it takes to create, maintain and restore them" (2001, p. 32).

Solomon and Flores provide an illustration of how the erosion of trust might occur in what they regard as "times of increasing uncertainty" (2001, p. 17). They state that in such times people may wonder, for example, whether they can "trust the market – or, rather, the leaders of our largest corporations – to act in the public interest" (2001, p. 17). They note in this context that "in corporations, the epidemic of downsizing and restructuring has taken its toll. Loyalty is often said to be an inappropriate or foolish virtue, and distrust and defensiveness … have all but destroyed the morale of many major companies" (2001, p. 20). They add that "in the fast-changing world of the 21st century, trust must be reconceived in terms of rapid change, converging societies in a global market, and a world that is (perhaps for the first time) self-consciously making its own history" (2001, p. 21). Considering the question of how we might operate in times of uncertainty, realist-oriented authors suggest that it is still incumbent upon inquirers to strive to find out the likelihood of certain conditions resulting in certain outcomes, so as to reduce levels of uncertainty as far as possible – cf. Barry and Elmes' discussion (1997) of this kind of outlook. However, constructivist-oriented authors emphasize that we do not necessarily earn trust in this fashion when others believe that there are still alternative ways of seeing (and attendant possibilities for acting) that have not been sufficiently accounted for.

For example, while leaders of corporations might refer to a professional way of running their organizations in the light of world competition, with attendant implications for job restructuring, this does not necessarily lead people to trust their manner of accounting for plans being made. An appeal to an informed understanding of "the realities" is not sufficient to earn trust. And indeed, leaders/managers (whether in private or public corporations) may recognize that pressures on them to try to "maximize performance" can create dilemmas that are not resolvable "neatly and efficiently" (Weil, 1998, p. 57). Maclagan points out, for example, that when moral concerns are brought into consideration (by managers and others), it becomes recognized that the notion of "organizational goals" (a notion used in both scientific and everyday discourses) is

> often ambiguous, and from an ethical standpoint other functions of an organization's activity, such as the provision of employment, service to dependent customers or pollution control, may be just as significant as return on investment for shareholders or benefits for local taxpayers in the case of a county council. (1998, p. 148)

Maclagan shows with reference to this and other examples that if we try to sanitize moral considerations out of the language of (management) "science", we fail to highlight for consideration the way in which people might develop their moral sensitivities in organizational and wider social existence. According to Maclagan, moral considerations are brought to bear in our thinking when we recognize "that there are several stories operating, not one objective story" (1998, p. 49). Part of the complexity of moral decision-making arises when we recognize that specific ways of framing "the realities" are themselves not without consequence. Or as Gummesson, citing Doctorow (1977), notes, "the world in which we live is still to be formed" (1991, p. 122). A trusting constructivist position suggests that it is preferable for people to indeed conceive "reality" as that which is in the process of being formed. Such a conception might encourage forms of discourse where people operate through a serious engagement with different stories as a way of expressing their involvement in reality.

As Gill comments, the advantage of focusing on theorizing as a function of human discourse, is that it opens possibilities for people to engage with one another's accounts without trying to resolve these in some uniform vision with reference to "the facts" (2001, p. 343). As she notes: "individual perspectives are brought forward and retained in the overarching narrative, maintaining the multiple levels of meaning" (2001, p. 343).

When working with a (trusting) constructivist view of the status of constructions, then, the task of "knowing" is not seen as linked to an attempt to "capture" the realities of life independently of human involvement therein. The commitment is instead to make some contribution toward enriching the discourses through which we generate ways of forming our lived world(s).

CONCLUSION

I have made a case in this paper for our reconsidering the way in which we define our knowing endeavors in the so-called knowledge age, rather than submitting to cultural pressures to define knowledge as representation of externally existing realities. I have suggested that our ways of defining the status of knowledge-constructions have implications not only for how we "know" but also for how we live. (See also Romm, 2001, pp. 96-97.)

I propose that in practice if one applies expectations of *discursive accountability* in the process of assessing others' way of defending visions and actions, then one desists from applying pressure on people to try to defend their decision-making with reference to an informed understanding of "the realities". One can then appreciate that people might be oriented to developing what Weil calls "responsible pathways through multiple realities" (1998, p. 44) through their engaging seriously with alternative visions and concerns that might be presented. I have suggested that realist-oriented ways of conceiving our knowing endeavors as striving to enable us to get closer to "the truth" about reality, may mitigate against the development of processes of trust earning and trust awarding in the terms in which I have explored (or rather outlined) in this paper.

REFERENCES

Banathy, B.A. (1999). "The difference that makes a difference: Incoming Presidential Address, delivered at the 42nd Annual Meeting of the *International Society for the Systems Sciences,* Atlanta, USA, 1998," *General Systems Bulletin, 28*: 5-8.

Barry, D., and Elmes, M. (1997). "On paradigms and narratives: Barry and Elmes' response," *Academy of Management Review, 22*: 847-849.

Davis, J. (1997). *Alternate Realities: How Science Shapes Our Vision of the World.* Plenum, New York.

Doctorow, E.L. (1977). "False documents," *American Review, 5*: 215-232.

Gill, P.B. (2001). "Narrative inquiry: Designing the processes, pathways and patterns of change," *Systems Research and Behavioral Science, 18:* 335-343.

Gregory, W.J. (2000). "Transforming self and society: A 'Critical Appreciation' model," *Systemic Practice and Action Research, 13*: 475-501.

Gummesson, E. (1991). *Qualitative Methods in Management Research.* Sage, London.

Hacking, I. (1999). *The Social Construction of What?* Harvard University Press, Cambridge, Mass.

Jackson, M.C. (1993). "Don't bite my finger: Haridimos Tsoukas' critical evaluation of Total Systems Intervention," *Systems Practice, 6*: 289-294.

Jackson, M.C. (2000). *Systems Approaches to Management.* Kluwer Academic/Plenum Publishers, New York.

Jervis, R. (1997). *System Effects: Complexity in Political and Social Life.* Princeton University Press, Princeton.

Keys, P. (1997). "Approaches to understanding the process of Operational Research: Review, critique and extension," *Omega*, 25: 1-13.

Laszlo, A. (2001). "The Epistemological foundations of evolutionary systems design," *Systems Research and Behavioral Science*, 18: 307-321.

Maclagan, P. (1998). *Management and Morality: A Developmental Perspective*. Sage, London.

McIntyre-Mills, J.J. (2000). *Global Citizenship and Social Movements*. Harwood Academic Publishers, Amsterdam.

Midgley, G. (1996). "What is this thing called CST?," in *Critical Systems Thinking: Current Research and Practice* (R.L. Flood and N.R.A. Romm, eds.), Plenum, New York.

Midgley, G. (2000). *Systemic Intervention: Philosophy, Methodology and Practice*. Kluwer Academic/Plenum Publishers, New York.

Romm, N.R.A. (1990). "Gouldner's reflexive methodological approach," in *Sociology and Society* (C. J. Alant, ed.), Southern Book Publishers, Johannesburg.

Romm, N.R.A. (1995). "Knowing as intervention," *Systems Practice*, 8: 137-167.

Romm, N.R.A. (1996a). "Inquiry-and-intervention in systems planning: Probing methodological rationalities," *World Futures*, 47: 25-36.

Romm, N.R.A. (1996b). "Systems methodologies and intervention: The issue of researcher responsibility," in *Critical Systems Thinking: Current Research and Practice* (R. L. Flood and N.R.A. Romm, eds.), Plenum, New York.

Romm, N.R.A. (1998). "Interdisciplinary practice as reflexivity," *Systemic Practice and Action Research*, 11: 63-77.

Romm, N.R.A. (2001). *Accountability in Social Research: Issues and Debates*. Kluwer Academic/Plenum Publishers, New York.

Solomon, R.C., and Flores, F. (2001). *Building Trust*. Oxford University Press, Oxford.

Tyler, T.R., and Degoey, P. (1996). "Trust in organizational authorities: The influence of motive attributions on willingness to accept decisions," in *Trust in Organizations* (R.M. Kramer and T.R. Tyler, eds.), Sage, London.

Weil, S. (1998). "Rhetorics and realities in public service organizations: Systemic practice and organizational learning as critically reflexive action research (CRAR)," *Systemic Practice and Action Research*, 11: 37-62.

Section Six

Social Systems Design

AUTOPOIESIS AND THE THEORY OF VIABLE SYSTEMS

John Brocklesby[1] and John Mingers[2]

[1]School of Business and Public Management
Victoria University of Wellington
Box 600, Wellington, New Zealand
john.brocklesby@vuw.ac.nz

[2]Warwick Business School
Coventry CV47AL
orsjm@wbs.warwick.ac.uk

INTRODUCTION

This paper examines the application and usage of the idea of autopoiesis – a theory of living systems - within the context of viable systems theory. The paper is part of a broader consideration of how, at the social and organisational level, the relationship between these two sets of ideas about might be rethought and reconfigured to produce more comprehensive insight into the nature of the relationship between systems and the environments in which they are embedded (see, for example, Brocklesby, 2001).

AUTOPOIESIS AND VIABLE SYSTEMS

At the outset it is important to acknowledge that there is a close connection between viable systems theory (Beer, 1979; 1981) and the concept of autopoiesis (Maturana and Varela, 1980), and there are some fairly obvious surface similarities. Since much of the inspiration for the development of the VSM arises out of Beer's interpretation of the operation of the human nervous system, and since the latter is a living system that we know a good deal about, this intersection between the two sets of ideas should not come as a major surprise.

Beer's choice of terminology also speaks of overlapping concerns. The term viability itself has biological connotations, and the vernacular use of the term 'survival' which is used extensively in the VSM literature has to do with maintaining life. The central characteristic of viability – maintaining an independent existence - is similar to Maturana and Varela's insistence that living systems are separate from and exhibit autonomy from the medium in which they exist. Hence all three authors speak of their

Edited by Ragsdell *et al.*, Kluwer Academic/Plenum Publishers, New York, 2002.

respective systems as being 'subject to their own laws' (Beer, 1981:103; Maturana and Varela, 1987:48).

It is also important to acknowledge that the idea of autopoiesis is already employed in the context of viable systems theory. Beer cites the concept as a useful set of ideas for the VSM, and for practical applications such as the well-known Cybersyn project; he endorses Maturana and Varela's work in the preface to their 'Autopoiesis and Cognition' (1980), and further claims that Maturana provides: " . . . the quintessence of this (i.e. viable systems) cybernetic thinking . . ." (Beer, 1979:405).

In addition, the term autopoiesis itself crops up from time to time in descriptions of the viable systems model (Beer, 1979:405-408; Flood and Jackson, 1991:96). Finally, VSM researchers such as Espejo and Harnden (see 1989, for example) have clearly been influenced by the phenomenological aspects of autopoietic theory (see Maturana, 1988), if not the formal biological aspects.

In extending the basic resonances between his thinking and that of Maturana, Beer remarks:

> "Naturally I had very closely compared the conditions for life as expounded by the theory of autopoiesis with the conditions for a viable system . . . to me they were complementary and mutually enriching . . ." (Beer, 1981:338)

Cementing the link between the two theories further, he says that we may argue that "a viable system is autopoietic (and) the autopoietic faculty . . . is embodied in the totality and in its systems one . . . " (1981:338).

In the light then of Beer's conviction that a viable system is an autopoietic system juxtaposing the two sets of theory involves assessing the grounds on which Beer makes this claim. Inevitably, since he claims that complex systems such as enterprises and social systems are also viable systems, this leads into territory that has been well-traversed by others (see for example, Zeleny and Hufford, 1992; Hejl, 1984; Mingers, 1992).

It is possible to use Maturana's distinction between the 'specification' of a system, the basic form (or 'organisation') that the specification implies, and the specific systemic forms ('structure') through which the basic form might be realised, to show why – in a very basic sense – autopoiesis and viable systems are certainly similar but fundamentally different entities. In his major phenomenological work, Maturana (1988) claims that objects are brought into existence through the 'doing' of the observer. On this basis when an observer specifies some real world entity as a 'viable system', he/she is carrying out an operation of distinction that implies a basic form in which Systems 1-5 and the various flows of information around the viable system facilitate the on going maintenance of an independent existence.

In contrast, specifying the entity 'living system' implies an organisation constituted through a specific dynamic or 'manner of relating' (Maturana, 1993) of molecules; namely one in which the molecules produced produce networks that produce more molecules. Through such means, the class identity autopoiesis specifies the entity living system, the canonical example of which is the single biological cell. In this sense autopoietic and viable systems are fundamentally different.

In fairness to Beer, it is easy to see where the confusion has arisen. In explaining this it helps to begin by drawing attention to Varela's (1981:103) claim that potentially there are many examples of systems that are 'capable of specifying their own laws', and therefore can be regarded as exhibiting autonomy. Under such a conception, autopoiesis and viable systems do indeed arise as autonomous systems i.e. as specialised subsets of a more general class entity. Yet, as we shall now aim to demonstrate, autopoietic and viable systems are clearly not autonomous in the same way.

Maturana's interest in the autonomy of living systems developed out of experimental work that he and others carried out in the 1960's. Subsequently he sought an explanation to support a conviction that the purpose of a living system is internal to it, and to counter an idea – apparently prevalent at the time - that human beings are relatively more independent from their environments than are other living systems. In this context, autopoiesis – components producing networks of components which produce more components – arises as the central mechanism that provides all living systems with a degree of autonomy and independence.

This, of course, is not to say that the environment is unimportant. The canonical example of the individually autonomous living system - the single biological cell - is not completely autonomous because it is embedded in a multi-cellular organism which in turn exists in, and is affected by, an environment. The cell's autonomy has to do with its bounded self-defined nature, not its complete insulation from the systems in which it is embedded.

In contrast, as one might expect for someone with a cybernetics background, when Beer speaks of autonomy - either in the body or in social systems - he means that the system is responsible for its own regulation. In complex systems, this means maintaining a stable internal environment – 'homeostasis' – in the face of environmental complexity, or more precisely, using Ashby's (1952) terms, as a means of absorbing 'environmental variety'. In complex environments it is necessary to have autonomous subsystems to have sufficient amplification to survive. Thus, the human body must have an autonomous nervous system because control mechanisms cannot be controlled consciously by the brain. In enterprises, the task facing management is to find the optimum level of operational autonomy that is necessary in order to survive but which still permits overall system cohesiveness. At higher levels of recursion the total system is autonomous from the system of which it is a part for the same reason.

It seems to us that Beer presents autonomy in functional terms, i.e. as something that is necessary for survival or task accomplishment. Viable systems are autonomous because their environments are complex, and the amount of autonomy ceded to subsystems varies in relation to complexity in the environment. The VSM can then be seen as a complex working out, in organisational or structural terms, of how – depending on circumstances – varying degrees of autonomy might be ceded and then managed so as to ensure viability.

For Maturana it is not a case of autonomy being necessary. Rather autonomy is a constituent condition; it is more fundamental. Whereas Beer comes up with a list of characteristics of viable systems, of which autonomy is one, for Maturana autonomy is the fundamental condition that constitutes a living being. Living systems are autonomous because they are autopoietic, i.e. because the components are the result of a self-producing dynamic.

On this view, autonomy is a more fundamental condition. And it is autonomy that provides living systems with a particular identity that can be maintained in the face of significant and enduring structural change. In the theory of autopoiesis, autonomy is not a variable; living systems are living because they are autonomous in the manner in which their molecules relate. If molecules stop producing networks to produce more molecules then the system dies.

This is not to say that Maturana discounts the importance of Beer's point that systems must respond to prevailing environmental circumstances. It is just that on Maturana's view, autonomy is not the mechanism through which this occurs. For Beer, autonomy is a mechanism for adaptation; for Maturana, the relevant mechanism is 'structural coupling' which is quite different. The point is that while autopoietic and viable systems are both autonomous, the nature of the autonomy in the two cases is quite different.

However, the situation becomes more confused since Maturana (1991) does acknowledge that autopoiesis can exist in the conceptual as well as in the molecular/physical domains. Thus arises the possibility that the VSM might be a conceptual autopoietic system, if not a physical/molecular one.

This widening of the applicability of autopoiesis beyond the physical/molecular domain is an inevitable consequence of how the term is defined. Maturana, for example, acknowledges that because autopoiesis describes a relational phenomenon that exists in the dynamic space, clearly it does not depend on any particular set of components, and is thus potentially applicable to other non-physical/molecular systems. He remarks:

> "When I first spoke about living systems as autopoietic systems, I was speaking of molecular systems. Later . . . I realised that it was necessary to make the molecularity of living systems explicit in order to avoid confusion. A computer model of an autopoietic system does not take place in a molecular space . . . and this is why we did not claim to have a living system in the computer." (Maturana, 1991:376. Also cited in Mingers, 1995:45)

In other words it is molecularity and autopoiesis that specifies living, i.e. " . . . a living system is an autopoietic system in physical space . . . " (Maturana 1981:22-23). Or more precisely: " . . . any such autopoietic system realised in physical space is a living system" (Whitaker, 1996:6).

A viable system then, is not a living system. But is it a conceptual autopoietic system? If our interpretation of what Maturana and Varela mean by the term self production is correct, then the answer to this question is no.

In Beer's preface to Maturana and Varela (1980), he latches on to the terms organisation and structure. In true cybernetic style, he focuses on Maturana and Varela's remark that an autopoietic system is a homeostat, i.e. a device for holding a critical systemic variable constant within certain limits. Beer then goes on to say that Maturana and Varela's 'definitive point' is that, in the case of autopoietic homeostasis,

> " . . . the critical variable is the system's own organisation. It does not matter, it seems whether every measurable property of that organizational structure changes utterly in the system's process of continuing adaptation. It survives". (Beer 1980:70).

In the same piece, he then goes on to say: "their (i.e. Maturana and Varela's) 'it' (i.e. autopoiesis) is notified precisely by its survival in a real world."

Elsewhere, Beer (1979:405) remarks,

> "The enterprise(s) . . . staff may come and go, its departments may be closed down or opened up . . . and still it has and retains its identity. In cybernetic terminology, this enterprise is called autopoietic."

In trying to make sense of this, the meaning that Maturana and Varela attach to the term organisation is pivotal. Recall that this term describes the manner in which a system's components relate that defines its class identity; it is the general that underpins the particular.

Now while we would agree with Beer that this distinction between organisation and structure is very important, in tying it to autopoiesis we submit that he misrepresents Maturana and Varela. He interprets them as saying that autopoiesis is the continuation and preservation of a system's identity in the face of change. Thus,

> "any cohesive social institution is an autopoietic system because it survives . . . and because it may well change its entire appearance and its apparent purpose in the process" (Beer, cited in Mingers 1995:119).

Moreover, Beer remarks:

> "Look at any great institution: a hospital, a university, a multinational company, a social service, a country. All these things change, in so far as their elements are replaced; all these things change, insofar as some features disappear while others are invented. But Guys Hospital, Oxford University, the steel industry, education, and Britain itself, are recognisably themselves. There is, as in our bodies, every kind of change. But there is no alteration . . . . In the concept of autopoiesis we have the final testimonial to viability. The viable system is directed towards its own production". (Beer, 1979:405)

All of this suggests that in latching on to Maturana and Varela's distinction between organisation and structure, Beer has substantially modified the meaning of the term autopoiesis. Autopoiesis is not, as he would have us believe, the continuation and preservation of identity in the face of change; it is a very specific type of organisation that constitutes the class identity 'living'. Molecular systems such as motor vehicles and items of footwear and clothing change constantly through use, and they maintain a specific identity; there is structural change with conservation of organisation.

But this does not make them autopoietic systems. The same is true of hospitals, universities (to use Beer's examples) and other human social systems. Beer is right in saying that these change constantly but retain a sense of (observer-defined) identity. But autopoiesis defines its own organisation; it is not open to observer-dependent assessments of what it is that is conserved. Autopoietic systems define their own identity; they do not require an observer to do this for them.

In Beer's examples it is a particular observer-defined organisation (Maturana's term) of the system that is conserved. When Beer says that Guys hospital and Oxford University have existed for hundreds of years, he is saying that something has been conserved. It could be the way people interact, it could be the manner of production, or it may just the name. The point is that something has been conserved, and it this that gives the system its identity to some observer.

That all of this is the case is not, as Beer seems to think, autopoiesis. All systems exist as long as there is conservation of whatever it is that defines them. All systems arise this way. But we cannot say, as Beer does, that because systems exist or are viable over time, then they are autopoietic. It is only when a particular organisation is conserved – when there is a conservation of a specific manner if relating of the components, that we can speak of autopoiesis.

Specifically we cannot see anything in the theory of viable systems that implies self-production of components. In specifying the entity 'viable system' an observer does not have to specify that the components of System 1 must be produced by the system itself. All that is specified is that they directly produce whatever it is that an observer decides the system produces. Such is the case with viable systems such as enterprises. If an observer were to attribute the purpose 'knowledge production' to a university then the various knowledge producers within the university and its component organisational units would fulfill System 1 functions. But there is nothing to say that such researchers must have been produced within the system, even though some may have been, nor that they are engaged in producing systems to produce more knowledge producers.

Again it is possible to see what might have led Beer to make the claims that he does. He is correct when he says that people can enter and leave an organisation and that departments may be closed down or opened up - yet it still has and retains its identity. Ostensibly this resonates with Maturana's claim that molecules enter into a living system, participate in the autopoietic dynamic and then leave. In both cases there is a flow of matter through the system. But it is only when the molecules participate in a specific dynamic – producing networks that produce more molecules – that we can speak of autopoiesis. Indeed it is only when the molecules participate in this dynamic that we can legitimately regard them as components. Participating in the production of any other specific identity is something different entirely. Autopoiesis is not just the conservation of any identity, it is the conservation of a particular identity – the class identity of living systems – self production of components.

In debating the distinction between viable systems and autopoiesis, the teleological nature of the former is significant. Whereas notions of purpose, function, and goals are pivotal to the theory of viable systems, these concepts are irrelevant to autopoiesis. No matter how you look at viable systems it is difficult to get away from the idea of purpose. Observer-dependent judgements on the identity of the system will often have recourse to notions about 'the basic purpose of the system'. In turn these rely on assessments being made about the relationship between the viable system and the wider environment: for example: "the purpose of a university is to prepare students for the world of work". This, of course is possible for an observer who can observe both the system and the environment and explain the former in relation to the latter.

Internally the various components of the VSM are defined functionally (or, 'allopoietically') in terms of the purposive role that they play in supporting the overall purpose of the system. For example the purpose of System 1 is to do what the system does, the purpose of System 2 is to coordinate the activities of operational units; the purpose of System 5 is to formulate policy, and so on.

Inevitably then when the VSM is applied to some sort of social system, it is brought forth in functional terms. In other words it is constituted as something that fulfils some external purpose and its internal structure is defined in functional terms.

All of which is very different to autopoiesis. Maturana acknowledges that in our culture we often have a need to speak about the function or purpose of systems. However he argues that these distinctions are misleading, because they obscure how the system operates, which is as a result of how they are made or built, their components and relations. It is not because of a purpose. The distinction purpose belongs in the domain of the observer; it is a tool that people use to explain their experiences, and it arises as an explanation when the observer can observe the interaction between the system and its environment.

Purposes are not to be found in nature. The basic function of a living system is to produce itself and no more (see also Capra, 1996:106). On this view the notion that internal components have purposes in relation to the system of which they a part is equally misleading. The components of biological systems have no conception of an 'outside'

Having said all of this, it would be wrong to suggest that Beer believes that the purposes of organisations are pre-given. Attributing one or more purposes to a viable system is one of the first tasks that an agent must carry out in a VSM intervention. And for Beer, the purpose of a viable system is what it does, as this is defined by an observer. Thus, " . . . purpose is a mental construct imported by the observer to explain what is really an equilibriul phenomenon of polystable stable systems" (Beer, in Maturana and Varela, 1980:67).

So to the extent that Beer does give credence to the role of the observer, he and Maturana share common ground. However this does not suggest that there is some fundamental point of similarity between a viable system and autopoiesis. Among the other points of contrast that we have discussed, the teleological nature of the viable system, and the non-teleological nature of autopoiesis identifies them as fundamentally different kinds of entity. This lead us to the conclude that while there can be abstract autopoietic entities, Beer's viable system is clearly not one of these.

REFERENCES

Ashby, W.R. (1952). *Design for a Brain*, Chapman and Hall, London.

Beer, S. (1979). *Heart of Enterprise*, Wiley, Chichester.

Beer, S. (1980). "Autopoiesis: the Organization of the Living", in *Autopoiesis and Cognition - The Realisation of the Living.* (H. Maturana and F. Varela), D. Riedel Publishing Company, Dordrecht.

Beer, S. (1981). *Brain of the Firm*, Wiley, Chichester.

Brocklesby, J. (2001) "A Biological Perspective on the Evolution of Competitive Capability", in *World Famous in New Zealand – How New Zealand's Leading Firms Became World Class Competitors,* (C. Campbell-Hunt, J. Brocklesby, S. Chetty, L. Corbett, S. Davenport and P. Walsh, eds.), Auckland University Press, Auckland.

Capra, F. (1996). *The Web of Life - A New Synthesis of Mind and Matter*, Harper Collins, London.

Espejo, R. and Harnden, R. (1989). "The VSM - an On-Going Conversation", in *The VSM : Interpretations and Applications of S Beer's VSM,* (R. Espejo and R. Harnden, eds.), Wiley, Chichester.

Flood, R.L. and Jackson, M.C. (1991). *Creative Problem Solving*, Wiley, Chichester.

Hejl, P. (1984). "Towards a Theory of Social Systems: Self Organisation and Self-Maintenance, Self-Reference and Syn-Reference", in *Self-Organisation and Management of Social*

Systems: Promises Dubts and Questions (H. Ulrich and B. Probst, eds.) Springer-Verlag, Berlin.

Maturana, H. (1988). "Reality: the Search for Objectivity or the Quest for a Compelling Argument", *Irish Journal of Psychology*, 9: 25-82.

Maturana, H. (1991). "Response to Jim Birch, " *Journal of Family Therapy*, 13:375-393.

Maturana, H. (1993). Videotaped Seminar on Language and Cognition. St. Kilda, Melbourne.

Maturana, H. and Varela, F. (1980). *Autopoiesis and Cognition: The Realisation of the Living*, Reidel, Dordrecht.

Maturana, H. (1981). "Autopoiesis", in *Autopoiesis: A Theory of the Living Organization* (M. Zeleny ed.), Elsevier-North Holland, New York.

Maturana, H. and Varela, F. (1987). *The Tree of Knowledge - The Biological Roots of Human Understanding*, Shambhala, Boston.

Mingers, J. (1992). "The Problems of Social Autopoiesis," *International Journal of General Systems,* 21(2):229-236.

Mingers, J. (1995). *Self-Producing Systems - Implications and Applications of Autopoiesis*, Plenum Press, New York and London.

Varela, F. (1981). "Describing the Logic of the Living: The Adequacy and Limitations of the Idea Autopoiesis", in *Autopoiesis: A Theory of the Living Organization* (M. Zeleny ed.), Elsevier-North Holland, New York.

Whitaker, R. (1996). "The Observer Web: The Internet Nexus for Autopoiesis and Enaction", http://www.informatik.umu.se/~rwhit/AT.html.

Zeleny, M. and Hufford, C. (1992). "The Application of Autopoiesis in Systems Analysis: Are Social Systems also Autopoietic Systems?", *International Journal of General Systems* 21(2):145-160.

RECONCILING COMPLEXITY SCIENCE IN ORGANIZATIONS AND CHRISTIAN SPIRITUALITY

Eric B. Dent

University of Maryland University College
3501 University Boulevard East
Adelphi, MD 20783
edent@umuc.edu

INTRODUCTION

A body of knowledge has recently come onto the scene which has produced some exciting new insights in the study of organizations. The ideas are found under headings such as *new sciences, chaos theory, emerging sciences*, and *complexity science*. *Complexity science* will be the term used in this paper. Many complexity science ideas have grown out of recent work in the natural sciences. The underlying rationale for applying to organizations ideas initially developed in the physical and life sciences is that if change processes, structural dynamics, cooperative and competitive dynamics, and other organizational phenomena work well in nature, then, perhaps, nature should serve as a guide for how organizations could function. It is commonly accepted that business organizations are complex adaptive systems (CAS), "composed of a diversity of agents that interact with each other, mutually affecting each other, and in so doing generate novel behavior for the system as a whole," (Lewin and Regine, 2000, p. 18) operating according to the principles complexity science is unearthing.

A large number of books have been written on this subject and the journal *Emergence: A Journal of Complexity Issues in Organizations and Management* is specifically devoted to this topic. One special topics issue of the journal reviewed 34 books that had been published circa 1998 and that number certainly doubled or tripled within a year or two. Examples of such books include *The Complexity Advantage: How the Science of Complexity Can Help Your Business Achieve Peak Performance* (Kelly and Allison, 1999) and *Competing on the Edge: Strategy as Structured Chaos* (Brown and Eisenhardt, 1998).

A striking feature of the new works on complexity science is that those which reference the deep implications of the work subtly or overtly further the scientific tradition of negating religious traditions by alleging that these scientific findings provide additional evidence that certain religious traditions are not valid. Although the

Edited by Ragsdell *et al.*, Kluwer Academic/Plenum Publishers, New York, 2002.

interpretation of these scientific findings have different manifestations (such as denying a spiritual dimension to life or denying creation by intelligent design), at their essence they are all a denial of the existence of God or Supreme Being. Well-read sources with this orientation include Waldrop (1992), Gould (1999) and Kaufmann (1995). Those researchers did not see God's hand in the new data. Looking at the same evidence, however, I will suggest that many complexity science philosophies and evidence strengthen, rather than weaken, the case for the existence of a supreme being and the religious traditions associated with such a belief.

Consequently, it will be argued here that the new organization theories based on complexity science are consistent, rather than at odds, with religious traditions. Although much of the case made in this paper is consistent with other multiple religious traditions, the focus here is on Christianity and some examples are uniquely Christian.

Scientists claiming to have made discoveries which refute teachings of the religious and wisdom traditions are not new to the scene. Such activity has been going on for hundreds of years. In recent years, though, a number of scholars and researchers have endeavored to bridge the gap between science and religion. Books such as *The Marriage of Sense and Soul* (Wilbur, 1998) and the activities of over 150 groups such as The Center for Theology and the Natural Sciences are working on the integration of science and religion. These efforts typically take the form of finding a common core of science and its methods which is acceptable to the religious traditions and finding a common core of religion which is in accordance with scientific practice. If complexity science is to be consistent with attempts to integrate science and religion, it must not fall into the "traps" which some argue are simply bad science, and others note are inconsistent with religious traditions. Science has a wonderful history of illuminating unseen and unheard phenomena ranging from electromagnetic fields to ultrahigh frequency sound waves to the inside of an atom. At the same time, one of the overreaches by science is acceptance of the philosophy of scientific materialism, the "conception of the universe as composed basically of matter (or matter/energy)...best studied by science, and science alone" (Wilbur, 1998, p. 10).

As with the evolution of any new discipline, at this stage of development there is no single complexity "science" and work in complexity science varies in the degree to which it ascribes to scientific materialism. In general, though, complexity science has been shown to make the assumptions of what can be called an emerging worldview (EWV) as opposed to a traditional worldview (TWV) which includes scientific materialism (Dent, 1999). The worldviews comprise different viewpoints on a variety of polarities (see discussion of polarities below) which include equality/patriarchy, mutual causation/direct causation, understanding/prediction and many others.

The discussion here will be limited to four representative dimensions of worldview in organizations. The first two of these four dimensions have been shown to be among the three primary determinants of worldview in organizations (Dent, 1999). The choice of these four allows for illustration primarily from natural sciences in two cases and primarily from social sciences in the other two.

In general, then, recent work in complexity science supports the philosophical assumptions of the EWV. When those interpreters use the findings to challenge religion, they often make the premise that religion holds TWV philosophical assumptions. For example, Gould (1999) accepts that religion cannot have a paradoxical element because God would not permit a "contradiction" between his words recorded in scripture and his works recorded in nature. In fact, as this paper will show, the EWV assumptions are very consistent with Christianity. Each of the four sections which follows will introduce the philosophical dilemmas manifested in organizational challenges, explicate some recent work in complexity science, and provide religious evidence that demonstrates consistency with that work. The contention of this paper is that the illustrations here, though limited, are representative of the religious philosophy as a whole.

PARADOX (POLARITY)/LOGIC

Organizational leaders are perpetually faced with a series of questions. In this major change effort I want to initiate, can the change be driven from the top down or must it have bottom up leadership? How do I make sure my organization is constantly innovating and at the same time delivering a standardized level of service? How do I encourage my top management group to work as a team and at the same time not lose my star performers? What these three questions all have in common is that they cannot be answered using solely logical methods. Yet, one of the bedrock principles of science is the universal applicability of logic. Recently, though, a number of organizational scholars have suggested that an understanding and a facility with paradoxes is as important, if not more important than understanding logic.

Complexity science suggests that paradoxes are not problematic. Rather, they "create a tension from which creative solutions emerge" (Regine and Lewin, 2000, p. 19). This realization can shake someone at the core of his being. Charles Handy (1994) for example, writes that "the important message for me was that there are never any simple or right answers in any part of life. I used to think that there were, or could be. I now see paradoxes everywhere I look. Every coin, I now realize, has at least two sides" (p. xi). Others see the concept of paradox as so important that they now define leadership as essentially the management of paradoxes (Lewin and Regine, 2000; Farson, 1996). Paradoxes are defined as simultaneous or interdependent opposites. The terminology in the literature is not yet consistent. Sometimes the same phenomenon is described as a polarity (Johnson, 1992) or a dilemma.

Although leadership is defined above as the management of paradoxes, paradoxes are not managed in the way that problems are. Paradoxes have to be constantly managed, for they are never "solved" like problems. Handy's (1994) words are again instructive:

> I used to think that paradoxes were the visible signs of an imperfect world, a world which would one day, be better understood and better organized... I no longer believe in A Theory of Everything, or in the possibility of perfection. Paradox I now see to be inevitable, endemic, and perpetual (p. 12).

Perhaps, surprisingly, some see the role of paradox critical to concepts such as integrity. Peck (1987) asserts that

> if a concept is paradoxical, that itself should suggest that it smacks of integrity, that it gives off the ring of truth. Conversely, if a concept is not in the least paradoxical, you should be suspicious of it and suspect that it has failed to integrate some aspect of the whole (p. 238).

Handy (1994) has identified nine principal paradoxes he feels are critical for anyone wanting to understand societies or organizations. These range from the Paradox of Intelligence to the Paradox of the Individual. Farson (1996) identifies as most significant an understanding of the paradoxes of human relations, communication, the politics of management, organizational predicaments, change, and leadership. In the paradox of change, for example, research suggests that when it comes to human relations, big changes are easier to make than small ones (p. 110).

Paradox has been at the core of Eastern religious traditions for millennia. Today the notion of a koan and questions such as "what is the sound of one hand clapping" have made it into the mainstream of American culture. It is the judgment of those both decidedly not Christian (Bateson and Bateson, 1987, p. 146) and those Christian (Peck, 1993b, p. 21) that Christianity as a religious system is more consistent than the world's other great religions in its paradoxical structure. Peck was not a Christian when he wrote his most famous book, *The Road Less Traveled.* Since then, though, he has made a public statement of his new found Christian faith (Peck, 1993a, 156-157).

Paradoxes in Christianity abound. Jesus is both human and divine "and not simply fifty percent one and fifty percent the other. Paradox does not divide into categories but transcends categories through a mystery that may never be fully comprehensible and yet is often more real than the purely logical" (Peck, 1987, p. 242). A second example of a paradox in Christian faith is the claim that only by totally surrendering self are people completely free. (John 8:31b-32). Also, "humans are made in God's image and therefore have the potential for enormous good, yet as fallen beings they are also selfish and rebellious toward God" (Daniels et al, 2000, p. 553). A final example offered here is that God is both in heaven (Matthew 6:9) and in human hearts (1 John 4:12b-13).

It is common today for CEOs to report that "all management is people management." Given that Christianity, also, could be described as being about "people management," it is not surprising that the work world and the religious world would share the phenomenon of paradox.

HOLISM/REDUCTIONISM

As a leader in an organization, how do I establish an organizational structure which adequately addresses the fullness of the mission in a changing, white water context? How do I measure the performance of a top manager when I think she did a great job but doesn't have the bottom line to prove it because of circumstances beyond her control? If I benchmark the industry leader, how do I account for the role of values and culture in their success? Each of these questions is intractable for people thinking reductionistically. Proponents of quantitative organizational methods and scientific materialism are wedded to the assumption of reductionism. Simply put, from that viewpoint, the whole is equal to the sum of its parts. Until the past few hundred years, science, philosophy, and religion all subscribed to the theory that there are levels of being starting with inert matter, and passing through organic matter, non-human life forms, and then within people - body, mind, and soul. Scientists have argued that all of the higher levels can be reduced to matter. The soul, for example, is presently described as a series of brain functions which are not yet fully understood. Wilbur (1996) has critiqued this argument by noting that "many cosmologies have a materialistic bias and prejudice: the physical cosmos is somehow supposed to be the most real dimension, and everything else is explained with ultimate reference to this material plane. But what a brutal approach that is! It smashes the entire Kosmos against the wall of reductionism, and all of the domains except the physical slowly bleed to death right in front of your eyes" (p. 19).

Science has proceeded almost exclusively by using methods which are empirical and objectifying, revealing the exteriors and surfaces of phenomena. This approach allows for far greater understanding of a sailfish on a dissector's table than it does for one swimming in the wild. In organization theory, the reductionist method is to use the organization chart as a framework for study rather than process flow charts, for example. Leadership, too, is simply the accumulation of attributes and functions such as intellectual stimulation, individualized consideration, planning, and delegating rather than a holistic phenomenon including values, determination, motives, and other ways of being. Occasionally, though, the information most needed is discovered by reduction. However, a far greater number of phenomena of primary interest are essentially inaccessible by these methods. These questions include: what is leadership or spirituality? How do we know for sure whether a customer will maintain allegiance with one company? Will the new formula for a soda be a flop or a hit?

Wilbur (1996) suggests that these questions require study of interior dimensions which "can only be accessed by communication and interpretation, by 'dialogue' and 'dialogical' approaches, which are not *staring* at exteriors but sharing of interiors. Not objective but intersubjective. Not surfaces but depths" (p. 87). Another important factor is the inclusion of context. Bateson's (1979) work suggests that animals (and graduate students) which are studied in experimental settings are cued to the "I'm being studied" stimulus by the details of the lab which may include its smell, the experimental stand on which the animal is supported, and so forth. Consequently, findings which were confirmed in an animal study in the lab were not substantiated when the same studies were conducted with the animal in its natural context. Any change in context can introduce a variable which alters the findings.

Perhaps the work in physics is causing the most rapid decline in reductionist thinking and inspiring holistic organizational researchers. Atomic physicists now believe that particles within the atom are not independently existing analyzable entities. Prevailing research in elementary physics works under the assumption that subatomic study is of a set of interactions and relationships, not of things (Gell-Mann, 1994). This view is captured in Bell's Theorem: no theory of reality compatible with quantum theory can require spatially separated events to be independent (Schwartz and Ogilvy, 1979, p. 33). These discoveries provoke the possibility of relationship being the central concept of organizational life. Work in cybernetics included development of a proof that a focus on optimizing each of the parts results in the suboptimization of the whole. This proof has implications for TQM, reengineering, and other organizational optimization methods.

Scientists now making the EWV assumption of holism are philosophically in sync with Biblical scriptures. That Christianity affirms a spiritual level and a human soul does not need to be belabored here. Jesus Christ's earthly ministry was a daily lesson in holism. In nearly every challenge that Jesus receives from the Pharisees or others, he points out that the question or challenge contains an inappropriate reduction. For example, in the Parable of the Good Samaritan (Luke 10) when asked "who is my neighbor?" Jesus redefines the concept of "neighbor" to include anyone seen in need. In Matthew 19 Jesus is confronted by the rich young man who asks for a list of what it takes to receive eternal life. Jesus's response is that eternal life is not achieved via checklist, but by a holistic commitment and attitude change. In fact, theologians would probably not quarrel with the suggestion that a major function of Christ's earthly ministry was to restore the wholeness of life reduced legalistically by the scribes and Pharisees.

Holism and reductionism form a polarity. In the past couple of hundred years, Western society has emphasized reductionism. At work it has been manifested by assembly lines and an organization-chart mindset. In religion it is manifested by ritualistic form over function. Recognizing holism as an equal partner in the polarity will make both organizational and religious practice (restored to the guidance of scripture) healthier and more productive.

INDETERMINISM (FREE WILL)/DETERMINISM

How can I confidently expect a suitable return on assets invested in an unpredictable world? How do I maintain a sense of control over my organization knowing that people work best when they are given some autonomy? Deming (1986) has described the most important information I need as unknowable - what do I do now? Determinism is the philosophical assumption of classical science. For centuries, scientists believed that the functioning of the entire world could be predicted if only an equation large enough could be modeled. More recently, Ilya Prigogine was awarded a Nobel prize for showing the existence of indeterministic events in chemical reactions. Prigogine and Stengers (1984) now see determinism and indeterminism not as irreconcilable opposites but "each playing its role as a partner in destiny" (p. xxiii). Many scientific and managerial concepts are being reconsidered with the new assumption that there are "permanent gaps in our ability to predict events" (Peacocke, 1995, p. 277).

Human beings take for granted a world of three spatial dimensions and one time dimension (for a total of four). Early in the 20th century, though, Albert Einstein concluded from his scientific work that there must be additional dimensions. The only other extant document at the time to refer to additional dimensions was the Bible, which had been written thousands of years earlier (Ross, 1996, p. 21).

Einstein's conclusion was contrary to the generally held belief of the day that the universe is eternal and that time is reversible. George Ellis, Stephen Hawking, and Roger Penrose offered a proof of Einstein's theory in 1966. In 1993, the Nobel prize in physics went to Russell Hulse and Joseph Taylor who established beyond a shadow of scientific doubt that Einstein was right and whoever or whatever caused the universe must possess at least one more dimension of time than the one in our universe (Taylor, et al, 1992; Penrose, 1994). Even more recent work has suggested that at 10^{-43} seconds after the creation event began at least 10 dimensions existed, and they split into two pieces - a four dimension piece that became our dimensions of length, width, height, and time and a six or more dimension piece that permanently ceased expanding and never produced matter (Ross, 1996, p. 30). These six or more dimensions remain curled up everywhere within the other four dimensions.

The discovery of these additional dimensions provides a glimpse of an answer to the age-old question of whether human beings have free will or whether God has predestined everything that happens. It is possible that both are true. Scientists have shown that if three time dimensions are operative, God could pre-determine every action of every human being while still sustaining the operation of human choice. As an individual expresses her free will, God can also express His will in sufficient degree to keep His "plan" intact (Peacocke, 1995).

To get an inkling of how God might work in additional spatial dimensions, it is helpful to compare the human situation with one that is more restrictive. Consider a "world" of two spatial dimensions and one time dimension. "Flat-people" living in two spatial dimensions would have no clue what three spatial dimensions are like. They could not fathom it. In the same way, humans cannot really fathom four spatial dimensions. If you want to try, think about turning a basketball inside out without making a break in any of the surfaces. In four dimensions, this task is easily accomplished. The Bible tells of Jesus using an extra spacial dimension to enter a room after His crucifixion without opening the locked doors and windows of the Disciples' hideout. Imagine how easily people can create the same effect for flat-people. If you were to poke your finger in and out of the flat plane, flat-people would see a circle appear, then disappear, then reappear in a way that they could not explain. Although people have a difficult time picturing four spatial dimensions, in mathematics additional dimensions are used routinely. A simple equation such as $3x + 2y + z + 4w = 17$ cannot be illustrated graphically. Organizational researchers using statistical tools such as multivariate regression equations regularly employ additional dimensions.

The work of Einstein, Prigogine, Taylor, Hulse, Penrose and other scientists squares very nicely with Biblical teachings. On the question of free will and determinism, over 120 scriptural passages speak to either human free will or predestination by God, sometimes in the same passage (Ross, 1996, p. 136). For example, Ephesians 1:11-13,

> All things are done according to God's plan and decision; and God chose us to be his own people in union with Christ because of his own purpose, based on what he had decided from the very beginning. Let us, then, who were the first to hope in Christ, praise God's glory! And you also became God's people when you heard the true message, the Good News that brought you salvation. You believed in Christ, and God put his stamp of ownership on you by giving you the Holy Spirit he had promised.

Christianity teaches, then, that the managerial questions at the beginning of this section cannot be answered with certainty. From a human, four-dimensional perspective, the world is probabilistic and uncertain. Although some risks can be mitigated, the notion of control is paradoxical, as discussed above.

PERSPECTIVAL OBSERVATION/OBJECTIVE OBSERVATION

How do I help resolve a conflict between two subordinates when their sides of the story sound equally right (and/or wrong)? How do we design the perfect product for a stay-at-home mother when no one who works here has that viewpoint? Why do people think of me as arrogant when I never feel as though I'm being arrogant? Perspectival observation is a term which means that phenomena or information in the world are dependent, to some extent, upon the method of observation, and that the definition of an appropriate perspective may change depending on what information is requested. Moreover, the phenomena or information may be changed by the act of observing. Some suggest that truth is subjective or relative. Although perspectival observation suggests that most phenomena are interpreted or mediated, and that the interpretation may differ, it does not preclude the existence of an absolute truth. Certainly religion asserts some absolute truths. At the same time, variation in Christian denominations, even those considered part of the orthodoxy, witnesses to the perspectival interpretation which helps to keep a historic document timeless in its application. The scriptures themselves assert the importance of perspective suggesting that it is best to have two or more witnesses to a claim (Matthew 18:16). Moreover, Christ's predominant teaching method, the parable, is an inherently perspectival approach.

As with the reinterpretation of paradox (above), the ascendance of belief in perspectival observation means that aspects of the Bible which were formerly criticized are now in vogue with current thinking. For example, the story of Jesus' earthly ministry is told in four different books. According to the TWV, this is problematic because if reality is objective, there should be only one true story. The approach of some scholars to discrediting the Bible is to point out that in Luke it says three women went to the tomb and found the stone rolled away, but in John it says only one woman. Complexity science recognizes that if four people tell the same story it will be told in four different ways with different emphases, purposes, and details.

Traditional organizational research has also been critiqued for another feature which has been at odds with Biblical teaching. An important element of the scientific method, taught to every school child, is that the scientist must be a detached, arms-length observer. Berman (1981) notes that "scientific consciousness is alienated consciousness: There is no ecstatic merger with nature, but rather total separation from it. Subject and object are always seen in opposition to each other. I am not my experiences, and thus not really a part of the world around me. The logical end point of this world view is a feeling of total reification: everything is an object, alien, not-me; and I am ultimately an object too, an alienated "thing" in an world of other equally meaningless things" (p. 17).

A researcher assuming perspectival observation essentially changes her research question. Rather than asking, what is leadership, she asks, what is the *human experience* of leadership? (Berman, 1981, p. 187). The organizational researcher is then free to undertake studies in which he becomes, or already is, an integral part of the organization under study. Seen this way, interpretive and phenomenological methods are not lesser than quantitative, positivist methods - they simply provide a different type of data.

Christianity has always been a religion in which subject and object are not separate and distinct. For example, Christianity offers a very personal relationship with the deity. Jesus notes in John 15:15 "I do not call you servants any longer, because a servant does not know what his master is doing. Instead, I call you friends, because I have told you everything I heard from my Father. Moreover, there are many Biblical passages which demonstrate the oneness of the deity with the devout human, e.g. John 15:4a.

Perhaps the ultimate expression of "ecstatic merger with nature," is the Christian experience of "taking communion" - sharing in the body and blood of Christ. Although the scientific community will probably never appropriate a term loaded with such religious connotation, the experience of "communing" is a wonderful expression for an organizational researcher "conversing intimately; exchanging thoughts and feelings" (American Heritage Dictionary) with organizational employees. For Protestant Christians, the experience of sharing in the body and blood of Christ through the symbols of bread and wine (or juice) is one of the most sacred of experiences. For Roman Catholics, the merger is seen as more complete through the doctrine of transubstantiation, that the bread and wine are transformed miraculously at the time of consumption into the true presence of Christ. The lack of separateness is clear in passages like Mark 14:22-24.

CONCLUSION

Although many writings in complexity science would promote a further divide between complexity science in organizations and spirituality/religion, the contention of this paper is that not only are there great consistencies between them, but that complexity science can best inform organizational practice if it proceeds in ways that foster integration of management, spirituality, and religion. Several publications are advancing this integration although many of them do not explicitly use the terminology of complexity science and/or spirituality. Some examples of these are listed below.

Paradox (Polarity)

Wacker, Taylor, and Means (2000). The Visionary's Handbook: Nine Paradoxes That will Shape the Future of Your Business.

Farson (1996). Management of the Absurd: Paradoxes in Leadership.

Handy (1994). The Age of Paradox.

Holism

Kaplan and Norton (1996). The Balanced Scorecard: Translating Strategy into Action.

Greenleaf (1977). Servant Leadership: A Journey into the Nature of Legitimate Power and Greatness.

Rummler and Brache (1995). Improving Performance: How to Manage the White Space on the Organization Chart.

Indeterminism

Merry (1995). Coping with Uncertainty: Insights from the New Sciences of Chaos, Self-Organization, and Complexity.

Modis (1998). Conquering Uncertainty : Understanding Corporate Cycles and Positioning Your Company to Survive the Changing Environment.

Harvard Business Review on Managing Uncertainty (1999).

Stacey (1992). Managing the Unknowable: Strategic Boundaries Between Order and Chaos in Organizations.

Perspectival

Jones (1995). Jesus, CEO: Using Ancient Wisdom for Visionary Leadership.

Zohar and Marshall (2000). Sq : Connecting with Our Spiritual Intelligence.

Cooperrider, Sorensen, Whitney, and Yaeger (eds). (1999). Appreciative Inquiry: Rethinking Human Organization Toward a Positive Theory of Change.

Sociologists such as Sorokin (1957) have observed that one story line of history is that of an oscillating pendulum swinging between religious and scientific extremes. He has labeled as "idealistic" the time when there has been a harmonious synthesis of faith, reason, and empiricism. Idealistic times have been rare in the last few millennia. Perhaps the recent efforts of many to focus on such a harmony will help usher in a very meaningful and productive idealistic world culture.

REFERENCES

Bateson, G. (1979). *Mind and Nature: A Necessary Unity*, Bantam Books, Toronto.

Bateson, G. and Bateson, M. C. (1987). *Angels Fear: Towards an Epistemology of the Sacred*, Macmillan Publishing Company, New York.

Berman, M. (1981). *The Reenchantment of the World*, Cornell University Press, Ithaca.

Brown, S. L. and Eisenhardt, K. M. (1998). *Competing on the Edge: Strategy as Structured Chaos*, Harvard Business School Press, Boston.

Daniels, D., Franz, R. S., and Wong, K. (2000). A classroom with a worldview: Making spiritual assumptions explicit in management education, *Journal of Management Education*. Vol. 24, (5), 540-561.

Deming, W. E. (1986). *Out of the Crisis*, MIT Ctr for Advanced Engineering Study, Cambridge.

Dent, E. B. (1999). Complexity science: a worldview shift, *Emergence: A Journal of Complexity Issues in Organizations and Management.* 1, (4), 5-19.

Farson, R. (1996). *Management of the Absurd: Paradoxes in Leadership*, Simon & Schuster, NY.

Gell-Mann, M. (1994). *The Quark and the Jaguar: Adventures in the Simple and the Complex*, W. H. Freeman, New York.

Gould, S. J. (1999). *Rocks of Ages: Science and Religion in the Fullness of Life*, Ballantine, New York.

Handy, C. (1994). *The Age of Paradox*, Harvard Business School Press, Boston.

Johnson, B. (1992). *Polarity Management: Identifying and Managing Unsolvable Problems*, HRD Press, Inc., Amherst.

Kauffman, S. (1995). *At Home in the Universe: The Search for the Laws of Self-organization and Complexity,* Oxford University Press, New York.

Kelly, S. and Allison, M. A. (1999). *The Complexity Advantage: How the Science of Complexity Can Help your Business Achieve Peak Performance*, McGraw Hill, New York.

Lewin, R. and Regine, B. (2000). *The Soul at Work: Listen... respond... let go: Embracing Complexity Science for Business Success*, Simon and Schuster, New York.

Peacocke, A. R. (1995). God's interaction with the world: The implications of deterministic 'chaos' and of interconnected and interdependent complexity, *Chaos and Complexity: Scientific perspectives on Divine Action.* R. J. Russell, N. Murphy, and A. R. Peacocke (eds), The Center for Theology and the Natural Sciences, Berkeley.

Peck, M. S. (1987). *The Different Drum: Community-making and Peace*, Touchstone, New York.

Peck, M. S. (1993a). *Further Along the Road Less Traveled: The Unending Journey toward Spiritual Growth*, Simon and Schuster, New York.

Peck, M. S. (1993b). *A World Waiting to be Born: Rediscovering Civility*, Bantam Books, Toronto.

Penrose, R. (1994). *Shadows of the Mind*, Oxford University Press, New York.

Prigogine, I. and Stengers, I. (1984). *Order out of Chaos: Man's New Dialogue with Nature*, Bantam Books, Toronto.

Regine, B. and Lewin, R. (2000). Leading at the edge: How leaders influence complex systems, *Emergence: A Journal of Complexity Issues in Organizations and Mgt,* 2(2)5-23.

Ross, H. (1996). *Beyond the Cosmos: What Recent Discoveries in Astronomy and Physics Reveal about the Nature of God*, Navpress, Colorado Springs.

Schwartz, P. and Ogilvy, J. (1979). *The Emergent Paradigm: Changing Patterns of Thought and Belief*, SRI International, Menlo Park, CA.

Sorokin, P. (1957). *Social and Cultural Dynamics*, Extending Horizons Books, Boston.

Taylor, J. H., et al. (1992) Experimental constraints on strong-field relativistic gravity, *Nature*, 355, 132-136.

Waldrop, M. M. (1992). *Complexity: The Emerging Science at the Edge of Order and Chaos*, Simon and Schuster, New York.

Wilber, K. (1995). *Sex, Ecology, Spirituality: The Spirit of Evolution*, Shambala, Boston.

Wilber, K. (1996). *A Brief History of Everything*, Shambala, Boston.

Wilbur, K. (1998). *The Marriage of Sense and Soul: Integrating Science and Religion,* Random House, New York.

Zohar, D. and Marshall, I. (2000). *Connecting with our Spiritual Intelligence* Bloomsbury, London.

TAKING ACCOUNT OF HUMAN RIGHTS

Trevor Pearce and Joyce Fortune

Centre for Complexity and Change
The Open University
Walton Hall
Milton Keynes
MK7 6AA
J.Fortune@open.ac.uk

INTRODUCTION

For the last few years a programme of research has been underway which is using systems thinking to look into aspects of modern policing. Unlike most applications of systems thinking to policing this work has placed particular emphasis on aspects of policing which can be described as 'function based', that is relating to the procedures and methods which the police service employs to carry out its activities rather than to organisational matters or specific issues.

The functions that have been looked at so far are: the investigation of major crimes (Pearce and Fortune, 1995a); command and control in the policing of disturbances and major events (Pearce and Fortune, 1995b); the use of informers in intelligence-led policing (Pearce, 1999); and tackling organized crime. In all of these studies the Formal System Model (FSM) (see Fortune and Peters, 1995) has been used to examine whether, in systems terms, the processes being carried out and the structures in which they operate meet the requirements of systems capable of purposeful activity without failure. Each of the functions listed above has thus been tested to see whether it possesses:

- a decision-making subsystem
- a performance-monitoring subsystem
- a set of subsystems and elements carrying out the tasks of the system
- a continuous purpose or mission
- connectivity between components
- an environment with which the system interacts
- boundaries separating the system from the wider system and the wider system from the environment
- resources
- some guarantee of continuity

In almost all of the cases use of the FSM has proved very fruitful. For instance, the study of organized crime groups yielded the results summarised in Table 1.

Table 1. Examination of Organized Crime Groups

Environmental disturbances	Law enforcement Government Legislation Public perception
Wider system	Absence – to be expected as organised crime is loosely connected. This assumption was tested by re-drawing boundaries and looking at the higher-level degree of organisation between organised crime groups.
System	Present – with a clear boundary between those engaged in the organised crime group
Decision making sub-system	Present and conceptualised as the head of the group
Deciding on transformations	These are the methods of the crime. Can include types of commodity based on profit and risk of enforcement.
Providing resources and legitimating activity	Organised crime needs to be funded and this will come from the group itself. Legitimation of activity may be based upon the ethos of the group. i.e. not to deal in harder drugs.
Makes known expectations	These revolve around profit and internal discipline
Transformation sub-system	Present – these are the lieutenants and foot soldiers of the organisation
Performance monitoring sub-system	Depending upon the distance of the top tier criminal from the activity it could be that a lieutenant-type individual plays this role. More often than not this role will be fulfilled within the decision making sub-system
Provision of performance information	As above this is likely to be part of the decision-making sub-system's duties – a good example would be sub-system dealing with money laundering.
Attempts to influence the environment	These are present – such as corrupting officials, legal challenges to policing techniques, technical responses and other disruptions to law enforcement, and use of violence.

However, in the study of the use of informers in intelligence-led policing, although the FSM provided some insights, particularly in relation to the structure of the management system and the interactions between the components, it seemed to provide a less than complete examination of the situation. Use of informers is an aspect of policing which is recognised as being fraught with difficulty and danger, and one where errors of judgement or unethical behaviour can have severe consequences such as the life of an informant being put at risk or an expensive and protracted investigation and prosecution being compromised because of procedural or ethical failings. At the same time, however, it is widely accepted that the informer plays a vital role in policing; indeed in the drive for a more efficient, effective and economic police service in the 1990's the potential to be gained from informant activity was highlighted by the Audit Commission in its report *Helping with Inquiries: Tackling Crime Effectively* (Audit Commission, 1993). (See also Dunnighan (1994) and Dunnighan and Norris (1999).) Ethical issues associated with the use of informers therefore needed to be an important aspect of this study but though they were touched upon in terms of the 'disturbances from' and 'attempts to influence' the environment and in the 'makes known expectations' aspect of the system (see Figure 1) the FSM did not force them to be examined with any great degree of rigour. In particular, the analysis seemed to take little account of the tension between the two competing drivers: achievement of high performance levels; and adherence to ethical standards.

At the time of the study this perceived weakness in the FSM was merely noted but soon afterwards the requirement to be able to investigate areas of policing where ethical issues are to the fore became much more pressing due to changes in UK legislation. These changes will be looked at in the next section.

HUMAN RIGHTS

The Human Rights Act 1998 came into force across the UK on October 2 2000. It gave effect to certain Articles from the European Convention on Human Rights (ECHR), a treaty of the Council of Europe, which was adopted in 1950 and signed by the UK in 1951. Prior to 1998 UK citizens who wished to uphold their rights under the convention had to take their cases to the European Court of Human Rights in Strasbourg. This was a very lengthy and often costly procedure. (For a detailed consideration of the legislative and jurisprudential issues see Wadham and Mountfield (1999).)

The rights incorporated by the Act are: the right to life; freedom from torture, inhuman and degrading treatment; freedom from slavery; freedom from arbitrary arrest and detention; the right to a fair trial; freedom from retrospective penalties; the right to privacy and family life; freedom of religion; freedom of expression; freedom of association; the right to marry and found a family; freedom from discrimination; the right to peaceful enjoyment of property; the right to education; and the right to free elections. The 1998 legislation also places the following significant duty upon public authorities: 'It is unlawful for a public authority to act in a way which is incompatible with a convention right'. Should a public authority, and the definition of such bodies is wide, fail to respond appropriately then it becomes liable for that breach. An example of this might be where a firm fails to provide adequately for the safety of its staff. Should a member of staff be killed at work then the firm would have failed to protect the individual's rights under Article 1 – the right to life - and thus not discharged its duty of care. Further, the Act places a responsibility upon public authorities to actively uphold the rights of the individual.

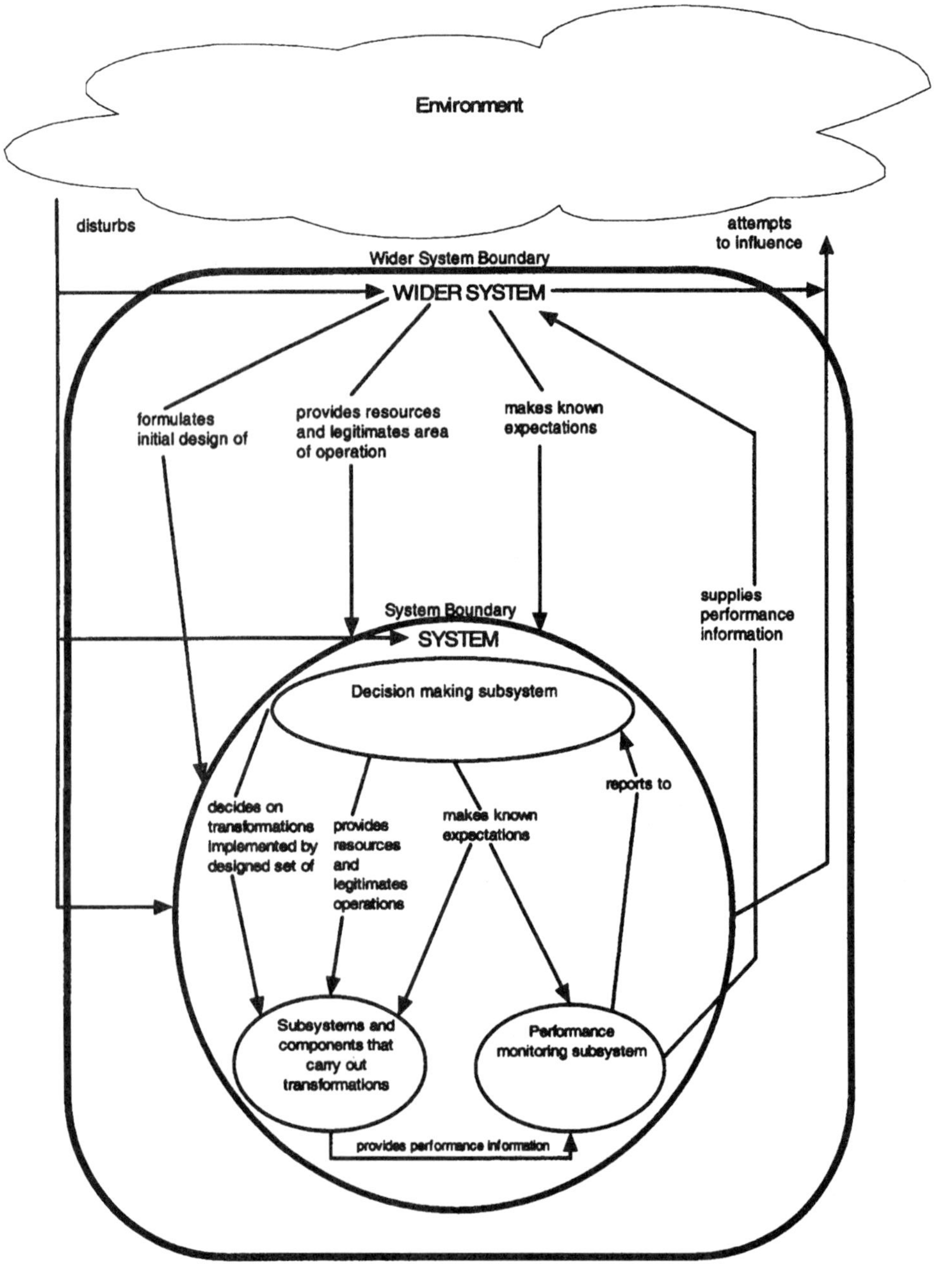

Figure 1 The Formal System Model

Some of the rights in the Act such as the right to free speech are 'qualified rights'. Where these are concerned, interference is allowed by the authorities under strictly controlled circumstances but such interference requires permissive legislation to be in place. This is a major departure from the principle that had previously been adopted in the legal system of England and Wales that anything was permitted unless prohibited by law.

It was for that reason that the Regulation of Investigatory Powers Act 2000 came into force in October 2000. If covert intrusive tactics and techniques were to be available to law enforcement and public authorities and such activities were to be Human Rights Compliant, it was essential that permissive legislation be put into in place allowing such activities. The Regulation of Investigatory Powers Act 2000 is this permissive legislation. It ensures that this form of law enforcement activity can be carried out without breaching Article 8, the right to privacy.

As the public authority most subject to public, media and governmental scrutiny, not to mention increasing litigation, it was vital for the police service to act in a pro-active fashion and be able to demonstrate it was ready to face the significant challenge ahead. After all, the impact of the legislation would be ranging across the entire ambit of policing activity from administration and duty of care to staff to the operational imperative. Therefore, the police service began to look closely at all its activities and formulate responses to the legislation. (A detailed appreciation of the response can be found in Neyroud and Beckley (2001).) They emphasise the increased complexity the legislation placed on the policing function: 'This, as we have seen, does not help us [as police officers] provide clarity in the policing role because the very nature of human rights, like the very nature of policing, is ambiguous and conflictual'.)

Human Rights and the Formal Systems Model

As has already been stated, the list of features of the FSM makes no specific reference to ethical or human rights issues. (Indeed Mansell (1996) has described the underlying logic of the FSM as cybernetic and the prevailing tone as functionalist.) The model does incorporate aspects such as 'makes known expectations' which can allow ethical or human rights issues to be considered, but no specific features are identified which force users of the model to take account of these or similar considerations. To look at it another way, it could be argued that of Jackson's three sets of interest (Jackson, 1991) it very definitely serves the technical interest, largely serves the practical interest but does not serve the emancipatory interest at all.

However, as the discussion on human rights above makes clear, in very many situations it is now essential to ensure that they are seen as an intrinsic part of the structures and processes in place and given the consideration they deserve. In the next section options for doing this within the context of policing are considered. It is suggested that though further work may be necessary to extend this discussion to ethical issues more generally much of what follows is equally relevant to other domains such as commercial activities and international development where there is great concern about ethical issues.

Human Rights within the FSM

One approach to forcing appropriate consideration of human rights would be to encompass ethical and human rights considerations within the list of features that need to be present if the components of a situation are to meet the requirements for purposeful action without failure. Thus consideration of Human Rights' would be a tenth bullet point at the end of the list on page 2 of this paper and lack of consideration of human rights' could emerge as a finding when comparing a situation viewed as a system with the FSM. This approach would have the benefit of being simple as far as the list of features is concerned but it would be problematical in a number of ways. First one cannot be certain that human rights considerations are genuinely different from other considerations such as other legislative requirements like health and safety to the point where they cannot be covered by other features on the list. Secondly, adopting this approach may make the process too static. A third reservation is that the list of features may start to become a checklist rather than a distillation of the model itself. Although it is very easy to add words to the list of features it is very difficult to see how the form of the FSM could be modified to reflect this addition.

Another option might be to retain the existing FSM and to regard the human rights legislation as an important component of the environment that is disturbing the wider system and the system. This would allow the requirements of the legislation to be considered in analysis to the extent of looking for a reaction to the disturbance but it may fail to probe in sufficient depth into the underlying structure and processes of the wider system and the system. A second drawback of this suggestion is that it might give a false view of the requirements placed upon the system. According to the FSM, the system responds to disturbances from the environment by trying to influence it. To suggest human rights' requirements are somehow negotiable is to misunderstand their real impact.

A further approach would be one which recognises the need for human rights to be a dynamic feature running through a system and encompassed within the 'guarantee of continuity'. This could be achieved through examination of the real world situation seeking to confirm that consideration had been given to such issues or that features that could be interpreted in terms of human rights issues were present. In particular the examination would focus upon the wider system and the decision-making sub-system with the expectation that human rights issues would be accounted for in the function 'makes known expectations'. It would be right and proper for both the wider system and the decision – making subsystem of the system to ensure that the fundamentals of Human Rights are complied with when delivering the system's purpose; therefore, it could be said that through this Human Rights issues transcend decision-making. Equally if a system is to operate in an ethical fashion taking into account Human Rights, then there would be an expectation that the 'legitimates operations function' would be expected to follow this convention.

A possible criticism of this approach is that the model incorporates no specific means by which human rights can be communicated through the wider system and systems expectations. To deal with this either a paradigm could be developed that would cut deeper into the comparison or alternatively a supporting set of heuristics could be used. If the latter the opportunity arises to tailor them to the specific situation under examination, thus providing a degree of flexibility and a response to the need to meet dynamic changes.

In the context of policing activity involving the use of informants or sensitive tactical options the following heuristics, drawn from within the human rights and ECHR jurisprudence, might have value:

Is there interference with someone's rights? *(Operational and subjective decision)*
Is there a positive obligation to take action?
Is the action lawful? *(Is there a law that says I can do this?)*
Is the action necessary? *(Could the same result be achieved by means, which did not interfere with someone's rights?)*
Is the action proportionate? *(Is the action unfair or arbitrary? Is it limited to that which is necessary to achieve the aim? How does the proposed action balance the rights of the individual against the community as a whole?)*
Is the action accountable? *(Is there a process for recording decisions and the reasons for it that will stand scrutiny?)*

CONCLUSION

Having used the Formal System Model to examine a range of policing activities the authors have identified one area where the model is lacking in rigour: it can fail to take sufficient account of ethical issues. This potential weakness has been exemplified by reference to the application of the requirements of the Human Rights Act 1998 to policing.

A number of options have been put forward for strengthening the FSM with respect to human rights considerations. These are: emphasising human rights legislation as an environmental disturbance; adding to the list of functions required of a purposeful system capable of operation without failure; searching for human rights as a dynamic feature embedded in the guarantee of continuity; and developing a set of heuristics. The last of these has been selected as the most promising and a set of possible heuristics has been identified. The next phase of this research strand will be to test these heuristics in order to determine their practical value in evaluating policing activities.

REFERENCES

Audit Commission (1993). *Helping with Inquiries: Tackling Crime Effectively*, HMSO, London.

Dunnighan, C. (1994). "Helping with inquiries", *Criminal Justice Matters*, 17:12.

Dunnighan, C. and Norris C. (1999). "The detective, the snout, and the Audit Commission: the real costs in using informants", *Howard Journal of Criminal Justice*, 38(1): 67-86.

Fortune, J. and Peters, G. (1995). *Learning from Failure*, Wiley, Chichester.

Jackson, M. C. (1991). *Systems Methodology for the Management Sciences*, Plenum Press, New York.

Mansell, G. (1996). "Book review", *Systems Research*, 13(1):501-503.

Neyroud, P. and Beckley, A. (2001). *Policing, Ethics and Human Rights*, Willan Publishing, Cullompton Devon.

Pearce, T. and Fortune, J. (1995a). "The investigation of major crime – a systems examination", paper presented to the Operational Research Society annual conference, University of Kent at Canterbury.

Pearce, T. and Fortune, J. (1995b). "Command and control in policing: a systems assessment of the gold, silver and bronze structure", *Journal of Contingencies and Crises Management*, 3(3):181-187.

Pearce, T. (1999). "The potential ethical dilemmas of the use of informers as an intelligence resource", unpublished paper for the department of Criminology, University of Cambridge.

Wadham, J. and Mountfield, H. (1999). *Blackstone's Guide to the Human Rights Act 1998*, Blackstone, London.

THE IMPACT OF COMPUTER HARDWARE THEFT ON ICT INTRODUCTION TO SOUTH AFRICAN RURAL COMMUNITIES: AN INTERPRETIVE ASSESSMENT THROUGH FOCUS GROUPS AND MORPHOLOGICAL ANALYSIS WITHIN A PROCESS-BASED RESEARCH FRAMEWORK

Jackie Phahlamohlaka and Hugo Lotriet

Department of Informatics
University of Pretoria
Pretoria, South Africa
jphahla@hakuna.up.ac.za & hlotriet@postino.up.ac.za

INTRODUCTION

The Universal Service Agency (USA) was established through the Telecommunications Act (103 of 1996) with the specific purpose of promoting access to Information and Communication Technologies (ICTs) for all people in South Africa. It was established on the assumption that ICTs are a basic right and a necessary aspect of social and economic functioning. Attempts have been made by both local and international institutions to introduce these technologies to South African communities. A disturbing trend of computer hardware theft is beginning to emerge from several rural communities, which threatens the goals of the USA and other related initiatives.

One initiative that is under threat, is an ongoing research-linked community ICT development project, conducted by the Department of Informatics, University of Pretoria, in partnership with the Siyabuswa Educational Improvement and Development Trust (SEIDET).

As part of this ongoing research and community development efforts, an action research study conducted under the title: *An investigation into community development imperatives brought about by the recent establishment of a local area network and access to the Internet at a rural South African community education centre* was initiated at the beginning of the year 2001. During the early phases of the project, a burglary took place at the community centre and most of the computers as well as some of the network equipment were stolen. As a result of this incident, both the research and the community ICT development efforts were immediately under threat, which prompted the researchers to ask questions about the possible implications of the theft. From subsequent enquiries it

Edited by Ragsdell *et al.*, Kluwer Academic/Plenum Publishers, New York, 2002.

emerged that several other institutions in the vicinity of the community centre also suffered from recent computer hardware thefts.

This background called for an immediate search for a better understanding of the theft trend within the context of the broad community development environment and in particular, the rural area of Siyabuswa and its surroundings. The aim of the study was therefore to enhance understanding of the dynamics and perceived motivators of the theft trend in order to:

1. Raise awareness in the academic and development communities, the affected community and the country about the theft trend;
2. Assess what the impact of this may be on ICT introduction in rural communities and;
3. If possible, suggest possible courses of prevention action.

This paper presents and discusses the results of three focus group discussions with the affected community groups on the perceptions they have about the impact of these thefts. An interpretive assessment of the perceived impact on ICT introduction in the respective rural communities is made using Morphological Analysis within a Process-based research framework. Further questions are raised in conclusion.

RESEARCH METHOD

We followed a qualitative research method designed to capture perceptions of directly affected people by the theft of computers from their rural communities. A Process-based research framework developed by Roode (1993) was used to generate a set of research questions. Focus group discussions were used to gather the perceptions and the results were analysed using a Morphological Analysis technique as well as the researchers' own interpretations, impressions and observations. A further interpretation of the results was made from a critical systems thinking perspective.

Background on a Process-Based Research Framework and Morphological Analysis

A Process-based Research Framework

A process-based research approach proposed by Roode (1993) in formulating the research questions was employed. Roode's process based approach is based on the taxonomic framework of Burrel and Morgan (1979), through which one consciously traverses the problem space (with its underlying ontological and epistemological assumptions) in order to develop a richer understanding of the nature of the concept under investigation. Roode's approach allows the researcher to deliberately pose different questions to explore different aspects of the problem or situation at hand. The researcher is not required to accept the assumptions associated with one question, but merely enquires about different facets of the research problem to obtain as much information about it as possible.

The research problem in this case was to capture the perceptions the various focus groups had about the impact of these thefts on ICT introduction in their respective rural areas. This research problem is interrogated from different perspectives through posing the following questions:

- **What is...?** The fundamental nature or essence of the research problem is first explored. The question aims at exposing the structure of the problem or the meaning of the underlying concepts or ideas. The fundamental assumption is that universally accepted descriptions for the concepts, ideas and problems do exist.
- **Why is...?** This question examines the real-life behaviour or characteristics of the problem. The focus is on determining relationships between aspects of and/or variables within the problem domain. The fundamental assumption underlying this question is that these relationships, when uncovered, can be used to generalise about the problem domain and causal consequences.
- **How does...?** The phenomenon or problem is directly observed and described as it manifests itself in reality.
- **How should...?** This question focuses on the conclusions, implications or normative aspects of the research results. It is an evaluation of the results or new insights obtained during the research. In some cases it might lead to prescriptive conclusions regarding the problem domain. In other cases it might enhance the understanding of the problem domain or redefine it. We discuss the specific questions raised in this way for our study in the research design.

Morphological Analysis

According to Ritchey (1997), Morphological analysis was developed by Fritz Zwicky as a general method for structuring and investigating the total set of relationships contained in multi-dimensional, usually non-quantifiable, problem complexes. The method is primarily based on defining a configuration space representing the variables and conditions of the problem complex being addressed (i.e. a morphological field), and on a special form of cross-impact analysis relying on internal consistency or coherence, rather than on causal relations. Ritchey (1997) uses morphological analysis for scenario development and risk management. In this context, he identifies three major problems for which morphological analysis is most suitable for application:

1. Many, if not all factors involved in analyzing complex policy fields and developing future scenarios are non-quantifiable, since they contain strong social-political dimensions and conscious self-reference among actors. This means that traditional quantitative methods, causal modeling and simulation are relatively useless.
2. The uncertainties inherent in such problem complexes are in principle non-reducible, and often cannot be fully described or delineated.
3. The actual process by which conclusions are drawn in such studies is often difficult to "trace. According to Ritchey, without some form of traceability we have little possibility of scientific control over results, let alone reproducibility.

As indicated earlier already, a morphological field is a complex configuration (e.g. matrix) defining a set of inter-related variables (dimensions), each with a range of (discrete) conditions, which it can express. The idea behind the morphological field is that no single variable is regarded as the dominant one, or "driver". According to Ritchey, any variable - indeed any single condition associated with a variable - could, given the right historical circumstances, become a dominant driving force for the evolution of the entire field. The number of possible states in a field is equal to the product of the number of conditions under each variable. The morphological approach was followed for this study up to the point of identifying the leading variables. Aspects of the morphological approach related to internal consistency of conditions were not deemed necessary for the problem at hand. The leading variables are classified as *themes* in our study and are briefly described in the research design. A brief description of the focus groups and their settings follows.

THE FOCUS GROUPS AND A BRIEF DESCRIPTION OF THEIR RESPECTIVE SETTINGS

Sibonelo Primary School

Sibonelo primary school is situated near the SEIDET community education centre in Siyabuswa. The school was selected by the National Department of Education to host a pilot multimedia project. Fifty computers, donated by the Netherlands government for this project were stolen. The school principal responded positively to a request to participate in this research initiative. At the time of the focus group session, the computers had been replaced by the National Department of Education.

SEIDET Community Education Centre

SEIDET has already been mentioned in the introduction. It is a community based educational project, which started in 1991, providing a wide range of supplementary tuition programmes to secondary school learners and the general community. It is largely through its partnership with the Department of Informatics, University of Pretoria that ICTs were later introduced in almost all its programmes. Twenty-four computers were stolen from its main computer laboratory in Siyabuswa.

The Primary Schools Science Project (PSSP)

The PSSP is a special partnership project between SEIDET, the provincial Department of Education and the Centre for Science Education of the University of Pretoria. At the time of the theft of computers at SEIDET, fifty-six primary school teachers from the KwaMhlanga district were busy with their final test on basic computer literacy training. They were thus directly affected by the theft.

RESEARCH DESIGN

Data was collected through the use of a combination of Nominal Group Technique (NGT, Delbecq *et al*, 1975) and focus group discussions. Three focus group sessions were held with the three focus groups already described.

The first was at Sibonelo primary school and was attended by sixteen people representing a wide spectrum of the community sectors. The sectors comprised the school principal and his deputy, representatives of the school governing body and learner representatives, the area police commissioner and other senior police officials, the circuit manager, the municipality councillor and representatives of the NGO sector. It was conducted on 19 June 2001. Forty-three teachers attended the PSSP session from the twenty-six participating schools, the curriculum implementer from the KwaMhlanga Education district, and representatives of the Centre for Science Education of the

University of Pretoria. It was conducted on 21 June 2001. The SEIDET session, attended by thirteen members of its executive committee, was conducted on 07 July 2001.

The procedure followed was the same in all the focus group sessions. We started by ensuring that the expectations were properly managed at the outset in each case. It was emphasized to the participants that our main aim was a better understanding of the perceived impact of the theft, explaining that we were interested in their perceptions and those of other community members they interact with. The participants were further asked to understand that at this stage no one had facts since the police would then have dealt with the theft problem accordingly. The need for this initial fundamental understanding was critical because of the nature of the problem.

Participants were then informed about the procedure to be followed. Each was asked to give one silent response written on the sticker and post it on the flipchart. This was followed by a full discussion based on issues raised individually by the participants. At each step, they responded to one question before the next one could be asked. They wrote responses on stickers and stuck them on the flipchart. Themes were identified through grouping of related stickers and the responses were discussed in group context. The themes were written on separate flipcharts during the discussion. Discussions were given sufficient time and members were encouraged to make notes for their own future reference and possible independent action where necessary and possible. The questions were asked in the following order in accordance with the process-based research framework:

- **Why is...?** I*n your own view, why were the computers stolen from Sibonelo/SEIDET?*
- **How does...?** *In your observation, how did the community react to the theft?*
- **How should...?** *In your own view, how should your community deal with the computer theft problem?*
- **What is...?** *What do you think the impact of the computer theft is on the introduction of ICT in your area?*

We present next the results, their analysis, interpretation and some observations.

ANALYSIS OF RESULTS AND THEIR INTERPRETATION

The results from each focus group were placed on to a Morphological field, in accordance with related themes. Because of space restriction, we are not able to present the separate Morphological fields per question and focus group. We are only able to show an integrated Morphological field of all the themes (leading variables) per question. These integrated results are shown in "**Table 1.** Table of integrated morphological results." The themes were ranked in accordance with the total number of times they were cited by the focus groups. (See the last column of the table.) The integrated themes corresponding to each questions are listed in the first column of the tables while the number of times each focus group cited such a theme are indicated in parenthesis in the second column.

A quick glance at the tables will give the reader a complete picture of the resulting themes. The ranking of the themes for question one show the three leading perceptions and motivators for the theft as *selling to get money, lack of security and high crime rate in the country* respectively. Similarly, the three leading perceptions with regard to how the community reacted to the theft are respectively, {*infuriated, angry, disappointed, devastated, saddened, shocked, frustrated, helpless}, feeling of being robbed of development opportunity, ignorance-no concern-mixed reaction.* An interesting theme ranked fifth is that of blaming the police. There is an implied aspect of blame, but at the same time an acknowledgement that co-operating with them could be the solution (response to question 3). With regard to the third question on how the community should deal with the computer theft problem, the three leading themes are *working with the police*, *educate people to respect community property* and *community needs to "become the police"*. For the last question that pertains to the perceptions on what the impact of these thefts were on the introduction of ICTs in their respective areas, *development progress delay/hindrance, negative development impact* and *donor discouragement* are the leading themes.

Table 1: Table of Integrated Morphological Fields

Themes	**No. of citations per Focus group [Sibonelo, PSSP, SEIDET]**	**No. of times cited combined**	**Rank**
Question 1 **In your view, why were the computers stolen from Sibonelo/SEIDET**			
Selling to get money	[3,5,5]	13	1
Lack of security	[6,5,1]	12	2
High crime rate in South Africa	[0,6,5]	11	3
Jealousy and selfishness	[2,5,2]	9	4
Establishment of own computer school/center	[2,3,1]	6	5
Lack of respect for community property	[3,2,1]	6	5
Unemployment and poverty	[0,4,1]	5	6
Syndicates	[1,1,1]	3	7
Own studies	[1,0,0]	1	8
Question 2 **In your observation, how did the community react to the theft?**			
Infuriated, angry, disappointed, devastated, saddened, shocked, frustrated, helpless	[4,17,7]	28	1
Feeling of being robbed of development opportunity	[0,9,3]	12	2
Ignorance-no concern-mixed reaction	[0,9,1]	10	3
Tried in vain	[5,2,1]	8	4
Blamed the police	[1,3,0]	4	5
The usual way of life – crime	[0,3,0]	3	6
Fear, criminals are within the community	[0,2,1]	3	6
Self-gain	[1,1,0]	1	7
Called traditional witch doctor	[2,0,0]	2	7
Future action/suspicion	[1,0,0]	1	8
Did nothing	[1,0,0]	1	8
Sympathy	[0,1,0]	1	8
Computers removed to home/police station	[1,0,0]	1	8
Security was improved	[1,0,0]	1	8

Table 1: Table of Integrated Morphological Fields (continued)

Themes	**No. of citations per Focus group [Sibonelo, PSSP, SEIDET]**	**No. of times cited combined**	**Rank**
Question 3 **In your own view, how should the community deal with the computer theft problem?**			
Working with the police	[4,9,5]	18	1
Educate people to respect community property	[2,10,1]	13	2
Community needs to "become the police"	[0,11,0]	11	3
Do not buy stolen goods	[0,7,2]	9	4
Community representatives do security reviews of premises	[3,3,0]	6	5
Improved security	[3,3,0]	6	5
Information for prevention	[0,0,4]	4	6
Working together between school/SEIDET/community	[3,0,0]	3	7
Pupil involvement	[0,2,0]	2	8
Report to higher authority	[0,1,0]	1	9
Follow-up on houses with hidden spaces underneath floors	[1,0,0]	1	9
Question 4 **What do you think the impact of the computer theft is on the introduction of ICT in your area?**			
Development progress is delayed/hindered	[0,16,6]	22	1
Negative development impact	[3,9,2]	14	2
Donors will become discouraged	[7,5,0]	12	3
Community is technologically disadvantaged	[0,10,1]	11	4
Bad image to community	[3,0,0]	3	5
Learners will suffer	[3,0,0]	3	5
Opportunity for community to work together	[1,1,0]	2	6
Positive impact/No impact	[0,1,1]	2	6
A wake-up call	[0,0,1]	1	7
Resource re-allocation	[0,0,1]	1	7

The interpretation of the results was informed by the work of Walsham (1995). In order to interpret these themes, we attempt to capture what the focus groups are saying in the following way: Because of insufficient security at the respective premises in which the ICT equipment are kept, the perceived ineffective policing and a high crime rate in the country, criminals are able to steal these equipment in order to sell them and make money. This result in the respective communities being deprived of development opportunities likely to be brought about by these technologies, leading to community anger, frustration, devastation and a sense of helplessness. In order to deal with this problem, the groups suggests that some form of "community policing", an enhanced co-operation with the police and a community education programme about the need for respect of community property could be considered. According to the focus groups, development progress is delayed/hindered and the donors would be discouraged by these thefts.

The manner in which the community takes action on these thefts would probably be informed by the prevalent perceptions regarding these thefts. In this regard, it would seem that the emotional reaction to the theft (probably those in the community that are intimately involved in the ICT projects) could be contrasted with the indifference (theme 3 of Question 2) of the rest of the community. The emotional reaction could serve as a driver to replacing the equipment, while the indifference may lead to an "acquired helplessness" in the face of the high crime rate in the country. The burden would therefore be on those involved in the ICT development process to take decisive action to ensure that development opportunities are not hampered.

We have scanned the literature on the problem under investigation, hoping to find that the developed world has ways to deal with situations where the demand for technology cannot be satisfied by available means, thus creating potential for theft. We discovered that, on the contrary, computer equipment theft is an international problem, prevalent even in highly developed countries like the United States of America and the United Kingdom. According to Stair and Reynolds (1999, p 634), in 1995, the theft of portable computers accounted for $640 million losses in the United States alone. They indicate that the statistics in the United Kingdom are just as staggering, with more than 30 percent of all reported thefts being computer related. This is contrary to our initial thoughts when we engaged in this study. It would seem that this might compel the researchers to look for a holistic and a critical systems thinking perspective to the problem. In this respect, the perspective of Midgley on critical systems thinking may be helpful. Starting from looking at the respective communities affected by these thefts as constituting elements within the primary boundary (see Jackson *et al*, 1991, p. 397), we come to the conclusion that we can neither exclude nor marginalize the wider national and even international contexts in terms of the issues addressed in this study. This compels us to conclude that the study results fit within a system that has wider primary boundaries than rural Mpumalanga in South Africa. Taking a systems perspective on this reduced system may lead to reduced understanding of the wider problem. Locating our study within a broader action research framework may result in substantial contributions to the problem within a wider context.

CONCLUSIONS

Bringing the focus groups together and discussing the theft trend, has been an exercise in raising awareness of the problem.

The communities' assessment of the impact of computer hardware theft were obtained and summarised in terms of themes. These themes seem to suggest that one of two development possibilities exist:

1. The community can adopt an indifferent stance of acquired helplessness, which may induce inaction, resulting in delayed introduction of ICTs and reluctance of donors to become involved.

2. Alternatively the community could channel the energy released through their anger and disappointment into taking preventative measures, such as becoming involved in the security management of ICT equipment and co-operating with the police.

The theft trend for this study has been considered within the context of rural ICT development. A consideration of worldwide trends in computer theft shows that the problem could be considered within a considerably expanded system.

The threat to our research in ICT introduction at SEIDET became an opportunity for a better understanding of the surrounding community. With this trend in computer theft, the question arises whether the goals of the Universal Service Agency will be realised, and if they are realised, how the theft problem can be effectively dealt with. The impact this theft trend could have on community development related to ICT introduction, as perceived by the focus groups, remains a challenge to the country as a whole and the international development community.

REFERENCES

Burrell, G. and Morgan, G. (1961). *Sociological paradigms and organizational analysis.* Heinemann, London.

Delbecq, A.L., van den Ven, A.H., and Gustafson, A.H. (1975). *Group techniques for program planning*, Illinois, Scott, Foresman.

Midgley, G. (1991). The sacred and profane in critical systems thinking, in *Systems Thinking in Europe,* (M.C. Jackson, G.J. Mansell, R.L. Flood, R.B. Blackham, S.V.E. Probert, (eds.)), Plenum Press, New York.

Ritchey, T. (1997). Scenario development and risk management using morphological field analysis: research in progress, in *Proceedings of the 5th European Conference on Information Systems*, (R. Galliers, S. Carlson, C. Loebbecke, C. Murphy, R. Hansen, R. O'Callaghan, (eds.)), Cork, Ireland.

Roode, J.D. (1993). Implications for Teaching of a Process-based Research Framework for Information Systems, in *Proceedings of the International Academy for Information Management Conference*, (L. Smith, ed.), Orlando, Florida.

Stair, R.M. and Reynolds, G.W. (1999). *Principles of information systems-a managerial approach.* 4th ed., course technology, Cambridge.

Walsham, G. (1995). Interpretive case studies in IS research: nature and method. *European Journal of Information Systems,*4 (2):74-81.

PARTICIPATION IN THAMES CHASE COMMUNITY FOREST: A MULTIMEDIA ELICITATION TOOL

M.W.J. Spaul, S.H. Evans and S.A. Unsworth

School of Design and Communication Systems
Anglia Polytechnic University
Victoria Road, Chelmsford, Essex, CM1 1LL
m.spaul@apu.ac.uk

INTRODUCTION

This paper describes the design thinking behind, and the outline structure of, a multimedia system to represent and elicit comments on some aspects of Thames Chase Community Forest in Essex. It is part of a programme of research projects being carried out at Anglia Polytechnic University intended to form links between the university and local communities and businesses. These projects are designed to be "dual use", with benefits for both the community and the academic work of the university. In the case of this project the aim is to provide Thames Chase with a multimedia facility for use in its schools outreach programme, and simultaneously - for the present authors - to be part of a wider research programme into the representation and consumption of countryside leisure sites. This paper is written principally from the latter, academic, perspective; it outlines the background to the development of community forests, a rationale for the research project, and the conceptual design of the multimedia system.

BACKGROUND

The concept of a community forest is a flexible one, implemented in different ways depending on geographical and development contexts; however, its core meaning is that of a forest managed and used principally for the benefit of a community living within, or close to, the forest. A further assumption is made that the management and use of forest resources is of a sustainable nature. In developing countries this concept is usually realised as a cooperative, community-based forestry enterprise (see, e.g., Ascher, 1994); but in the developed context of the UK the reality is more elastic and complex.

In the UK the development of community forests was driven by central government, through the Countryside Commission (1989) as an instrument for

Edited by Ragsdell *et al.*, Kluwer Academic/Plenum Publishers, New York, 2002.

addressing several interlocking problems: the dereliction of areas of the urban fringe around major cities as a result of the upheavals in mining and heavy industry in the early 1980's; changes in the economic environment of agriculture, with incentives for large tracts of land to be taken out of arable production; and the negative image of forestry (and the Forestry Commission) with its imposition of regimented coniferous landscapes on the countryside. Areas that had been left in a state of dereliction by the loss of mining, gravel extraction and heavy industry were targeted for regeneration projects based on broad-leafed forestry. These projects were deemed to offer multiple benefits in terms of the development of leisure resources and attendant service industries, light industries based on woodland products, a rise in property prices as the image of the areas changed, and the development of a focus for community life in the wake of the destruction of communal bonds based on lost industries. As an additional benefit, large-scale afforestation was seen as a means of helping to meet global commitments on climate change and biodiversity.

The model of governance of community forests is based on partnerships between a range of public and private bodies: the lead agencies of the Countryside and Forestry Commissions, local councils, private landowners, and a range of voluntary groups focused on conservation and countryside interests. The realisation of the forests as active community projects is based on a range of routes being available for public representation and participation - principally through the activities of voluntary conservation groups, direct involvement in planning exercises centred on forest design, planting initiatives, etc.

Thames Chase is one of twelve community forests in the UK. It was set up in 1990 and covers approximately 40 square miles of South Essex and East London, tracing an arc from Brentwood in the north, through South Ockendon, to Hornchurch in the west. The forest takes in pre-existing woods, country parks, and outdoor activity centres, with extensive new planting on reclaimed gravel quarries and disused farmland. Thames Chase has an active community outreach programme, with volunteer groups, planting events, schools liaison, participatory planning and a calendar of recreational events at its various centres.

Despite this activity it is reasonable to raise questions over the levels of public participation in the development and use of the forest, and whether the profile of this participation corresponds to the community-wide involvement envisaged in the initial policy document (Countryside Commission 1989, pp.1-18). A study of participation (Evans, 2001) showed that, even amongst those who expressed an active interest in the forest, the majority were only concerned passively to receive information; levels of participation in planning exercises was lower than that in directed planting and conservation work. Interest in Thames Chase was shown to be correlated with prior membership of wildlife and conservation organisations, and with profiles of age and social status typical of countryside leisure activities - owner occupiers (more heavily concentrated in the north of the Thames Chase area) were disproportionately represented. Investigations in under-represented areas (those with higher concentrations of local authority housing, to the south of Thames Chase) elicited responses which indicated a preference for built forms of development that might offer more employment opportunities. Such results throw into question the concept of the forest as an instrument of community development, as it seems to fail to engage with the life experience of large groups for whom it was intended to be a significant focus.

THE MORAL GEOGRAPHY OF PLANNING IN THE URBAN FRINGE

One way of approaching the participation problems experienced by Thames Chase, but one which is difficult to explore in any precise manner, is that of the broad boundary judgements inherent in the implementation of the community forest concept in the UK. Two broad, and interlinked, sets of assumptions that have conditioned the projects may be discerned. Firstly, there is the relatively fixed planning framework within which the forests had to be developed: green belt planning restrictions, a climate of approval for multi-agency, public/private partnership ventures, and the need to enlist the cooperation of private landowners. Secondly, there are the assumptions underlying the idea that large-scale afforestation may act as an instrument of social change: the landscape aesthetics enshrined in UK planning culture, and the more general cultural association between the countryside and social harmony. Shoard (2000, pp.74-76) has pointed out that for planners, as for much of the population, the urban fringe is an unsightly "no-place"; lacking the comprehensible characteristics of the urban or the rural environment, these "edgelands" - a patchwork of poor farmlands, straggling paddocks and shrubland, abandoned mineral workings and remnants of derelict industrial buildings - are seen, in the light of a dominant planning aesthetic, as only fit for eradication.

In the case of green belt land, as in Thames Chase, this eradication means a return to some image of the rural; this leads to the development of a gamut of woodland centres, country parks, outdoor pursuit centres, and golf courses interspersed with private woodlands to soften the overall landscape. These changes were foreshadowed in the policy document (Countryside Commission, 1989, pp.8-13) with an artist's impression of a scarred urban fringe wasteland transformed into a wooded leisure landscape, with the crucial added visual suggestion that such an environmental transformation would carry with it a social transformation - delinquency, despair and poverty giving way to a community happily enjoying the genteel pleasures of the countryside, and bolstered by a new prosperity. Implicit in this "visual argument" is a highly selective moral geography (see Matless, 1998, pp.40-43) and the assumption that this moral geography could become the defining sentiment of a community (in the strong sense of "community" - that of a group who share a moral and cultural outlook).

The aesthetics of this planning vision stand in a long tradition of English equations between landscape and psyche, and the community forest concept is explicitly linked with Howard's "garden city" (Hall 1996, 86-135), in which landscape and lifestyle interlock to expunge the social ills bred by Victorian urbanisation. This interlinking of rural landscapes and harmonious social relations is part of a rural utopianism fuelled by traditions of landscape gardening (Hunt, 1994; Williamson; 1995), literature (Williams, 1975), and art (Barrell, 1980). A ruralised, but decidedly not a concretely rural, existence is seen as the natural state for the English psyche. Images of the rural drawn from these traditions, and dressed in modern guise, are used liberally in policy documents, forest plans and interpretative materials. They can thus appear as a plausible, aestheticised justification for a landscape that, in broad terms at least, is the inevitable outcome of the planning environment that surrounds community forest development.

EXPLORING THE VISUAL CULTURE OF THAMES CHASE

The exploration of the boundary judgement that assumed the intrinsic attractions of the community forest initiatives, the aesthetic and cultural judgement that woodland landscapes and leisure activities organised around them are, or could become through an educational process, an important feature of the life of a community poses an interesting challenge. Faced with a situation in which a large fraction of the population appears to find no resonance with the community forest idea, is it possible to find some form of non-reductive explanation of why this is so, and might such an explanation be of some use in a wider strategy for improving rates of participation? It was of interest that such a project could also overlap with one that pursued Shoard's (2000, p.91) contention that the urban fringe should be explored and celebrated in its own terms; and that the "edgeland" and its culture must be assumed to have a validity and worth, rather than simply being viewed as a landscape awaiting a desirable replacement. Having had some experience of exploring and constructing multimedia representations of the urban fringe landscapes of Chelmsford in a teaching context, it seemed possible this experience might be put to use in a study of aspects of the visual culture of Thames Chase.

The underlying moral geography of Thames Chase is, in common with other community forest initiatives, strongly articulated in visual terms - as a set of favoured images, landscapes, and the graphical designs and logos used to give the forest an identity. It thus seemed appropriate to focus on the visual in the analysis of this moral geography, using some of the techniques becoming available from the expanding field of visually-based social research (see, e.g., van Leeuwen and Jewitt, 2001; Rose, 2001). A structured visual exploration encoded in the new media (multimedia representations, plus the necessary communication facilities to collect responses) was seen to be a possible way of eliciting images, anecdotes and experiences from a "captive" cross-section of the population via the schools outreach programme, to whom we had been offered access. Students of upper-school age were regarded as an ideal group to target, since they had been shown to be under-represented in Thames Chase participants, and they could be assumed to be comfortable - via the IT curriculum and computer games - with multimedia technology and representations. Our experience with first-year university students in multimedia (some of whom live in the target area), although clearly not representative, led us to expect reasonable degrees of visual literacy. It was regarded as vital that this new media representation should try and encode, and prompt responses to, a wide diversity of ways of seeing and using the urban fringe area that constitutes Thames Chase. A model for this development was offered by the web-based "Three Cities" project (Balshaw, *et al.*, 2000). This used the new media simultaneously as a means of structuring an investigation of a large historical archive, representing and disseminating the results, and eliciting responses from the academic community for consideration in future revisions of the material. The use of such a strategy seemed to be feasible with a segment of the Thames Chase population used to interacting with information technology.

An important insight from the "Three Cities" team was that the process of researching and connecting materials in, and presenting results in, a range of media has its own logic and is not simply the pursuit of text-based research by other means. Mindful of this experience our research strategy was, however, structured on an established approach to visual cultural analysis outlined by Lister and Wells (2001). In this approach an image, or group of images, is regarded as the visible centre of a system of social and cultural processes that serve to endow it with meaning - a processual meaning that is inevitably shifting and contested. Rather than questioning what an image means in itself, this approach seeks to articulate aspects of the social and cultural system surrounding the image. This is a special case of the kind of material cultural analysis practised in various spheres (see, e.g., Benjamin, 1999; Williams, 1975; Matless, 1998). In Lister and Wells' formulation the visual image is pursued in four inter-related dimensions: the provenance of the image, i.e., its production and history; the audiencing practices that bring people into the presence and consumption of the image; the wider cultural meanings that may be attached to the image via established visual codes and traditions; and the social identities that may respond to, and be reinforced by, contact with the image. The end result of such an analysis is not any kind of synoptic statement of the meaning of an image - rather the meaning is implicit in the dense web of connected images, social practices, historical facts, and explicit or implicit responses from audiences uncovered during the analytical process. For this reason, a new media representation is well-suited to guide and express the results of such an analysis.

The concept of visual image central to this form of cultural analysis is a broad one. The most obvious instance is that of an image both intentionally produced and sited for an audience; in the case of the current project this embraces such obvious items as the visual materials in planning documents, leaflets and other interpretative documents, but also the signage of sites, the architecture and layout of visitor centres, and - of course - the designed landscape itself. Visual analysis may be usefully extended to a range of images which may be neither intentionally produced nor sited: chance configurations of signage, objects in the landscape, material cultural items discarded at sites, landscapes produced by random historical and natural processes, etc. All may be gathered together in the detective work of tracing the meanings that are attached to landscapes.

The rationale for this approach to the aesthetic and cultural assumptions underlying the community forest initiatives is that the reasons for the persuasive or antagonistic powers of an image are not readily available for introspection and discourse. It is a common planning practice to use visual forms in validating a design, and the new media are increasingly used in this way: an audience drawn from a target population is asked to comment on, choose between, or perhaps modify a design on the basis of building fly-throughs, or photo-realistic representations of landscapes. Such approaches can neglect certain aspects of the experience of place, in that the lived experience of a place may be different from an initial visual impression, and that this experience may lie below a level of consciousness that subjects are able directly to articulate (de Certeau, 1984; Lefebvre, 1991). An elicitation process based on visual cultural analysis is based on prompting subjects to share valued images, anecdotes and experiences within an environment offering multiple "hooks" to which they might connect their lived experience, and expressive resources through which such responses might be shaped.

This approach to elicitation based on multimedia is an invitation to a form of social dialogue; but one of a quite asymmetric kind, since it is based, in the first instance, on a reading of place constructed by the designers. Subsequent revisions take account of the perspectives of target groups, but these are, once again, incorporated into a designed framework. This is not the straightforward reflection of an open dialogue in an electronic medium (of the type common in community forums and bulletin boards), but rather a "dialogical" approach in Bakhtin's sense (see Bakhtin, 1981; Hirschkop, 1999). The central design idea is the orchestration of the different "social voices" visually detectable within the landscape of Thames Chase into a structure that makes their differences apparent, and makes them available for comment and criticism. This cannot be a neutral process, since there is no uninvolved standpoint from which the selection and arrangement of materials may begin. Such neutrality as can be achieved is through designing in the qualities that Bakhtin identified as the essence of the novel - for him the apex of a democratic culture. A communicative structure that reflects, and in turn reflects upon, a form of social life (Hirschkop, 1999, p.5).

CONCEPT DESIGN

The basic structure of the multimedia system is constrained by the context for which, and audience for whom, it is intended. As it is embedded in the Thames Chase schools outreach programme it must be seen initially by the user as representing the forest in a manner broadly consistent with the other interpretative devices used in the programme. Any distinct breaks in perspective from established to "alternative" representations are managed within the system, where they can be set in context.

The "front end" of the system thus presents Thames Chase in a conventional and accessible way: as a bounded area on a map, as sites with different characters and activities, as a repertoire of woodland images, and as a well-recorded developmental history. Similarly, the means adopted to structure and navigate this material are those that have become standard in tourist kiosks and related web sites: interactive maps of the forest and of individual sites, navigable tours around sites, and linking themes such as recreation or tree types. In addition, it was perceived as necessary to use sufficient "high-end" technology to keep a young audience engaged, using the paraphernalia of dynamic HTML, Flash animations, panoramas, and movie clips (the exact technical details are not crucial to the concept). This material, and the remainder described below, is accompanied by a set of standard "response boxes" inviting anecdote or reaction from the user.

This conventional front-end leads, however, to more multivalent material, and invites users to "take another look". In addition to the topography and development of Thames Chase held within this framework there is a more subtle human geography encoded in the gaps between these representations. This structure is reachable as a set of named themes through the main navigation and via lower-level links from appropriate topographical material. These themes are intended to disrupt established perceptions by approaching Thames Chase through a set of provocative themes: "under new management" (a study of the way in which aspects of the landscape are being remoulded and given a "brand identity"), "signs and signals" (a study of the official and unofficial signage of Thames Chase, from notice-boards to graffitti, and their chance juxtapositions), "boundaries, bridges and barriers" (the impediments to experiencing Thames Chase as a geographically or communally unified entity), and "the car in the landscape". This latter theme will be expanded to give an example of the experience of "taking another look".

Thames Chase, as both a concept and an area, has an ambivalent relationship with the car. Although cars seldom appear in the planning and promotional material - they make way for the more visibly sustainable transports of walking and cycling - they are experienced as a key feature as one moves around the forest. To walk the lanes that link the sites of the forest is an alienating and at times frightening experience. The forest is bounded (approximately) by major routes - including the iconic M25 - with very high traffic densities, and the small lanes that cross the area carry fast-moving, "rat-running" traffic. Planting schemes attempt to screen the Thames Chase sites from the sights and sounds of the road but more indirect influences are equally real. Local traffic includes, on occasion the vehicles of the Thames Chase management; beside their role as transport they act as travelling billboards for the forest, emblazoned with multiple logos. The major sites such as Belhus or Thorndon (popular dog-walking and picnic territory) are most commonly accessed by car, and their visitor catering and interpretation facilities are constructed with this fact in mind. To enter a site via the "back door" of a footpath is to have a very different encounter with the landscape. Unplanned influences may also be seen: the Baldwin's farm site is a common dumping ground for burned-out, unwanted or stolen cars; also the car is the enabler of the practice of 'fly-tipping', that turns lanes and car-parks into a contemporary archaeological trail of household waste. As Shoard (2000, p.89) has observed, the urban fringe is a visible manifestation of social process, and the multimedia design of this themed zone attempts to present what might be seen as intrusion and delinquency in a neutral light; as features of a landscape to be read and understood.

CONCLUDING REMARKS

At the time of writing, the project has reached the stage - in a process modelled on that of Lynch and Horton (1999) - of content creation and implementation, with a prototype being due for evaluation at the end of 2001. A complete version, starting the first cycle of feedback is due by the early summer of 2002. Wider contact via the route of the new media with other elements of the Thames Chase community has been considered, but such a project is hampered by the problems of differential access and facility with technology, which would compromise any results obtained.

REFERENCES

Ascher, W. (1994). *Communities and Sustainable Forestry in Developing Countries*, ICS Press, Oakland.

Bakhtin, M. (1981). *The Dialogic Imagination*, University of Texas Press, Austin.

Balshaw, M., Kennedy, L., Notara, A. and Tallack, D. (2000). *City Sites: Multimedia Essays on New York and Chicago 1870s - 1930s*, University of Birmingham Press, Birmingham.

Barrell, J. (1980). *The Dark Side of the Landscape*, Cambridge University Press, Cambridge.

Benjamin, W. (1999). *The Arcades Project*, Belknap Press, Cambridge, Mass.

Countryside Commission, (1989). *Forests for the Community*, Countryside Commission, Cheltenham.

De Certeau, M. (1984). *The Practice of Everyday Life*, University of California Press, Berkeley.

Evans, S. (2001). *With the People or For the People: The Role of Participation in Community Forest Creation*, Ph.D. Thesis, Anglia Polytechnic University, Chelmsford.

Hall, P. (1996). *Cities of Tomorrow*, Blackwell, Oxford.

Hirschkop, K. (1999). *Mikhail Bakhtin: An Aesthetic for Democracy*, Oxford University Press, Oxford.

Hunt, J. (1994). *Gardens and the Picturesque*, MIT Press, Cambridge, Mass.

Lefebvre, H. (1991). *The Production of Space*, Blackwell, Oxford.

Lister, M. and Wells, L. (2001). "Seeing Beyond Belief: Cultural Studies as an Approach to Analysing the Visual," in *Handbook of Visual Analysis*, (T. Van Leeuwen and C. Jewitt, ed.), Sage, London.

Lynch, P. and Horton, S. (1999). *Web Style Guide: Basic Principles for Creating Web Sites*, Yale University Press, New Haven.

Matless, D. (1998). *Landscape and Englishness*, Reaktion Books, London.

Rose, G. (2001). *Visual Methodologies*, Sage, London.

Shoard, M. (2000). "Edgelands of Promise," *Landscapes*, 1(2):74-93.

Van Leeuwen, T. and Jewitt, C. (2001). *Handbook of Visual Analysis*, Sage, London.

Williams, R. (1975). *The Country and the City*, Paladin, St. Albans.

Williamson, T. (1995). *Polite Landscapes*, Sutton, Stroud.

Section Seven

Business

A SYSTEMS APPROACH TO MARKETING

Declan P Bannon

University of Paisley
Paisley, PA1 2BE
declan.bannon@paisley.ac.uk

INTRODUCTION

The objective of this paper is to evaluate a systems approach to marketing in the light of developments in Marketing Theory over the past decade. In particular, contrasts are drawn between traditional transactional based marketing and the presently hailed new paradigm, that of relational exchange based marketing.

The concept of Marketing as a set of activities that stimulate and satisfy demand has been developed over thousands of years. In fact elements of marketing are observed and demonstrated by the behaviour of 'the oldest profession'. Marketing traditionally was viewed as a process of transactions for the exchange of goods between producer and consumer. Successful marketing was perceived as having occurred if sales revenue and market share were increasing. However the Marketing Concept has been widened and encapsulates all forms of exchange including industrial marketing, services marketing, public sector marketing and not for profit sector marketing. These developments challenged the existing paradigm of marketing as a purely commercial activity and plunged the discipline into a mid-life crisis.

The 1960's saw the rapid growth in the academic discipline of Marketing. Research was characterised by a theoretical focus on industrial markets, with papers and academic texts on Marketing Systems not uncommon. In the 1970's considerable academic effort was directed towards the not for profit sector, whilst the 1980's were dominated by Services Marketing. Nowadays, standard Marketing text books do not refer to Marketing Systems and virtually nothing has been published on Marketing Systems since the late 1960's.

In the 1990's, marketing academics were smitten by a new concept, that arguable was not new but carried with it a logical and noble view of the future of marketing, that of Relationship Marketing. 'Relationship Marketing concerns attracting, developing and retaining customer relationships' (Jackson, 1985). 'Relationship Marketing is a customer centred approach whereby a firm seeks long-term business relations with prospective and existing customers' (Morgan and Hunt, 1994). Relationship Marketing is a systems-

Edited by Ragsdell *et al.*, Kluwer Academic/Plenum Publishers, New York, 2002.

orientated approach, however little or no emphasis is placed on utilising general systems theory to organise the complexities of the interrelationships that are central to Relationship Marketing.

This paper critically evaluates the benefits and limitations of the two dominant approaches to marketing and argues that hybrid approaches can and should be developed as Marketing Systems. Relationship Marketing is systems-orientated, yet it includes managerial aspects (Gronros, 1994). A systems approach to Marketing encompasses all relevant actors, environmental influences, and even the process nature of marketing.

MARKETING SYSTEMS

A systems approach to marketing is nothing new. As early as the 1960s marketing academics were exploring the possibilities of marketing systems theory opening up new avenues for investigation. The systems school of thought came about due to the huge explosion in interest in applying systems theory to operations management and other related management fields. In pursuing an understanding of the complexities and inter-relations that is an organisation, it is habitually argued that the organisation should be viewed as a system (Boulding, 1956; Kuhn, 1963; Fisk, 1967; Bertalanffy, 1968; Katz and Kahn, 1966; Dowling, 1983).

> 'General Systems Theory can organise the bewildering interrelationships between production, marketing and consumption into a coherent and unified perspective from the standpoint of the consumer, the marketing channel commander, the business manager, or the social welfare of the nation. General systems Theory provides an analytical framework of systems levels for classifying problems. Marketing problems as a class belong to systems characterised by communication and adaptation in social organisations' (Boulding, 1956).

A system requires a set of interacting elements, that whilst complex, form an organised whole, with the intention of creating outputs. A social system is more complex than a mechanical or biological system simply because you are introducing elements that are complex such as subjectivity, values, beliefs, attitudes, symbolism etc.

Bell (1966) provides an excellent summary of social systems as related to marketing:

1. A system is a group of interrelated components. Elements outside this relationship are outside the boundaries of the system and are components of the environment.
2. Components both inside and outside the system can effect the system.
3. Marketing systems by their nature are open systems.
4. The major focus of systems analysis is on the process of the system and how it changes as conditions change.
5. Systems are dynamic i.e. made up of variables that change. A component that does not change is not part of the system.
6. In marketing all components are dynamic.

Laser and Kelley (1962) discuss the component elements of marketing systems:

1. A set of functionally interdependent marketing relationships among people and institutions in the system i.e. the producer of the product or service, the distribution network, facilitating agents (advertisers, PR, advocates) and the complex buying and consumption elements.
2. The interaction between all individuals in the exchange relationship.
3. Objective setting and the establishment of activities to attempt to achieve such goals.
4. A customer focused environment where exchange takes place within a framework of constraints such as legal requirement, competitive forces etc.
5. The infrastructure for communication, financial transaction, distribution and marketing research are in place (Lazer and Kelley, 1962)

> 'It seems apparent that we need to devote considerable conceptual research to marketing as a system versus marketing systems' (Sheth, *et al*, 1988).

Marketing Systems are the aggregate representation of procedures required to operate efficiently within an exchange environment. Between the parties in a network system, exchanges occur that directly and indirectly impinge on members of that system. A flow of goods, information, finance and social exchanges take place in the network system (Hakanasson, 1982; Johanson and Mattsson, 1985; Kock, 1991).

> 'In such a network system the role and forms of marketing are not very clear. All exchanges, all sorts of interactions, have an impact on the position of the parties in the network. The interactions are not necessarily initiated by the seller - the marketer according to the marketing mix management paradigm and they may continue over a long period of time, for example, for several years' (Gronroos, 1994).

Figure 1 depicts a set of interacting elements that form a representation of the Marketing System.

Marketing has never had a general theory Sheth, *et al.* (1988); Hunt (1983); Baker (2000); Gronroos (1994). However, the Management School with its marketing mix management paradigm and its essentially transactional-based marketing has in the past been regarded as such. Some authors argue that the Management School of thought has had its day or should be confined to the marketing of consumer goods and that has no place in modern marketing e.g. Baker (2000); Gronroos (1994); Sheth *et al.* (2000); Hunt (1994); Webster (1994); Payne *et al.* (1995) and Kotler (1992). The majority of authors are advocating the Relationship School of Marketing as the modern saviour in Marketing's mid-life crisis. Others believe that the way forward is a hybrid form of marketing that incorporates the best of the transactional school and relationship school or that the two can and should co-exist Bannon (2000); Pels (1997) and Gronroos (1994).

> 'The post-industrial era has seen substantial development towards relationship marketing, both in practice and in managerial thinking. Marketers started realising the need to supplement a transaction orientation with an orientation that shows more concern for customers' (Sheth and Parvatiyar, 2000).

Figure 1. The Marketing System

RELATIONSHIP MARKETING AS A SYSTEMS-ORIENTATED APPROACH

'Relationship Marketing is systems-orientated, yet it includes managerial aspects. A systems approach is well suited as a basis for a general theory of marketing, because it makes it possible to include all relevant actors, environmental influences, and even the process nature of marketing' (Gronross, 1994).

Relationship building is the cornerstone of *relationship marketing*. The interaction and network approach clearly views marketing as an interactive process in a social context where relationship building and management are a vital cornerstone (Sheth *et al*, 1988; Bagozzi, 1974; Webster, 1994). They are in some respects clearly related to the systems-based approaches to marketing of the 1950s. Customers appreciate and value the integration of products and services which provide augmented products. This is hardly surprising; they wanted to buy the benefit of the combined bundle not the individual pieces. The systems approach was then adopted to an extent by the consumer goods and service sectors (Shapiro and Posner, 1979).

Baker (2000) describes the essential differences between the Marketing Management School and the Relationship School as follows. The Marketing Management School 'is based around a model that seeks to understand consumer needs better so that it can manipulate the elements of the marketing mix more effectively and so bend demand to the available supply. It tends to emphasis the transaction and sees exchange largely as a zero sum game in which there are inevitably winners and losers. Needless to say it is the seller who intends to win. In other words it is concerned with what marketers can do to buyers. By contrast, relationship marketing reflects the marketing concept much better. Relationship marketing is built upon the creation and maintenance of mutually satisfying exchange relationships. It enjoys a win-win perspective and sees the role of a producer as doing things for customers.'

Relationship Marketing as first discussed by Berry in 1983, is not a new concept but a redefining of the purpose of marketing itself. Relationship Marketing recognises the importance of sustaining mutually satisfying exchanges and building customer relationships as central to generating loyalty and repeat business. 'The concept suggests that instead of the narrow, transactional, one sale at a time view of marketing, marketing should emphasize relationships between the organisation and its markets more strongly' (Payne, *et al.*, 1994).

IMPLEMENTATION OF THE MARKETING CONCEPT

Most writers in the field of Marketing agree that implementation of the Marketing Concept is more difficult than it sounds and that many companies are simply paying lip service to the concept (Lambin, 2000; Baker, 2000; Jobber 1995; Bradley, 1995; Payne *et al*, 1994; Dibb, *et al.*, 2001. The marketing concept is integrated differently in organisations as demonstrated by Table 1.

Table 1. The Changing Role of Marketing

Marketing Theory

Marketing in Practice	Action	Analysis	Culture	Timescale
Passive Marketing	-	-		
Transactional Marketing	Yes	Limited	managerial	months
Strategic Marketing	Yes	Systematic	professional	years
Market-driven	Yes	MR# driven	manipulative	short/medium
Relationship Marketing	Yes	Sociological	altruistic	long term

 # Marketing Research

Relationship Marketing, integrates the marketing philosophy into the marketing function forming a marketing system that encompasses the entire organisation. Kotler (1997) argues that most companies do not embrace the marketing concept until they are driven to by combinations of sales decline, slow growth, changing buying patterns and increased competition increasing marketing expenditures. Table 2 contrasts transactional marketing with a relational marketing approach and attributes factors to each.

Table 2. Factor Comparative Matrix

Factor	Transactional Marketing	Relational Marketing
Time Horizon (Gordon 1998)	Short/ medium-term (Christopher *et al*1991)	Long –term (Anderson, 1982)
Strategy	Offensive/ Defensive	Relationship building (Dibb *et al* 2001)
Activities	New Business Generation (Christopher *et al* 1991)	Retention/ Customer Loyalty (Doyle, 1998)
Focus	Operational/ Functional (Gordon, 1998)	Philosophical (Dibb et al 2001)
Relationship	Proxy/ Professional Service	Representative/ Trusting/ Partnership (Gordon, 1998)
Scope	Restricted (Gronroos 1994)	Inclusive (Gordon 1998)
Attitude	Manipulative (Van Waterscoot, 1999)	Partnership (Gordon, 1998)
Communication	Spin/ Soundbites	Message Integrity
Style	Authoritative/ Patronising/ Mechanistic	Paternal/ Altruistic/ Craftlike (Mintzberg, 1994)
Marketing Ethos	Transactional Exchange (Gronoos, 1994)	Exchange Relationship (Van (Waterscoot, 1999)
Exchange Type (Bagozzi 1974)	Transactional (win-lose) (Christopher *et al* 1991)	Relational (win-win) (Baker, 2000)
Business Issues	Profit/Self interest	Quality of relationship (Brassington and Pettitt 1997)
School of Marketing	Management School (Baker, 2000)	Relationship School (Payne *et al* 1998)
Branding	Superficial/ Patronising	Brand Equity

CONCLUSION

The 21st century is already being referred to as the Knowledge Age and Marketing is going to have to operate within the Knowledge Economy. Relationship Marketing requires systems to enable the marketing process to develop and become more sophisticated. The use of hardware and software system has revolutionised the functional practice of marketing. However the nature and types of systems require to manage relationships, particularly the softer aspects of relationships is unclear. What is clear is that sustainable competitive advantage can be derived from the managing of relationships better than competitors.

This paper has compared and contrasted two approaches to the implementation of Marketing i.e. the traditional transactional approach and the relational exchange approach.

1. Transactional marketing which is focused on single discrete transactions, short in timescale and provides low levels of customer care and commitment.
2. Relationship Marketing as a systems approach focusing on the importance of customer retention, high levels of customer service and commitment over a long-term strategic horizon.

A systems approach to Marketing should be able to add value to and possibly lead to a better understanding of Marketing and the complexity of all its elements. Further research into systems methodology and application to Marketing is a strand of theoretical marketing that will be both fascinating and hopefully fruitful.

REFERENCES

Anderson, P. (1982). "Marketing, strategic planning and the theory of the firm", *Journal of Marketing*, 46(2):15-26.

Bagozzi, R. (1974). "Marketing as an organised Behavioural System of Exchange" Vol. 38 Oct. 1974, 77-81

Baker, M. (2000). *Marketing Strategy and Management* (3rd ed..) Macmillan Business Press, Basingstoke.

Bannon, D. (2000). "Political Marketing: Plotting the development of political activity into an evolutionary framework", in *Proceedings of the UK Academy of Marketing Special interest group on Political Marketing*, Loughborough University.

Bathie, D. (1994). "The Marketing Cycle" unpublished manuscript

Bell, M. (1966). *Marketing : Concepts and Strategy,* Houghton Mifflin Company, Boston.

Bertalanffy, L. (1968). *General Systems Theory*, George Braziller, Inc, New York.

Boulding, K. (1956). "General Systems Theory", *Management Science*, 2(April):197-208.

Berry, L. L. (1983). "Relationship Marketing". In Berry, L. L., Shostack, G. L. and Upah, G.D. (eds.) *Emerging Perspectives on Services Marketing* American Marketing Association, Chicago, pp.25-28

Bradley, F. (1995). *Marketing Management,* Prentice Hall, Harlow.

Brassington, F. and Pettitt, S. (1997). *Principles of Marketing,* Pitman Publishing, London.

Christopher, M., Payne, A. and Ballantyne, D. (1991). *Relationship Marketing* Oxford: Butterworth-Heinemann. The Marketing Series for the CIM pp.4

Dibb, S., Simkin, L., Pride, W. and Farrel, O. (2001). *Marketing: Concepts and Strategies,* (4th ed.), Houghton Mifflin Company, London, 14-15.

Dowling, G. (1983). "The application of General Systems Theory to an analysis of Marketing Systems", *Journal of Macromarketing*, 3(Fall):22-32.

Doyle, P. (1998). *Marketing Management and Strategy,* (2nd ed.), Prentice Hall Europe, Harlow.

Drummond, G. and Ensor, J. (1999). *Strategic Marketing*, Butterworth and Heinemann, Oxford.

Fisk, G. (1967). *Marketing Systems: An introductory Analysis*, Harper and Row, New York.

Gordon, I. (1998). *Relationship Marketing*, John Wiley and Sons, New York.

Gronroos, C. (1994). "From the marketing mix to relationship marketing", *Management Decision*, 32(2):4-20.

Hakanasson, H. (1982). *International Marketing and Purchasing of Industrial Goods*, John Wiley and Sons, New York.

Hunt, S. (1983). *Marketing Theory*, Irwin Series, Illinois.

Hunt, S. (1994). "On rethinking Marketing: our discipline, our practice, our methods", *European Journal of Marketing*, 28(3):13-25.

Jackson, B.B. (1985). *Winning and keeping industrial Customers*, Lexington Books, Lexington, Mass.

Jobber, D. (1995). *Principles and Practice of Marketing*, McGraw-Hill, London.

Johanson, J. and Mattsson, L. (1985). "Marketing Investments and market investments in industrial networks", *International Journal of Research in Marketing*, 4:185-95.

Katz, D. and Kahn, R. (1966). *The Social Psychology of Organisations*, John Wiley and Sons, New York.

Kock, S. (1991) "A Strategic process for gaining external resources through long-lasting relationships", Swedish School of Economics, Helsinki/Vasa, Finland.

Kotler, P. (1997). *Marketing Management*, (9th edition), Prentice Hall, London.

Kuhn, A. (1963). *The Study of Society: A unified approach*, Dorsey Press, Homewood, Illinois.

Lambin, J-J. (2000). *Marketing driven Management*, Macmillan Business Press, Basingstoke.

Lazer, W. and Kelley, E. (1962). "The systems approach to marketing" in *Managerial Marketing: Perspectives and viewpoints*, Richard D. Irwin, Homewood, Illinois.

Mintzberg, H. (1994). "The rise and fall of strategic planning", *Harvard Business Review*, (Jan-Feb):107-114.

Morgan, R.M. and Hunt, S. (1994). "The commitment-trust theory of Relationship Marketing", *Journal of Marketing*, 58:20-38.

Payne, A., Christopher, M., Clark, M. and Peck, H. (1995). *Relationship Marketing*, Butterworth-Heinermann, Oxford.

Pels, J. (1997). "Actors' exchange paradigm and their impact on the choice of marketing models", in *Proceedings of the 22nd Macro Marketing Conference*, Vol. 2 (A. Falkenberg and T. Rittenburg, eds.), Bergen.

Shapiro, B.P. and Prosner, R.S. (1979). "Making the major sale", *Harvard Business Review*, 59(6):103-110.

Sheth, J., Gardner, D. and Garret, D. (1988). *Marketing Theory, Evolution and Evaluation*, John Wiley and Sons, New York.

Sheth, J. and Parvatiyar, A. (2000). "The Evolution of Relationship Marketing" in *The Handbook of Relationship Marketing*, (Sheth, J. and Parvatiyar, A. eds.), Sage, London.

van Waterscoot, W. (1999). "The marketing mix", in *Encyclopaedia of Marketing*, (M.J. Baker. ed.), International Thomson Business Press, London.

Webster, F. (1994). "The changing role of Marketing in the corporation", *Journal of Marketing*, 56(4):1-17.

SYSTEMIC REQUIREMENTS FOR THE DEVELOPMENT OF COMPETITIVE CAPABILITY: LESSONS FROM A NEW ZEALAND STUDY OF COMPETITIVE ADVANTAGE

John Davies

School of Business and Public Management
Victoria University of Wellington
PO Box 600, Wellington, New Zealand
john.davies@vuw.ac.nz

INTRODUCTION

This paper reports on an ongoing study of the evolution of competitive capability in a set of exemplar New Zealand enterprises. The project has involved case-research (Eisenhardt, 1989) and historiographic studies (Goodman and Kruger, 1988) of a set of New Zealand enterprises over the full span of their history. These organisations have displayed a consistent ability to thrive over several decades of their history under different economic regimes, and include companies that have leading global positions in their specialist markets (See Appendix 1 for a list of firms).

The research project complements other work which reflects the resource-based view of the firm (Barney, 1991; Peteraf, 1993). Whilst that work examines the sources of a firm's long-run competitive advantage, particularly intangible resources such as reputation and relationships (Kay, 1993), organisational knowledge and learning (Nonaka and Takeuchi, 1995), and socially-complex, path-dependent configurations of 'activity systems' (Miller, 1996; Porter, 1996), this research seeks to build a complementary understanding of the evolutionary paths and processes by which configurations of advantage develop.

A large multi-disciplinary team, based at Victoria University of Wellington, has been working since 1998 to develop theory on this topic, funded by a major grant from New Zealand's Foundation for Research Science and Technology. The team involves specialists in decision sciences, human resource management, international marketing, operations management, organisation behaviour, organisation theory, strategy, and technology management. Their perspectives have variously been biological in nature to better understand learning processes; been based on network notions to examine processes of co-evolution or to examine how firms collaborate with suppliers and customers to internationalise; been drawn from cultural frameworks; or based on other conventional frameworks giving rise to the resource-based view of the firm. The organisations were selected by an advisory panel of business and policy leaders, and are mostly of medium size, employing between 100-1000 people.

The intent has been to describe and examine the evolution of organisations and the paths that the organisations have followed, and to do so by developing, linking and juxtaposing different perspectives that emerge from alternative disciplinary approaches underpinned by, and representative of, different paradigms (Gioia and Pitre, 1990). As such, our approach spans and accords with the regulatory dimension of Burrell and Morgan's (1979) paradigmatic framework in that our work is primarily founded in the interpretivist paradigm seeking to develop insights and understanding of the evolutionary development of competitive capability, and of the context of that development, aware that that those insights may be used with functionalist instrumentality by other organisations to better diagnose their competitive situation as a precursor to informed decision-making. Adopting Gioia and Pitre's definition (1990) of theory as "any coherent description or explanation of observed or experienced phenomema", our use of different theory-building approaches grounded in different paradigms is regarded as an appropriate way of fostering necessary comprehensive "portraits" of the complex phenomena of building competitive capability.

The study has involved semi-structured interviews with senior managers having a minimum of ten years experience in the selected organisations; the compilation of documentary information, both internal documentation and that available in the public domain; the writing of case histories; and feedback workshops with the organisations concerned to confirm facts and understanding. Such emergent understanding was also exposed to critique at a series of seminars with managers and practitioner organisations.

The remainder of the paper will briefly outline some of the general or composite findings of the study and then indicate how the use of a viable systems perspective has provided insights that have contributed to the development of the findings.

GENERAL FINDINGS OF THE STUDY

The study is ongoing, but significant components of a theory of the evolution of competitive capability are emerging. First, these exemplar firms display portfolios of capabilities that are unusually broad in scope, relative to what is known of the foundations of competitive advantage in New Zealand firms at large (Campbell-Hunt and Corbett, 1996). The deliberate leveraging of one source of advantage into another has also played a key role in creating broadly-based portfolios of advantage. The process has also been driven and moderated by coherences that emerge between one competitive capability and another.

The evolution of these broad portfolios of capability is strongly influenced by factors contingent on the context from which they were developed. For these New Zealand firms, this means that attributes of market isolation and distance, and of smallness, have helped develop capabilities which the firms have later found to be powerful competitive differentiators during their international growth. The roots of advantage in these firms are specific to the context and the time in which they were developed, but have been repeatedly re-configured to meet the evolving competitive challenges faced by them as enterprises. One such important and typical development challenge relates to the processes of internationalisation observed in the New Zealand organisations which processes follow a different pattern from those models manifest in large northern hemisphere economies.

The history of these firms also suggests that their portfolios of competitive capability develop two further properties: of balance - the strength of a capability to appropriately match the demands placed upon it by other capabilities and by the firm's overall strategic direction; and coherence - the ability of the firm to bring together all of its diverse activities and capabilities into a coherent whole, delivering an integrated bundle of value to the ultimate consumer. Indeed, balance and coherence play important roles in the evolution of competitive capability. Disturbances to such balance and coherence are considered to be what challenge and catalyse a firm to re-configure its capability portfolio. These disturbances are produced as often by the dynamic of a firm's own growth and development as by external changes. They are likely to be unique and often transformational, and new competitive capabilities are often the result of successful resolution of these disturbances and challenges. These exemplar firms have successfully navigated several of these transitions in their history. Some such transitions or journeys have been intentional, guided to some degree by recognised goals and by a degree of foresight. Others may be construed to be more driven by internal dynamics and to be the product of a more widespread process of 'structural coupling' (Maturana and Varela, 1987) as the organisation strives to recapture coherence in its internal processes, often without any awareness or clarity of the external forces acting on the enterprise.

A VIABLE SYSTEMS PERSPECTIVE

In this section, the activities of the case organisations are examined and discussed using the viable systems framework and notions of systemic function developed by Beer (1979; 1981; 1985). Beer's approach to issues of organisational design, effectiveness and capability is not pre-occupied with structural determinants. Instead, Beer focuses on the systemic functions which enhance organisational viability, and which provide a basis for adaptive learning about what are effective organisational behaviours and goals in a climate of complexity and change (Davies, 1999). Beer's research (1979; 1981; 1985) has demonstrated that certain systemic features or functions are necessary to any system's viability or survival. Knowledge of these systemic functions can be used to examine the systemic strengths and weaknesses in organisations, and/or to examine the design of organisations as they change or evolve to provide required systemic features and capabilities. Beer's cybernetic framework for analysing organisational systems is known as the Viable Systems Model (VSM). The use of cybernetic science to underpin the design of self-regulating, adaptive technical systems that can maintain required outputs, and work within established norms, is well known and predates Beer's attempts to extend its use to organisational systems (Francois, 1999; Shenhav, 1995). However, Beer's deliberate use of a definition of cybernetics as "the science of effective organisation - the science of communication and control" (1985: ix) goes beyond emphasising the applicability of cybernetics to natural or technical systems. It also emphasises the importance of communication - building communication channels, generating information flows and installing information feedback mechanisms - as part of organisational systems design to enhance learning and adaptive response, that is, to develop requisite variety in action and action choices for the organisation. Beer conceptualises all viable systems as a network of communication channels bonding five complementary sub-systems. The sub-systems, whose effective functioning and communication links are necessary to any system's viability, comprise - an operational system, S1, of autonomous units that act on and act out the identity and purpose of the overall system, and a supporting management meta-system comprising four other sub-systemic functions: S2 - effecting overall coordination of the autonomous units; S3 - operational planning, monitoring, control and audit functions relating to the autonomous units; S4 - intelligence and strategy development serving the whole organisation's future; and S5 - the creation and promulgation of identity, vision, direction, purpose and mission, throughout the organisation and its environments (Brocklesby *et al.*, 1995).

The success, over decades, of the chosen NZ organisations is testimony to their long term survivability and viability. It would seem appropriate then to use Beer's viable systems framework to examine aspects of each organisations' design, functioning and effectiveness to help determine any common systemic characteristics and any communalities in their environments or relationships with them. As such, the viable systems framework can help identify and interpret systemic function and dysfunction within an organisation or in its relationships with others in the wider system.

Here, our attention has been drawn to the S5/S4/S3 meta-systemic functions of the case organisations. In particular, we find S5 behaviour that builds and fosters identity linked to core values; that is visionary and also has missionary qualities; that is relationship oriented, identifying, and identifying with key constituencies; that is proactive, leading, creating rather than approving, reviewing, monitoring etc. Similarly, we find S4 functionality endorsing forward thinking and the long term view; by encouraging/developing an external focus, building toward responsiveness; by valuing intelligence and developing an awareness of wider system issues.

We note that many of our organisations have survived over long periods, responding, adapting and evolving to changes of huge complexity in their environments. In some cases, the evolution is akin to inter-generational change in a family, with the "inheritance" of key behavioural values and physical characteristics impacting on whether and how the organisation determines a new purpose or identity for itself, or even "re-invent" itself in a way that is congruent with its physical and financial assets and its conserved or inherited core values. In this sense, Beer's systems framework admits to a notion of adaptation that is more than a learning process of response to flux and change in the environment. It is a matter of the organisation's S4 capacity to model and understand itself, recognising "what I am", "what my strengths are", and what new purpose is congruent with the recognised strengths, accepted identity and character – a manifestation of structural coupling (Maturana and Varela, 1987).

These notions in themselves, are often considered congruent with a purposeful system which is visionary in nature, exclusively ends-oriented and aligned with strategic goals and objectives (Cummings and Davies, 1994). However, the purpose and motivation for an organisation's continuing existence may also reflect missionary characteristics of individual and strategic behaviour embedded in a value system that helps mutually define identity. In one sense, we see motivation, behaviour and adaptive learning linked by cognitive processes to ends-oriented purpose. In another sense, we see missionary purpose, behaviour and adaptive learning determined by the organisation's structural and value systems. We thus see a necessary interaction between the S5 function of deciding purpose as vision and mission, and the S4 functioning involved in recognising one-self, one's values and strengths. Effective interaction of this nature leads to a different dimension of systemic balance and coherence observable in the New Zealand organisations that have exhibited adaptive learning and changed. In some circumstances, we have evidence of single-loop learning activity (Argyris and Schon, 1978) driving organisations purposefully, though episodically, towards strategic goals:

> "Everything we did was about achieving those ends; everybody in the company knew, they always had this idea in mind," Rick Wells, Formway

... and in other cases, we have evidence of the necessary double-loop learning required as organisations experience what may be termed habitat change or the 'internal" organisational change that reflects restructuring, down-sizing, merger or acquisition, and which may involve re-inventing purpose and/or identity or self.

> "There is no ultimate design solution, just continual change," Rick Wells, Formway

> "I would debunk another myth. Many people will say a successful business starts from a business plan ... know where you are going and you know what you are going to do. That's rubbish ..." Angus Tait, Tait Electronics

We may also comment on how several organisations maintained balance and systemic coherence in terms of the right mix of autonomy and control, innovative free-thinking and focus. Such coherence not only reflects the systemic balancing act between Beer's 'here and now' S3 function and its S4 future oriented, strategising and intelligence function, but also the balanced interactions between the S5/S4/S3 management meta-system and the S1 operational units. In many cases, we note that even where decisions were seemingly outcome driven, weighted against purpose, this really reflected enhanced S2/S3 functionality, with clear policies and adopted norms leading to effectively coordinated and synergistic initiatives. The S5 functioning involved in promoting identity within and without the organisation, establishing and fostering ethical and behavioural values, has an impact on how coordination is effected throughout the organisation. Strong identity and widespread adoption of 'company' values allows for a diminution of formal control mechanisms and associated thick policy manuals, and a consequent increase in trust and autonomy ceded to the S1 operational units which dominate interactions and interfaces with those customers and suppliers etc. who make up the wider environment and system. However, it would be inappropriate to suggest that the promotion and adoption of 'shared values' is just a deliberate top-down S5 function. Ethical signals, for example, can emerge anywhere in an organisation, and values can be transmitted in many informal ways across boundaries at different logical levels as members engage in cross-functional projects or social activities. Similarly, those values can be transmitted or conserved on an intergenerational basis as individuals 'survive' major organisational change or the transition from one organisational form to another. Our organisations provide several examples of such S5 functioning.

> " ... we invest in all the capital resources needed to develop the land, and to create wealth for New Zealand. And that is a strong philosophy in this company ... we feel ourselves very much a New Zealand company,"
> Peter Anderson, Montana Wines Ltd.

> "We didn't stuff money in our back pockets, we built a business with it. I have enjoyed the opportunity of putting virtually all that money back into the business. ... What is the advantage of a bag full of gold ... over a structure that provides careers for lots of people and puts products into the world that are well respected?" Angus Tait, Tait Electronics.

As such, where organisations have stable, long term employment, and S2 recruitment and induction processes serve to endorse S5 core values and imprint identity, the mutually reinforcing behaviours of employees emerge as S2 coordinating norms that guide choice in the organisations' interests at all levels.

> "You basically have to have a philosophy which you are following and identify the routes that could bring that philosophy to pass. I am a great believer in that if something is right it will eventually happen.... you do not have to force it."
> Fred Holland, Nuplex

The need for S1 operational units that can 'respond' or act appropriately, rapidly, innovatively and effectively, even intuitively to environmental complexity is thus made possible by ceding autonomy in a climate where trust exists about expected S1 behaviours reflecting organisational values and norms.

> "Some people seize opportunities ... when you have 250 people always looking for a new idea ... that is a creative force," John Williams, PEC(NZ).

> "Everyone in this organisation is looking for better ways of serving the customer," Fred Holland, Nuplex.

> "The sort of people we employ give their best shot ... they're not driven by money all that much." Colin Campbell, Formway.

Of course, the S1 units also need to be supported appropriately by the organisation's meta-system. Market research and R&D activities are examples of how S4 intelligence functioning can be relevant to our organisations. Additionally, operational planning and financial resourcing represent S3 support for the S1 activities that are closest to the customer and to the market. There is evidence that our NZ organisations provided such tangible support as well as the intangible feel-good factor that builds from effective S5 leadership in values and behaviour.

> "...researching the US market ... get into the market place, talk to people at the ground level ... the key movers of product within the industry ... there is no substitute for doing it yourself." Tom Macky, Criterion.

The notion that the systemic design and operational characteristics of a viable system are immutable is questioned by the research findings. Beer has conceptualised all viable systems as a network of communication channels bonding the five complementary sub-systems. As such, we consider communication functions and processes of building communication channels, generating information flows and installing information feedback mechanisms as a necessary part of organisational systems design. We note that at Formway, and elsewhere, there is no place for autocratic managers.

> "The object is to allow ideas from any quarter a chance to flourish in a receptive atmosphere." Colin Campbell, Formway.
>
> " ... any good idea from any PEC staff member will always be given a chance to be evaluated." John Williams, PEC.
>
> "Problems can be aired without fears of repercussion." Andy Lusti, Svedala.

Such processes enhance learning and effect adaptive response. They help develop requisite variety in action, and may even effect change to other aspects of organisational design, including S5 purpose and identity, that is, they may contribute to the re-invention of organisation noted in this study. For example, PEC(NZ)'s open communication channels and willingness to listen to "insiders" and "outsiders" and to pickup on new product ideas, resulted in a major change of company direction and purpose, moving the company from electronic petrol pump products to security card systems. It is important to note though that PEC(NZ)'s change or evolutionary development, which was triggered unexpectedly from within the wider system, may only have been possible because its then existing structure and capabilities accommodated such change. Similarly, we note that Formway's success in reinventing itself can be attributed not only to its ability to recognise fundamental changes in market characteristics as a trigger for change in itself, but also to its ability to build and evolve its technical and manufacturing design capabilities to capitalise on such changed market circumstances. At Criterion, much of the change can be described as cultural change, with heightened recognition of the potential arising from individual development being sensed as changes in management style, but leading to the ceding of authority to self-managed teams and more appropriate levels of S1 autonomy. In these cases, it is suggested that the organisations maintained their S4 capacity for critical self-reflection even whilst redefining S5 purpose and direction. As an example, we may consider how Nuplex harnessed the benefit of government export incentives to redefine S5 purpose, to exploit S4 strategic opportunities and to bolster their S3 resourcing of development, but did so without becoming reliant upon what they regarded as a temporary phenomenon.

SUMMARY

In summary, we note these exemplar organisations highlight the importance of environmental change and opportunity as possible triggers for organisational learning, responsiveness and change. They also highlight the importance of S4 sense making of such change, the necessary acceptance of S5 directional change within the system, and an accommodation of change which may be facilitated or limited by existing capabilities and internal structures, as determinants for the evolution of competitive capability.

The multi-disciplinary makeup of the project team accords with Gioia and Pitre's (1990) approach to theory building and with their description of what constitutes theory. Whilst the project team has attempted to accommodate different paradigmatic assumptions and the consequent multiple perspectives that emerge from use of the different disciplinary frameworks underpinned by those assumptions, this paper has merely outlined some of the findings that have emerged from a viable systems perspective and how they have contributed to a more comprehensive or synthetic view of the organisational processes and circumstances involved in the evolution of competitive capability. Whether Beer's framework, which has been mainly used in other studies to effect snapshot assessments of organisational design and effectiveness, is the most appropriate systems vehicle for the examination of the learning and adaptive processes involved in the evolution of organisational capability has been considered in other work (Brocklesby, 2001) but is not discussed here. Nevertheless, we affirm that Beer's framework has been useful in providing insight about the systemic competitive capabilities that are conserved, emerge or evolve as organisations' attempts to survive are impacted upon by internal structures, and internal and external dynamics and forces.

Appendix 1 – The Sample Organisations

The Criterion Group	Ready-to Assemble Furniture
Formway Furniture Ltd	Office Furniture Systems
Freightways Express Ltd	Goods and Packet Distribution
Gallagher Group	Electronic fencing and gates
Kiwi Dairies	Cooperative dairy factory, milk products
Montana Wines Ltd	Wine maker
Nuplex Industries Ltd	Chemical industry - Australasia's largest
PEC (NZ) Ltd	Electronic petrol pumps, POS systems
Scott Technology Ltd	Manufacturing systems
Svedala Barmac Ltd	Quarry Equipment manufacturer
Tait Electronic Ltd	Mobile radio systems

REFERENCES

Argyris, C., and Schon, D.A. (1978). *Organizational Learning: a Theory of Action perspective*, Addison-Wesley, Reading, MA.

Barney, J. (1991). "Firm Resources and Sustained Competitive Advantage," *Journal of Management*, 17:99-120.

Beer, S. (1979). *The Heart of the Enterprise*, Wiley, Chichester.

Beer, S. (1981). *Brain of the Firm,* (2nd edition), Wiley, Chichester

Beer, S. (1985). *Diagnosing the System for Organisation*, Wiley, Chichester.

Brocklesby, J., (2001). "The Evolution of Competitive capability: a Biological Perspective", in *World-famous in New Zealand: How New Zealand's Leading Firms became World-Class Competitors*, (C. Campbell-Hunt et al. eds), Auckland University Press, Auckland NZ..

Brocklesby, J., Cummings, S., and Davies, J. (1995). "Demystifying the Viable System Model as a Tool for Organisational Analysis", *Asia Pacific Journal of Operational Research*, 25(1):65-86.

Burrell, G., and Morgan, G. (1979). *Sociological paradigms and organizational analysis*, Heinemann, London.

Campbell-Hunt, C., and Corbett, L.M. (1996). *A Season of Excellence?*, New Zealand Institute for Economic Research, Wellington.

Cummings, S., and Davies, J. (1994). "Vision, Mission and Fusion," *Long Range Planning,* 27(6):147-150.

Davies, J. (1999). "The Effective Organisation of National Sports Bodies in NZ - a Systems Perspective," in *Sports Management in New Zealand,* (L. Trenberth, and C. Collins, eds), Dunmore Press, Palmerston North.

Eisenhardt, K.M. (1989). "Building Theories from Case Study Research," *Academy of Management Review,* 14:532-550.

François, C.O. (1999). "Systems and Cybernetics in a Historical Perspective," *Systems Res & Behavioural Sci,* 16(3):203-220.

Gioia, D.A., and Pitre, E. (1990). "Multiparadigm Perspectives on Theory Building," *Academy of Management Review,* 15:584-602.

Goodman, R.S., and Kruger, E.J. (1988). "Data Dredging or Legitimate Research Method? Historiography and its Potential for Management Research," *Academy of Management Review,* 13:315-325.

Kay, J. (1993). *Foundations of Corporate Success,* Oxford University Press, Oxford.

Maturana, H., and Varela, F. (1987). *The Tree of Knowledge - The biological roots of human understanding,* Shambhala, Boston.

Miller, D. (1996). "Configurations Revisited," *Strategic Management Journal,* 17:505-512.

Nonaka, I., and Takeuchi, H. (1995). *The Knowledge Creating Company,* Oxford University Press, Oxford.

Peteraf, M.A. (1993). "The Cornerstones of Competitive Advantage," *Strategic Management Journal,* 14:179-191.

Porter, M.E. (1996). "What is Strategy?," *Harvard Business Review,* Nov-December:61-78.

Shenhav, Y. (1995). "From chaos to systems: the engineering foundations of organization theory, 1879-1932," *Administrative Science Quarterly,* 40:557-585.

RETHINKING BUSINESS STRATEGY WITH COMPLEXITY THEORY

Wai Ming Mak

Department of Management
The Hong Kong Polytechnic University
Hung Hom, Kowloon, Hong Kong
mswmmak@polyu.edu.hk

INTRODUCTION

Business strategy is regarded as the capstone subject in management education. The subject has developed since the early sixties in the time of a comparatively stable business environment. However, as time goes by, it has found that the strategy concepts cannot explain new phenomena facing business managers recently. As we enter the knowledge age, the subject requires some new ideas to enrich its sustainability. After the incoming-tide of the game theory in the nineties, it is time to rethink business strategy and its application from the perspective of complexity theory in the new millennium.

WHAT IS STRATEGY?

Business strategy provides the right direction for a company to move forward. The origin of the word strategy is the Greek *strategos* that means leader of an army. Drawing on the principle of competitive exclusion advances by a Russian biologist Gause in 1934, Henderson (1989) concludes that business survival in a competitive environment depends on strategies of differentiation. To put it in simple term, no two companies can make their livings in an identical way. Take a real life example, if there are two shops that sell the same product, at the same price, and in the same place, one of them will fail.

Porter (1996) explains strategy in a competitive concept, "it means deliberately choosing a different set of activities to deliver a unique mix of value." He says that the essence of strategy is choosing to perform activities differently than rivals do. At the same time, McMaster (1996) concludes that, "the strategy of a corporation should be focused on what can be done in the present to increase its ability to operate in future."

Edited by Ragsdell *et al.*, Kluwer Academic/Plenum Publishers, New York, 2002.

From my point of view, business strategy is to manage change. The company has to plan and execute strategic decisions on the present in order to shape the future. A future that is suitable and favourable for the company to survive, develop, and prosper.

Strategy in the Industrial Age

Rumelt *et al.* (1994) trace the origin of business strategy to Chandler (1962), Sloan (1963), and Ansoff (1965) in the early sixties. As the Chairman of General Motors, Sloan (1963) explains, "the strategic aim of a business is to earn a return on capital, and if in any particular case the return in the long run is not satisfactory, the deficiency should be corrected or the activity abandoned." As an academic, Chandler (1962) conducts case research in four companies, namely, Du Pont, General Motor, Sears Roebuck, and Standard Oil of New Jersey (now Exxon) and concludes that organisational structures must be adapted to changes in strategy, that is, structure follows strategy. It was a good description of how companies grow in a stable environment. Subsequently, Ansoff (1965) introduces the concept of corporate planning that strategy means creating a match between the organisation and its environment.

At the same time, Harvard Business School offers a subject on Business Policy for their MBA students with Learned *et al.* (1965) as the textbook. Later on, Andrews (1971) advances SWOT analysis as a framework to prescribe a match between internal strengths and weaknesses and external opportunities and threats. The subject develops gradually in the seventies.

Porter (1980, 1985, and 1991) applies the industrial organisation economics concepts to the problem of strategy formulation. He suggests three generic strategies, namely, cost leadership, differentiation, and focus, to represent and consider a company's competitive position. He also proposes to use the concept of five forces of industry, that is, threat of new entrants, bargaining power of suppliers and buyers, availability of substitute product or services, and the rivalry of competition. The two models provide useful analytical tools for the senior management to plan the business strategy of the companies.

The concept of game theory came to the front of business strategy in the nineties. McMillan (1992) defines game theory as the study of rational behaviour in situations involving interdependence. Interdependence means any player in the game is affected by what the others do. Moreover, rational behaviour implies no moral judgement and the players are trying to do the best they can. Hence, it provides a better way to understand the principles of strategic decision making. They conclude that most business situations are interactive. As a result, business strategy becomes the interaction among the competitors in the same industry.

Like playing chess, good business decisions require each decision-maker to anticipate the actions of the others. The decisions are made under uncertainty because actions are anticipatory. In terms of business strategy, we must anticipate each move of our competitors before we take our own.

Brandenburger and Nalebuff (1995) argue that successful business strategy is about actively changing the rules of the game companies play, not just play the game they find. They propose that game theory can be called the science of strategy and provide a systematic way to develop strategies when one person's fate depends on what other people do. Hence, if we cannot change the rules of the game, we may have to change the players of the game or even withdraw from the game. To conclude, in the industrial age, business strategy assumes organisable industry boundaries, manageable competition, and knowable future.

Brown and Eisenhardt (1998) criticise that game theory offers a more dynamic approach to business strategy. However, it can only answer the first question of business strategy, "Where do we want to go?" However, the second question of business strategy, "How are we going to get there?" is still not answered.

Strategy in the Knowledge Age

As we enter the knowledge age, the unprecedented rate of change makes our traditional strategy concept obsolete. As the future is unpredictable, more and more companies are resorted to short-term financial return. The time and effort to prepare a five-year plan is not worthwhile. The only way to move forward is trials and error method.

Back in the eighties, Mintzberg and Waters (1985) discuss that strategy making is not only deliberate, but also emergent. However, most people view it as short-term approach and it cannot break into the mainstream of business strategy. To speak the truth, the world is not what it used to be. The problems we are facing in the knowledge age are unorganisable industry boundaries, unmanageable competition, and unknowable future. Strategy as emerged has become more practical in the knowledge age.

A decade ago, Hamel and Prahalad (1989) state, "Regaining competitiveness will mean rethinking many of the basic concepts of strategy." Schoemaker (1990) argues that, "if the purpose of strategy is to gain competitive advantage, then by implication theories of strategy should be continually in a flux." Stacey (1996) agrees that if one accepts the premise that the dynamic of success is chaotic, all forms of long-term planning are completely ineffective. Beinhocker (1997) argues that traditional strategy framework can explain the success of companies retrospectively. However, it is difficult to use them to look forward in industries undergoing rapid change.

To develop a better mindset for the knowledge age, complexity theory is a good starting point for us to rethink and reinvent the concepts of business strategy. Ralph Stacey of the University of Hertfordshire is one of the pioneers introducing complexity theory to strategic management (Stacey, 1996). Hull University Business School is the first university in the United Kingdom that teaches Complexity Theory and Strategic Management as a subject in its DBA programme (Murray, 1997).

THE RISE OF COMPLEXITY THEORY

What is complexity? Flood (1987) answers, "In general, we seem to associate complexity with anything we find difficult to understand." Complexity is about the interaction of people and thing, including people and people as well as thing and thing. Pippenger (1978) points out, "The most important lesson of complexity theory is the demonstration of the diversity of phenomena that can arise through the interaction of simple components." In fact, complexity theory was first developed in the field of biology and physics. Coveney and Highfield (1995) define, "Complexity is the study of the behaviour of macroscopic collections of such units that are endowed with the potential to evolve in time." Irreversibility and nonlinearity are two ingredients necessary for complexity.

In the real world, there are three types of situations facing managers. Order and chaos are two extreme positions. Order represents a stable situation that managers can take their actions based on standard regulations and procedures. Chaos means disorder. It is turbulence or dynamic situation those managers cannot control the situation no matter what they do. The third situation is somewhere between order and chaos. It is called complexity or the edge of chaos. In this situation, managers must find the hidden order and balance the two extremes between order and chaos. It is not only searching a trend but also a pattern of what are happening around us.

The Santa Fe Institute's Stuart Kauffman believes that there is continuum from order to complexity to chaos. Using an ice cube as a metaphor, Sherman and Schultz (1998) explain, "order would be represented by the solid phase, complexity by the liquid, and chaos by the gas."

LESSONS FROM SANTA FE

Santa Fe is the capital of the state of New Mexico in the United States. In terms of art, it is the third most important city, after New York and Los Angeles. The place originally belongs to the Red Indians with a rich Indian culture. With an altitude of 7,200 feet above sea level, we can see the clear blue sky and the starry night. I thought that it was only when people left the city centre and stay out of the noise of their neighbours, that they had a better environment to think and search for something new, both creative and innovative.

The Santa Fe Institute is a private, non-profit, multidisciplinary research and education centre. It was founded in 1984 and has devoted itself to creating a new kind of scientific research community, and pursuing emerging science. In the past seventeen years, distinguished scholars from different disciplines build a new discipline of complexity theory together. For examples, Arthur (1996) advances the law increasing returns, Gell-Mann (1994) discusses simple and complex, Gleick (1987) explains chaos, Holland (1995) depicts the hidden order, Kauffman (1995) outlines self-organisation, Waldrop (1992) describes complexity, etc.

The Santa Fe Centre for Emergent Strategies is a spin-off of the Santa Fe Institute. Its main purpose is to introduce complexity theory to the business community. In order to learn more about complexity theory, I attended a Seminar on Complexity and Business organised by the Santa Fe Centre. The seminar was held at the beautiful Sol y Sombra retreat, from 14 to 16 July 1999. There were three facilitators, namely, Bruce Abell, Howard Sherman, and Mike Simmons, in the seminar. After three days learning, all participants are ready to go back to their companies with a new set of tools on complexity.

BASIC CONCEPTS OF COMPLEXITY THEORY

Complexity theory provides us new lens to see the world around us. It helps us to explain new phenomena that cannot be explained by the traditional strategic concepts.

As the telecommunications industry experiences accelerating rate of change in the past decade, it is used as real life example to illustrate the basic concepts of complexity theory.

The Butterfly Effect

Gleick (1987) explains butterfly effect has a technical name: sensitive dependence on initial conditions. Edward Lorenz says that given a slightly different starting point, the weather should unfold in a slightly different way. He sees there must be a link between the unwillingness of the weather to repeat itself and the inability of forecasters to predict it: a link between aperiodicity and unpredictability. Berreby (1998) describes butterfly effect as a well-worn cliché to describe nonlinear systems. The image is of a butterfly in, say, Beijing, flapping its wings and causing a typhoon in, say, Hong Kong. For example, the divestiture of American Telephone and Telegraphs in 1984 has created a huge impact on the telecommunications industry at that time and today, in the United States and around the world. We must be aware of what is happening around us because it may affect our business.

The Strange Attractor

Attractor means behaviour passively following a pattern. If the industry structure is stable, key players compete until one dominates the market and the industry reach equilibrium. Gleick (1987) describes two kinds of attractor. Fixed-point attractor represents behaviour that reached a steady state while limit cycle attractor repeats itself continuously. However, in the knowledge age, the industry is facing a strange attractor, which is changing all the time without a limit cycle. Hence, if the industry structure is dynamic, no one player is powerful enough to stay on the top forever. When the barrier to entry, such as, monopoly, is removed and new products and services come out everyday, the market keeps moving all the time. As a result, a focus strategy is insufficient.

The Red Queen Effect

The Red Queen effect is coined by Van Valen. It is based on the comment by the Red Queen to Alice, "It takes all the running you can do, to keep in the same place." Kauffman (1995) explains, "Coevoluation also occurs between predator and prey species. ... In a coevolutionary arms race, when the Red Queen dominates, all species keep changing and changing their genotypes indefinitely in a never-ending race merely to sustain their fitness level." In the industrial age, we fight our competitors by enhancing the quality of our service or reducing the price of our product. With an economy of scale, the market leader can kill all competitors. However, in the knowledge age, the economy of scale becomes less important. Whatever the market leader can do, the challengers can copy. Hence, a price war between two companies will make one of them fail in the short run. In the long run, both companies will suffer. The more we think we are solving the problem, the more problem we create.

The Law of Increasing Return

Arthur (1996) explains that, "the style of competition in the increasing returns arena is more like gambling." He explains that in casino gambling, it is important to choose which game to play and playing them with which skill. Moreover, the art of playing on the casino table is primarily a psychological one.

Hence, the law of diminishing return in the industrial age is not applicable in the knowledge age. In the industrial age, once we dominate the market, there is no way for the challengers to compete with us. In that case, the chance for the success of new company is slim. However, VHS successfully captures the market by providing a cheaper platform.

The Rugged Landscape

Borrowing the concept from biology, Kauffman (1993) explains the rugged fitness landscapes. Subsequently, Beinhocker (1997) puts the ideas in real life application. In the industrial age, the industry structure is stable. There is a fitness landscape with one high mountain. The strongest player climbs up to the top and stay there. However, in the knowledge age, the industry structure is dynamic, even there is a high mountain in the landscape, it might collapse and create different mountains and peaks around. To compete in the rugged landscape, the players cannot just stay at the top, they must explore the environment all the times in order to survive in the long run. In the past, company follows a single strategy of producing one major product. However, if there is no more demand in the market, the company will fail. For example, although Microsoft invests in windows in the early nineties, it also invests in other systems as well. In case windows fail, it still has a fall back strategy. Beinhocker (1999) explains instead of asking, "What was Microsoft's strategy?" we should ask, "What were Microsoft's strategies? Hence, do not put all eggs in one basket.

CONCLUSION

Fairtlough (1995) concludes, "Our strategic thinking, and our organisational responses to changes in strategy, need the help of new mental models. ... that biology can be a fruitful source of mental models, of new conceptual schemes." Later on, Lissack (1997) proposes that complexity theory metaphors add value by allowing managers new ways of thinking and the use of a different language. He summarises, "complexity metaphors when accepted within the vocabulary of an organisation, can change both the way managers manage and the problems they choose to manage." Black (1962) explains metaphor is used, "to bring two separate domains into cognitive and emotional relation by using language directly appropriate to the one as a lens for seeing the other."

The fact is that what we learnt from the industrial age could not help us to tackle the problem we encounter in the knowledge age. The classical economics theory cannot explain the phenomena. There is a new economy in the knowledge age. Arthur (1999) summaries that the new economy is created by the minds of its agents, namely, consumers, managers, and decision-makers. Instead of a physical economy, it is a cognitive one. Hence, we need a new mindset to meet the challenges ahead. In order to succeed in business strategy, we have to rethink the concept with complexity theory.

Arthur (1996) concludes that, "strategy in the knowledge world requires CEOs to recognise that a different kind of economics is at work. CEOs needs to understand which positive and negative feedback mechanisms are at play in the market ecologies in which they compete." Berreby (1998) admits that, "the atmosphere of complexity work is of a construction site, not a completed building. ... No one can yet define complexity or say exactly how to manage a company in accordance with its principles." More researches on complexity theory is needed in future.

However, complexity theory cannot tell us what to do in future. It can only help us to make a better sense of what is happening around us. Using a metaphor, we are all walking in the dark. Complexity theory is like a torch that can help us to have a better understanding around us so that we can find our way forward. However, it is only a tool. In the end, we still have to make our strategic decisions: Where are we? Where are we going? And, how can we get there?

REFERENCES

Andrews, K. (1971). *The Concepts of Corporate Strategy*, Irwin, Homewood, IL.

Ansoff, H.I. (1965). *Corporate Strategy*, McGraw-Hill, New York.

Arthur, W.B. (1996). "Increasing Returns and the New World of Business," *Harvard Business Review*, 74(4):100-109.

Arthur, W.B. (1999). "The End of Economic Certainty," in *The Biology of Business: Decoding the Natural Laws of Enterprise*, (J.H. Clippinger III, ed.), Jossey-Bass, San Francisco.

Beinhocker, E.D. (1997). "Strategy at the Edge of Chaos," *The McKinsey Quarterly*, (1):24-39.

Beinhocker, E.D. (1999). "Robust Adaptive Strategies," *Sloan Management Review*, 40(3):95-106.

Berreby, D. (1998). "Complexity Theory: Fact Free Science or Business Tool?" *Strategy and Business*, (1):40-50.

Black, M. (1962). *Models and Metaphors: Studies in Language in Philosophy*, Cornell University Press, Ithica.

Brandenburger, A.M., and Nalebuff, B.J. (1995). "The Right Game: Use Game Theory to Shape Strategy," *Harvard Business Review*, 73(4):57-71.

Brown, S.L., and Eisenhardt, K.M. (1998). *Competing on the Edge: Strategy as Structured Chaos*, Harvard Business School Press, Boston, MA.

Chandler, A.D. (1962). *Strategy and Structure: Chapters in the History of the Industrial Enterprises*, MIT Press, Cambridge.

Coveney, P., and Highfield, R. (1985). *Frontiers of Complexity: The Search for Order in a Chaotic World*, Faber and Faber, London.

Fairtlough, G. (1995). "Biological Models and Business Success," *Business Strategy Review*, 6(3):27-40.

Flood, R.L. (1987). "Complexity: A Definition by Construction of a Conceptual Framework," *Systems Research*, 4(3):177-185.

Gell-Mann, M. (1994). *The Quark and the Jaguar: Adventures in the Simple and the Complex*, Freeman & Co., New York.

Gleick, J. (1987). *Chaos: Making a New Science*, Viking, New York.

Hamel, G., and Prahalad, C.K. (1989). "Strategic Intent," *Harvard Business Review*, 67(3):63-76.

Henderson, B.D. (1989). "The Origin of Strategy", *Harvard Business Review*, 67(6):139-143.

Holland, J.H. (1995). *Hidden Order: How Adaptation Builds Complexity*, Perseus Books, Cambridge, MA.

Kauffman, S.A. (1993). *The Origins of Order: Self-Organisation and Selection in Evolution*, Oxford University Press, New York.

Kauffman, S.A. (1995). *At Home in the Universe: The Search for Laws of Self-Organisation and Complexity*, Oxford University Press, New York.

Learned, E.P., Christensen, R.C., Andrews, K., and Guth, W. (1965). *Business Policy: Text and Cases*, Irwin, Homewood, IL.

Lissack, M.R. (1997). "Of Chaos and Complexity: Managerial Insights from a New Science," *Management Decision*, 35(3):205-218.

McMaster, M.D. (1996). *The Intelligence Advantage: Organising for Complexity*, Butterworth-Heinemann, Oxford.

McMillan, J. (1992). *Games, Strategies, and Managers: How Managers Can Use Game Theory to Make Better Business Decisions*, Oxford University Press, New York.

Mintzberg, H., and Waters, J.A. (1985). "Of Strategies, Deliberate and Emergent," *Strategic Management Journal*, 6(3):257-272.

Murray, P.J. (1997). *Study Guide on Complexity Theory and Strategic Management*, The University of Hull.

Pippenger, N. (1978). "Complexity Theory," *Scientific America*, June:114-125.

Porter, M.E. (1980). *Competitive Strategy: Techniques for Analysing Industries and Competitors*, Free Press, New York.

Porter, M.E. (1985). *Competitive Advantage: Creating and Sustaining Superior Performance*, Free Press, New York.

Porter, M.E. (1991). *Competitive Advantage of Nations*, Free Press, New York.

Porter, M.E. (1996). "What is Strategy?" *Harvard Business Review*, 74(6):61-78.

Rumelt, R.P., Schendel, D.E., and Teece, D.J. (1994). "Fundamental Issues in Strategy," in *Fundamental Issues in Strategy: A Research Agenda*, (Rumelt, R.P., Schendel, D.E., and Teece, D.J., eds.), Harvard Business School Press, Boston, MA.

Schoemaker, P.J.H. (1990). "Strategy, Complexity and Economic Rent," *Management Science*, 36(10):1178-1192.

Sherman, H., and Schultz, R. (1998). *Open Boundaries: Creating Business Innovation through Complexity*, Perseus Books, Reading, MA.

Sloan, A.P. (1963). *My Years with General Motors*, Sedgewick and Jackson, London.

Stacey, R.D. (1996). *Strategic Management and Organisational Dynamics*, Second Edition, Pitman Publishing, London.

Waldrop, M.M. (1992). *Complexity: The New Science at the Edge of Order and Chaos*, Viking, London.

Section Eight

Information Systems

INFORMATION SYSTEM RISKS AND COUNTERMEASURES IN SUPPLY CHAIN INTEGRATION AND MANAGEMENT FOR AN ENTERPRISE

Joseph Akomode

School of Computing and Mathematical Sciences
Liverpool John Moores University
Byrom Street Liverpool L3 3AF
England UK
Email: j.o.akomode@livjm.ac.uk

INTRODUCTION

Manufacturing may be viewed as the foundation for economic growth and modern comfort. The increasing application of the internet and electronic commerce (e-commerce) coupled with market forces has the potential to lead an enterprise into increased business risks in the domain of supply chain integration and management (SCIM) in manufacturing. Consequently, there is a need for a generic evolutionary framework for optimising activities in SCIM and for supporting managers in decision making. In manufacturing, *supply chain* represents a system of services (or a network of activities) and distribution alternatives that combine to perform the tasks of: (i) procurement of input materials, (ii) transformation of the input materials into intermediate and final-stage products and (iii) the delivery of these products to customers who may be business buyers or individual consumers. The explanation given above about the supply chain may be substantiated by the work of some experts and commentators such as: Lee and Billington (1995); Ganeshan and Harrison (1995); Swaminathan *et al* (1996). Supply chain as a network of activities is not exclusive to manufacturing but does exist in the service industry and firms, often with different levels of complexity. The research findings presented in this paper form part of an ongoing research effort towards the development of a generic (or useful) evolutionary framework for optimising SCIM activities. The sections below discuss: (1) conventional and contemporary supply chain mechanisms; (2) supply chain integration and management; (3) manufacturing activities and potential risks in SCIM; (4) possible countermeasures for improving security.

CONVENTIONAL AND CONTEMPORARY SUPPLY CHAIN MECHANISMS

Figure 1 represents a conventional supply chain arrangement in the domain of manufacturing. Generally this set-up contains serial processes as indicated in the block diagram. Basically, activities such as: purchasing of raw material, planning and manufacturing, distribution, marketing and sales to the end-consumer (business buyer or individual customer) are independently operated along the supply chain. In Figure 1, typically the *retailer* interacts directly with the *end consumer* through the sale of products. When the level of products is low or more specifically low inventory level, the retailer places orders with supplier(s), in this case the *distributor,* to replenish the inventory. In turn when the inventory level of the distributor is low, s/he places orders with the supplier (in this case the manufacturer) to replenish the inventory. The manufacturer will consume their inventory of raw materials and place orders with each raw material supplier for replenishment. At the same time the manufacturer builds up his/her inventory of output products. In turn, the raw material supplier will place orders with his (or her) suppliers and so on. Logically, the situation results in a sequential process and the flow of material or products is serial as shown in Figure 1. In this conventional arrangement, we consume inventory and when the inventory level is low we send information to our supplier for replenishment. In other words, each block in Figure 1 has its *individual cycle of operation* that forms the serial approach to supply chain management. The individual cycle of operation involves the repeating process of: consume inventory – place order – replenish inventory (CI-PO-RI). While information flow in such a repeating process may precede or follow inventory request, the situation may lead to inefficiency and major business risks, such as: (1) slow customer response, (2) reduced information sharing between partners and (3) delayed replenishment of inventory. Such drawbacks (or risks) may hamper competitiveness and customer satisfaction, especially in enterprises with little or no vertical integration.

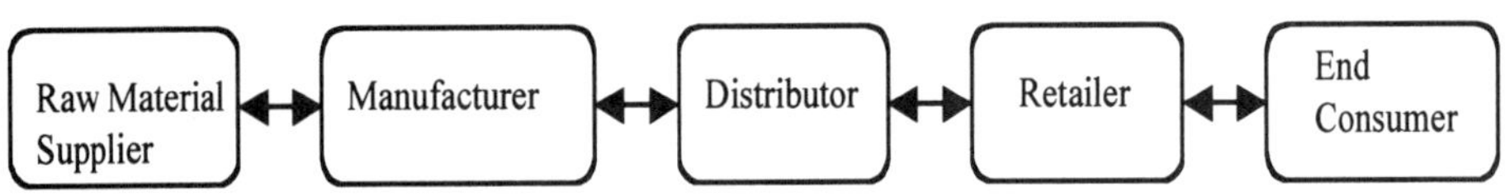

Figure 1. Conventional View of Supply Chain Activities in Manufacturing

Due to changing technology and competitiveness, there is a move from the conventional supply chain approach to a modern approach simply referred to here as the contemporary supply chain mechanism. It gives a different perspective of supply chain and involves the application of a computer network, more appropriately the internet (or e-commerce). Figure 2 represents the e-commerce based perspective of a supply chain. As an internet enabled infrastructure it supports increased information sharing between supply chain partners and facilitates faster responses for inventory replenishment. The end consumer in the lower block of Figure 2 can make direct contact with the: (1) retailer, (2) distributor, (3) manufacturer and (4) raw material supplier. Also, any of the partners can make contact with the other directly through, for example: (1) an interactive web-site, (2) electronic mail or (3) a special channel of EDI (Electronic Data Interchange). Such contact and response is often faster compared to the conventional serial process of CI-PO-RI in a supply chain as discussed earlier. The great advantages of the contemporary supply chain mechanism include its support in: (1) increased cycle time for supply chain transactions, (2) continuous replenishment suggestions, (3) quicker customer response for effecting replenishment, (4) on-line evaluation of suppliers to meet changes, (5) faster co-ordination between the different participants in the supply chain, (6) internet/intranet (or client/server) solution. An aspect of an internet-enabled supply chain is discussed in www.delboy.com/iwp4.htm. Despite the huge advantages offered by the internet perspective of supply chain there are also disadvantages in the form of business risks, details are discussed below.

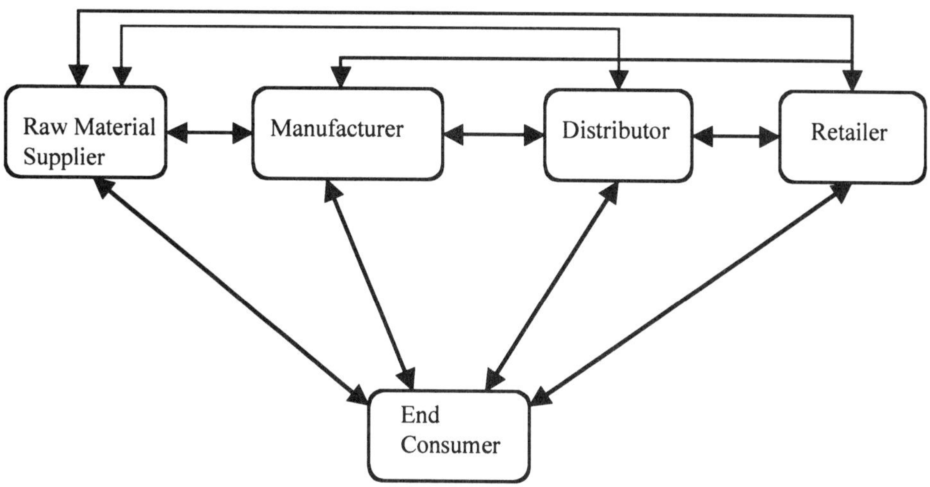

Figure 2. E-Commerce View of Supply Chain Activities in Manufacturing

SUPPLY CHAIN INTEGRATION AND MANAGEMENT

Fundamentally, a suitable co-ordination of the various participants and processes in the supply chain is a *sine qua non* to effective and efficient management, especially with regard to manufacturing. The term '*supply chain integration*' implies coordinating and optimising the supply chain mechanism for better operations and effective management. This may involve a systemic amalgamation of major decision areas that impact on the supply chain, such as: (1) location, (2) manufacturing (or production), (3) inventory and (4) shipment. For a manufacturing enterprise, the consideration and assessment of different functions of the supply chain is also necessary and this may include: policies, processes, purchasing contracts, systems (e.g. logistics, information communication technology (ICT) and associated databases), marketing and business strategies often with diverse trading associates (i.e. suppliers and customers). The involvement of suppliers and customers is a significant factor in achieving an effective integration result. On the other hand '*supply chain management*' implies a strategy through which supply chain integration could be achieved and operated effectively. Consequently, the goal of supply chain integration and management (SCIM) is essentially to: (1) improve responsiveness, (2) reduce lead times, (3) improve quality of services/products (4) reduce inventory level and (5) minimise business risks. Also, the purpose is to improve the effectiveness of the whole supply chain from suppliers of raw materials through internal manufacturing to delivery of the finished products to the customers. The explanation and view of supply chain integration and management presented above may be corroborated with the work of some other authors and commentators (cf: Freeman (1988); Lewis *et al* (1998); O'Connor (1998); Teigen (1997). The continuous monitoring and control of the supply chain components and processes are essential as part of a plan for effective management, especially in a dynamic competitive environment.

The different functions mentioned above in SCIM may be carried out by managers of some enterprises without adequate holistic consideration of all the functions and without appropriate overall risk assessment of the associated components, for example:

1. ICT may not be adequately coordinated with processes in the manufacturing plant, suppliers/customers and security issues
2. purchasing contracts may be arranged on the basis of historic buying patterns with little or no information based on the complete spectrum of the other functions
3. enterprise policies and business strategies may not involve adequate risk assessment of the dynamically changing (a) technological environment and (b) competitiveness
4. inventory levels may not be adequately co-ordinated with shipment from vendors to the manufacturing process and from the manufacturing plant to major customers, for example, to distributors let alone individuals.

The above mentioned situations may lead to severe business risks, such as: (a) delays in responding to customers, (b) poor quality of products/services to customers, and (c) a decrease in the competitiveness of a manufacturing enterprise. In this regard there is a need for an *improved supply chain mechanism* based on adequate risk assessment of the dynamic business environment with the aim of obtaining a single integrated plan (or a systemic arrangement) that is effective and useful to an enterprise.

MANUFACTURING ACTIVITIES AND POTENTIAL RISKS IN SCIM

The management of an organisation (for example, a manufacturing enterprise) is responsible for over 80% of quality improvement of its products/services and for competitiveness (cf. Deming, 1985, 1993; Taguchi and Clausing 1990). In most organisations, managers are responsible for the strategic and operational activities that encompass the minimisation/sharing of risks coupled with improvement of quality of products/services, in order to cope with dynamic changes in the marketplace. Also, managers often find that supply chain integration (SCI) risks associated with their own supply-chain have a deleterious effect on the health of their manufacturing goals and competitiveness. Business risks associated with SCIM are often compounded with the impact of the application of modern technologies such as intranet/internet and e-commerce, which are rapidly changing the way humans live, communicate and do business. In manufacturing, the implication of the changes in marketplaces that are propelled by modern technologies, is the increasing globalisation of the selling/buying process. Consequently, the *selling/buying situation is dynamic* such that vendors and purchasers (also in a supply chain condition) may not necessarily know each other in advance of the experience of doing business together. Such globalisation and evolution in SCIM may breed increased business risks and may require a different solution approach. As a result, manufacturing risk assessment decision-making models in SCIM should not be 'static' but should be *dynamic and adaptive* to support managers effectively in coping with evolutionary changes in supply-chain activities. This ongoing investigation covers current activities of SCIM with regard to small-to-medium-sized enterprises, the analysis of data/information obtained and the development of 'conceptual' models that will in turn lead to the development of a generic evolutionary risk assessment framework for SCIM.

A manufacturing organisation can be viewed as a place for carrying out various activities that include the following.

1. *Delivery of suppliers' products/services*: a delivery point for products/services of suppliers, from which contractors and their string of subcontractors send their output products/services that in turn serve as inputs to the manufacturing process.
2. *Production of products*: a place where output products are made and where associated strategies and services for such products are co-ordinated.
3. *Delivery of manufactured goods*: a source for the delivery of manufactured products coupled with services to: wholesalers/distributors, business firms and other types of consumers.
4. *Inventory centre*: a store of raw materials (at a reasonable level) to the manufacturing process and for finished goods (output products) from the manufacturing process.
5. *Information gathering, processing and dissemination centre*: an information gathering, information processing and dissemination point with regard to; products/services, staff/suppliers, competitors, benchmarking, quality standards, technologies, supply chain issues, customers and their satisfaction levels.
6. *Social interaction*: a social environment where people interact and carry out different activities towards stated goals and objectives.

The above mentioned activities in manufacturing with an internet (or e-commerce) enabled SCIM set-up have inherent business risks, such as: (a) inappropriate communication of requirements/demands for input and output materials; (b) weakness in the continuous replenishment arrangement in the supply chain; (c) unsuitable supply chain integration (e.g. due to lack of adequate vendor/inventory management); (d) lack of adequate decision making (e.g. regarding supply chain coordination/optimisation activities); (e) lack of delivery precision (or delay in response); (f) misalignment of supply-chain infrastructure or logistics.

As a result of the intranet/internet (or e-commerce) enabled SCIM, business risks associated with information system (IS) assets of an enterprise may include the following: (1) risk in access control to the IS asset; (2) risk in processing control; (3) risk in physical controls; (4) risk of losses from hackers and crackers due to a breach of security; (5) invasion of the right to privacy; (6) on-line defamation; (7) losses from cyber-thieves; (8) disgruntled employees and breach of security from the inside;(9) viruses/bugs;(10) cyber terrorism and information warfare, Akomode (2000).

POSSIBLE COUNTERMEASURES FOR IMPROVING SECURITY

Risk management for improving security in computer network and in an internet related business (e.g. e-commerce) involves a combination of factors as explained below.

1. *Physical protection for the premises*: adequate measures should be put in place to secure appropriate protection for the premises where the information technology asset is kept (e.g. the e-supply-chain mechanism of the enterprise). For example, security cameras should be installed in strategic positions, security personnel should be employed to monitor the premises.
2. *Human resource management and monitoring of dissatisfaction of staff:* as already pointed out, disgruntled and bribed employees are responsible for the majority of security breaches in most establishments. In order to improve security, it is necessary to put in place a proper human resource management structure that include the monitoring of angry staff, suppliers/customers (e.g. Akomode 2001; Cronbach, 1951, 1984; Hammersley, 1992; Kirk and Miller, 1986; Litwin, 1995; Nunnally, 1967; Silverman, 1994) and to apply suitable measures to motivate staff/suppliers, in order to nullify (or properly handle) any possible anger or rebellious attitude from any employee.
3. *Legal issues:* when new personnel are recruited (or when new suppliers are engaged) they should sign a legal document in their contract to emphasis the serious outcome of security breaches, for example, that the manufacturing company will take all necessary measures to locate such staff/supplier or former staff and bring him/her to justice.
4. *Technical security:* suitable forms of technical security measures should be installed within budget, for example: (a) firewall; (b) encryption technology; (c) intruder detection systems capable of intercepting intruders in case of any attack and for the detection of the attacker's identity. A combination of the technical security measures may be used instead of a single type of system.
5. *Insurance provision:* as another form of security for the information technology asset (or e-supply-chain mechanism) of an enterprise, it is prudent to secure reasonable insurance cover for the physical equipment and against possible damage to information resources which may be due to intrusion from both internal and external sources.
6. *Incessant watchfulness:* despite the above precautions perpetual attention is still required as an additional security measure.

CONCLUSION

The core of SCIM is to ensure that all aspects of the supply chain work together, instead of a scenario of uncoordinated activities. Therefore, in manufacturing (at least) the notion of SCIM involves the philosophy that deals with the time-related positioning and control of input materials/services and output products/services, in order to achieve appropriate quality (fitness-for-use) level and minimise business risks/waste while improving and ensuring the satisfaction of customers. For SCIM the issues of identification and mitigation of risks are often emphasised by managers of most enterprises, researchers and other interested individuals. In this on-going investigation the findings presented are aimed at: (i) supporting managers and researchers with a strategy for risk mitigation associated with the internet (or e-commerce) enabled SCIM, in order to satisfy customers' product/service quality requirements; (ii) providing a basis for the development of a generic evolutionary risk assessment framework. It should be noted that the development of security mechanisms for the protection of e-supply-chain is still undeveloped with regard to both technology and management frontiers. Also, hackers will continue to have interest in confronting (or attacking) any new type of technical security system for protecting information technology assets, since they are likely to see it as a new challenge in the digital race. An attempt has been made to investigate the potential business risks and countermeasure associated with information system supported supply chain integration and management activities. It is expected that the research findings presented above will lead to increased awareness in e-business risks, demand chain, and improved decision making in the domain of SCIM enabled by e-commerce potential.

REFERENCES

Adams, J. (1998). Hacker Ideology http://vallejo.phil3.uni-freiburg.de/1998/hackers.html

Akomode, J. (1999). "Project Risks and Issues of Security Associated with Information Technology Systems in an Enterprise," in *Proceedings of the 9[t] Annual BIT Conference*, 3rd/4th November 1999, (Generative Futures), (R. Hackney, ed.), (CD).

Akomode, J. (2000). "Potential Risks in E-Business and Possible Measures for an Enterprise," in *Proceedings of the 10th Annual BIT Conference*, 1st/2nd November 2000, (E-Futures), (R. Hackney, ed.), (CD).

Akomode, J. (2001). "Measuring the Satisfaction of Customers to Minimise Business Risks and Improve Quality in New Products" *in Proceedings of the 5th World Multiconference on Systemic, Cybernetics and Informatics (SCI 2001)*, (N. Callaos; S. Long and M. Loutfi, eds.), Orlando, Florida, USA, July 22 –25, 2001, (xvii):575-580.

Briney, A. (1999). Got Security? www.infosecuritymag.com/july99/cover.htm

Cronbach, L. J. (1951). Coefficient Alpha and The Internal Structure of Tests, *Psychometrika*, 16(3), September 1951, 297-334.

Cronbach, L. J. (1984). *Essentials of Psychological Testing*, Harper & Row, London.

Deming, W. E. (1985). Transformation of Western Style of Management, *Interfaces* (May/June), (15):6-11.

Deming, W. E. (1993). *The New Economics for Industry, Government, Education*. Massachusetts Institute of Technology.

Freeman, B. (1998). http://www.infochain.org/quarterly/Spring98/Survey.html

Ganeshan, R. and Harrison, T. P. (1995). "An Introduction to Supply Chain Management. Penn State University," http://silmaril.smeal.psu.edu/misc/supply_chain_intro.html).

Graham, B. (1999). www.computeruser.com/newstoday/99/11/09/news9.html

Hammersley, M. (1992). *What's Wrong with Ethnography: Methodological Explorations*, Routledge, London.

http://www.delboy.com/iwp4.htm

Info-sec.com (1999). http://www.info-sec.com/internet/99/internet_080399d_j.shtml

Kirk, J and Miller, M. (1986). *Reliability and Validity in Qualitative Research*, Qualitative Research Method Series, No. 1, Sage, London.

Kovacich, G. L. (1999). "I-Way Robbery: Crimes on the internet," *Computers & Security*, 18(3):211-220.

Lee, H. L. and Billington, C. (1995). "The Evolution of Supply Chain Management Models and Practice at Hewlett-Packard". *Interfaces* (25):42-63, 5 September-October, 1995.

Lewis, J.; Naim, M. Wardle, S.; Varity, L. and Williams, E. (1998). "Quick Scan Your Way to Supply Chain Management," *Control UK*, 24 (5):14-16.

Litwin, M. S. (1995). *How to Measure Reliability and Validity*, Sage, London

Nunnally, J. C. (1967). *Psychometric Theory*, McGraw-Hill, London

O'Connor, B. (1998). "Supply Chain Challenges," *Control UK*, 24 (6):23-26.

Proctor, P. E.(1998) Information Risk Management. www.centraxcorp.com/techpubs.html

Silverman, D. (1994) *Interpreting Qualitative Data: methods for analysing talk. Text and interaction,* Sage, London

Swaminathan, J. M; Smith, S. F. and Sadeh, N. M. (1996). A Multi Agent Framework for Modelling Supply Chain Dynamics. Technical Report, The Robotic Institute, Carnegie Mellon University

Taguchi, G. and Clausing, D. (1990) Robust quality, *Harvard Business Review*, Jan-Feb, pp.65-75.

Teigen, R. (1997). http://www.eil.utoronto.ca/profiles/rune/node5.html

The Times (1996). 26th March, UK.

Turkle, S. (1984). *The Second Self: Computers and the Human Spirit*, Simon and Schuster.

Warren, M. J. (1998). Cyber Terrorism, in *Proceedings of the IFIP TC 11 14th international Conference on Information Security* (31 August – 2 September, 1998), Vienna, (A. Papp and R. Posch, eds.), Austrian Computer Society

AN EVALUATION FRAMEWORK FOR INFORMATION STRATEGIES AT HIGHER EDUCATION INSTITUTIONS

Yongmei Nie Bentley and Steve Clarke

Department of Finance, Systems and Operations
Luton Business School, University of Luton
Park Square, Luton, Bedfordshire LU1 3JU, UK.
E-mail: yong.nie@luton.ac.uk

INTRODUCTION

An effective information strategy is becoming an increasingly important component of organisations. Specifically in the field of higher education, a number of institutions have developed institutional information strategies during the last few years. A major problem to be faced is how to measure the success and effectiveness of such strategies. In response, through a review of the theoretical underpinning, particularly critical social theory and theories of corporate strategy; and through empirical research conducted in higher education institutions, this research develops a framework for the evaluation of information strategies of higher education institutions. The overall purpose of the framework is to provide a generalised tool by which the benefits and effectiveness of information strategies in higher education institutions may be assessed, and thus, to help higher education institutions ensure accountability in the strategic management of their information.

THEORETICAL UNDERPINNING - CRITICAL SOCIAL THEORY

Critical social theory (CSoT) sees 'our understanding of the world as determined by a *priori* conditions which are uncritically accepted' and seeks to 'release human beings from their *false consciousness* to a position from which true potentiality can be attained' (Clarke, 2001, p66). CSoT can be traced from the work of Kant, through Marx and the Frankfurt School. One of the most widely accepted modern theorists is Habermas (1971), and it is to him that management science turned in the 1980s in order to develop a more human-centred view of its domain.

CSoT applied to the field of information strategy is appealing for its denial of the natural scientific principles on which study has largely hitherto been based. Seen through

Edited by Ragsdell *et al.*, Kluwer Academic/Plenum Publishers, New York, 2002.

a scientific framework, information strategy appears as a rational plan for a system to satisfy objective, verifiable requirements which are the same for all involved since they are independent of human opinion. CSoT refutes this, seeing our understanding of the world as determined by a *priori* conditions which are uncritically accepted. Critical theory seeks to expose these, and thereby release human beings from their 'false consciousness' to a position from which true potentiality can be attained.

From the roots of CSoT came the development, in the domain of management science, of critical systems thinking (CST).

CST is an approach to systems practice that was first developed in the early 1980s, and has since been debated widely in the literature (Midgley, 1995b). Authors who have reflected on significant contributions to this debate include: Flood and Jackson (1991b), Schecter (1993), Midgley (1995a), Flood and Romm (1997), and Clarke (2001). The various views of the authors mentioned above are reflected in the development of the evaluative framework for information strategy in HEIs.

TOWARDS A CRITICAL APPROACH FOR THE FRAMEWORK DEVELOPMENT

The foundation of a critical systems approach to organisational studies owes much to the work of Churchman (1968), which was built on foundation laid by Singer, and has been continued by Ackoff and other adherents to the systems school (see Britton and McCallion, 1994).

The work of Jackson and Keys (1984) proved a major turning point in the development of a critical framework which is true to the commitments of critical systems thinking. By exploring the relationship between the different problem-solving methodologies that had arisen, as a guide to intervening in problem situations and to understanding the strengths and weaknesses of the different methodologies. Jackson and Keys (1984) developed a grid with four boxes, representing four different types of perceived problem context, and then aligned different systems methodologies with each of them. These four boxes were later expanded to six by Jackson (1987b), and the authors called the resulting grid of contexts "Systems of Systems Methodologies" (SoSM).

A number of developments have followed this initial work, see, for example, Oliga (1988), Flood and Jackson (1991a), and Midgley (1992, 1995b), and this research is based on those developments.

THEORIES OF CORPORATE STRATEGY

Mintzberg (1987) represents corporate strategy as the 'strategy as a plan' versus 'strategy as a pattern' debate; Quinn (1980) as the distinction between planned and incremental strategies. Planning approaches to strategy are to be found in the work of the 'design school' (Ansoff, 1964), which is further traceable to scientific reductionism. The process is objective, and lacks attention to human activity.

Clarke (2001, p32) characterises the 'design school' in the following terms: 1.Complete strategy formation always precedes implementation. 2. Responsibility for strategy rests with senior management, who are placed in a command-and-control position in relation to the environment. 3. Strategies emerge from the design process, fully formulated and ready to be chosen. 4. There is no room for an emergent view.

Mintzberg (1987, p67) argues that 'virtually everything that has been written about strategy-making depicts it as a deliberate process', whilst the evidence shows this not to be the case, with strategy emerging from the organisation without there having been a deliberate plan. This leads Quinn to challenge the standard 'rational-analytical' approach, which he sees as normative. The goals and objectives of strategic planning are seen by Mintzberg and Quinn (1998) as determining what is to be achieved and when, but not how the results are to be achieved. This distinction between the 'how' and 'what' of strategy becomes important when looking at the strategic context of information. Information strategy has the broader remit of addressing the 'what' questions, echoing the foundation already determined in interpretivism and radical humanism.

The above is only a brief summary of the literature reviewed for this study. The research carried out to date has led to a clear focus in critical social theory as a grounding for information strategy - a grounding which appears equally well fitted to both the *information* content and the *strategy* content.

EMPIRICAL RESEARCH AND FINDINGS

For this research, following the literature review, primary data have been collected and analysed from two pieces of action research (AR). The first investigated the student records system (SRS) at a UK university; the second looked at participation in the development of an information strategy at a UK university. Information was also drawn from the information strategy workshops facilitated by the HEFCE Joint Information Systems Committee (JISC), through which contacts were made and interviews conducted with representatives of a great number of HEIs.

The empirical research has yielded valuable findings, which call into question the over-emphasis on operational issues; give guidance to HEIs on information strategy formulation and implementation based explicitly on pragmatic judgements; and raise a question regarding the nature and potential value of a theoretical grounding to the domain.

Based on the first piece of AR, the main issues have been identified as:

1. The SRS lacked strategic focus, and exhibits significant operational deficiencies. Because of the large incidence of error, the system frequently produces incomplete, inconsistent, ambiguous, or even completely inaccurate information about students. It fails to fit within the university's overall strategic plan.
2. The system users were insufficiently informed of system development, and had failed to use the system to its full potential.
3. Effort had been wasted on duplication of information at different levels.
4. Users had no clear understanding as to ownership and management of the system. It was not clear to students where they should go to register changes.
5. There was no systematic approach to reporting and resolving problems.

The problems found have further focused attention on human issues such as the lack of user involvement with the design and implementation process, the diversity of interests and agendas in the organisation, and lack of communication.

From the second piece of AR, the benefits of having an information strategy have been identified as: elimination of unnecessary duplication, removal of bureaucracy, release of time, an informed and knowledgeable community, and recognition and sharing of good practice.

The findings from the research conducted at JISC's workshops have further supported the view that information strategy may be enhanced by investigating the domain from the perspective of social theory. In particular:

1. Most respondents interviewed were not clear what an information strategy should consist of, how to produce one, or what the real benefits had been to the pioneering universities who had developed their institutional information strategies. Many seemed to see an information strategy as synonymous with an IT Strategy and/or an IS strategy.
2. IT people from higher education institutions (HEIs) had much less sympathy with the idea of developing an information strategy. They did not think the final document, the information strategy itself, would work.
3. Although some universities have developed an information strategy, early indications have shown that they do not abide by them.
4. In the development of an information strategy, many who claimed to have used human-centred methods had in fact followed rather traditional functionalist and goal-oriented approaches.

EVALUATIVE FRAMEWORK FOR INFORMATION STRATEGIES

The data from the different types of empirical work have been triangulated to enhance the validity of the research; in addition, the findings have been combined with knowledge gained from the theoretical and empirical literature review. Gaps identified in the research to date show that planning approaches to information strategy gain precedence over alternative methods. All these research activities have enabled the formulation of a framework for evaluating information strategies in HEIs. As indicated in Figure 1 at the end of the paper, the concept of an ‘information strategy’ has been broken down into three broad categories – control structure, external environment analysis, and internal environment analysis, which are then further divided into certain elements. More detailed information on each of the framework elements, including relevant theoretical and empirical support are specified in Table 1.

The framework has been given an initial trial by means of a case study dealing with assessing the successes or failures of the development and implementation of the information strategy at a specific HE institution. Based on the findings from the case study, a revised version of the framework has been formulated (see Figure 2 at the end of the paper).

Table 1. Elements of the Proposed Framework for Information Strategy Evaluation

Framework Element	*Theoretical Support*	*Empirical Support*
Control structures	Framework for IS Strategic Management (Clarke, 2001); Analysis, choice, implementation (Johnson and Scholes, 1993)	Student record system - very 'operational' and lack of strategic controls
Information analysis	Information Needs (Baets, 1992)	Information needs paramount (JISC).
Evaluative structures	Context, process, form & content, effects (Pettigrew, 1987)	JISC pilot sites point to need for participation; Action research 2 - information strategy embedded in 8 principles.
Strategic alignment	The Strategy Concept (Mintzberg, 1987); Align information strategy with corporate strategy (Baets, 1992); Strategic alignment (Henderson & Venkatraman, 1993).	Information strategy must support all organisational activities (JISC).
Competitive advantage	Value chain/five forces (Porter, 1990); The strategic advantage to be gained from information (Porter, 1990)	Cast study: competitive advantages gained by strategic management of information.
Information / IT synthesis	The 'what' and 'how' of information (Earl, 1989)	Six pilot sites point to need for participation (the 'what'); JISC participants see Information and IT as synonymous.
The internal environment	Strategic grid (McFarlan, 1989); SWOT analysis; Strategies for Managing the Technology Infrastructure (Ward and Griffiths,1996)	SRS participation was weak, but problems were attributed to human issues.
The external environment	Business Strategy (Campbell and Stonehouse, 1999); Value chain/five forces (Porter, 1990)	Internal and external contexts are important (JISC).
Organisational culture	The organisational paradigm Cultural web (Johnson and Scholes, 1993)	Need to move to more inclusive, participative culture (JISC and SRS).
Organisational structure	Systems Strategies for Information Management (Jackson, 1987a); Mechanistic to adhocratic (Mintzberg, Quinn et al., 1998)	Incorporation of 'user' views is weak (JISC).
Managing strategic change	Issues in change management (Johnson and Scholes 1993; Managing Change in Organisations (Carnall, 1990)	Overall, management of information strategy is weak, with over emphasis on operations, lack of long-term vision and human issues poorly addressed.
Resources	Means, ends and resource planning (Ackoff, 1981); Business Strategy (Campbell and Stonehouse, 1999)	Strategic development seen to be resource constrained (Case study).
Strategic choice	Choice models: TSI (Flood and Jackson, 1991a); Complementarist framework (Flood, 1995); Knowledge constitutive interests (Habermas, 1971)	Choice of models informed from other practice.

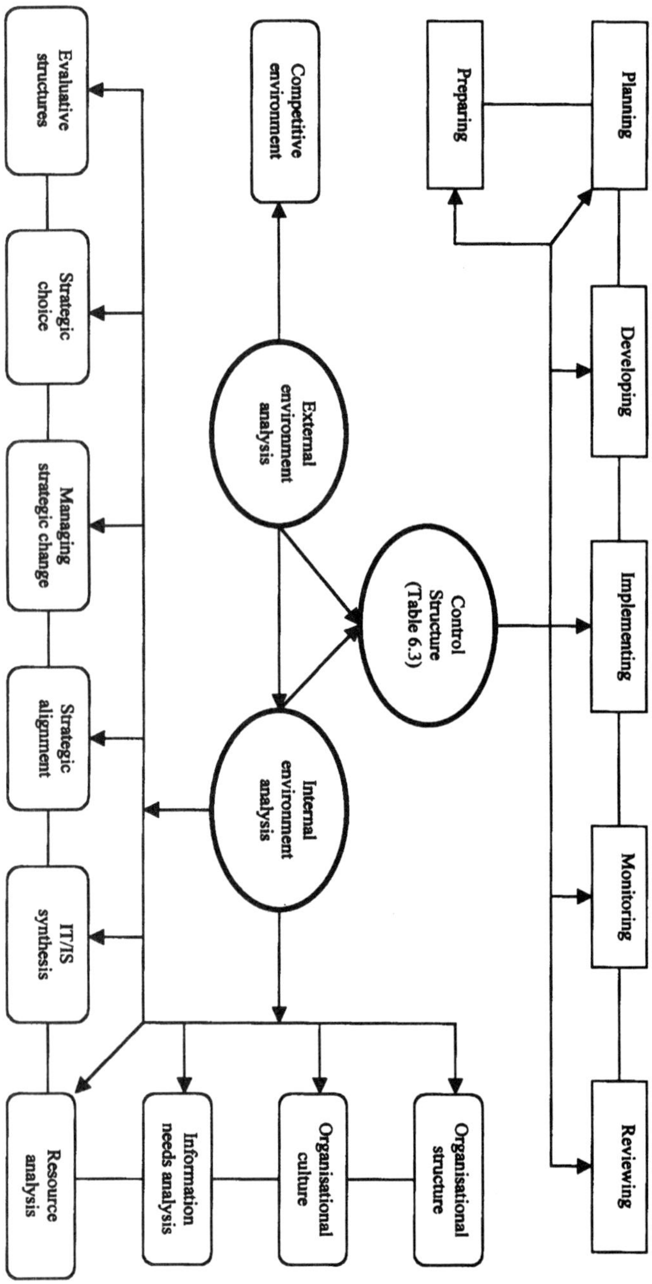

Figure 1. Framework for Information Strategy Evaluation

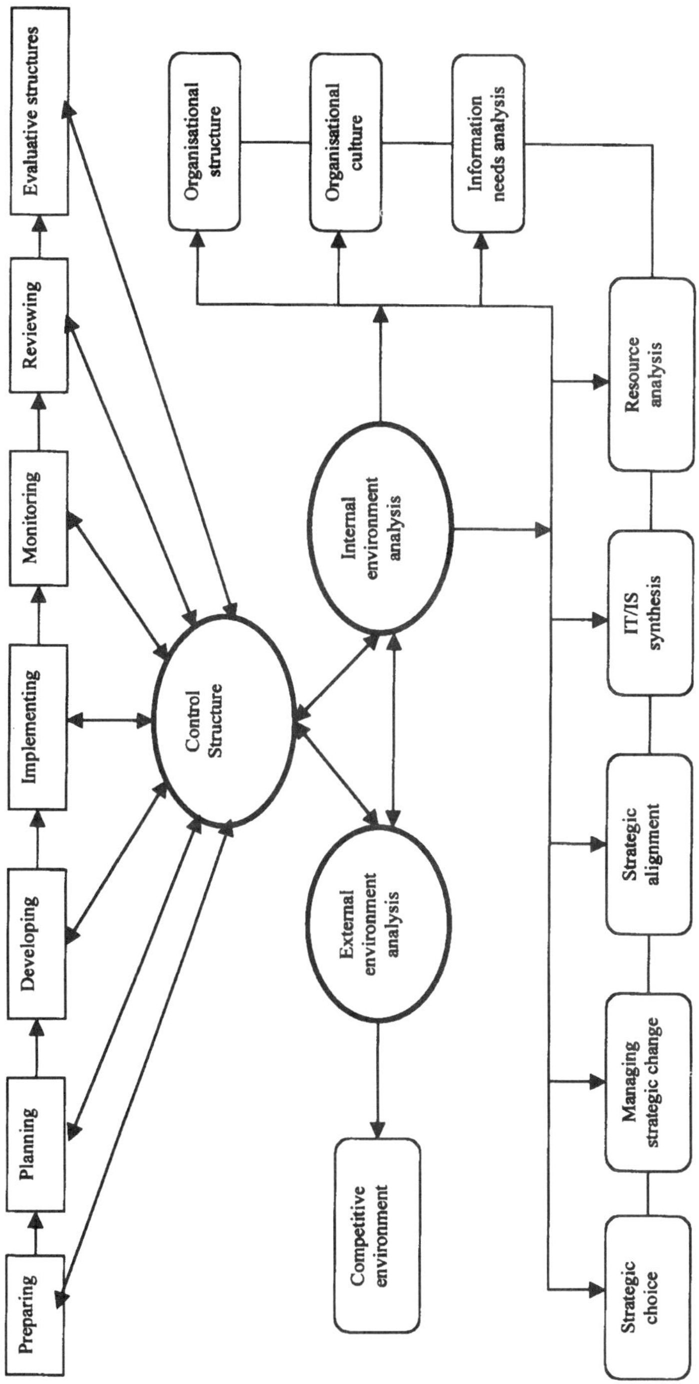

Figure 2. Revised Framework for Information Strategy Evaluation

REFERENCES

Ackoff, R. (1981). *Creating the Corporate Future*. Wiley, New York.

Ansoff, H. (1964). "A Quasi-Analytical Approach of the Business Strategy Problem." *Management Technology* (IV): 67-77.

Baets, W. (1992). "Aligning Information Systems with Business Strategy." *Journal of Strategic Information Systems* 1(4): 205-213.

Britton, G. and McCallion, H. (1994). "An Overview of the Singer/Churchman/Ackoff school of thought." *Systems Practice* 7(5): 487-522.

Campbell, D. and Stonehouse, G. (1999). *Business Strategy: An Introduction*, Heinemann, London.

Carnall, C. (1990). *Managing Change in Organisations*. Prentice-Hall, New York.

Churchman, C. (1968). *The Systems Approach*. Dell, New York.

Clarke, S. (2001). *Information Systems Strategic Management: An Integrated Approach*. Routledge, London.

Earl, M. (1989). *Management Strategies for Information Technology*. Prentice Hall, London.

Flood, R. (1995). "An Improved Version of the Process of Total Systems Intervention (TSI)." *Systems Practice* 8(3):329-333.

Flood, R. and Jackson, M. (1991a). *Creative Problem Solving: Total Systems Intervention*. Wiley, Chichester.

Flood, R. and Jackson, M. (1991b). *Critical Systems Thinking: Directed Readings*. Wiley, Chichester.

Flood, R. and Romm, N. (1997). From Metatheory to "Multimethodology," in *Multimethodology: The Theory and Practice of Combining Management Science Methodologies*, (J. Mingers and A. Gill, eds.), Wiley. Chichester.

Habermas, J. (1971). *Knowledge and Human Interests*. Beacon Press. Boston.

Henderson, J. and Venkatraman, N. (1993). Strategic alignment: leveraging information technology for transforming organisations. *IBM Systems Journal* 32 (1), 4-16.

Jackson, M. (1987a). "Systems Strategies for Information Management in Organisations which are not Machines." *Intl Journal of Information Management* 7: 187-195.

Jackson, M. (1987b). "New directions in management science". In Jackson, M. C. and Keys, P. (eds.), *New directions in Management Science*, Gower, Aldershot.

Jackson, M. and Keys, P., (1984). "Towards a System of Systems Methodologies." *Journal of the Operational Research Society* 35(6): 473-486.

Johnson, G. and Scholes, K. (1993). *Exploring Corporate Strategy*. Prentice Hall, Hemel Hempstead.

McFarlan, F. (1989)."Portfolio Approach to Information Systems". in *Software Risk Management*. (B. W. Boehm, ed.), IEEE Computer Society Press: 17-25, Washington.

Midgley, G. (1992). *Unity and Pluralism*, Ph. D. thesis, City University, London.

Midgley, G. (1995a). "What is this thing called Critical Systems Thinking?" in *Critical Issues in Systems Theory and Practice*, (K. Ellis, A.J. Gregory, B.R. Mears-Young, and G. Ragsdell, eds.), Plenum, New York.

Midgley, G. (1995b). "Mixing Methods: Developing Systemic Intervention." *Hull University Research Memorandum* No. 9.

Mintzberg, H. (1987). "Crafting Strategy." *Harvard Business Review* 65(4): 66-75.

Mintzberg, H. and Quinn, J. (1998). *The Strategy Process: Revised European Edition*. Prentice Hall, Hemel Hempstead.

Oliga, J. (1988). "Methodological foundations of systems methodologies". *Systems Practice*, (1): 87-112.

Pettigrew, A. (1987). "Context and action in the transformation of the firm". *Journal of Management Studies* 24(6), 649 - 670.

Porter, M. (1990). *The Competitive Advantage of Nations*. Macmillan, London.

Quinn, J. (1980). *Strategies for Change - Logical Incrementalism*. Homewood, Il., Irwin.

Schecter, D. (1993). "In defence of the System of Systems Methodologies: some comments on the Mingers/Jackson debate." *Jrnl of the Operational Research Soc.*, **44,** 205-306.

Ward, J and Griffiths, P. (1996). *Strategic Planning for Information Systems*. Wiley, Chichester.

SYSTEMS THINKING IN THE THEORY AND PRACTICE OF STRATEGIC INFORMATION SYSTEMS PLANNING

Irwin Brown

Department of Information Systems, University of Cape Town
Private Bag Rondebosch, 7701, Cape Town, South Africa
Email: ibrown@commerce.uct.ac.za

INTRODUCTION

Information systems planning has been defined as the process of identifying a prioritised portfolio of information systems that are efficient, effective and/or strategic in nature, together with the necessary resources (human, financial and technical), management of change considerations and organisational structure necessary for implementation (Baker, 1995). Strategic information systems planning (SISP), the topic of this paper, differs from lower levels of planning in that, in terms of scope, it is organisational, in perspective, that of top management, in terms of level of abstraction, conceptual, and in time frame, long (Segars *et al*, 1998).

The overall aim of the paper is to highlight the contribution of systems thinking to the theory of SISP and to demonstrate its appropriateness and relevance to SISP practice. The paper will therefore proceed by firstly examining how systems ideas have contributed towards SISP theory. This will lead into a discussion on what has been termed an approach to SISP. The most effective approach, characterised by a high degree of both rationality and adaptation, is then discussed. It is shown that the underlying principles of rational adaptation are highly congruent with those of soft systems thinking and a soft systems-based approach to SISP is then described.

DEVELOPMENTS IN SISP THEORY

The focus of early research in SISP was on the derivation of methods and techniques for carrying out SISP. However, many of these were developed in the absence of a sound theoretical underpinning, and partly as a result of this, there were many reports of methodological problems and plan implementation failures (Lederer and Sethi, 1988). More recently, attempts have been made to fill this theoretical gap by conceptualising SISP in terms of an input-process-output model, or system (Baker, 1995; Lederer and

Salmela, 1996). The SISP dimensions identified from these studies include the internal and external environment, planning resources, informational input, the planning process, planning methods, the information plan, plan implementation, and planning outcomes.

It has thus become evident that describing SISP in terms of method only is clearly simplistic. A multi-dimensional conceptualisation captures the richness and complexity of SISP as a result of the interactions between, and the interrelatedness of the various dimensions. Thus, attention has shifted to what have been termed approaches. An approach is much broader than method, derived from a consideration of characteristics relating to process, method, implementation, and resources (Doherty *et al*, 1999). A recent empirical study provides evidence that an approach characterised by a high degree of both rationality and adaptation is the most effective in achieving SISP outcomes (Segars and Grover, 1999). Rationality is reflected in a highly comprehensive, formalised process, with a top-down flow, and a focus on control. High consistency and broad participation profiles reflect adaptation (Segars *et al*, 1998). The characteristics are defined in "**Table 1**. SISP Characteristics."

Table 1. SISP Characteristics (Segars *et al*, 1998; Doherty *et al*, 1999)

Characteristic	Definition
Comprehensiveness	The extent to which an organisation attempts to be exhaustive or inclusive in making and integrating decisions.
Formalisation	The existence of structures, techniques, written procedures, and policies that guide the planning process.
Flow	The locus of authority or devolution of responsibilities for strategic planning (top-down, bottom-up, interactive)
Focus	The balance between creativity and control orientations inherent within the strategic planning system.
Consistency	The frequency of planning activities or cycles, and, relatedly, the frequency of evaluation/revision of strategic choices (occasional vs. continuous).
Participation	The breadth of involvement in strategic planning (narrow vs. wide)

THE CASE FOR SOFT SYSTEMS THINKING IN SISP

Given that much of the research focus has been on examining characteristics of alternative approaches, detailed descriptions of these have received little attention. Some insight has, however, been given into the schools of thought underlying them (Segars and Grover, 1999). That which underlies rational adaptation is called the learning school and seems to bear some resemblance to the systems thinking approach to dealing with problem situations, as embodied in the soft systems methodology (SSM). The term methodology as used here is more akin to what Iivari *et al*, (1998) call an approach than it is to method or technique. They define an approach to be the set of goals, principles and concepts underlying methodology. SSM is essentially a process of inquiry for dealing with complex, ill-structured problem situations. It has traditionally been represented as a seven stage cyclical model, whilst recognising that these stages need not be followed in any particular order, and that it is possible to iterate between stages if needs be. These stages are as follows (Checkland, 1981):

1. Problem situation: unstructured.
2. Problem situation: expressed.
3. Formulation of root definitions of relevant systems.
4. Conceptual model building (human activity models).
5. Comparison of 4 with 2.
6. Identification of feasible, desirable changes.
7. Specification of actions to improve the problem situation.

To provide more substantial validity to the claim that SSM is highly congruent with rational adaptation, it will be examined with respect to the characteristics of comprehensiveness, flow, formalisation, focus, consistency, and participation. These characteristics have been defined within the context of SISP, but will in this analysis be generalised in order to assess their presence within a typical SSM study.

Comprehensiveness in SSM

Comprehensiveness in SSM is evident in several ways, as is outlined below (Checkland and Scholes, 1999):

1. Finding out about the problem situation may include the drawing of rich pictures, which enable the complex interrelationships inherent in the situation to be more clearly presented.
2. To complement the logic-based enquiry of the seven stages, a cultural stream of enquiry may be employed to analyse the roles, the social system, and the political system in the problem situation.
3. Several alternative root definitions of relevant systems may be formulated, each based on a different underlying worldview, rather than formulating only a single root definition.
4. In building a conceptual model, 7 +/- 2 activities are identified. Each activity may in turn be a candidate for further analysis, whereby a root definition is formulated for it, and a lower level conceptual model is developed (Flood and Jackson, 1991).
5. The comparison of the conceptual models with the real world proceeds via a debate about possible changes that may entail backtracking to earlier stages to gain more insight.
6. Changes ultimately agreed upon must then meet the criteria of culturally feasibility and systemic desirability.
7. The actions identified to improve the problem situation are thus arrived at after an exhaustive and inclusive (i.e. comprehensive) process.

Formalisation in SSM

SSM provides a formalised structure for dealing with messy, ill-structured problem situations, and recommends specific methods and techniques to be used at each stage. The most well known include rich pictures, root definitions, conceptual models, gap analysis, and an assessment of cultural feasibility, and systemic desirability (Checkland,

1981). Various other techniques may also be employed, either in lieu of the aforementioned, or to complement them. Examples include cognitive mapping for finding out about the problem situation (Ho and Sculli, 1994), and decision variable partitioning for conceptual model building (Ledington and Ledington, 1999). There have even been attempts at providing computer support for SSM (Zhang *et al*, 1997). Given that SSM is participative, requiring debate at various stages, workshops are typically used as a means of carrying out the study (Galliers *et al*, 1994).

Flow in SSM

SSM was developed through a process of action research that attempted initially to apply hard systems thinking to the socially complex problem situations faced by managers. It is thus essentially a process for managing, where the real world is seen as a changing flux of events and ideas, and managing as making sense of parts of that flux, and then deciding upon a course of action (Checkland and Scholes, 1999). This implies that SSM is a top down process, as the types of problem situations being addressed are those typically faced by managers. It is managers that will authorise any intervention, and SSM requires that they participate, if they are in any way involved with the situation. That the flow in SSM is decidedly more top-down than bottom-up is furthermore borne out by critics who note that in assessing changes for cultural feasibility, they will more than likely be assessed against the dominant culture, which is often that of the most powerful (Flood and Jackson, 1991).

Focus in SSM

Both control and creativity orientations are present within SSM. In terms of creativity, it is fostered through the use of techniques such as rich picture building (Ho and Sculli, 1994). Several alternative root definitions based on different worldviews are formulated in a bid to broaden thinking about possible ways forward. Conceptual models are a way of employing creative systems thinking, without being constrained by the detailed context of real world operations. These models are then compared with the real world to generate debate and creative ideas about possible changes. Indeed if the root definitions and conceptual models are found to be lacking in insight, they can be reformulated in a search for more creative and innovative ways forward.

A control focus is evident when building conceptual models, as for each model a monitoring and control sub-system should be included, in which efficiency, effectiveness, efficacy, ethicality and elegance may be assessed (Checkland and Scholes, 1999). There are thus in effect two major sub systems to the conceptual model – the operational sub-system and the monitoring and control sub-system.

There is no conflict in a rational adaptive approach to SISP of having both control and creativity orientations present. In fact, in an empirical study examining alternative SISP approaches in the UK, it was found that the most successful approach was characterised by a creativity focus (Doherty *et al*, 1999). This contrasts with the finding by Segars and Grover (1999) that a rational adaptive approach has a control focus, and may be an indication that having both orientations is advantageous.

Consistency in SSM

Consistency is one of the hallmarks of SSM, as it is essentially a never-ending learning cycle. It seeks to learn about problem situations in order to generate commitment to actions for improvement (Checkland, 1981). The resultant new situation that emerges as a result of these actions can again be examined by repeating the cycle. Learning cycles are also apparent within SSM between stages, as ideal conceptual models are compared with the real world and debate about changes takes place, sometimes necessitating a recycling back to earlier stages.

Participation in SSM

Participation is another defining characteristic of SSM, whereby all stakeholders concerned about a problem situation should be involved (Flood and Jackson, 1991). Participation allows for alternative worldviews to be surfaced and for debate to take place between stakeholders that helps generate shared understanding and accommodations between sometimes-conflicting views (Checkland and Holwell, 1998). Ultimately the sense of ownership engendered through this process helps to gain commitment for actions to be taken to improve the situation (Flood and Jackson, 1991).

Rationality and Adaptation in SSM

SSM thus rates highly in terms of both rationality (high comprehensiveness, high formalisation, top-down flow and control focus) and adaptation (high consistency and broad participation). It can therefore be used as a basis for designing a rational-adaptive approach to SISP. The application of SSM to the field of information systems is not new, however (Checkland and Holwell, 1998). Deriving any SSM-based approach to SISP therefore requires an examination of the literature in this domain, and is the subject of the next section.

SSM AND SISP

Any approach to SISP grounded in SSM needs to take into account its paradigmatic perspective with regards to information and information systems. Information is defined as data plus the meaning attributed to it in some context (Iivari *et al*, 1998). Attribution of meaning is recognised as being a distinctly human activity, which gives rise to the concept of an information system as having two major sub systems: the serving sub-system (typically a computer-based data manipulation system), and the served sub-system (human activity system). Any development of the serving system would firstly require a prior declared account of the system to be served (Winter *et al*, 1995).

Thus conceptual models in SSM may serve as the basis for deriving information flows and information requirements in support of the human activity they imply. A comparison of these information flows and requirements with current information provision in the real world, together with the feasible and desirable changes that are agreed upon helps to identify potential information systems to be developed (Checkland and Holwell, 1998). Galliers (1993) describes and illustrates an example of an SSM-based approach to SISP that incorporates these ideas. Using the Galliers (1993) model as a guide, a similar model is derived by incorporating the major phases and stages expected of an SISP study into the SSM cycle. These phases and stages are listed in "**Table 2**. Phases and Stages in SISP", together with the corresponding SSM stage that they would fit into. The assumption being made is that although there may be a recommended set of SISP phases and stages, the approach to carrying them out is what differentiates SISP outcomes (Segars and Grover, 1999).

Table 2. Phases and Stages in SISP (amended from Mentzas, 1997)

SISP Phase	SISP Stages	SSM Stages
Preparation	Planning for planning (pre-planning).	One
Organisation analysis	Analysis of internal business environment. Analysis of internal IS environment.	Two
External environment analysis	Analysis of external business environment. Analysis of external IS environment.	Two
Strategy conception	Scanning the future. Identification of alternative scenarios including information flows and requirements for each. Scenario elaboration.	Three/Four Five
Strategy formulation	Formulation of agreed organisational recommendations. Formulation of information architecture. Synthesis and prioritisation of portfolio of information systems to be developed.	Six
Strategy implementation planning	Definition of action plan elements. Elaboration of action plan. Definition of follow-up and control procedures.	Seven

TOWARD A RATIONAL ADAPTIVE APPROACH TO SISP

The resultant model that emerges from incorporating the SISP phases and stages into the SSM cycle, is displayed in Figure 1 on the next page. The model illustrates clearly that the approach is grounded in SSM. For it to be truly rational and adaptive requires that it be conducted in a highly comprehensive formalised manner, with a top-down flow and a focus on control. There should be frequent iterations between stages where necessary, as well as a continuous learning process, such that once a planning cycle is complete, both the process and outcomes of the study are evaluated, and these results fed back into the next planning cycle (Baker, 1995). Broad participation of all major stakeholders should furthermore be ensured.

CONCLUSION

In conclusion, then, systems theory has made a contribution towards a better conceptualisation of SISP in terms of an input-process-output model or system (Lederer and Salmela, 1996). This in turn has given rise to multi-dimensional studies focused on approaches to SISP. Recent empirical evidence suggests that an approach to SISP characterised by rationality and adaptation is the most effective (Segars and Grover, 1999). However, there is little in the way of descriptions of these. It has therefore been shown that SSM (a systems thinking approach to problem solving) exhibits the required characteristics of rationality and adaptation, and that SSM-based approaches to SISP are examples of rational adaptation. This is not to say that they are the only examples. Other problem solving approaches may also be examined to assess whether they too possess the required characteristics. If so, they may also form the basis for design of this type of SISP approach.

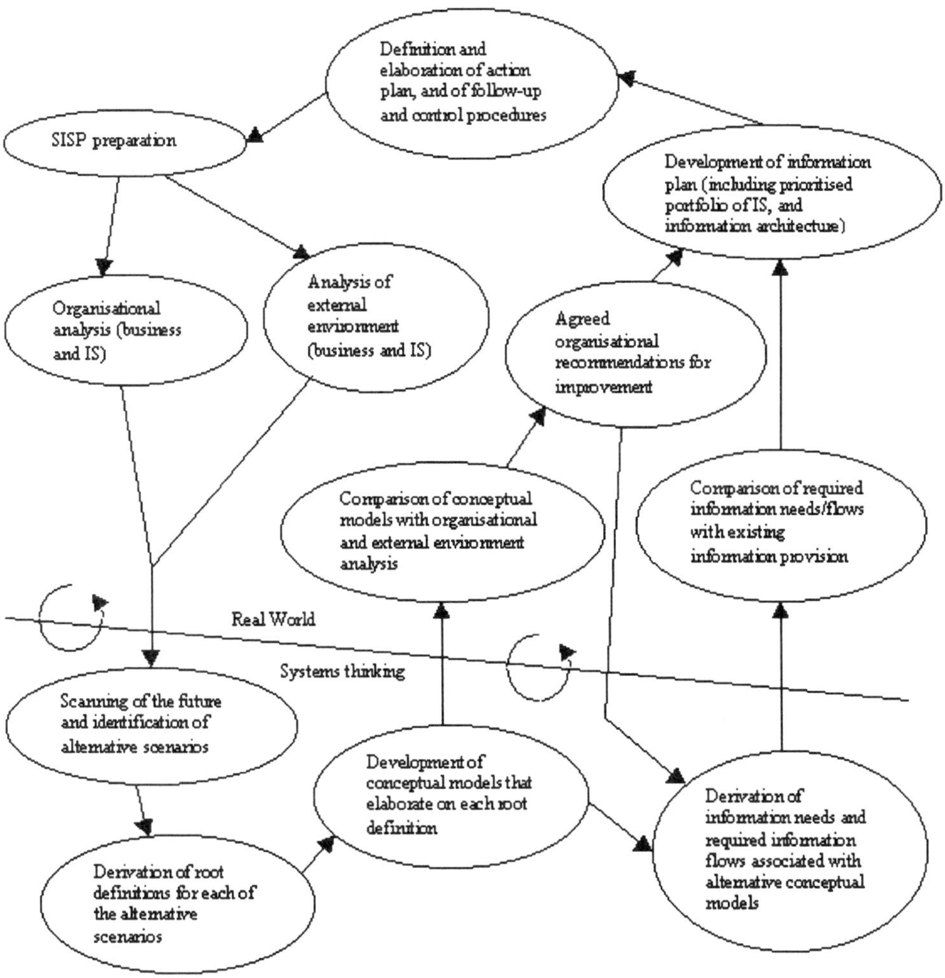

Figure 1. A Rational Adaptive Approach to SISP (amended from Galliers, 1993)

REFERENCES

Baker, B. (1995). "The role of feedback in assessing information systems planning effectiveness," *Journal of Strategic Information Systems*, 4(1):61-80.

Checkland, P. (1981). *Systems Thinking, Systems Practice*, Wiley, Chichester.

Checkland, P. and Holwell, S. (1998). *Information, Systems and Information Systems – making sense of the field,* Wiley, Chichester.

Checkland, P. and Scholes, J. (1999). *Soft Systems Methodology in Action,* Wiley, Chichester.

Doherty, N., Marples, C. and Suhaimi, A. (1999). "The relative success of alternative approaches to strategic information systems planning: an empirical analysis," *Journal of Strategic Information Systems*, 8(3):263–283.

Flood, R. and Jackson, M. (1991). *Creative Problem Solving: Total Systems Intervention*, Wiley, Chichester.

Galliers, R. (1993). "Towards a flexible information architecture: integrating business strategies, information strategies and business process redesign," *Journal of Information Systems*, 3(3):199-213.

Galliers, R., Pattison, E. and Reponen, T. (1994). "Strategic information systems planning workshops: lessons from three cases," *International Journal of Information Management*, 14(1):51-66.

Ho, K. and Sculli, D. (1994). "Organisational theory and soft systems methodologies," *Journal of Management Development*, 13(7):47–58.

Iivari, J., Hirschheim, R. and Klein, H. (1998). "A paradigmatic analysis contrasting information systems development approaches and methodologies," *Information Systems Research*, 9(2):164–187.

Lederer A. and Salmela, H. (1996). "Toward a theory of strategic information systems planning," *Journal of Strategic Information Systems*, 5(3):237-253.

Lederer, A. and Sethi. V. (1988). "The implementation of strategic information systems planning methodologies," *MIS Quarterly*, 12(3):445-461.

Ledington, J. and Ledington, P. (1999). "Decision variable partitioning: an alternative modelling approach in soft systems methodology," *European Journal of Information Systems*, 8(1):55–64.

Mentzas, G. (1997). "Implementing an IS strategy - A team approach," *Long Range Planning*, 30(1):84-95.

Segars, A. and Grover, V. (1999). "Profiles of strategic information systems planning," *Information Systems Research*, 10(3):199-232.

Segars, A., Grover, V. and Teng, T. (1998). "Strategic information systems planning: planning system dimensions, internal coalignment, and implications for planning effectiveness," *Decision Sciences*, 29(2):303-345.

Winter, M., Brown, D. and Checkland, P. (1995). "A role for soft systems methodology in information systems development," *European Journal of Information Systems*, 4(1):130–142.

Zhang, J., Smith, R. and Watson, R. (1997). "Towards computer support of the soft systems methodology: an evaluation of the functionality and usability of an SSM toolkit," *European Journal of Information Systems*, 6(2):129–139.

PRACTICAL APPROACHES TO AN INFORMATION SYSTEMS ARCHITECTURE IMPLEMENTATION AT GLAXOWELLCOME

Andreas Rassau[1], Martin Sykes[2] and Misha Hebel[3]

GlaxoSmithKline
Programme Management Group
Global Manufacturing & Supply IT
Stockley Park, Uxbridge,
MIDDX, UB11 1BD
E-mail:axr93984@GSK.com

[2]Head IT Strategy
Global Manufacturing & Supply IT
GlaxoSmithKline
Priory Street, Ware
E-mail: mjs46934@gsk.com

[3]Brignition Ltd,
c/o 54 Hillingdon Hill,
Uxbridge,
MIDDX, UB10 0JD
E-mail: mishah@brignition.com

INTRODUCTION

For some years GlaxoWellcome was one of the world's leading research-based pharmaceutical and healthcare companies. Over the last year it merged with a company of similar profile, SmithKline Beecham, to form GlaxoSmithKline (GSK). This merger of two large, diverse and global holding companies evidently presents many challenges from an Information Systems Architecture perspective. This paper considers aspects of architectural work conducted within the GlaxoWellcome Research Department, that might be considered preparatory to the merger. The focus in this paper is on the difficulties that GlaxoWellcome experienced in moving towards an architected approach to systems development and in particular the specific challenges of managing the required organisational changes within the IS department.

As we move into the 21st century, the pharmaceutical industry finds itself in the midst of a period of intense and wide-ranging change. Rapid advances in technology have lead to a proliferation in chemical and biological information. At the same time, and in response to the escalating costs of drug discovery and development, there is a trend towards consolidation amongst the large players in the pharmaceutical industry. Indeed, Price Waterhouse Coopers (PWC) has projected that by the end of 2005, there could be as few as 13 industry giants (PWC, 2000). In this context it is necessary to consider mergers and acquisitions as essential components of the business environment for which there must be active planning. Specific challenges, which the large pharmaceutical companies face as a consequence of this market conditions, are: increased demand for tools to facilitate global collaboration; requirements for highly flexible information systems architectures and an ongoing requirement for systems integration. As a consequence many large pharmaceutical organisations struggle to keep their information systems up to date, avoiding duplication and the excessive costs of maintaining interoperability solutions.

In June 1998 a report was published internally in GlaxoWellcome Research outlining a 10 year strategic plan for drug discovery research (GlaxoWellcome, 2000). Key issues that were highlighted in this plan included the fragmented nature of the existing infrastructure and duplication of data across the organisation. In this context it became obvious that there was a clear business need for a wide-ranging information systems project to meet the businesses' long term strategic needs. Over the next six months project sponsorship and management were assembled for an initiative, which came to be known as **HAL,** the Hits and Leads Project. Here the terms "hits" and "leads" refer to key stages in the process to identify potentially useful pharmaceutical compounds. The HAL programme was formally launched in May 1999 with a stated objective:

> "To develop a global informatics infrastructure and toolset to support the lead identification and lead optimization processes in the GlaxoWellcome Research Department." (HAL project, 2000)

From the very outset, Systems Architecture was deemed to be a vital component of the HAL initiative. In this context Systems Architecture was understood to mean the models of the business and of the IS systems as they are now (as-is) and as we wish them to be at some point in the future (to-be). Embedded within this concept of architecture was an understanding that it is also necessary to describe an acceptable path to bridge the gap between the as-is and the to-be representations, by means of standards and intermediate implementations.

On a conceptual level the advantages of a fully integrated systems architecture are indisputable. By developing a consistency of representation and removing redundancies, integrated systems architectures have the potential to create organisations which can operate on a truly global scale and are able to cope more effectively with a rapidly changing business environment (Cook, 1996). Offset against these clear motivations, however, are the significant practical difficulties of migrating from an organisation within which systems are developed in isolation (as had previously been the case in GlaxoWellcome), to one in which systems are developed in the context of an overall integrated architecture. It is our contention that this shift in approach requires significant changes in the overall structure, required skill-sets and desired behaviours within the

Information Systems Department. Indeed, this requirement for internal change within the IS department is very seldom acknowledged in the literature.

In this paper we will draw on experiences from the HAL architecture project within the GlaxoWellcome Research organisation. Using this case as a framework for discussion, recommendations have been developed on how best to implement an integrated systems architecture, from the perspective of managing IS department organisational change.

GLAXOWELLCOME RESEARCH, HAL ARCHITECTURE PROJECT

In the first five months, the HAL programme focused on clarification of business benefits and global requirements. Five opportunities were initially selected as the focus for HAL Stage 1, one of which was the development of an overall Information and Systems Architecture. In March 1999, one of the co-authors, Martin Sykes was appointed as the global project manager for Architecture and between May and September 1999 a core architecture team was formed. The idea was that this team would work in support of the business area project teams but would also remain distinct from them. By August 1999 the team had decided to use the META Group architecture methodology (META Group, 2001), as represented in Figure 1.

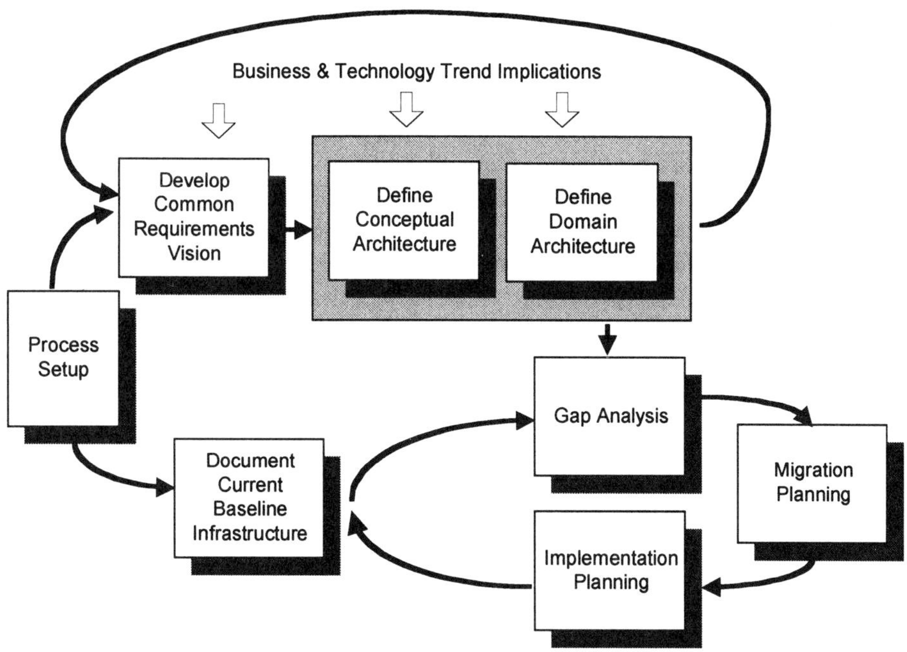

Figure 1. Diagram Summarising the MetaGroup Methodology for Enterprise Wide Technology Architecture Development (META Group 2001)

As part of the GlaxoWellcome Research IS architecture implementation, the *Conceptual Architecture* and *Common Requirements Vision* were developed between November and December 1999. Even at this early stage, however, the first signs of a rift between the architecture team and the wider IS department had started to emerge. The reasons for the emergence of this resistance will be examined in the following sections. After defining a conceptual architecture the next stage in the META Group process is to define so called domain architectures which provide architectural guidance in specific subsets of the architecture. In December 1999 two high priority domains, data integration and middleware, were selected as the initial areas of focus for the HAL architecture initiative. The initial target was that both domains should be completed by the end of February 2000. In the first quarter of 2000, however, the HAL Architecture initiative started to experience real resistance. In an attempt to address the growing rift between architects and developers, some of the individuals most vehemently opposed to the concept of an integrated systems architecture were included within the domain teams. It was hoped that by involving them at this point they could be encouraged to buy into the overall initiative. Ultimately, however, this led to further resistance and divisiveness.

Between March and June 2000, following increasing disquiet within the department as to the status and direction of the HAL architecture project, a series of reviews were conducted. The overall purpose of these reviews was to determine whether the HAL Architecture Team was on-track to deliver a well-defined, achievable and sustainable architecture in the timeframe required by the HAL programme (HAL, 2000a). The review process included interviews with representatives from the architecture management, the architecture team, primary architecture customers (project managers of other HAL projects), management external to HAL, sponsors, business managers and technical staff. To provide objectivity and add credibility the review process was conducted by external consulting agencies and IS professionals from areas of Glaxo Wellcome outside of the HAL programme (HAL, 2000b).

In this review it was found that the ultimate benefits of the architecture initiative were generally well understood. Furthermore, it was acknowledged that the initiative had clearly promoted discussion between the Research Department and Research IS. It was noted, however, that the project teams could not bridge the gap between the architectural vision and their day to day project work. In fact, their main experience of the overall architecture was the impact that the architecture schedule was having on their own time-scales. Furthermore it was identified at this stage that there was a clear risk that if architecture was not integrated fully into the project teams, then they will design their own architectures to meet their own time-scales (HAL, 2000c).

BUILDING SUPPORT FOR AN ENTERPRISE ARCHITECTURE INITIATIVE: KEY LESSONS FROM HAL

In a project of this scope, lessons can be derived on many different levels. This paper focuses on the softer issues which affected the acceptance of the architectural initiative into the IS projects organisation. In part these issues were touched upon by Cook (1996) who stated that:

> "An architecture does not at first seem to be in the interest of the internal information systems departments with their own proprietary approaches to systems development and there may be extreme resistance from these organisations, especially decentralised teams who have had complete freedom in the past" (p1).

As we will see below, however, these factors identified by Cook are just some of the reasons for resistance to develop. Although related on some level to specific practical or technical concerns, many of the causes for opposition are entirely generic and as such insights can be drawn from conventional change management theory.

Understanding the Reasons for Resistance

In the course of an initiative to integrate systems architectures into a single overall view, the shift in the mind-set is fundamental: from a locally optimised response to a globally optimised plan. It seems, in this context, almost inevitable that there should be some resistance (Hebel, 1998a & 2000). This resistance may manifest itself in many different ways, for example by an increased sense of project urgency being used as an excuse for failing to adhere to standards or, as occurred in the case of HAL, by product zealots becoming increasingly entrenched and vocal. In order to trace the causes of such resistance it is necessary to understand the response to change on a fundamental human level. Clarke (1994) identifies seven major factors that can lead to resistance to change. As should be apparent from the following discussion, all are relevant in the case of an architecture implementation.

1. **Loss of control**: The more that people feel that they are the victims of change the more likely they are to oppose it. The HAL architecture review revealed that many developers felt that their views had been ignored in formulating the original architecture strategy and hence this was a significant source of resistance (HAL, 2000c).
2. **Why are we doing it?**: Often the need for the change is not at all clear. For many people working in the project teams there was a gap of experience between these long-term goals and their day-to-day experience.
3. **Uncertainty and ambiguity:** Both are inevitable in an integrated systems architecture initiative. From the responses recorded during the HAL architecture review it was clear that many people were extremely uncomfortable with this ambiguity (HAL 2000c).
4. **Springing Surprises**: In general it would be inadvisable to develop an integrated systems architecture in a vacuum with the intention of introducing it to the IS projects organisation at some later stage.
5. **Loss of face**: There is a danger that in implementing an architecture initiative we are seen to be making a tacit value judgement on systems that had previously been delivered within the organisation. This could induce an emotionally rather than rationally motivated defensiveness towards existing techniques and practices.
6. **Can I cope?**: In any major change initiative steep learning curves are required in order to develop the competencies necessary to cope with the change. In the case of an architecture initiative staff with deep skills in a particular technical area may find that their skill-set is no longer required. It was a key concern among some members

of technical staff that they would have to build new skills following the changes brought about by the architecture initiative (HAL, 2000c).

7. **More Work**: It can be difficult to convince people of the long term time savings which should accrue as a result of the integrated approach to systems architecture.

In addition to these generic factors, there is an additional source of resistance specific to the context of an organisation moving to an integrated systems architecture. One of the key goals of architecture as a discipline is to optimise overall performance by selective sub-optimisation of certain of the component parts (Armson and Paton, 1994). This is familiar concept from the field of engineering however there is a crucial difference in that, in the case of enterprise architectures, we are dealing with human being rather than mechanical components. Sub-optimisation in this context seems to present a direct challenge to the western culture of individualism, whereby the ingenuity of the individual is prized above all else. As pointed out by Cook (1996), there is a clear danger that in this context the architecture comes to be regarded more as a policing function than as a valued infrastructure.

Managing the Architecture Implementation

In order to effectively manage an architecture implementation, it is necessary to have an understanding of the context for change from the systems perspective as well as the wider cultural context. It is then necessary to consider how to manage the change, taking into account these conditions and anticipating the likely human responses. In the following sections we outline some of key considerations in managing such changes.

Understand the Context of the Change Initiative

Before starting an integrated architecture initiative it is important to develop an understanding of existing conditions. As such, it is vital to give early consideration as to how to adapt existing processes and frameworks. It is necessary to develop an understanding of the existing culture, norms and practices before any architecture can be designed. This is not a topic always covered in architecture texts perhaps because it suggests the possibility of architecture not being the answer. This is due to the implication that the resolution of data management problems may simply be too difficult to conceptualise (Goodhue *et al*, 1988).

As well as understanding the existing frameworks, processes and architecture of the organisation it is vital to have an appreciation of the underlying value systems. It is these value systems which determine how the initiative will be regarded within the organisation and they should be taken into account when planning both the implementation and the associated communications. As a development on considering early conditions Hebel (2000) offers a number of methods for modelling the values and value systems present in an organisation. In particular it is worth noting that people hold a relatively small number of values in common use and these rarely change except in priority. The design and introduction of architectures however offers opportunities for people and groups to share values or experience their conflict. There should therefore be an awareness in planning an architecture implementation that this sort of dialogue will be an intrinsic part of the implementation. This implies that an architecture implementation cannot be planned in purely technical terms and must also take account of the impact on human activity systems.

Prepare the Foundations for Change

At the outset of any major change initiative, it is important to mitigate the inevitable resistance to change. By building a shared understanding of what the initiative will actually entail many of the causes of resistance outlined in the previous section, especially the feelings of loss of control, will be limited (Hebel, 1998b).

Streubel (1998, p139) points out that:

> "Employees and organisations have reciprocal obligations and mutual commitments both stated and implicit that define their relationship... corporate change initiatives whether proactive or reactive alter these terms".

Streubel goes on to suggest that unless managers define new terms and persuade employees to accept them, it is unreasonable to fully buy into changes in the status quo.

In general, the natural response to lack of "terms for change" is to immediately address the need for a formal compact. Arguably however the first stage should be to start from the social and psychological dimensions in order to build a sense of ownership for the new approach. At the start of the HAL architecture initiative there was a general feeling that there was a need for change. There may well have been a significant opportunity missed at this point. By failing to get the project teams from HAL together at the outset of the project an opportunity was missed to get to their buy-in at this stage. Certainly it is difficult to say whether this sort of "soft compact building" can be effective once impressions have already built up around an initiative of this nature.

Drive the Architecture Initiative

Having established the foundations for change it is important to maintain the momentum of the process. This is a delicate balance between effective inclusive communications and retaining an overall drive. It is crucial in this context to ***visibly*** maintain the connection to the original business benefits. In the case of the HAL Architecture initiative, it was widely noted that there was an apparent disconnect between the business requirements and the technical architecture (HAL, 2000b; Downey, 2000). This apparent gap appeared to have had more to do with communications, in that the people were not provided with sufficient information to see how the technical architecture was derived from the business requirements (Cook, 1996).

CONCLUSIONS

In carrying out an enterprise architecture initiative it is important to acknowledge the effect that this change will have on members of staff within the IS organisation. In common with any other major change initiative, a proper foundation must be laid by forming a compact on the need for change between the individuals and the organisation (Hebel, 2000). Having set the change initiative in motion momentum must be maintained by retaining focus on the fundamental business drivers. Much of the analysis of architecture work to date has focused on the considerable amount of work required to identify, develop and communicate models. There is as much effort to be expended on a change in the processes and behaviours of the IS community.

REFERENCES

Armson, R. and Paton, R. (1994). *Organisations, cases, issues, concepts*, (2nd edition), Paul Chapman Publishing, London.

Clarke, L. (1994). *The Essence of Change*, Prentice Hall International, New Jersey.

Cook, M.A. (1996). *Building Enterprise Information Architectures*, Prentice Hall, New Jersey.

Downey, K. (2000). "Cap Gemini Review of the HAL Architecture Initiative", (internal company document).

GlaxoWellcome (2000). "Lead Discovery 2000: GlaxoWellcome Discovery Research Strategic Business Plan", (internal company document).

Goodhue, D.L., Quillard, J.A. and Rockart, J.F. (1988). "Managing The Data Resource: A Contingency Perspective", *MIS Quarterly*, 12(3):373-392.

HAL (2000a). "Architecture Review Proposal Document", (internal company document).

HAL (2000b). "Architecture Review Executive Summary", (internal company document).

HAL (2000c). "Architecture Review Interview Responses", (internal company document).

HAL Project (2000). "Phase I Summary", (internal company document).

Hebel, M. (1998a). "Human Values in MIS - Fit or MISFfit?", *Systemist*, 20:140-152.

Hebel, M. (1998b). "Values Systems - A way to greater understanding", *Systems Practice and Action Research*, 11(4):381-402.

Hebel, M. (2000). "Human Values and the Management of Technological Change", *Cognition, Technology & Work*, 2:106-115.

META Group (2001). "Presentation made to GlaxoWellcome Research IS on process model", META Group.

PWC (2000). Price Waterhouse Coopers, "Pharma 2005: An Industrial Revolution in R&D", available from http://www.pwcglobal.com/gx/eng/about/ind/pharm/phpubl_pharma2005.html (15/12/00).

Streubel, P. (1998). "Why Do Employees Resist Change", *Harvard Business Review on Change*, Harvard Business School Press, Boston MA: 139-158.

THE ROLE OF SURPLUS VISION IN KNOWLEDGE SYSTEMS

Robert Stephens

School of Information Systems
University of the West of England
Frenchay, Bristol, UK BS16 1QY
Robert.Stephens@uwe.ac.uk

INTRODUCTION

Although it is topical to distinguish knowledge-based organizations, work or economies as a progression from the industrial, factory era, knowledge has of course always been key to organizational success. However, partly because of the availability of appropriate technology and partly because of the IT enabled structural change in organizations the idea of knowledge systems within institutions has gained currency. This follows a long tradition of attempting to make knowledge robust and explicit through procedures and technologies, and a more mixed history of support for cultures of knowledge (Dixon, 2000) or communities of practice (Seely Brown and Duguid, 2000). As Zuboff (1989) observed, information systems and technologies can augment either, but not both of these traditions. Recent trends in networking, internet technologies, standards and markup languages such as ontologies and XML, support the former, but evidence suggests (see Kling, 2000) that on their own they will not informate the knowledge activities in organizations.

The social capital of organizations is now very much identified with its knowledge resources, which are increasingly inseparable from information systems. This paper describes and analyses the role of 'surplus vision' within such knowledge systems (broadly construed). Surplus vision is a concept derived from Bakhtin's writings (1981; 1990) on dialogism, and describes a particular kind of participative knowing. Specifically, a community will to some extent perceive and experience a common environment, but each participant will also have unique perspectives. The paper explains how the meeting of these perspectives in properly formed dialogic circumstances creates 'an excess of seeing' (surplus vision) within the discourse situation itself, which promotes reflexive knowledge systems.

Surplus vision is relevant to ideas of knowledge work and organizational learning, as well as information systems practice, technologies and teaching. The notion of a "reflexive worker" is appropriate here: one who is located and locating themselves within

Edited by Ragsdell *et al.*, Kluwer Academic/Plenum Publishers, New York, 2002.

a range of information genres, able to translate and accommodate discourses in ambiguous and dilemmatic situations. At issue, however, is the possibility of genuine dialogic environments. Dialogism is based on people interacting and positioning in language, but may be compromised by communication media that either condense, replace or disengage such activity.

The paper is the product of ongoing research into dialogic aspects of information systems. Systems readers will probably be unfamiliar with the theoretical background of dialogism so we will begin by attempting to place this in an organizational context, before the paper's central topic of surplus value is discussed.

THEORETICAL BACKGROUND

Famously, in launching their *Invent* campaign, chairperson of Hewlett Packard, Carly Fiorina, announced *"if only we knew what we know at HP"*. Fittingly for an IT company, this was taken as a call for knowledge management technology in the shape of software and procedures to facilitate coordination and communication throughout the organization. In common with most knowledge management initiatives and thinking, the assumption is that organizational knowledge is distributed and fragmented, when it should be networked and accessible. However, a different reading of *if only we knew what we know* is possible. This concerns knowledge that is implicated but not articulated in collaborative situations, and suggests that students of organizational knowledge may well look beyond the Cartesian coordinates of time and space when identifying the territory for their analyses and technological interventions. The distinction, therefore, is between providing a channel for the communication, sharing and sourcing of knowledge on the one hand, and supporting the stewardship and (re)systemization of what is implicitly known on the other.

In this paper I shall seek to distinguish this type of learning situation from the mere sharing of information by drawing on work done by Yuri Lotman (1990) on the *functional dualism of texts*, and on the *dialogic* theory of Mikhail Bakhtin (1981). Although not aware of the concerns of information systems researchers, both were engaged with a central problem in IS scholarship - how to reconcile the formalization of communication associated with the growth of organizational information technologies while preserving the conditions required for the growth of knowledge.

According to both these authors, learning was unlikely to occur in situations of information assimilation, but took place where texts or speech acts (both of which may include information technologies) functioned as 'thinking devices' which promoted new understanding or perspective. Instead of being merely extended in time and space, knowledge systems operate in communities of practice where a dynamic tension between perspectives ensures there are always different readings of any event or situation.

In systems terms, this may be termed a redundancy or surplus of meaning or perspective, and as such is not a novel insight. Indeed, this is the logic behind single and double-loop learning and is shared by many scholars of organizational knowledge (e.g. Dixon, 2000). However, following Bakhtin and Lotman, another approach is to focus on the interaction of language genres in speech and literate situations, rather than on ideas or viewpoints that are 'cognitively completed'. Thus, while we may be quite familiar with the Socratic idea of the growth knowledge through a dialogue that entails a dialectic of people's perspectives, here we are considering Bakhtin's notion of dialogue as the interplay of language genres. The approach, therefore, is non-psychological and non-cognitive, but rather attempts to apprehend how texts work at a *cultural* level. Although still at a tentative stage, because this work shares a contemporary assumption that for some purposes it is warranted to treat infornation systems as *texts* (e.g., Zuboff, 1989; Nardi and O'Day, 1999), it is expected that dialogic theory will offer a distinctive insight on contemporary IS problems.

DIALOGISM AND INFORMATION SYSTEMS

Dialogue here is considered to be an epistemological condition of knowledge (or better, the knowing person), but concerns more than a Socratic exchange. What is significant is not so much the social fact of several people exchanging words, as the idea that each word itself contains diverse, discriminating, often contradictory 'talking' components because it is populated with the intentions of both speaker and listener. The more often a word (or better 'utterance') is used, the more contexts it accumulates and the more its meanings proliferate. In this way, utterances do not forget, but themselves convey the plurality of meaning or perspective coveted by systems thinkers, and this is the origin of the term 'surplus vision'. Surplus vision therefore, refers to the plasticity of the utterance that allows it to inform beyond the contexts of its issue.

Understood like this, dialogue is a model of a creative process ministered in language, and surplus vision is a characteristic of dialogic communication. However, not all communication is dialogic, much is 'univocal' or 'monologic', that is, it is resistant to the intentions of speaker or listener and therefore unable to embody more than a single voice or consciousness.

A typical organization will display both modes of communication, with the more bureaucratic or mechanical showing a univocal preference. While information systems technologies are often designed to support a formalization of communication, and often supposed by critics to achieved this (cf. Habermas, 1984), the technologies themselves are notoriously ambiguous. Often known as 'interpretive flexibility' information technologies have a variety of use values that usually far exceeds their specified uses. The most celebrated examples of this are the reconfiguration of the personal computer from a desktop toy designed for executive amusement to an essential general purpose device, and the short-message-service (SMS) technology, again designed for executives, but adopted by the youth for text-messaging. This points to a powerful, but often unacknowledged, sensual attribute of information systems technology that are deployed as media rather than instruments in an information-control system. A distinction first made by Marshall McLuhan is helpful here. Hot and cool, or high and low definition media invoke different degrees of personal intellective engagement (McLuhan, 1964). High definition media are loud, bright, clear, and importantly, fixed, while the presentations of low definition media are soft, shadowy, blurred and changeable. The psychological logic of this distinction is that we are obliged and seduced to work harder - get more involved - to fill in the gaps with the lower profile, less complete media (Levinson, 1999).

Low definition media have become pervasive since the second half of the twentieth century and quietly ventilate culture, attitudes and interpersonal behaviour. But, as

Zuboff (1989) observed, the degree to which low definition media, such as information systems, may transform behaviour by informating activities is conditioned by the distribution of responsibility and authority within an organization. She also warns that the formalization of communication as an objective on its own poses a severe threat to human comprehension and critical judgement. This is also the conclusion Bakhtin arrived at through his dialogic theory (no doubt influenced by his own experiences of the Soviet system), although in his view univocal communication invoked no consciousness. This is not to valorise univocal or dialogic communication, but merely to observe that not all communication or social structures produce knowledge. The question to be answered therefore from this theoretical standpoint is, can, and to what extent can formalized communication support dialogic interaction? A key component of dialogic interaction is the presence of multiple voices in communicated items (utterances). Such multiple voices are considered to convey a *surplus vision* in any properly formed dialogic context, and thereby support learning and knowledge development.

SURPLUS VISION

In dialogism, aesthetic events can take place only when there are two or more participants present; it presupposes two non-coinciding consciousnesses'. The meaning generated by this event is a 'self'. A single consciousness could not generate a sense of self; only the awareness of another consciousness outside the self can produce that image. Behind this notion is an affirmation of the importance of being-in-the-world, particularly Bakhtin's very strong sense of the physical and spacial materiality of bodily being. He writes:

> "This other human being whom I am contemplating, I shall always see and know something that he, from his place outside and over against me, cannot see himself: parts of his body that are inaccessible to his own gaze (his head, his face and its expression), the world behind his back .... are accessible to me but not to him. As we gaze at each other, two different worlds are reflected in the pupils of our eyes .... to annihilate this difference completely, it would be necessary to merge into one, to become one and the same person. This ever present excess of my seeing, knowing and possessing in relation to any other human being is founded in the uniqueness and irreplaceability of my place in the world". (Bakhtin, 1990: 23).

He continues to assert that we can only exist as knowing beings through the 'borrowed axiological light of otherness'. Since each of us occupies a unique time/space, we can see and experience things others cannot, within our sphere of self-activity. And the reverse is equally true: the other can visualize and apprehend things that we are unable to. Hence, the other has a 'surplus of seeing' with regard to ourselves, and vice versa, a scenario that corresponds directly to what Merleau-Ponty calls the 'reversibility of perspectives' (Gardiner, 1998: 138).

Bakhtin argues that this co-participation cannot occur solely through the medium of 'cognitive discursive thought', but that authentic thinking and acting requires an engaged, embodied relation to the other and the world at large. Dialogism attempts to reconcile prosaic life (bound to our physical bodies) and cognitive or theoretical thinking (free to move as it pleases) in a concept of the answerable and accountable act which is forced upon us by our unique position as persons in history and location. At the same time that we are implicated by our historical being and our language in social interaction, we answer to the world for this active presence. Answerability links theory with action, art with life, and forcloses the possibility of a dichotomous relationship between them. Bakhtin finds both aesthetic and theoretical thinking problematic precisely because they abstract what they imagine to be 'important' from actual events located within real time and space, and gain life of their own within this abstract realm of singular thought. In doing so they deny a dialogical (answerable) relationship with the concrete other who is displaced by an abstract category within a monologic theoretical framework.

A central axiom of dialogism is that the voice of the other is genuinely independent of the author's own voice: an authentic polyphony exists when the author permits his characters to have the status of another I, standing over against the claims of his own authorial view. It is this independent status of the other which permits the author to disclose an extra-spatial realm. When two people meet each other at a particular place and at a particular moment in time, they see, to some extent, a common environment. At the same time they see things which the other does not see. This aspect of the situation which the one person sees and the other doesn't is termed 'surplus vision', and is introduced by Bakhtin in order to underscore the independence of author and character as spatially and uniquely located partners in a dialogical relationship.

However, there is a particular subtlety in the dialogic approach to participative thinking. Surplus vision operates not at the cognitive-discursive level, but at the level of the word itself, or more correctly, the utterance (which is chosen to draw attention away from linguistic form in favour of the use of language in action and communication). Meanings in dialogism are generated and heard as social voices anticipating and answering one another. But in properly formed dialogic contexts, a latency of meaning exists because of the animation of many voices, not just those of the author, that form a simultaneity in the utterance. Thus, discourse does not become dialogic just because speakers take turns; discourse is dialogic to the extent that each utterance is dependent on another and that the tension between voices creates new meaning. Surplus vision is the non-articulated but latent part of the dialogue.

THE RELEVANCE OF SURPLUS VISION

To explain the relevance of surplus vision, I would like to recruit Bourdieu's (1998) critique of academic practice. Bourdieu argues that academic descriptions of action and reason tend to reify the product of such activity as *the* components of the activity at the expense of the full set of resources that could be marshalled by real agents. Such 'official accounts' produce distorted understandings of the situation in question, by reifying and over-emphasizing ideals, norms, values, etc., and placing closure on what in the real world is likely to be an open-ended and contested situation. Bourdieu identifies in particular, elements integral to practice such as 'style', 'tact', 'dexterity', 'saviour-faire' (skill, etiquette), but especially, 'improvisation', 'misfires', and 'strategies' that are omitted from the academic model. Bourdieu's point is that authentic inquiry must estimate the set of dispositions available to an actor, rather than just those that were enacted, and recognise those enacted as mediating constructs, rather than determining ones.

Now, if we consider a learning situation or an interpretation as an aesthetic event, according to Bourdieu's formulation a number of dispositions or the agents' 'social capital' will mediate the event. However, academic requirements demand an account of such dispositions that is frozen in time in order to provide a singular interpretation. This effectively renders them timeless, and the singular academic interpretation or monologue denies the human agency of the 'subject' because that agency is founded on an irreducible otherness.

So, while the academic model may be timeless, real agents and real knowledge systems are extended in time-space because they are located in the unfinalizability and openendedness of ordinary life (Bakhtin, 1981). Bakhtin notes that we are always tempted to view our own horizons as extended in time and space, yet to regard others' horizons as finite and completed. For perceivers, their own time is forever open and unfinished; their own space is always the centre of perception, the point around which things arrange themselves as a horizon whose meaning is determined by wherever they have their place in it. By contrast, the time in which we consider others is perceived as closed and finished, and the space in which others are seen is never a significance-charged surrounding, but a neutral environment, ie, the homogenizing context of the rest of the world.

The way that time and space are constructed and theorized is particularly significant for information systems (e.g., DeSanctis and Poole, 1994; Giddens, 1994). For example, most knowledge management technology is designed to repair a perceived time-space disjunction that occurs in modern organizations, as in the case of aircraft fitters who work in London and Paris or who work different shifts, or both. This is in common with paradigmatic sociology which treats co-presence, consociality or social interaction as based on spatial and temporal *presence* (Sandywell, 1998). Thus this type of IS objective is one of a surrogate presence.

'Time' in Bakhtin's work, can be viewed as a dialogically mediated construction that is inseparable from a society's general communicative and interactional strategies (Sandywell, 1998: 205). The speech genres that make proper communication dialogical can be viewed as social frameworks of meaning mediating biographical and historical experience. Thus the narratives used to map the course of a life are drawn from historical rhetorics which are in principle available to others in representing their own 'location' in space and time. But this is not limited to spoken discourse: societies can create space-time relations *in* and *as* sociotextual inscriptions across a whole range of cultural practices. For example, inscriptive technologies such as writing systems, art forms and electronic media are central to the construction of different forms of temporal consciousness. The contrasting experience of time between pre-literate and literate societies has long been observed by anthropologists, and time may provide the key distinction between industrial and informational societies (Eriksen, 2001).

The relevance of the notion of surplus vision to IS becomes apparent when we adopt the idea of a 'community-of-practice' as a key analytic unit (Wenger, 2000; Seely Brown and Duguid, 2000). The social capital of such communities is increasingly identified with its knowledge resources, and how these are articulated and integrated with information systems. Information systems will conventionally attempt to extend these resources across time and space, but it has been argued that a further dimension exists when the information system 'text' has a dialogic orientation such that it can convey 'surplus vision'.

TWO TYPES OF ORGANIZATIONAL COMMUNICATION

Formal and bureaucratic organizational structures are designed for allocating resources and for accountable decision-making, but it is well known they are not conducive to the development, application and spread of knowledge. The informal nature of communities of practice (Wenger, 2000) is much more fertile environment for learning and knowledge development. From a dialogical perspective a further distinction concerns the *nature* of communication and the conditions for learning: communication can be *univocal* or *dialogical* (Stephens 1997, 1999b). The former concerns the exchange of information in conditions of pure intersubjectivity through the 'conduit' of language, and, because a shared alterity or representation is assumed, learning is single-loop. Double-loop learning occurs in the latter model because communication involves the intersection of two non-coinciding consciousnesses, hence a surplus of vision, and information is the product rather than the condition of the process.

Clearly, as characterized, these are both *ideal types*. The univocal type is typified by the Shannon and Weaver communication model, while the dialogical type resonates with the `biology of language' advanced by Maturana and Varela (1980). No modern system can operate by full-scale dialogue alone, all kinds of arrangements or 'relief mechanisms' are required to "condense or replace mutual understanding in language" (Habermas, 1984:181). However, the distinction is instructive for the prospects of 'informated' organizations. Lotman argues that communicative texts may serve two functions - univocal and dialogical - and one or the other is dominant in a given context. The univocal function has gained most attention, even though the dialogical function, according to Lotman (1988), is vital for understanding and for creative thought:

> "In an overall cultural system, texts fulfil at least two basic functions: to convey meaning adequately, and to generate new meanings. The first monologic function is fulfilled best when the codes of the speaker and the listener most completely coincide and, consequently, when the text has the maximum degree of univocality .... Since it is the aspect of a text that is most easily modeled with the means at our disposal, this aspect of a text has been the most noticed." (1988, p 34-5).

Both Lotman and Bakhtin eschew a universal communication model: "The second dialogic function of a text is to generate new meanings". In this respect a text ceases to be a passive link in conveying stable information between senders and receivers, but functions as a 'thinking device'. Whereas in the monologic or univocal case a difference between the message at the input and at the output of an information circuit occurs as a result of a defect, and is to be attributed to the technical imperfections of this system, in the dialogic case such a difference is the very essence of a text's function.

CONCLUSION

This paper represents the first stages of a research programme to theorise information technology using concepts derived from dialogism. Previous research has focussed on methodological issues (Stephens, 1997; 1999a; 1999b). A requirement to think beyond time-space coordinates of the distribution of information is likely to be more keenly felt as IS moves its concerns from information structures to information content (Willcocks and Lester, 1999.), and dialogism has some ready made conceptual tools available to scholars. This paper has outlined the relevance of surplus vision, but work in progress concerns the functional dualism of texts and an investigation of the 'chronotope' as a marker for constructing temporal relations.

A dialogical interpretation is interesting and useful for IS because of its insistence upon the urgency of maintaining maximum openness between speech communities, lest the language and the community of speakers become frozen in a debilitating inbreeding of existing ideas and speech genres that are eventually emptied of meaning. We know from Habermas that modernity readily sacrifices *ideal* speech situations for communicative efficiency and utility. Bakhtin's ideas can keep us on our guard.

REFERENCES

Bakhtin, M. (1981). *The Dialogic Imagination: Four Essays by M.M. Bakhtin*, (M. Holquist, ed.), (trans. C. Emerson and M. Holquist), Univ. of Texas Press, Austin.

Bakhtin, M. (1990). "Author and Hero in Aesthetic Activity", in *Art and Answerability* (M. Holquist and V. Liapunov, eds.), (trans. V. Liapunov), Austin, Univ. of Austin Press.

Bourdieu, P. (1998). *Practical Reason: On the Theory of Action*, Polity Press, Cambridge.

DeSanctis, G. and Poole, M.S. (1994). "Capturing the complexity in advanced technology use: adaptive structuration theory", *Organization Science*, 5(2):121-147.

Dixon, N.M. (2000). *Common knowledge: how companies thrive by sharing what they know*, Harvard Business School Press, Boston.

Eriksen, T.H. (2001). *Tyranny of the moment: fast and slow time in the information age*, Pluto Press, London.

Gardiner, M. (1998). "The incomparable monster of solipsism", in *Bakhtin and the human sciences,* (M.M. Bell and M. Gardiner, eds.), Sage, London.

Giddens, A. (1994). "Foreword," in *Now here space, time and modernity* (R. Friedland and D. Boden, eds.), University of California Press, Berkeley.

Habermas, J. (1984). *The theory of communicative action II,* (trans. T. McCarthy), Polity Press, Cambridge.

Kling, R. (2000). "Learning about Information Technologies and Social Change: The Contribution of Social Informatics," *The Information Society*, 16(3):4-19.

Levinson, P. (1999). *Digital McLuhan: a guide to the information millenium*, Routledge, London.

Lotman, Y.M. (1988). "Text within a text", *Soviet Psychology*, 24(3):32-51.

Lotman, Y. M. (1990). *The Universe of Mind: A semiotic theory of culture*, (trans. A. Shukman), I.B. Tauris and Co Ltd, London.

Maturana, H.R. and Varela, F.G. (1980). *Autopoiesis and Cognition*, Reidel, Dortrecht.

McLuhan, M. (1964). *Understanding Media*, Mentor, New York.

Nardi, B.A. and O'Day, V.L. (1999). *Information ecologies: using technology with heart*, MIT Press, Cambridge.

Sandywell, B. (1998). "The shock of the old: Mikhail Bakhtin's contributions to the theory of time and alterity," in *Bakhtin and the human sciences,* (M.M. Bell and M. Gardiner, eds.), Sage, London.

Seely Brown, J. and Duguid, P. (2000). *The Social Life of Information*, HBS Press, Boston.

Stephens, R.A. (1997). "Dialogism: A Bakhtinian perspective on communication," in *Systems for Sustainability: People, Organizations and Environments*, (F.A. Stowell, R.L. Ison, R. Armson, J. Holloway, S. Jackson and S. McRobb, eds.), Plenum, New York. 513-518.

Stephens, R.A. (1999a). "Authenticity and Ethical Practice in IS: Technicism, Reflective Practice and Answerability", *Proc. AIS 99*, Milwaukee. AMCIS.

Stephens, R.A. (1999b). "Shared alterity in interpretive research: a case of mistaken identity?", in *Synergy Matters: Working with systems in the 21st century*, (A.M. Castell, A.J. Gregory, G.A. Hindle, M.E. James, and G. Ragsdell, eds.), Plenum, New York. 469-474.

Wenger, E. (2000). *Communities of practice: learning, meaning and identity*, Cambridge University Press, Cambridge.

Willcocks, L.P. and Lester, S. (1999). (eds.), *Beyond the IT Productivity Paradox*, Wiley, Chichester.

Zuboff, S. (1989). *In the Age of the Smart Machine: The Future of Work and Power,* Heinemann, Oxford.

CROSS-CULTURAL ISSUES AFFECTING ELECTRONIC TRANSFER OF PERSONAL HEALTH INFORMATION

Richard Whiddett[1] and Jocelyn Handy[2]

[1]Department of Information Systems and [2]School of Psychology
Massey University,
Private Bag 11222,
Palmerston North, New Zealand
Email: r.j.whiddett@massey.ac.nz

INTRODUCTION

For some time now the health sector in developed countries has been seeking to utilise information and communications technology to improve both the quality of patient care and organisational efficiency (Hovenga *et al.*, 1996). One strategy has been the introduction of electronic medical records (EMR) systems. An EMR is an integrated database system that consolidates all information relating to the healthcare of an individual patient. Its aim is to improve the quality of patient care by ensuring that complete and up to date information is readily available to all caregivers (Dick and Streen, 1997).

As more healthcare providers are storing their records online and are becoming connected to communications systems, it is becoming feasible to develop a 'virtual' lifelong electronic medical record of a patient (Mount *et al.*, 2000). With this technology a complete medical record no longer needs to be stored in one location, instead, a virtual record could be constructed by piecing together the jigsaw of a patient's history of contact with the health sector from information that is scattered throughout various information systems. With such a system, a healthcare professional anywhere in the world could, theoretically, simply enter a patient's identification and access their entire medical history.

The technical and ethical issues raised by such systems are currently being examined by various organisations such as the European Standards Committee (CEN) and the HL7 consortium (HL7). This work is now being consolidated by the International Organisation for Standardization (ISO) Technical Committee on Health Informatics (ISO-TC215). A series of technical international standards relating to the coding, storage and communication of patients' information are expected to emerge over the next few years which will facilitate the exchange of personal health information on a

Systems Theory and Practice in the Knowledge Age
Edited by Ragsdell *et al.*, Kluwer Academic/Plenum Publishers, New York, 2002.

global scale. To date, these developments have tended to focus on technological and security issues. Less attention has been directed towards the ethical and privacy issues that are associated with transfer of patients' information between different environments.

The sensitive and confidential nature of patients' health information is widely recognised (Goldberg, 2000; Sadan, 2001) and is often addressed within the privacy legislation and medical codes of ethics of various countries such as the UK Data Protection Act 1984 and the New Zealand Health Information Privacy Code (Slane, 1994). However, relatively little attention has been paid to identifying the issues that need to be considered or the controls that are required when patient information is being transferred between systems that are located in environments that have different organisational structures, legislative controls or cultural norms. This paper will highlight some of the problems of current access control mechanisms and identify the issues that need to be resolved to ensure that future EMR systems can meet the requirements for confidentiality and will be acceptable to both patients and medical professionals.

PRIVACY OF PERSONAL HEALTH INFORMATION

Although the potential benefits of health information systems are widely accepted, there are also serious problems concerning the potential threats to patient privacy (Carter, 2000). Health records can contain sensitive information concerning issues like drug and alcohol problems, sexual behaviour, mental health or a genetic predisposition towards certain diseases. The consequences of its inappropriate release can therefore be extensive and may impact on a person's ability to gain employment, to maintain their marriage or to obtain loans or life insurance (Chadwick, 1999; Woodward, 1995).

Two major problems arise in the move from paper-based medical records to EMRs. Firstly, individual records are potentially more vulnerable to breaches of privacy since they can be accessed from a remote location, and secondly, computer based systems facilitate sophisticated searching and record matching operations which can breach privacy in ways that would not be practical with paper based systems. For example, a banker in the US obtained a list of cancer patients from the local hospital, correlated it with his customer files and called in any outstanding loans to the patients (Gorman, 1996). These threats to personal privacy are not unique to healthcare and have been recognised for some time now. As long ago as 1980, the OECD published a series of guidelines to control of the flow of personal information of any kind across international borders (OECD, 1980). The basic principles of these rules have been adopted into the legislation of many western nations (e.g. Australia, Canada, New Zealand and the United Kingdom), although the details of the implementation vary. Even with legal protection, concerns over patient privacy remain, to the extent that in 1997 the British government commissioned the Caldicott Committee Report (1997) which examined the privacy issues related to the exchange of patient information between organisations within the UK, which resulted in recommendations to improve patient privacy. Furthermore, some countries still do not have sound laws to protect people's privacy, and the judicial and cultural rules that apply to the release of the information can vary widely between different cultures. Therefore the transfer of personal health information from one legislative or cultural setting to another which has different or less stringent controls can place a patient's privacy at increased risk.

Confidentiality is essential to the patient - physician relationship. Unless a patient can be sure that personal information will not be distributed against their wishes, they may be reluctant to disclose sensitive information that may in fact be crucial to their correct treatment. The security of health information systems and comprehensive access control mechanisms are therefore particularly important. Unfortunately, the protection of health information is particularly difficult due to the technical complexity of the data and the organisational complexity of the healthcare setting. EMRs can be very large (particularly with the storage of medical imaging systems), long-lived (being relevant for many years), with a complex structure which may try to represent the logical patterns of medical processes and episodes of care. The data is often fragmented and distributed across many systems, which are operated by the many separate and independent stakeholder organisations in the healthcare sector, such as hospitals, primary care surgeries, diagnostic laboratories and pharmacies. Each of these stakeholder organisations will also employ a variety of staff with different levels of information needs. It is therefore difficult to define an appropriate mechanism that can balance the information needs of the stakeholders with the privacy needs of the patients.

CURRENT ACCESS CONTROL MECHANISMS

In general, access control mechanisms for computer based information systems use three subsystems, a user-management system, an access control system and a monitoring system (ISO, 2000). The purpose of the user-management system is to identify the legitimate users of the system and to ensure that only legitimate users can gain access by implementing some authentication system, usually by using a username/password combination. Authenticated users are then permitted access to different parts of the application system, depending on their status, under the control of the access control system. Monitoring systems usually maintain some audit trail or log to record users' activities so that any breaches of security can be investigated. Unfortunately, there are problems with the application of these techniques to EMR systems.

First, there is the problem of managing the large and constantly changing pool of potential users of an EMR system, many of whom may work for various organisations, often on short-term contracts. To some extent the problem of authenticating users from other organisations can be tackled by using public-key infrastructures and digital certificates, but there remains a significant burden of user-registration and administration to be done by some trusted authority. An example of such an infrastructure is the New Zealand Health Intranet, a Virtual Private Network (VPN) that is sponsored by the New Zealand Ministry of Health (NZHIS, 2001), access to which is restricted to members of the healthcare community authorised by the Ministry. The giant American healthcare organisation, Kaiser, is planning a similar approach (Gattenbein and Stepanek, 2000).

The emerging Role Based Access Control systems (Park and Sandhu, 1999; Sandhu *et al.*, 1996), which are found in modern network operating systems such as Novell Netware and Microsoft Windows NT, also attempt to tackle some of the problem of managing large user populations. In this approach, access permissions are associated with 'roles', and then users are assigned to appropriate roles, thus simplifying the allocation of access privileges. The major problem with such Role-Based systems is that they tend to locate control at the application level or file level which is too coarse grained for the needs of EMRs. With role-based systems it is possible to enable or disable access to complete categories of information from users with particular roles, but it is not then possible to further restrict access to a limited range of patients. For example, the system should not only be able to restrict access to information about HIV status of all patients, but access to that information by a particular user should be granted on a patient by patient basis.

Ideally, for EMR systems, access control should operate at the level of elements within a patients' record. A study commissioned by the British Medical Association came to this conclusion, recommending that "*Each identifiable clinical record should be marked with an access control list naming the people or groups of people that may have access to it*" (Anderson, 1996). Other writers have gone even further, for example Beznosov (1998) suggested that access should not only be controlled by the user's role, but also by other factors such as their location, affiliation and relationship to the patient. These sorts of facilities would allow the implementation of sophisticated restrictions such as limiting the access medical staff to the records of patients who are on a particular ward. Mandl *et al.* (2001) further argue that the patient should have complete control over the access of any part of their record. Whilst such restrictions would seem to represent an ideal level of control, they are largely unproven and little is known about the practicality of implementing and of maintaining systems with such fine grained control.

MATCHING SYSTEMS TO ORGANISATIONS AND CULTURES

Whilst there is widespread agreement about the need to apply stringent controls to the access of EMRs, the details of these controls will vary and need to be matched to the particular organisational, legislative and cultural structures of the country in which they are located. Because of these differences it is not possible to define a universal access control policy, instead, a flexible mechanism is required which can be tailored by each country to implement a policy that will suit its own unique environment. Several factors may influence the details of the policy that is required for a particular environment.

Firstly, the sensitivity of the information contained within different parts of the EMR and the appropriate degree to which information may be shared is defined by the surrounding culture. The contentious issues relating to access to information often arise where there is a conflict between the interests of the individual patient and the interests of other individuals, a group of people or society as a whole. Examples of such areas are the rights of adopted children and their parents, immunisation records, HIV status and DNA fingerprinting. In each of these areas, a different policy might be adopted in a culture depending on the weighting that the society places on rights of the individual to their privacy verses the interests of others. For example, traditional New Zealand Maori regard relationships with the extended family, or whanau, as having much greater obligations and responsibilities than is usual in a conventional western society. Consequently, the members of the whanau sometimes expect to have a greater degree of access to the health information of other members of the whanau than is permitted under New Zealand's western-based privacy laws. This can cause problems if Maori healthcare providers follow cultural norms rather than legal requirements.

Furthermore, certain issues, particularly in the areas of sexual or social history, which are extremely contentious in one culture, may be regarded as less so in another. This means that the categories of information within a patient's record that need selective control and the access rights that are allocated to various roles may be different for different cultures. An example of this kind of situation was recently reported by the New Zealand Privacy Commissioner (2001) in a case where a Fijian student studying in New Zealand was cut off by her family when aspects of health information were leaked to her family by a distant relative who worked in the hospital where she was a patient. This situation arose partly because of the different cultures of New Zealand and Fiji, which meant that the hospital did not fully appreciate the degree of privacy that was required.

Secondly, the roles that exist or the access requirements for a particular role may differ between institutions or countries because of the ways that healthcare is delivered. For example, different countries give nurses and midwives different levels of responsibility. Their requirements for access to information will therefore vary depending on the level of responsibility that they are expected to assume.

Such examples show that it is not possible to have universal access control policies and the technology must be adaptable so that it can be tailored or customised to accommodate different cultures or organisational structures. Any access control mechanism will therefore need to be designed so that it can be customised to define the categories of information at a fine level of detail, to define a number of access roles and then to allocate access privileges to these roles that are appropriate to the local culture and customs. This is most important for the acceptance of EMR systems by both patients and medical practitioners. In a recent study of New Zealand primary care practitioners, respondents were clear that they would partially boycott an inter-organisational EMR system if they believed that it would not meet patients' privacy requirements (Handy *et al.*, 2001).

Unfortunately, even if some mechanisms can be designed so that systems can be built to meet the specific requirements of various cultural contexts, this flexibility will itself cause difficulties when individual records are moved between environments. Each time information is transferred between significantly different environments it may be necessary to translate or change the access control rules to the new context to ensure that an appropriate level of privacy is maintained. For example, it may be necessary to make adjustments for the differences in the responsibilities of healthcare professionals or some issues may become more salient and require more security. Although the above case of the Fijian student did not arise from the inter-organisational transfer of information, it did effectively arise from an inter-cultural transfer, and similar cases are much more likely to occur as personal information becomes more readily accessible from remote locations.

In conclusion, the different healthcare structures and varying cultural norms that exist within different societies have two major implications for the design and use of EMR systems. Firstly, in order that individual information systems are capable of being adapted to suit the needs of different cultures, they need to be designed with flexible and robust access control mechanisms that can be tailored to meet the confidentiality requirements of the various stakeholders to a very fine level of detail. There also needs to be a clear identification of these requirements within different cultures in order that culturally-appropriate policies can be implemented. Secondly, mechanisms need to be developed at the interfaces between systems to ensure that patient's interests are protected when information is transferred out of a system and across a cultural boundary. Appropriate mappings need to be defined between the access control protocols of different cultures to ensure not only that there is no loss of privacy when information is transferred, but also to ensure that increased security is provided in situations where it is required.

REFERENCES

Anderson. R.J. (1996). *Security in Clinical Information Systems*, BMA, London.

Beznosov, K. (1998). "Requirements for Access Control : US Healthcare Domain". In *Proceedings of the 3rd ACM Workshop on Role-Based Access Control, Fairfax, Virginia, October*, ACM, New York

Caldicott Committee. (1997). *Report on the review of patient-identifiable information,* NHS Executive, London. http://www.doh.gov.uk/confiden/crep.html

Carter, M. (2000). "Integrated electronic health records and patient privacy: possible benefits but real dangers", *MJA*, 172:28-30.

Chadwick, R. (1999). "The Icelandic Database – Do modern times need modern sagas?" *BMJ*, 319:441-4.

CEN, European Standards Committee http://www.centc251.org

Dick, R. and Streen, E. (1997). *The Computer Based Patient Record: an essential technology for healthcare*, National Academy Press, Washington.

Gattenbein, D. and Stepanek, M. (2000). "Kaiser takes the cyber cure", *Business Week*, Feb 7:EB80,EB84,EB88.

Goldberg, I. (2000). "Electronic Medical Records and Patient Privacy", *Health Care Manager*, 18(3):63-69.

Gorman, C. (1996). "Who's looking at your file?" *Time* 6 May: 60.

Handy, J.A., Hunter, I. and Whiddett, R. (2001). "User Acceptance of Inter-Organisational Electronic Medical Records", *Health Informatics Journal* , 7(2):103-107.

HL7 Consortium http://www.HL7.org

Hovenga, E., Kidd, M. and Branko, C. (eds.) (1996). *Health Informatics: an Overview*, Churchill Livingstone, Melbourne.

ISO, International Org. for Standardization (ISO) http://www.iso.ch/iso/en/ISOOnline.frontpage

ISOTC-215, ISO Technical Committee on Health Informatics http://www.astm.org/COMMIT/ISO/tc215.html

ISO (2000). *Information Technology – Code of Practice for Information Security Management ISO/IEC 17799:2000(E)* ISO, Geneva.

Mandl, K.D., Szolovits, P., Kohane, I.S., Markwell, D. and MacDonald, R. (2001). "Public standards and patient control: How to keep medical records accessible but private", *BMJ* 322(3): 283-287.

Mount, C.D., Kelman, C.W., Smith, L.R. and Douglas, R.M. (2000). "An integrated electronic health record and information system for Australia?", *MJA*, 172:25-27.

New Zealand Privacy Commissioner. (2001). *Health Waikato Investigation*, http://www.privacy.org.nz/shealthf.html

NZHIS (2001). New Zealand Health Info. Service http://www.nzhis.govt.nz/intranet/index.html

OECD (1980). *Guidelines on the Protection of Privacy and Transborder Flows of Personal Data.* OECD http://www.oecd.org/dsti/sti/it/secur/prod/

Park, J.S. and Sandhu, R. (1999). "RBAC on the Web by Smart Certificates". In *Proceedings of the Fourth ACM Role-Based Access Control Workshop. Fairfax, Virginia, 28-29 October*, ACM, New York

Sadan, B. (2001). "Patient Confidentiality and Patient Rights", *Int. J. Med Informatics*, 62(1)41-49.

Sandhu, R., Coyne,E., Feinstein, H. and Youman, C. (1996). "Role-Based Access Control Models", *IEEE Computer*, 29(2): 38-47.

Slane, B. (1994). *New Zealand Health Information Privacy Code*, Privacy Commissioner, Auckland.

Woodward, B. (1995). "The Computer-Based Patient Record and Confidentiality", *New England J. Medicine*, 333:1419-22.

COMMENT ON 'HOW' TO CONDUCT AN ACTION RESEARCH STUDY IN THE DOMAIN OF INFORMATION SYSTEMS DEVELOPMENT

Casey McQuinn Wilson

Sheffield Hallam University
School of Computing and Management Sciences
City Campus, Howard Street, Sheffield, S1 1WB

INTRODUCTION

In recent years there has been much interest in applying an action research framework to Information Systems Development, particularly Information Systems Development which resides in the interpretivist paradigm. Consideration of the action research literature, indicates that many descriptions of action research do not comment on the practicalities of 'how' to conduct the action research study, but rather focus on the nature of action research and its merits as an alternative to positivist research (Checkland, 1991; Stowell *et al.*, 1997).

In this paper, a piece of action research, which investigated the practical application of interpretivist methods of analysis to Information Systems and their Development is presented. Upon reflection of the field study, a number of lessons have been drawn which support the adoption of an action research framework, as well as yielding practical advice on how to conduct an action research study.

ACTION RESEARCH AND INTERPRETIVIST INFORMATION SYSTEMS DEVELOPMENT

If an interpretative approach is to be used in Information Systems Development there should be a framework of ideas that includes reflection, learning and articulation of the learning (Hughes and Wood-Harper, 1999) within which the work can be conducted. In recent years there has been much interest in applying an action research framework to Information Systems Development. The design, development and implementation of Information Systems has been likened to research conducted within an action research framework (Avison and Wood-Harper, 1990; Stowell and West; 1994, Stansfield, 1997), and action research has been considered as a means of conducting Information Systems research.

In taking the view, presented in this paper that an organisation is a social construct, there is a strong argument for an Information System to be developed in such a manner that a close 'fit' with the organisational context is achieved (Pliskin *et al.*, 1993; Walsham, 1993). In order to achieve this the analyst must have a comprehensive understanding of the organisational processes the Information System is being designed to serve (Jackson, 1997). In adopting systems ideas an organisation is viewed holistically and an understanding of the situation is gained from the views and opinions of the individuals involved, and through understanding relevant action is taken (Avison, 1989; Checkland and Casar, 1986; Stowell and West, 1994; Lewis, 1995). Action research provides the vehicle by which such complexity can be investigated

Many descriptions of action research do not comment on 'how' a study may be conducted, but rather discuss the nature of action research and its merits as an alternative to positivist research (Checkland, 1991; Stowell *et al.*, 1997; Checkland and Holwell, 1998). Checkland (1991; 1998) provides a more practically oriented view of 'how' an action research study may be conducted by arguing that for a researcher to conduct an action research study, there has to be a declaration in advance of the intellectual framework in which results are to be explained and of the methodology which is to be used, otherwise the action researcher leaves the study open to criticism. The declaration in advance of an action researcher's chosen intellectual framework and methodology is an essential aspect of action research if rigour is to be maintained, but it is not sufficient to ensure the credibility of an action research study (Stowell *et al.*, 1997), because it is difficult to support an argument which is based upon the subjective interpretation of the researcher's experience within a situation. It is important when conducting an action research study to declare the intellectual framework, because the conclusions drawn from the research experience will be based upon the mind set of the researcher. The selection of methods in advance presents a problem in the sense that a method may only be recognised as inappropriate for a particular setting through reflection during the action research study. A method may be considered inappropriate for a number of reasons, for example the clients find it difficult to use or understand, or the chosen method may not have the 'facilities' to cope with a particular aspect of a setting. Attempting to adhere to an inappropriate method may have negative consequences for the rigour of the research process, while changing methods may invalidate early record keeping, since if the method itself was inappropriate then so might its recommended record keeping practices. In either case the rigour of the action research study could be undermined. Action research requires a flexibility of methods, because the appropriateness of a method will be defined by the setting (Jonsson, 1991), and it will not be possible to assess in advance which method will meet the needs of the setting.

In the following section an action research study is presented. The study documents an Information Systems project in which interpretivist ideas were conducted and evaluated.

THE ACTION RESEARCH STUDY - THE CARE OF THE ELDERLY DAY HOSPITAL

In order to investigate the practical application of interpretivist approaches to Information Systems Development an action research study was conducted in which an Information System was developed for a Care of the Elderly Day Hospital. The field study was conducted over a period of 18 months, during which an interpretivist (Burrell and Morgan, 1979; Myres, 1997) view of information systems development was conducted and evaluated.

The notion of a Rich Picture was used as a medium for promoting discussion amongst the day hospital staff. The reason for this choice was, that during the early discussions with the Day Hospital staff, the researcher noticed the staff experienced difficulties in articulating their understanding of the Day Hospital situation and that the individual disciplines had limited understanding of the roles and activities of their colleagues. Consequently it was felt that any attempt to describe the situation in a narrative form was likely to be incomplete, and it would be tempting for the researcher to add their own interpretation in order to produce a 'well written' account. In order to create the Rich Pictures, qualitative semi-structured interviews were conducted with prominent staff from the Day Hospital team. The interviews were tape recorded and from the interview transcripts the researcher created the first Rich Picture. This picture was then discussed, as a group, with all the Day Hospital staff. The result of tape recording this group discussion was the creation of a second, enriched Rich Picture. The enriching process added depth of context to the issues which had only been touched upon in the interviews, identifying for both the researcher and the Day Hospital staff areas of concern which were emerging as a consequence of the development of the clinical record.

The favourable response to the Rich Pictures led to further semi-structured interviews being conducted, and the process of discussing and enriching the Rich Pictures was repeated twice more during the study. The result was a series of pictures which not only were used to enable communication between the Day Hospital staff and the researcher, but also provided the researcher with a complete history of events and attitudes of those involved in the field study.

Data Flow Diagrams were also employed by the researcher, as the Day Hospital were still considering the option of implementing a technological component to the clinical record. For this reason the researcher felt that a detailed specification of the Day Hospitals data requirements should be obtained and Data Flow Diagrams were considered to be an appropriate medium by which to achieve this. The way in which the Data Flow Diagrams were constructed was influenced by the researcher's understanding of 'soft' systems ideas, consequently the Day Hospital staff were actively encouraged to participate in the construction of the diagrams.

The results of employing Rich Pictures, Data Flow Diagrams and the group discussions during the development process was to shift the focus from the clinical record as a lone entity to the Day Hospital situation as a whole. Consideration was no longer restricted to what data might be required by the disciplines but extended to identifying and understanding the changes required to enable the move from uni-disciplinary to multidisciplinary working. The Day Hospital staff were encouraged to discuss actively the problem situation, so that the researcher and the staff could learn about the problem situation and one another's perspectives. Through understanding, the Day Hospital staff and the researcher could consider, and come to an agreement, on how best to implement the changes required to support the new mode of working. In this way the development of the Day Hospital Information System was underpinned by interpretivism.

CONDUCTING AN ACTION RESEARCH STUDY - THE LESSONS DRAWN

Throughout the field study the researcher maintained a field diary. At the end of the study, the diary material was collated together along with the minutes recorded in the formal meetings. Analysis of the field data and the collection of Rich Pictures have enabled the researcher to draw several lessons relating to 'how' an action research study could be conducted.

Lesson 1 -Reflection on Action

> Consideration of the 'actions' during a field study enables problems to be addressed which might otherwise not be recognised.

In the early stages of the field study, the researcher by reflecting upon the 'action' within the situation, began to realise that the pre-arranged meetings were a contributory factor toward the poor communication between the researcher and the Day Hospital staff. By re-evaluating the situation, the researcher realised that the best way to improve communication was to talk directly to those working with the clinical record. This was achieved through the introduction of group discussions within the Day Hospital with the staff members who were using the clinical record on a daily basis. While the increased contact between the researcher and the Day Hospital staff aided communication, it also focused the group's attention onto the issue of identifying data fields for the clinical record, rather than their requirements for the computer based clinical record. In re-evaluating this situation at the end of the research, there is the suggestion that had group discussions not been undertaken then the Day Hospital staff would have resisted any changes made to their situation, because these changes would have been controlled by the medical consultant and a perceived outsider, the researcher.

Lesson 2 - Secrecy

> It is important that the researcher is aware that the client group may not report all of their feelings, attitudes and opinions about the situation.

In any particular situation what appears to be 'open' communication between the group members and the researcher may only be 'selective' communication, leaving the researcher uninformed as to certain aspects of the situation (Argyris *et al.*, 1982; Jonsson, 1991). This lesson may be supported further by considering the reaction of the researcher, who had the 'feeling' that problems existed within the Day Hospital but had no evidence to support what the problems were. It was the unease of the researcher combined with previous experience of interpretivist ideas that prompted the use of 'soft' systems ideas and further investigation of the problem situation.

Lesson 3 - Modes of Action Research

> The concept of two modes of Action Research is a useful means of analysing the research situation both when immersed in the study and in retrospect.

Upon reflection of the research in terms of the notional cycle of action research it can be seen that prior to the researcher considering the use and application of 'soft' Systems ideas, the reflective process revolved around the 'action' where reflection informed further 'action'. The application of 'soft' ideas led the researcher to reconsider the research and begin to realise that the development of the clinical record was merely part of a wider, more complex Information System. The reconsideration of the 'theory' led to a deeper investigation of 'systems' ideas which in turn informed further action. This may be seen within the field study where Rich Pictures were created from interviews and then used to promote debate amongst the client group. Such debate enriched the picture being discussed and led to action being taken by the Day Hospital staff.

The use of Rich Pictures and Data Flow Diagrams proved to be beneficial for promoting communication and enabling both the researcher and the Day Hospital staff to gain an appreciation of the problem situation and of one another's perceptions. The differences between the pictures raised the researcher's awareness of the radical changes the professionals were experiencing in their working practices. While the process of developing each picture as a collaborative effort between the researcher and the Day Hospital staff, provided the opportunity for potential problems to be addressed whilst still in their infancy By reflecting, at the time upon the use of Rich Pictures the researcher realised their potential for facilitating communication and enabling appreciation. In this way the Rich Pictures proved to be invaluable in enabling the progression of 'action' in the research.

Lesson 4 - Documentation

> Tools and techniques used to enhance communication can also act as a means of documenting the action research study.

Reflection upon the field study produced two further lessons. The first concerns the employment of Data Flow Diagrams to gain an understanding of the data flows in the situation. In hindsight this process, and the way the Data Flow Diagrams were constructed was influenced by the researcher's understanding of 'soft' systems ideas. The Data Flow Diagram, because the Day Hospital staff were involved in their creation, offered the staff a medium by which they could learn and understand their data to an extent that they were able to identify areas which they felt could be changed and initiate those changes. From this event the lesson that what may be considered to be a 'positivist' tool may be incorporated into an interpretivist study was drawn. In this case Data Flow Diagrams were not used as a tool to obtain specifications, but as a tool for communication which enabled the clients to appreciate the problem situation in terms of client perceived data structures. At the time the Rich Pictures were used to enable communication, but consideration of the 'set' of Rich Pictures at the end of the field study provided the researcher with a complete history of events and attitudes of those involved in the field study. When considered in retrospect the documentation provided by the Rich Pictures enabled a number of lessons to be drawn from the research while at the same time ensuring the rigour of the action research process.

COMMENTS ON CONDUCTING ACTION RESEARCH

Following on from the lessons drawn from the field study, comment can be made concerning 'how' an action research study could be conducted.

The first comment relates to the importance of maintaining relevant documentation in order to ensure the rigour of the study, and how this may be achieved. Documenting events and ideas in action research is hard to achieve at the time such events occur. The reasons for this is that the researcher has to constantly maintain a balance between satisfying the needs of the clients and investigating the research ideas. The researcher is in a position where they are actively collaborating with the clients, making it difficult to write down every word of an intense group discussion without missing subtleties of body language. This research experience indicates that the researcher needs to adopt a means by which an accurate record of events may be obtained, for example obtaining copies of the formal minutes of any meetings or using tools which enable the researcher to communicate and appreciate the problem situation, while at the same time creating a permanent record of the events within the situation. In the case of this research Rich Pictures and Data Flow Diagrams, while used at the time as learning and communication tools for the client group, upon reflection have enabled a permanent record of the research's evolution to be created, reflecting attitudes, ideas and perceived structures both at the time they emerged and how they were addressed and changed over time.

The second comment which can be made concerning action research is related to the potential difficulty of keeping the research question in mind, and being able to address research issues. Difficulties can arise if the needs of the clients come to dominate. This may be seen by considering the initial research question in this study, which was to standardise a clinical record to promote changes within a particular client group in order to bring that client group in line with regulations outlined by their governing body. This presented problems for the researcher who was placed in the position of being both a consultant, in the sense that an 'end product' would be delivered, and a researcher. The research initially followed the consultancy route and this led to difficulties in thinking of the project as a piece of research. This may be illustrated by considering that it was the clinical consultants of this particular setting who were approached and invited to collaborate in this research and allow the presence of a researcher to address a perceived need, and to promote change within the situation. As the project began to unfold the researcher began to realise both the complexity of the situation and the nature of the intervention required to enable the project to progress as a piece of research. This realisation emerged through the researcher reflecting at the time upon the 'action' and identifying a number of problems, such as poor communication between the Day Hospital staff and the researcher, and amongst one another. This identification of the problems led to the decision to further investigate the literature. These elements combined led to a shift in the research ideas and a reforming of the research questions. The domain, while pre-selected by the initial instigators of the project, was recognised by the researcher as a suitable vehicle for research.

The initial emphasis of the project was upon addressing the perceived problems of the clients, and may be considered to have been in 'consultancy mode' (Stowell *et al.*, 1997). As the researcher's ideas began to emerge and the client group came to an understanding of how the research might benefit them, the study moved towards a 'field study mode' (Stowell *et al.*, 1997), in which priority was given to answering the research question. If the research had remained in 'consultancy mode' (Stowell *et al.*, 1997) it would have been difficult to extract lessons which contributed to the body of knowledge within the domain of Information Systems. This is not to say that lessons would not have been drawn, but that they would have been related to the requirements of the clients and their particular problems. In reflecting upon the study, and utilising the concept of 'modes' of action research, it may be possible for a researcher to assess the degree of balance between research aims and client needs. Consideration of the research in terms of the diagrams can help the researcher shift emphasis from 'action' to 'theory' and back again, so that both the concerns of the researcher and of the clients can be met. In addition the diagrams provided a means of highlighting the importance of lessons learned from reflection on both 'action' and 'theory'. In this study the realisation of the shift toward what may be considered 'consultancy mode' (Stowell *et al.*, 1997) prompted a re-evaluation of the research, and steps were taken in an attempt to re-focus on the research question.

Finally within the action research literature it is unclear whether it is possible to have a number of iterations of either 'action' or 'theory', before one may inform the other. An example of this from this study can be seen in the way that Rich Pictures were used. Reflection upon the positive response of the client group to the first Rich Picture, and the debate which this generated, prompted the researcher to re-assess the use of Rich Pictures and continue to use them as a medium for enabling communication and appreciation. This may be considered to be reflection upon 'action' re-informing further 'action'. At the end of the field study, reflection upon the use of Rich Pictures throughout, led to the realisation that they were a medium by which the research process could be documented, and informed 'theory'.

CONCLUSION

Overall action research has been found to beneficial for information systems development in this setting. Lessons have been drawn and constructive comment on 'how' to conduct action research has been presented. An important aspect of this field study which has indicated that action research is beneficial is that the clients gained a practical outcome from their involvement, which they managed.

REFERENCES

Argyris, C., Putnam, R., McLain-Smith, D. (1982). *Action Science Concepts Methods and Skills for Research and Intervention,* Jossey-Bass, San Francisco.

Avison, D.E, and Wood-Harper, A.T. (1990). *Multiview: An Exploration in Information Systems Development,* Blackwell Scientific Publishers, Oxford.

Avison, D.E. (1989). "An Overview of Information Systems Development Methodologies", in *Systems Prospects: The Next Ten Years of Systems Research*, (R.L. Flood, M.C. Jackson, and P. Keys, eds.), Plenum Press, New York.

Burrell, G., and Morgan, G. (1979). *Sociological Paradigms And Organisational Analysis*, Heinemann, London

Checkland, P.B. (1991). "From Framework Though Experience to Learning: The Essential Nature of Action Research", in *Information Systems Research. Contemporary Approaches and Emergent Traditions*, (H.E. Nissen, H.K. Klein, and R. Hirschheim, eds.), North-Holland, Amsterdam.

Checkland, P.B., and Casar, A. (1986). "Vicker's Concept of an Appreciative System: A Systemic Account", *Journal of Applied Systems Analysis*, 13:3-17.

Checkland, P.B., and Holwell, S. (1998). *Information, Systems, and Information Systems: Making Sense of the Field*, John Wiley and Sons, Chichester.

Hughes, J., and Wood-Harper, A.T. (1999). "Systems Development as a Research Act". *Journal of Information Technology*, 14:83-94.

Jackson, M.C. (1997). "Critical Systems Thinking and Information Systems Research". in *Information Systems: An Emerging Discipline*, (J. Mingers, and F.A. Stowell, eds.), McGraw-Hill, Maidenhead.

Jonsson, S. (1991). "Action Research", in *Information Systems Research*, (H.E. Nissen, H.K. Klein, and R. Hirschheim, eds.), North Holland, Elsevier Science Publications B.V, Amsterdam.

Lewis, P. J. (1995). "New Challenges And Directions For Data Analysis And Modelling", in *Information Systems Provision: The Contribution of SSM*, (F.A. Stowell, ed.), McGraw-Hill, Maidenhead.

Myers, M., (1997). "Interpretative Research in Information Systems", in *Information Systems: An Emerging Discipline*, (J. Mingers, and F.A. Stowell, eds.), McGraw-Hill, Maidenhead.

Pliskin, N., Romm, T., Lee, A.S., and Weber, Y. (1993). "Presumed versus Actual Organisational Culture: Managerial Implications for Implementation of Information Systems". *The Computer Journal*, 36(2):143-151.

Stansfield, M.H. (1997). "The Use of Computer Based Technology in Supporting a Subjective Method of Inquiry". *Ph.D. Dissertation*, University of Paisley (unpublished).

Stowell, F.A., and West, D. (1994). *Client-Led Design: A Systemic Approach to Information System Definition*, McGraw-Hill, London.

Stowell, F. A., West, D., Stansfield, M. (1997). "Action Research as a Framework for Information Systems Research", in *Information Systems: An Emerging Discipline*, (J., Mingers, and F., Stowell, eds.), McGraw-Hill, Maidenhead.

Walsham, G. (1993). *Interpreting Information Systems In Organisations*, John Wiley and Sons, Chichester.

Section Nine

Modelling

COMPLEX CLUSTERS IN LOCAL ENVIRONMENTS: MULTI-LAYERED MULTI-MEDIA LEARNING

P. Baker and K. Fielden

Patrick Baker, MASTERS Institute, 51 Carr Road, Mt Roskill, Auckland, New Zealand
Kay Fielden, UNITEC Institute of Technology, Carrington Rd Mt Albert, Auckland, New Zealand
Patrick.baker@masters.ac.nz
Kfielden@unitec.ac.nz

INTRODUCTION

This research investigates the early stages of forming a community of multi-media teachers. The aim of the study is to capture the complex clustering patterns of triggering factors associated with teachers' migration from a state of confusion to a state of knowledge as they became proficient. CDCM (Context Dependent Cluster Model) has been derived from Soft Systems Methodology (SSM) (Checkland and Scholes, 1990) and Complexity Theory (CT). (Stacey, 1996) CDCM is used to develop and interpret a conceptual model of the system being investigated at the micro level at the initial phase of community formation for multi-media teachers. CDCM is used as the interpretation tool for data gathered from a five-level Technical Action Research (TAR) project. TAR is considered the most appropriate approach for this study for testing an intervention based on a pre-specified theoretical framework. The researchers are questioning whether the selected intervention could be successfully applied in a practical setting (Holter and Schwartz-Barcott, 1993). In the first level of TAR teachers are trained in the use of a commercial package called Illuminatus Opus to develop strategies for using in instructional design, particularly with respect to the application of student-centred multimedia projects (SCMP+T) in their own classroom multi-media projects. In the second level of TAR these teachers implemented student-centred, project-based multi-media projects in their own classrooms. The third level of TAR involves an exhibition of the students' project work. The fourth level involves the formation of an on-line learning community for these multi-media teachers and the final level is the analysis of questionnaires, observations and informal interviews all with the teachers. The data is then interpreted using CDCM to elicit emergent possible and multiple directions for change.

Edited by Ragsdell *et al.*, Kluwer Academic/Plenum Publishers, New York, 2002.

RESEARCH METHODOLOGY

The research methodology described is in two parts: (1) Technical Action Research (TAR), the methodology employed in the study for considering the impact of multimedia training as a technology intervention strategy for linking New Zealand curriculum outcomes to ICT (Information and Communication Technology) skills; and (2) Context Dependent Cluster Model (CDCM) used to situate the study within the problem domain and to interpret the qualitative data resulting from the level one and level two of the TAR study. TAR is embedded within CDCM at step 2, Problem Situation Structured.

Technical Action Research (TAR)

The five stages of TAR are: diagnosing, action planning, action taking, evaluating, and specifying learning (Nereu *et al.*, 2000). This provides a research model suited to supporting teachers' observations of the application of student-centred multimedia projects (SCMP+T) in their own classrooms. Teachers are required to use this enquiry process to consider the effectiveness of this approach. Characteristics of this method include: a project initiated by an individual seen to be an expert or authority figure; the focus of the project should be to promote more effective or efficient practices; and the project, though product directed, leads to personal involvement by the target group in the pursuit of improved performance (Masters, 1997). The researchers have chosen a mentoring and support approach to professional development based on the assumptions regarding TAR that mentoring and support are viewed as essential to successful implementation of the project. Previous work suggests that this distinctive characteristic of action research (mentoring and support) leads to the development of a stronger linkage between organisations and research centres and to organisational development and improvement (Nereu *et al.*, 2000).

TAR Level One

The training workshops incorporate three distinct areas:

1. Teachers trained in the use of *Illuminatus Opus*. The researcher uses a theoretical, conceptual framework adapted from *The Challenge 2000 Multimedia Project* (Means and Golan, 1998, p.3) to develop a multimedia training process.
2. Workshops to train teachers in general instructional systems design using "Multimedia Design Model". The model includes four major functions: 1) analysis, 2) design, 3) production, and 4) evaluation. Each function is divided into a set of specific activities (MIME, 1998).
3. Teachers trained to use and assess multimedia projects. This component of the training focuses on; defining project-based learning with multi-media and outlining goals and objectives for student projects, (Penuel *et al.*, 2001, pp.61-63) defining assessment strategies for projects and classroom management techniques, (Hill, 2000, pp.14-15) and completing project plans (Penuel *et al.*, 2001, p.63). The two lead school clusters receives six training sessions of four hours each of six months. This will include time for a pre-training, open-ended interview, a post-training questionnaire and a post –implementation questionnaire.

TAR Level Two

Over the following six months teachers implement student-centred multimedia projects (SCMP+T) in her/his own classroom to develop an interactive CD-Rom. Teachers use the same training process to teach students to use "Illuminatus Opus" to develop a project based multimedia CD. Student behaviour is monitored and recorded to answer those research questions relating to the introduction of student-centred multimedia projects and technology (SCMP+T) supports into classrooms.

TAR Level Three

During this level a multimedia exhibit is held to exhibit student's work. Students' multimedia projects are aligned to determined strands and learning outcomes on the New Zealand curriculum Framework (Te Kete Ipurangi) (Harris, 2001).

TAR Level Four

Teachers use the *Te Kete Ipurangi* website (the official New Zealand Ministry of Education website) to share their knowledge and experiences with other schools.

TAR Level Five – Data Collection

Table 1. TAR Data Collection Matrix

Data collection methods	Research Design Level	Sources of information	Person Responsible for data collection
Survey	Pre -Implementation	Teachers	Researcher
Questionnaire One	Pre -Implementation	Teachers	Researcher
Open –ended Interview	Level One	Teachers	Researcher
Observation	Level One	Teachers	Researcher
Observation	Level Two	Students	Teachers
Questionnaire Two	Level One	Teachers	Researcher
Questionnaire Three	Level One	Teachers Students	Researcher
Multimedia Projects }	Level Three *Not part of data collection*	Students *Not part of data collection*	Teachers

The Data Collection Matrix Table 1 indicates the different proposed phases of the research project and the data collection method proposed for level one and two. An open-ended interview is conducted after the researcher has collected and collated survey and questionnaire data about technology needs and teacher attitudes regarding the use of multimedia as an ICT intervention strategy in classroom teaching. The researcher observes teachers use of technology and supports and assists individual teachers where necessary to ensure mastery of the software. Observations of teacher needs and training needs is noted as part of improving the training process in the use of Illuminates Opus as a multi-media teaching tool for classroom use. Observations are also used to help design a student training handbook, which then forms part of the development of a collaborative approach to teacher training in the use of "Illuminatus Opus". A post-training questionnaire is conducted to determine the possible change in teacher need, attitude and proficiency regarding the use of multimedia as a teaching tool. A post-implementation questionnaire is conducted at the end of implementation phase to determine the possible change in teacher need, attitude and proficiency regarding the use of multimedia as a teaching tool.

Data Interpretation Model - Context Dependent Cluster Model (CDCM)

CDCM is underpinned by both Soft Systems Methodology (SSM) (Checkland and Scholes, 1990) and Complexity Theory (CT) (Stacey, 1996). The following summarises the changes to the 7 SSM steps:

1. Problem situation unstructured remains the same.
2. Problem situation expressed remains the same.
3. Finding root definitions for relevant systems is applicable for a macro view of systems. In CDCM this step is modified to explore local clusters of responses.
4. Extract conceptual models from the root definitions of relevant systems also needed to change to align with these local clusters. In CDCM local peaks of responses to perceptions are regarded as individual conceptual models. The formation of these local clusters has been informed by an interpretation model based on the idea that clusters occur in local peaks within complex systems. More importantly, core systems concepts - communication, control, structure and emergence (Checkland, 1984) also underlie these local clusters.
5. Interpretation of step 2 with the model devised in step 4.
6. Results of this interpretation expose the emergent and possible changes of step 6.
7. Possible multiple actions to improve the problem situation some of which may be in conflict. This is described in CT (Stacey, 1996) as 'walking on the edge of chaos'.

Research Context - The Problem Situation Unstructured (step 1)

A New Zealand Ministry of Education report (ERO, 2001) states that although schools have well developed ICT infrastructures, many schools and teachers have yet to demonstrate a shift in their thinking from '*How do we get ICT in place?*' to '*How do we use ICT successfully to improve outcomes for our students?*' In many New Zealand schools ICT is not well integrated into the curriculum, and there are few demonstrable ways in which it has improved teaching and learning (other than in ways related to the use of the technology itself). A recent literature review regarding the use of multimedia as a teaching tool indicates that teachers need to move through different stages of ICT proficiency. Teachers also need to understand that the introduction of ICT into the classroom can be used to change classroom dynamics as well as have an impact on classroom management (Hill, 2000, p.19). This suggests a definite paradigm shift in approach to teaching and teacher attitudes towards the use of ICT in the classroom. This research describes a multimedia teaching model to assist New Zealand teachers in making the shift from a teacher-centred to a student-centred approach (Eadie, 2000; Riley, 2000, p.4; Tearle, 1998, p.13). Such a paradigm shift acts as a catalyst for changing teaching and learning in New Zealand classrooms by supporting teachers in their efforts to design and implement student-centred projects wherein students demonstrate their ideas and knowledge through multimedia presentations (Means and Golan, 1998, p.1). This is however not a miraculous remedy and should be seen for what it is, one of many teaching tools available to teachers to improve both teaching and learning outcomes.

The Problem Situation Structured (step 2)

The five levels of TAR as described above are embedded in step 2. Major stakeholders are; The Ministry of Education, the **2 lead schools** selected for the study, all teachers undergoing multi-media training, all students in classrooms receiving and contributing to MMT and the researchers. Release time for teachers to undergo MMT is always a major source of conflict between teacher trainees and school administrators. Relationships are enriched and learning deepens as MMT progresses using this application of TAR.

Research Methodology - Forming Relevant Theoretical Systems (step 3)

In CDCM each individual is considered within her/his own local domain. This is a major departure from developing root definitions for relevant systems as is done in SSM. In CDCM it is more appropriate to consider individuals in their local context as opposed to the macro level for HAS. (Fielden and London, 2001)

Looking at the use of CT applied to Human Activity Systems (HAS) (Stacey, 1996) led to considering the use of multiple factors as triggers. Chait (2000) suggests that CT is more suited to considering systems at the micro level. Applying content analysis (Potter, 1999) to clusters of words and/or phrases in the data allowed inferences to be made by the writers about these new MMT learners. Emotional state, level of learning maturity, position in the continuum from disengaged to engaged in the MMT process and the number and nature of obstacles and triggers to engaging in MMT emerged from the content analysis. Earlier work (Fielden, 1996) on depth learning and intrinsic motivators also informed the theory formation and interpretation. The content analysis performed on groups of words and phrases showed that it is clusters of intrinsic motivators that are the triggers for engaging in the MMT process. An individual's cluster is dependent on her/his local MMT community, and may be different from one person to the next both within local environments and in the population of new researchers as a whole. McKenna (1999) suggests that individuals draw their own inner complexity maps. This is consistent with individual clustering of obstacles and engagers. When the total number of clusters is greater or equal than the total number of obstacles for an individual then engagement has happened.

Conceptual Models from Complexity Theory and Systems Core Concepts (step 4)

In the original SSM Checkland (1984) draws on formal systems concepts (control, communication, hierarchy and emergence) and other systems thinking to extract the core ideas from step 3 to develop conceptual models. In CDCM conceptualisation is informed by the application of CT to HAS as well as core systems theory (ST) concepts. The most important core concept from both ST and CT is emergence since it is emergence that flows through into subsequent steps in CDCM. Group rules for emergence that have been observed in this study include: generating collaborative energy or trust; openly sharing information and learning; deep personal commitment to aligning the individual with the group; co-ordinating co-evolution; self-discipline and self-responsibility and emerging synergy. Chait (2000) identified that CT has more application at the micro level. In this study on new MMT learners for this particular problem situation it was discovered that at the micro level there were local peaks of responses that could be extracted from step 3 for the emergence of both engagers and obstacles to the learning process. For this particular problem situation it was discovered that there were local peaks of responses that could be extracted from step 3 for the emergence of both engagers and obstacles to the learning process.

Findings - Interpretation of the Problem Situation Expressed with Theoretical Conceptual Models (step 5)

In CDCM the problem situation expressed is interpreted with the conceptual model developed at step 4 rather than comparison of step 2 with step 4 as in the original SSM. The MMT learning process was triggered when:

1. The number of engagers on a cluster are greater than the number of;
2. Recognising the importance of engaging with initial confusion;
3. Becoming aware of the learning process from disengaged to engaged as a complex process;
4. Recognising that clusters of engagers are context dependent;
5. Recognising that while there are context dependent patterns that an individual's cluster may lie outside the context-dependent cluster;
6. Clusters of intrinsic motivators are the engagers; and
7. Individual as well as whole class engagement is important to boost self-esteem in a complex learning process.

Emergent and Possible Changes (step 6)

In CDCM Step 6 becomes emergent and possible rather than feasible and desirable changes. This allows for diversity at the micro-level. Emerging and possible changes are to:

1. Expose new MMT learners to chaotic situations as part of the learning strategy. For instance, expose new MMT learners to many MM tools, techniques and strategies;
2. Raise the level of awareness for individual new MMT learners to chaotic states in the early stages of learning;
3. Facilitate early stages of MMT through the stages outlined in figure 4 from overwhelm through confusion to frustration;
4. Normalise negative emotional states to chaotic states in the facilitation process.

Experiencing chaos is essential for the research process to start. Engaging in chaos-inducing exercises, a supportive learning environment and developing skills to deal with negative emotional states all form part of the new MMT learner's individual tool kit to start to engage. New MMT learners thus develop their own cluster of intrinsic motivators and the own cluster of obstacles. Full and rich engagement with the complexity of the learning process occurred for those new MMT learners who engaged in multiple ways (had larger clusters of engagers) and who learnt in community with others. Those MMT learners who remained in isolation and who did not utilise the multiple communication channels provided had fewer intrinsic motivators and they were more likely to stay in a state of chaos overlaid with negative emotions.

Future Implications - Possible Multiple Actions for Problem Improvement (step 7)

In CDCM the emphasis is on possible multiple actions for problem improvement. These may be in conflict and may induce or encourage states of confusion and chaos. In CT (Stacey, 1996; Merry, 1995; Plsek *et al.*, 1997) all stress the importance of initial conditions for multiple actions to emerge. In establishing an environment conducive to allowing the learning process to be triggered, exposure to chaos through process design is essential. From the facilitator's point of view it is important to realise that educators, new to technology or the use of multimedia will be overwhelmed, confused and frustrated. They will have negative emotions. This is a normal part of the process. Being aware of the complex patterns of engaging in multimedia design allows for these early stages to emerge. Stacey (1996) also stresses the importance of allowing the shadow (informal) system to operate. This is supported in a community of multimedia developers or educators learning to use multimedia as a teaching and learning strategy by an environment in which multiple communication patterns may occur. Whole-class discussions and small group interactions are all provided within this environment as a safe medium in which the shadow can operate. Realising that each learning community has context-dependent clusters of triggers to engage means that the facilitator can adapt both the learning style and environment to encourage local emergence.

IMPLICATIONS FOR FURTHER RESEARCH

It is the authors' belief that CDCM is a generic model for interpreting (Wolcott, 1994) – as opposed to analysing – qualitative research data. CDCM attempts to capture the richness and complexity of context-dependent data. CDCM allows for possible and emergent changes to a problem situation that may be in conflict. CDCM is predictive rather than prescriptive. Applying CDCM to further sets of complex qualitative data is needed to substantiate this claim.

TAR provides a research model suited to facilitating teachers proficiency and skills in implementing student-centred multimedia projects (SCMP+T) in their own classrooms. What this research highlights is that CDCM provides a framework that can be applied to different research models (such as SCPM+T) using appropriate methodologies (such as TAR) to assist practitioners in understanding that the different stages of development from a state of confusion to a state of clarity in application is a natural progression applicable in a number of research situations that can successfully be applied to any theoretical study or practical setting (Holter and Schwartz-Barcott, 1993, pp.298-305). This synergy between CDCM and SCPM+T can provide a new theoretical framework that can be used with a variety of research designs to provide richer research data within pre-defined parameters.

CONCLUSION

This research investigated the early stages of forming a community of multi-media teachers. Preliminary results indicate that there are unique complex clustering patterns of triggering factors for each teacher as they migrate from a state of confusion to a state of knowledge. These individual sets of triggering factors are context dependent and are clustered around local peaks within complex patterns. In this paper CDCM has been developed in which there is better alignment at the micro level for a HAS than in the original SSM model. Exciting results from applying this modified SSM to a problem situation suggest that becoming engaged *in the MMT learning process* is complex, context-dependent and happens in community with multiple communication channels. Four important findings are that intrinsic motivators are primary engagers, entering into a state of confusion should be viewed as an important early step in engaging in *MMT learning,* engagement is more likely to be triggered in community and individual, context-dependent clusters of triggers are required.

REFERENCES

Chait, H.N. (2000). "The Soul at Work: Listen...Respond...Let Go: Embracing Complexity Science for Business Success", *Personnel Psychology*, 53(4):1011-1013.

Checkland, P. (1984). *Systems Thinking, Systems Practice.* John Wiley and Sons Ltd, Chichester.

Checkland, P. and Scholes, J. (1990). *Soft Systems Methodology in Action.* John Wiley and Sons, Chichester.

Eadie, G. (2000). *The Impact of ICT on Schools: Classroom Design and Curriculum Delivery. Study of Schools in Australia, USA, England and Hong Kong, 2000.* Wellington.

ERO. (2001). *The Implementation of Information and Communications Technologies (ICT) in New Zealand Schools*. http://www.ero.govt.nz/Publications/pubs2000/implementationICT.htm.

Fielden, K. (1996). "Extending Human Potential in a Technical Learning Environment". PhD Thesis, University of Western Sydney.

Fielden, K. and London, K. (2001). "Complex Clusters in Local Environments: Investigating Triggers for Engagement," to be presented at ANZSYS2001, Perth, November.

Harris, D. (2001). *Information and Communication Technology in the New Zealand Educ. Context* http://www.minedu.govt.nz/web/document/document_page.cfm?id=3840&p=1003.1024.

Hill, R. (2000). Managing a multimedia classroom. *Computers in New Zealand Schools, 12*(2)13-19.

Holter, I. M., & Schwartz-Barcott, D. (1993). Action research: What is it? How has it used and how can it be used in nursing? *Journal of Advanced Nursing, 18*(3), 298 - 304.

McKenna, S. (1999). "Learning Through Complexity", *Management Learning.* 30(3):pp301-320.

Masters, J. (1997). *The History of Action Research in Hughes, I. Action Research Electronic Reader*. URL:http://www.beh.cchs.usyd.edu.au/~arow/Reader/rintro.htm.

Means, B., and Golan, S. (1998). *Transforming Teaching and Learning with Multimedia Technology. Report 1998.* Menlo Park: Silicon Valley Network.

Merry, U. (1995). *Coping With Uncertainty: Insights from the New Sciences of Chaos, Self-Organisation and Complexity*. Praeger, New York.

MIME. (1998). *Multimedia Development*. URL:http://mime1.marc.gatech.edu/MM_Tools/.

Nereu, F., Kock, J., McQueen, J., & Scott, L. (2000). *Can action research be made more rigorous in a positivist sense? The contribution of an iterative approach.* http://www.scu.edu.au/schools/gcm/ar/arr/arow/kms.html.

Penuel, B., Korbak, C., Yarnall, L., & Pacpaco, R. (2001). *Silicon Valley Challenge 2000: Year 5 Multimedia Project Report.* San Jose: Silicon Park Network.

Plsek, P., Lindberg, C. and Zimmerman, B. (1997). "Some Emerging Principles for Managing Complex Adaptive Systems". Working Paper version Nov 25.

Potter, J.W. (1999)." Rethinking Validity and Reliability in Content Analysis", *Journal of Applied Communication and Research*, 27(3):258-284.

Riley, R. (2000). *The National Educational Technology Plan. Report.* Washington: U.S. Department of Education.

Stacey, R. (1996). *Strategic Management and Organizational Dynamics*. (2nd ed.), Pitman, London.

Tearle, P. (1998). *Core Curriculum for Telematics for Teacher Training.* Luxembourg and Brussels: University of Exeter.

Wolcott, H.F. (1994). *Transforming Qualitative Data: Description, Analysis and Interpretation.* Sage Pub Inc, USA.

HANDLING COMPLEXITY IN NETWORKED MULTIMEDIA SYSTEMS: THE CONTRIBUTION OF SEMIOTICS

John H. Connolly and Iain W. Phillips

Department of Computer Science
Loughborough University
Loughborough, LE11 3TU, UK.
J.H.Connolly@lboro.ac.uk, I.W.Phillips@lboro.ac.uk

INTRODUCTION

Networked multimedia systems are commonplace in the world of today. The most obvious example is the world-wide web, but other such systems exist besides this, for instance the intranets of commercial organisations, which may or may not have internet connections. Furthermore, there is no doubt that web-sites and intranets have the potential to reach very high levels of complexity. One reason for this is that they can have a very intricate structure, attributable at least in part to the use of hyperlinks. Another reason is that they may be internally very diverse. They may contain text, tables, pictures, diagrams, animations, video and audio recordings, and so forth, in almost any combination.

This complexity poses a problem for those who have to engage with the systems concerned. The people we have in mind here include those who have the task of specifying, designing, implementing and/or evaluating networked multimedia systems, and also those who need to analyse or describe such systems for other purposes, for instance academic researchers investigating the relationship between the structure of a web site and its effectiveness in meeting expectations of user satisfaction or commercial success.

It is our contention that a useful handle on this problem of complexity can be provided by Semiotics. The field of Semiotics is concerned with communication, and in particular how this process is conducted by means of what are called 'signs'. The latter include symbols, such as numerals or letters of the alphabet (usually combined into longer texts), and also 'icons', which, unlike symbols, actually resemble (to some extent) what they represent. For example, a picture of a house is an iconic representation of the building concerned, as it bears a degree of physical resemblance to that building (even though it may be smaller than the real house), whereas the *word* 'house', being symbolic in nature, bears no such resemblance to the object that it denotes. A sketch of a

Edited by Ragsdell *et al.*, Kluwer Academic/Plenum Publishers, New York, 2002.

document folder, as typically seen on a computer screen, or a diagram of an electrical circuit, constitute further examples of icons in the semiotic sense of the term. For a survey of the field of Semiotics, see for instance Fiske (1990).

To appreciate the applicability of Semiotics to the matter in hand, it is necessary to understand a multimedia display as a type of communication system. For instance, on a web page, the purpose of any text is generally to convey some message to the reader, while the purpose of any picture will be, at least in part, to communicate the physical appearance of someone or something (real or imaginary). True, some material may be included purely for decoration, but as long as there is some message or set of messages to be communicated by the multimedia display, then the latter can be treated as a semiotic phenomenon. What is more, even background material can have semiotic significance, in conveying a particular mood or style. For previous work on the application of Semiotics to multimedia, see for instance Clarke (2001), Kress and van Leeuwen (1996; 2001), Purchase (1998; 1999) and van Leeuwen (1999).

Because Semiotics is concerned with all kinds of signs used in communication, it provides the breadth of coverage that is required in order to cope with the diversity of multimedia. This is a key property in enabling it to offer a coherent framework within which multimedia can be analysed and described. The remainder of the paper will be concerned with outlining such a framework.

In order to be compatible with Systems Theory, any analysis must avoid reductionism or atomism. In the present paper, therefore, we shall be concerned not merely with the analysis of multimedia systems into their components, but also with considerations of how these components interrelate to form larger entities which are not simply aggregates of their constituent parts.

SEMIOTIC ANALYSIS AND DESCRIPTION

One approach to the analysis of multimedia displays is to employ the distinction between symbols and icons, mentioned above. In furtherance of such a purpose, both of these semiotic classes can be sub-classified. Symbols can be either alphanumeric (letters, numbers, punctuation marks and such like) or non-alphanumeric (for example, notes on a musical stave, or arrows to help people find their way). Icons, for their part, can be either relatively concrete (as in the case of pictures) or relatively abstract (as in the case of diagrams or of those icons, such as folders, buttons, and so forth, that are characteristic of graphical human-computer interfaces). For more on the sub-classification of signs, see for instance Andersen (1997: 216-229), Fischer and Nänny (1999) and Purchase (1998; 1999). It is true that this sub-classification of signs is not watertight; for instance, a graphic can be built up out of alphanumeric characters, thus forming an icon out of symbol-type components. Nevertheless, borderline cases of this kind are in the minority, and the classification is generally applicable.

Let us take the Loughborough University home page as an example: http://www.lboro.ac.uk as at 5 October, 2001, organised as shown in Figure 1.

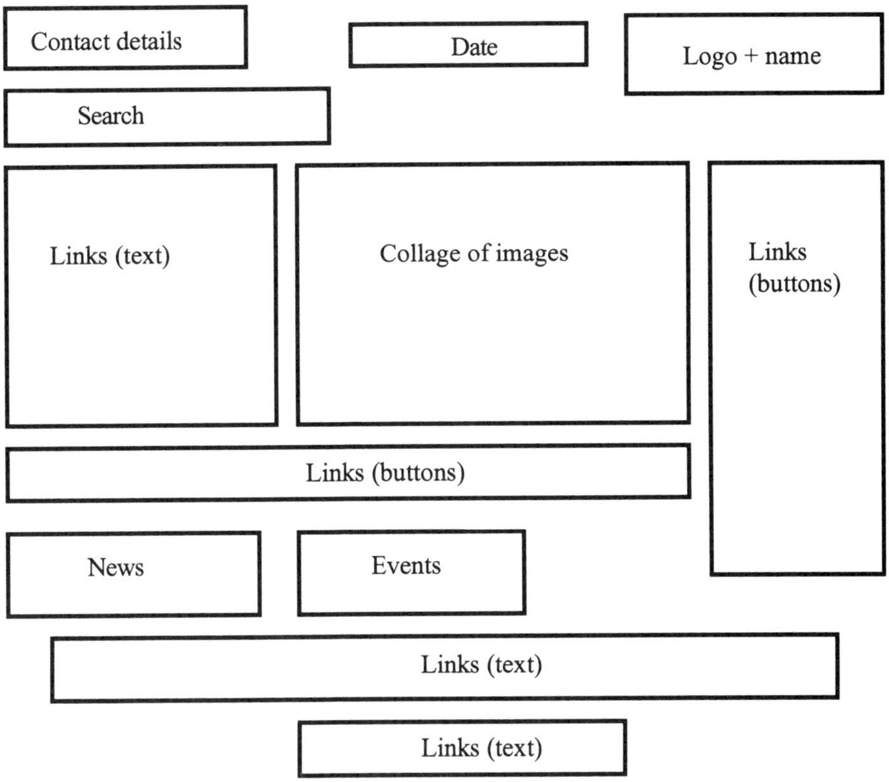

Figure 1. Layout of the Loughborough University Home Page

We see that the page is composed largely of text, which is alphanumeric and therefore symbolic in nature. The majority of this text comprises links to other pages in the same web-site. Some of these links are placed on coloured buttons, which are abstract-iconic in character. The page also contains a collage of colour images, featuring students in the foreground and a hall of residence in the background. These images are concrete-iconic in nature.

The icons in this example are all static. However, dynamic icons, which change or move, are also possible. Video and animations are the obvious examples of this.

Signs can also be sub-classified according to the sense modality through which they are perceived. Visual signs include written or printed alphanumeric characters/texts, pictures, diagrams, and so on. Auditory signs, on the other hand, are sub-classified by van Leeuwen (1999) into speech, music and sound effects. An example of a sound effect would be a sequence of tones emitted by a computer to indicate the arrival of an email message.

Given, then, the possibility of analysing a multimedia display into components, some or all of which are signs, the next question from the system-theoretic perspective is how the signs in question interrelate within the broader system to which they belong. In fact there are various possible relationships among signs within a multimedia system. These relationships include the following:

- Combination
  - Composition
    - Nesting
  - Dependency
  - Order
- Alternation

First of all, there is the relation of combination. This is simply the putting together of two or more signs to form a larger semiotic unit, which is consequently termed a composite sign. The web page described above is an obvious example. Within a composite sign we may find different types of relationship among the components, notably composition, dependency and order.

Composition is the part-whole relationship whereby a composite sign is composed of (or consists of) its component units. For instance, individual alphanumeric symbols are usually combined into larger, composite units, as when letters combine to form words, words to form phrases, phrases to form clauses, and so on. From the system-theoretic perspective it is significant that the composite unit may have emergent properties which the smaller units of which it is composed do not. For example, consider the clause 'the library will close next week', which consists of three phrases: 'the library', 'will close' and 'next week'. The first phrase, 'the library', is the subject of the clause, the second phrase, 'will close' is the verbal element of the clause, and the third phrase, 'next week', is an adverbial element. Hence, the clause has the structure subject + verbal element + adverbial element, abbreviated SVA. This SVA combination is a property of the clause. By contrast, none of the phrases within that clause has an internal structure of a similar type. This is because it is a structure of a kind which emerges only when phrases are combined into the larger structure of the clause.

It is not only units of language that can manifest the relationship of composition. For instance, a collage is composed of pictures which are constituents of the whole image; and the property of being a collage is an emergent one, since clearly none of the individual pictures can be a collage on its own. Moreover, a web page will often be composed of several different signs, again as in the example above. If each individual sign happens to be single-medium, then one of the emergent properties of the whole page will be its multimedia character.

Nesting is a special type of compositional relationship, namely the embedding of a structure within an encompassing structure of the same type, the latter being termed the superordinate, while the nested structure is described as the subordinate. For instance, we could take the clause 'I have been told at last' and the clause 'who stole the money', and embed the latter into the former to yield the sentence 'I have been told who stole the money at last'. To take another example, a composite multimedia sign can have other such signs nested within it, as can be seen when, within a hierarchically organised web-site, a page in which text and graphics are combined provides access, via a link, to a subordinate page in the hierarchy which likewise contains a combination of those two media.

Signs which co-occur in combination with one another often (though not always) exhibit relationships of dependency. For instance, if the definite article 'the' is to appear in a phrase, then it must be combined with another word (usually a noun) to which it can apply, and which thus acts as its target. Accordingly, we may construct a phrase like 'the computer' or 'the monitor', but not 'the' on its own. Hence, there exists a relationship of dependency here. The occurrence of the word 'the' depends on an accompanying target word (such as 'computer' or 'monitor') also being present. Without the presence of the target word, the word 'the' would not itself be permitted to appear; and so when the two words do co-occur, the definite article is dependent upon its target.

For further exemplification of dependency, let us consider a labelled graph displayed on a screen. Each label is applied to a target, namely some vertex or edge in the graph, and is dependent on that target, inasmuch as the occurrence of a label would be unwarranted if it did not apply to anything. Conversely, if the graph does not make sense without the labels, then the former must depend on the latter for its interpretability. Hence, in such circumstances there is a mutual dependency between the labels and their respective targets.

When signs co-occur but do not occupy exactly the same time and space, then it is possible to speak of the order in which they occur. This order may be temporal or spatial, and in the latter case it may involve more than one dimension. A simple example is seen in the fact that the definite article 'the' has to precede its target word. Even if two signs overlap, we can still speak of ordering in respect of their respective starting points and their respective ending points.

Order can be a problematic issue in relation to multimedia displays such as web-sites. It is true that if we consider a visual display occupying just a single screen, then the placement of items along the horizontal and vertical dimensions can normally be described straightforwardly enough. On the other hand, because of the possibility of jumping around within a web-site, or between web-sites, by means of hyperlinks or bookmarks, the sequence in which a particular set of signs is encountered may not be predictable. If there also exist dependency relationships among those signs, such that some are uninterpretable unless certain others have already been encountered, then this may impede the communicative goals of the authors of the web-sites involved.

Finally, let us deal with alternation. This is a relationship among two or more signs, only one of which can be present in a particular place at a particular time. For instance, an on-line multimedia illustrated catalogue of goods might afford the user the option of displaying the prices in pounds sterling or American dollars. In that case, the prices in sterling and the prices in dollars would be alternatives. (If both prices were displayed together, then they would be in combination rather than alternation. Of course, the display of this combination could itself be regarded as an alternative to displaying just the one or the other!) Alternation is an emergent property that arises from the existence of choices; it cannot apply where there is no option.

LEVELS OF ORGANISATION

Another important way in which Semiotics can contribute to the handling of complexity in networked multimedia systems is by providing a framework that enables us to describe the internal organisation of such systems in terms of a series of different levels of abstraction.

The most highly elaborated framework of this nature is that proposed by Stamper (1991), comprising the following six levels:

1. The social world.
2. The level of pragmatics.
3. The level of semantics.
4. The level of syntactics.
5. The level of empirics.
6. The physical world.

This framework indicates that communication takes place not only within the physical world but also within a world of social and cultural conventions, expectations and assumptions. The signs that serve to facilitate the communication process can be described in terms of their form or structure (a description at the level of syntactics) and in terms of their meaning (a description at the level of semantics). When these signs are deployed in actual situations of communication, the factors involved in this practical communicative activity can be described both from the point of view of the social/psychological interaction involved (a description at the level of pragmatics) and from the point of view of the physical patterns which signal the message (a description at the level of empirics).

For instance, imagine that a sports club has a web page containing a button as shown in Figure 2.

Apply to join

Figure 2. A Button on a Sports Club Web Page

This button has the (syntactical) form of a box enclosing a short piece of text, and the (semantic) meaning that by clicking on the button, an application for membership will be initiated. If someone actually clicks on the button, then they embark on a (pragmatic) interaction, via the system, with the committee of the club in question, who are then expected to consider the membership application. This interaction is supported by observable (empiric) patterns of physical appearance and physical activity. The physical appearance of the button is characterised by properties such as the colour and geometric shape of the box, and the colour, typeface and size of the text. The physical activity includes both the display of the button on the screen and the transmission of messages across the network.

The six-level semiotic framework makes a substantial contribution to the handling of complexity in networked multimedia systems, in that it enables us to concentrate on one aspect at a time without being distracted by other aspects that are not of direct relevance. For instance, ensuring the adequate flow of signals across the network is a matter of concern at the empiric level, and is not directly influenced by the precise semantic content of the messages that will be conveyed over it. On the other hand, in conformity with Systems Theory, the framework allows us to speak meaningfully of the interrelationships among the different levels. For example, the design of the (syntactical) organisation of a web page may be influenced by the (empiric) properties of the browsers of the anticipated audience of that document, inasmuch as the author of the page will not want it to be rearranged by the time it appears on the reader's screen.

CONCLUSION

We have seen, then, that Semiotics provides a broad and usable framework for the analysis and description of networked multimedia. Importantly, the analytical framework is not reductionist, but accommodates the relationships that exist among the internal components of multimedia systems.

REFERENCES

Andersen, P.B. (1997). *A Theory of Computer Semiotics*, Cambridge University Press, Cambridge.

Clarke, R.J. (2001). "Towards a Systemic Semiotic Approach to Multimedia Interface Design," in *Information, Organisation and Technology: Studies in Organisational Semiotics*, (K. Liu, R.J. Clarke, P.B. Andersen and R.K. Stamper, eds.), Kluwer, Boston: 247-270.

Fischer, O. and Nänny, M. (1999). "Introduction: iconicity as a creative force in language use," in *Form Miming Meaning: Iconicity in Language and Literature*, (M. Nänny and O. Fischer, eds.), Benjamins, Amsterdam: xv-xxxv.

Fiske, J. (1990). *Introduction to Communication Studies*, second edition, Routledge, London.

Kress, G., and van Leeuwen, T. (1996). *Reading Images: the Grammar of Visual Design*, Routledge, London.

Kress, G., and van Leeuwen, T. (2001). *Multimodal Discourse: the Modes and Media of Contemporary Communication*, Arnold, London.

Purchase, H.C. (1998). "Defining Multimedia," *IEEE Multimedia*, 5(1):58-65.

Purchase, H.C. (1999). "A Semiotic Definition of Multimedia Communication," *Semiotica*, 123: 247-259.

Stamper, R.K. (1991). "The Semiotic Framework for Information Systems Research," in *Information Systems Research: Contemporary Approaches and Emergent Traditions*, (E. Nissen, H.K. Klein and R. Hirschheim, eds.), North-Holland, Amsterdam: 515-527.

van Leeuwen, T. (1999). *Speech, Music, Sound*, Macmillan, Basingstoke.

SOME CONSIDERATIONS FOR A SEMANTIC ANALYSIS OF CONCEPTUAL DATA SCHEMATA

Wei Hu and Junkang Feng

School of Information and Communications Technologies
University of Paisley
High Street, Paisley PA1 2BE, United Kingdom
wei.hu@paisley.ac.uk
junkang.feng@paisley.ac.uk

INTRODUCTION

The now widespread use of computers for data-processing in large and distributed organizations means that it is often the case that one organization will hold data in a variety of information technology based information systems. The choice of an appropriate representation of data is one of the most crucial tasks in the entire systems development process (Moody, 1998). Although the data modelling phase represents only a small proportion of the total systems development effort, its impact on the final result is probably greater than any other phase (Simsion, 1994). The last 50 years have been characterized by a tremendous growth in the number and importance of database applications. Databases are essential components of information systems. Designing databases has become a popular activity. Unfortunately, database design methodologies are not very popular; most organizations and individual designers rely very little on methodologies for conducting the design of databases, and this is commonly considered one of the major causes of failure in the development of information systems. Many of these problems are due to a lack of clarity in understanding the exact nature of data at an abstract and conceptual level. The conceptual level has been introduced in the context of a three-schema framework for database management systems. In this context, a conceptual schema is defined as "a unique central description of the information that may be in the database." Nowadays, a conceptual schema plays a key role in the development of a large variety of systems (Loucopoulos, 1992).

In this paper we will present some of our ideas on how a conceptual data schema should be analysed in order to arrive at an adequate conceptual design of the data aspect of an information system. First we will review related work; then we will define what we mean by 'semantic analysis'; then we will present our basic ideas for our semantic analysis. The analysis is conducted at the 'information' and 'data' level respectively, so we will present inference rules for the former and our principle for the latter. When the information level is considered, we have to first of all identify information requirements for and from an application domain. The ideas and methodological work of soft systems thinking plays a central role in that stage.

RELATED WORKS

Some conventional methods have been applied on the analysis of conceptual data schemata. For the purpose of measuring the qualities of a conceptual schema, some criteria have been formally defined in the literature. Batini (1992) used completeness, correctness, minimality, expressiveness, readability, self-explanation, and extensibility to indicate the qualities of a conceptual schema. Most work has been done on the validation and verification of a specification from a real world reality. In this paper, we will not consider the problems of extracting and producing a conceptual data schema from inputs to the design process, namely requirements, but the existing schemata.

For the purpose of improving and optimising a conceptual data schema, the act of reshaping a schema is adopted. The kind of act is said to be a conceptual schema transformation. A knowledge of schema transformations helps us see what different design choices are possible (Halpin, 1995). The goal of applying transformations is to restructure the schema to produce a better version in terms of some quality criteria. In order to achieve the completeness, correctness and minimality of a conceptual data schema, much work and many approaches have been found based on the syntactic level. Eventually, schema transformations apply to an input schema S_1 and produce a resulting schema S_2. Some literature has discussed transformations in terms of how they change the information content of a schema. It is very difficult to formally define the information content of the schema or to prove that the information content of two schemata is identical (Batini, 1992). A means for comparing the information content of two schemata S_1 and S_2 is to compare their ability to reply to queries: S_1 and S_2 have the same information content (or they are equivalent) if for each query Q that can be expressed on S_1, there is a query Q' that can be expressed on S_2 giving the same answer, and vice versa. However, because of some connection traps existing in the schema, this means cannot exactly provide us a sound and complete result regarding the comparison of S_1 and S_2 (Batini, 1992).

In addition, nature language check and syntactic level check on schemata have also been used to ensure and determine whether the schemata effectively represent the data requirements and whether there are any other constraints or assumptions that the designer has made cannot be expressed through the schemata. But in the literature, little seems to have been provided for a systematic semantic analysis on data schemata.

SEMANTIC ANALYSIS

In the field of computer science, semantic analysis is normally utilized on analysis of software languages. The aim of applying semantic analysis is to guarantee proper execution of programming languages and enhance efficiency of execution of the translated program. It involves both the description of the analysis to perform and implementation of the analysis using appropriate algorithms. Following the rules of programming languages, software engineers can optimise the execution of their applications.

In designing an information system, the method of semantic analysis provides a means for modelling a business system by devising a canonical formalism with focus on the responsible agent and his repertoire of behaviour (Liu, 2000). This method covers a range of activities in information modelling such as requirements elicitation, analysis, specification and representation. The objective of this method is to elicit and specify user requirements. One of its theoretical foundations is organizational semiotics (Stamper *et al.*, 1997). Compared with Liu's semantic analysis, we will carry out a study on the data representation and transformation of two conceptual data schemata. The difference is schematically presented in Figure 1 (a) and Figure 1 (b).

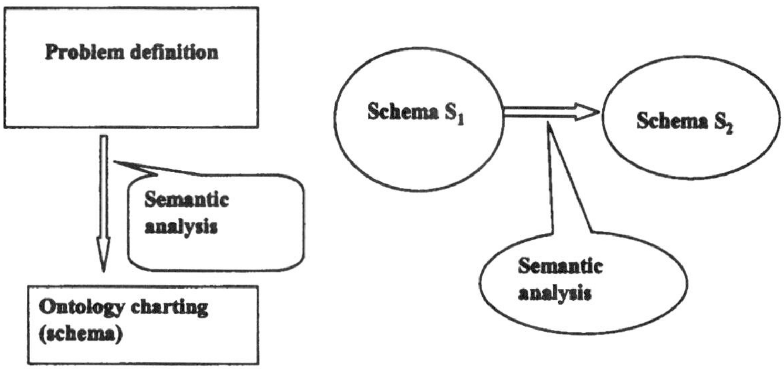

Figure 1(a) Liu's Semantic Analysis **(b)** Our Semantic Analysis

THE BASIC IDEA

The hardest problem in schema design is to define what is a good schema and what is a bad schema (Loucopoulos, 1992). Our goal is to articulate a systematic technique for creating a minimal data schema for a given set of information requirements. The central idea is the separation between information and data, and data bears (represents) information. So there are two levels, as shown in Figure 2 below. X and Y mean the set of states of affairs. A and B mean data constructs or data schemata.

Similar to our approach to IBC (Information Bearing Capacity) of a data construct (Xu and Feng, 2001), we should analyse a conceptual data schema from both the *what* and *how* perspectives.

For the *what* aspect, the starting point and the basis for our analysis is the conditions for data to represent information.

The issue of *granularity* requires particular attention. The granularity of IBC in Xu and Feng (2001) is individual data construct such as an entity and a path, and individual *semantic relation*. It is neither *data schema*, which is a collection of a number of *data constructs*, and a set of semantic relations, nor instances of a data construct and instances of a semantic relation.

To develop our ideas for the semantic analysis of the *what* aspect of a conceptual data schema, we draw on soft systems thinking (Checkland, 1981) and formal approaches to normalisation. We will describe our ideas in some detail in the following two sections.

INFORMATION LEVEL

Information requirements are seen as a set of semantic relations, or a set of states of affairs. The information content of a state of affairs might include another state of affairs. Moreover, two states of affairs may have an "informational relationship" as long as one's existence changes the probability of the existence of the other (Dretske, 1981).

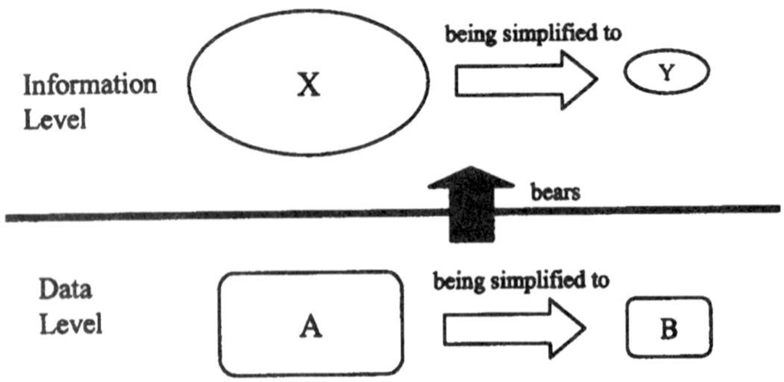

Figure 2. Information and Data Levels

Information Requirements Identification

To identify information requirements for an ill-structured application domain is not a trivial task. Many writers advocate a soft systems approach to resolving this problem, such as Checkland (1981), Checkland and Holwell (1998), Stowell and West (1994), and Wood-Harper and Fitzgerald (1982). We subscribe to this approach and these ideas, and have developed a method that combines the soft systems methodology (SSM) (Checkland 1981) and what we call semantic theories of information, namely the theories and ideas from Devlin (1991), Dretske (1981), and Mingers (1995). With this method, we elaborate the 'conceptual model' for a 'notional system' in using SSM by means of identifying and formulating relevant 'situation types' (Devlin, 1991). Information requirements are subsequently the 'info units' in the 'data storage situation type' and the 'basis situation type'. For details, the reader is referred to Feng (1999).

Information Requirements Simplification

Information requirements identified by using the aforementioned method, or other intuitive means are likely to be subject to redundancy, duplication, and inconsistency. In this section, we will present a set of inference rules for reasoning about the information requirements by exploiting Dretske's notion of the "informational relationship" between states of affairs (Dretske, 1981). The use of these inference rules should help remove redundancy and duplication, and identify inconsistency (to describe how these can be achieved however would require a whole set of notions and algorithms, which is beyond the scope of this paper unfortunately). We base our thinking for discovering and formulating these inference rules on Dretske's definition of the informational content of a sign. We restate his definition as follows:

Supposing that X, Y, W and Z are "states of affairs" and k denotes what we know about the information source including its scope. Let X, Y, W and Z be one of many possible states of affairs of the information source. The informational content of X includes Y, if given X and k, the probability of Y is 1, given k alone, the probability of Y is less than 1.

Inference rule 1

$$X \supset Y \models I(X) \supset Y$$

This rule means that if a state of affairs X includes another (smaller) state of affairs Y, then the informational content of X includes Y.

Inference rule 2

$$W \supset Z, I(X) \supset Y \models I(WX) \supset ZY$$

This rule means that if a state of affairs W includes another (smaller) state of affairs Z and meanwhile the informational content of X include Y, then the informational content of the union of W and X includes the union of Z and Y.

Inference rule 3

$$I(X) \supset Y, I(Y) \supset Z \models I(X) \supset Z$$

This rule means that if the informational content of a state of affairs X includes Y and meanwhile the informational content of a state of affairs Y includes Z, then the informational content of X includes Z.

Inference rule 4

$$I(X) \supset Y, I(X) \supset Z \models I(X) \supset YZ$$

This rule means that if the informational content of a state of affairs X includes another two states of affairs Y and Z superlatively, then the informational content of X includes the union of Y and Z.

Inference rule 5

$$I(X) \supset YZ \models I(X) \supset Y, I(X) \supset Z$$

This rule means that if the informational content of a state of affairs X includes the union of Y and Z, then the informational content of X includes Y and Z separately.

DATA LEVEL

Our goal here is to simplify the set of data constructs that bear the information requirement set X or Y into its minimal cover. A minimal cover can be defined as a data schema that does not have any redundant data construct for a given set of information requirements (we call them a set of state of affairs). This is a process of schema simplification.

The basic idea is this: let a1 be a data construct in A, if (A – a1) bears Y, then a1 is redundant in relation to Y, and it should be removed from A. More precisely, a1 being redundant means that: One, if a1 is sound and complete for bearing $y1 \subset Y$, then in (A – a1) there is at least one data construct that is also sound and complete for bearing y1; Two, if a1 is not sound and complete for bearing any $y1 \subset Y$, then removing a1 from A does not damage any other data construct in A that is the only sound and complete data construct for a particular $y1 \subset Y$. Figure 2 illustrates that set A of data constructs is simplified into its minimal cover B.

FUTURE WORK

Based upon *inference rules*, we can define the notions of *closures*:

1. The closure of an individual semantic relation a that a data construct can represent, perhaps denoted with a^+, is the collection of all semantic relations implied by a plus a itself.
2. The closure of a set X of semantic relations, perhaps denoted with X^+ is the collection of all semantic relations implied by X plus X itself.

Then we can have the notion of "X covers Y" for sets of semantic relations that can be represented by data constructs as follows: Assume that X and Y are two different sets of semantic relations that can be represented by data constructs, X cover Y if $X^+ \supset Y$. Furthermore, *equivalence* can be defined as follows: For two sets of semantic relations X and Y that can be represented by data constructs, if X covers Y, and vice verse, then X and Y are equivalent; for two data constructs α and β, if α covers β, and vice verse, then α and β are equivalent. Further, we can define the notions of 'minimal sets and covers'.

Using above notions, we can achieve the optimisation of data schema transformation.

SUMMARY

In this paper, we have reviewed the relevant literature on the quality of conceptual data schemata and semantic analysis in the field of information systems design. We have presented some inference rules for reasoning at the information level. The information requirements are identified first by using a method that combines SSM and semantic theories of information. We have also described the principle for simplification at the data level. We have also outlined some immediate future work based upon what is presented here. With those rules to be proved and used in the future, the optimization of conceptual data schemata can be achieved.

REFERENCES

Batini, C., Ceri, S. and Navathe, S. B. (1992). *Conceptual Database Design: An Entity-Relationship Approach*, The Benjamin/Cummings Publishing Company, Inc. US.

Checkland, P. (1981). *Systems Thinking, Systems Practice*, John Wiley & Sons, Chichester.

Checkland, P. and Holwell, S. (1998). *Information, Systems and Information System*, John Wiley & Sons, Chichester, New York.

Devlin, K. (1991). *Logic and Information*, Cambridge University Press, Cambridge.

Dretske, F. I. (1981). *Knowledge and the Flow of Information*, Basil Blackwell, Oxford.

Feng, J. (1999). "The use of Semantic Theories of Information in Identifying and Formulating Information Requirements for Information Systems Design", in *proceedings of BIT'99*, Manchester.

Halpin, T. (1995). *Conceptual Schema & Relational Database Design, Second edition,* Prentice Hall of Australia Pty Ltd, Australia.

Liu, K. (2000). *Semiotics in information systems engineering*, Cambridge University Press, Cambridge.

Loucopoulos, P. and Zicari, R. (1992). *Conceptual Modelling Database, and CASE*, John Wiley & SONS, Inc, Canada

Mingers, J. (1995). "Information and meaning: foundations for an intersubjective account", *Information Systems Journal*, 5:285 – 306

Moody, L.D., (1998). "Metrics for Evaluating the Quality of Entity Relationship Models", in proceedings of 17th International Conference on Conceptual Modelling, ER'98, Springer, Berlin.

Simsion, G.C. (1994). *Data Modelling Essentials*, Van Nostrand Reinhold, New York.

Stamper, R. (1997). "Organisational Semiotics", in *Information Systems: An Emerging Discipline?*, (J.M. Mingers and F. A. Stowell), The McGraw-Hill Companies, London.

Stowell, F. and West, D. (1994). *Client-led Design*, McGraw-Hill, London.

Xu, H and Feng, J. (2001). "Towards a Definition of the Information Bearing Capability of a Conceptual Data Schema", *Computing and Informtion Systems*, 8:129 – 133, University of Paisley.

Wood-Harper, A.T. and Fitzgerald, G. (1982). "A taxonomy of current aproaches to systems analysis", *The Computer Journal*, Vol. 25, No. 1.

PRODUCT AND SYSTEMS DESIGN THROUGH LINGUISTIC MODELLING

Janus Korn

London School of Economics
116 St Margarets Road
Edgware, HA8 9UX, UK

INTRODUCTION

The term 'design' refers to events and conditions which lie in the future (Korn, 1996a). More specifically, we have an object's physical condition such as 'The door of the house is *blue*' or a mental condition of a living object such as 'This child appears to be *unhappy*' which are for some reason considered unsatisfactory. We can envisage other conditions yet non-existent: 'The door of the house is *red*' or 'This child now shows *signs of happiness*' which are to be realised at some future time. The specification or description of an unsatisfactory initial and that of a final condition or state is called the 'design problem'. The 'design solution' is concerned with creating plans for a 'product' and/or schemes for a 'system' or for an 'assembly of interacting objects' with the task of delivering the product to the object undergoing a change of state. The delivery is specific and monitored, thus, the design solution invariably involves *purposive activity* (Korn, 2001a,b).

Accordingly, the envisaged final state, the product and assembly are unobservable. However, the initial state of a 'changing object' is observable otherwise it could not be described as unsatisfactory. Also, the assembly operates in the context of existing 'environmental objects' that affect and/or can be affected by design. The changing object, product, the assembly and the environmental objects together are called a *situation or scenario*. At time equals zero only the observable parts of a scenario exist. Unobservability allows the free flow of imagination and creativity essential in design, especially in the fields of fashion, architecture and art.

There is no comprehensive theoretical framework for design. Design thinking is different from analytical thinking (Roozenburg, Eekels, 1995). Engineering design except in conventional control systems engineering is restricted to components (Pahl, Beitz, 1984, Hubka, Eder, 1990). The problem solving approach of soft systems methodology (Checkland, 1981) is an attempt at design involving human activity situations. Methods for generating requirements is a step towards design in information systems (Flynn,

Edited by Ragsdell *et al.*, Kluwer Academic/Plenum Publishers, New York, 2002.

1998). The observabilty of the initial state in the design problem and that of the relevant characteristics of the changing and environmental objects called *quiescent properties* into which a future product and assembly must fit, enable us to develop a scheme for a design method. The scheme can be expressed in specific terms through linguistic modelling (Korn, 1996, 2000, 2001c). The objective of this paper is to describe the basic elements of this scheme.

GENERAL FEATURES OF THE DESIGN METHOD

Before going into details, we outline the general features of the design method that may help its critical appraisal.

1. Design or problem solving as an intellectual activity is applied in everyday activities such as preparing a meal as well as to complex technical and social scenarios and their combinations. Designing a technical control system for maintaining the flight of an aircraft such as an autopilot, a social system such as a home for the elderly and a sociotechnical system like a production line are examples. Fashion and textile design, art and craft also involve creation and innovation in the quest for something *not-yet-existing*.

A design method that claims to be more comprehensive should thus be applicable wholly or partly in all fields where *intentional* changes are envisaged to occur. The basic ideas behind the design method to be considered in this paper have been described in the INTRODUCTION briefly which suggest that the method is likely to be domain-independent.

2. A design method to be specific in suggesting changes and objects as candidates for the role of products and assemblies should use 'properties' (Korn, 2001a) as qualifiers expressed as adjectival and adverbial phrases and clauses. Properties that fit into an inferential structure can be used to predict the occurrence of events.

3. For complete description of an object we need an infinite number of statements carried by declarative sentences of the subject-predicate pattern incorporating concrete and/or abstract properties (Korn, 2000). A number of these statements are used to define the object, others are *contingent* or situation-dependent. When an object plays a part in a scenario a number of contingent statements emerges. Also, as the object moves from one scenario to another different statements emerge. As a simple example, some of the contingent properties of a pencil when used for writing are: 'it is sharp', 'it is long enough', 'its lead is of the correct hardness' and so on. When the same pencil is used as a hook for holding a small bucket of water by its handle, we have: 'it is strong enough', 'it is held horizontally' and so on.

Thus, the suitability of an object for use is governed by its contingent properties. The emergence of a set of suitable properties in a design method is ensured by *requirements*. Basically requirements are generated from properties involved in change of state as we show later (Korn, 2001a).

4. Modelling a scenario involves identification of objects thought to be associated with the scenario, their contingent properties and interactions. Any such identification is subjective to a certain extent. However, identification may appear to be a straightforward matter in problems met in conventional science (physics in particular) as in any of its branches (electricity, for instance) the objects, properties and interactions are well defined. The same applies to multidisciplinary problems (Korn, 1995).

When activities by living, in particular human, agents are involved identification of objects, properties and interactions may be subject to debate as neither of these is well defined in terms of its involvement in a scenario, its nature and description (Checkland, 1981). In general, any design method involves debate concerning alternatives, shapes, sizes etc due to lack of precise information and for creativity to operate. The design method described here is more structured yet it allows, and needs, debate to take place concerning objects, properties and interactions involved in an emerging scenario to cater for the legitimate need of the interpretive view (Jackson, 2000).

In addition, when we deal with generation and transmission of information mental properties of objects are involved which are not directly observable, they can either be received in terms of information or can be inferred from concrete properties (Korn, 2000, 2001a).

PROBLEM SOLVING SCHEME

The basis of the design method is the consideration of an envisaged change from an existing initial to a final state expressed in terms properties carried by a *changing object* as shown in Fig.1. *One property* selected for a projected change of a changing object is called the *changing property*. The rest of the properties of this object and of the environmental objects are assumed to remain *quiescent* for the moment. When the property to be changed is *abstract* (Korn, 2000), its change can be accomplished:

indirectly through changing concrete properties which are contained in the abstract property such as quality of material of skin, geometry and number of specific deformations are contained in 'wrinkled', or

directly through information (Korn, 2001a) in case of *mental properties* such as 'happy'.

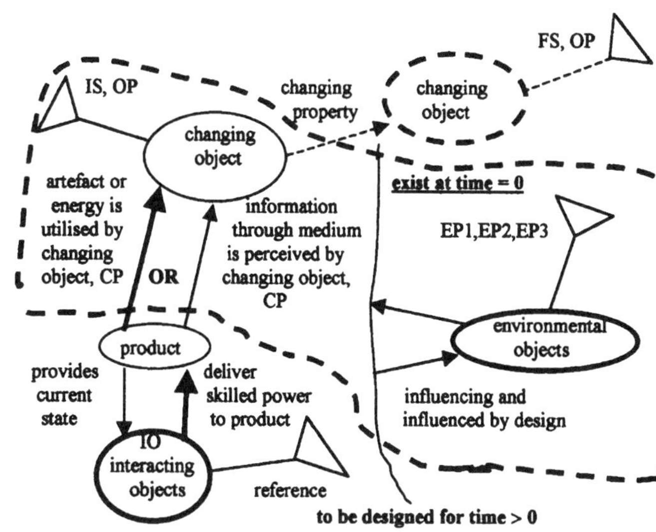

Figure 1. Diagram of a Purposive Situation and Change of State

Before and after change, there is an *initial* (IS) and a *final* (FS) state. In general, change may take place by *chance* or *in accordance with a purpose* (Korn, 2000). The design scheme operates purposively as shown by the feedback path from product to mutually interacting objects. IS denotes a description/specification of a property which is in some way regarded as *unsatisfactory or needing change*. FS means a description/specification of a property that is *desired or envisaged*. The two properties are *consistent* in the sense that there is an operation described as a *dynamic verb* which can convert an IS into a FS. For instance, there is a specific operation called 'painting' which can turn a '*white*' sheet of paper into '*blue*' but not into '*warm*'.

No change can take place *by itself*. Interaction in terms of *physical power* carrying the appropriate *energy* or of *influence* with theoretically zero power carrying the appropriate *information*, is required. An interaction symbolised as a *dynamic verb*, is capable of accomplishing a change described by consistent adjectives (Korn, 2000, 2001a,b,c).

Product is a set of properties delivered by an appropriate interaction to a changing object to bring about a corresponding change of state. The following considerations are intended towards the development of a *unified treatment of product*. To form a product, properties enter into relations:

1. *Stative artefacts*: These are purpose built objects consisting of concrete, stative properties *delivered* to a changing object by *interaction* designated by a *dynamic verb* with energetic property. A 'reciprocating knife *moving* in bread', a 'lawn mower with a rotating blade *pushed* along grass' or a 'flat stone for covering a part of pavement *lifted* into position', satisfy this criterion of product.

2. *Energy* : Energy is defined in physics and chemistry in terms of energetic properties. For example, the energy in a compressed or extended spring may be described in terms of its material and stative, energetic property (displacement).

3. *Information plus primary medium* : Information has been defined in linguistic modelling as 'A context-free, simple sentence with qualified constituents attached, as a *subordinate clause*, to a *verb* which designates influence interaction (Korn, 2000, 2001a,b,c). 'Data' are the *numerical part* of a qualifier used for achieving precision. Without a qualifier data by themselves although meaningful carry no information since data lack reference to an object which is given by the qualifier.

For example, in the sentence 'The fireman shouted to the other that the corrosive liquid in the vessel stands at a level of 2 m', 'fireman' is the *sender*, 'other' is the *receiver* and the verb 'shouted' designates *influence interaction*. The context-free sentence in the subordinate clause is 'Liquid stands'. 'Liquid' is qualified by 'corrosive', an adjective, and the verb is qualified by adverbial phrases 'in the vessel' and 'at a level' where the latter is further qualified by data '2 m'.

The structure and effectiveness of information is discussed in detail in (Korn, 2001a). We have introduced the notions of *'meaning', 'message carrier' and 'medium'* as the constituents of information (Korn, 2001c). 'Medium' refers to the physical realisation of information in terms of concrete properties called 'informatic'. We have *two categories of medium*: 'primary' which carries symbols such as *natural language* and 'secondary' that can be constructed of any object intended to convey a message such as a work of art, a hand gesture or other forms of body language. Any object or its change, an event, can serve as secondary medium. *Medium with encoded information* constitutes the *product* that is intended *to change the mental state of* a changing object capable of receiving and decoding it.

Relation between Properties

We have identified a changing object, its environmental objects and a product that operate as a part of a purposive situation depicted as a diagram in Fig.1. Such a situation has to be *designed* it is unlikely to occur *by chance*. The physical state of a *user* or the mental state of a *receiver* both called the changing object, is to change as a result of application of product. The final state of a changing object is either *given* having noted an unsatisfactory initial state or postulated as a result of *envisaging an as-yet-non-existent state* arrived at by inspiration, a flash of imagination or insight, a *matter of creativity*. We consider the scheme in Figure 1 in four stages through a simple example.

I. Primary representation of scenarios

Artefact – 'After a meal, dishes owned by a lady get dirty with remainder of food and need to be cleaned'

Energy – 'The angular speed of a rotating inertia subject to friction is to be increased'

Information – 'A gardener working in a neighbourhood with others is offered money for a job'

II. Analysis of change of state and synthesis of dynamic requirement

Changing objects– dishes owned by a lady (environmental object)
inertia subject to friction (environmental object)
gardener working with others (environmental object)

IS – dishes are stained with remainder of food
inertia is stationary
gardener is unaware of offer

FS – dishes are free from remainder of food
inertia is rotating at angular speed of 100 rad/s
gardener is aware of offer

Changing property -- dislocating remainder of food until dishes become free of such food
angular speed
gardener becomes aware of offer

First step: 'Reminder of food is *removed* from dishes in 45 min'
'Inertia is *accelerated* from angular speed zero to 100 rad/s in 10 s'
'Gardener is notified that an offer is to be made'

Second step: 'Remainder of food is removed from the dishes in 45 min' *requires* 'Means to remove remainder of food from dishes in 45 min' (*dynamic requirement*)

'Inertia is accelerated from angular speed zero to 100 rad/s in 10 s' *requires* 'Means to accelerate inertia from angular speed zero to 100 rad/s in 10 s' (*dynamic requirement*)

'Gardener is *notified* that an offer is to be made' *requires* 'Means to notify gardener that an offer is to be made' (*dynamic requirement)*

III. Synthesis of quiescent requirements

We use the 'dishes' example to demonstrate in terms of a few, assumed characteristics how the *quiescent requirements* and the enabling/hindering properties can be deduced.

Third step: CP -- 1. force to remove remainder of food from the dishes is applied at an acute angle

2. force to remove remainder of food is sufficient

OP – 1. dishes are fragile

2. there are 30 dishes sized between 10 and 30 cm in diameter

EP1 -- 1. owner of dishes is concerned about breakage

EP3 – 3. there is no supply of electricity

Fourth step: Adverbial phrases

CP,1 -- 'Force to remove remainder of food from the dishes is applied at an acute angle' *requires* 'Means is to exert force' and 'Means is to be applied at an acute angle'

CP,2 – 'Force to remove remainder of food is sufficient' *requires* 'Means is to operate with sufficient force to remove remainder of food'

OP,1 -- 'Dishes are fragile' *requires* 'Means is to operate gently'

EP,1 -- 'Owner of dishes is concerned about breakage' *requires* 'Means is to operate to reassure owner'

Adjectival phrases related to changing object

OP,2 -- 'There are 30 dishes sized between 10 and 30 cm in diameter' *requires* 'Means is to hold 30 dishes sized between 10 and 30 cm in diameter'

Adjectival phrases related to IO

EP,3 – 'There is no supply of electricity' *requires* 'Means is to operate in removing remainder of food from the dishes without electricity'

IV. Selection/invention of product and selection of IO

After selecting an alternative or particular object or product, we go on to the next step.

Fifth step: 'Means to remove remainder of food from dishes in 45 min' ⊃ 'Jet of water is (to impinge on dishes so as to remove remainder of food in 45 min)'

'Means to accelerate inertia from zero angular speed to 100 rad/s in 10 s' ⊃ 'Shaft is (to rotate inertia so as to accelerate it from angular speed zero to 100 rad/s in 10 s)'

'Means to notify gardener that an offer is to be made' ⊃ 'Stream of air, medium, with information is (to hit gardener's ear so as to notify him/her that an offer is to be made)'

Sixth step: 'Jet of water is (to impinge on dishes so as to remove remainder of food in 45 min)' *requires* 'Jet of water is (to be pointed so as to impinge on dishes)'

'Shaft is (to rotate inertia so as to accelerate it from angular speed zero to 100 rad/s in 10 s)' *requires* 'Shaft is (to be supplied so as to accelerate inertia from angular speed zero to 100 rad/s in 10 s)'

'Stream of air with information is (to hit gardener's ear so as to notify him/her that an offer is to be made)' *requires* 'Stream of air is (to be blown so as to hit gardener's ear)'

The sentences generated by the 1st, 2nd, 5th and 6th steps of the design method can be represented as a semantic diagram depicted in Fig.2. This diagram is an expression of the result of considerations implied by the scheme in Fig.1.

In Figure 2 the properties, ip(4,3) and ep(4,4), ep(5,5), enabling/hindering interaction and change of state respectively (Korn, 2000, 2001abc) must be specified. This is done by the quiescent properties and worked out through *linking* the third, fourth and seventh steps.

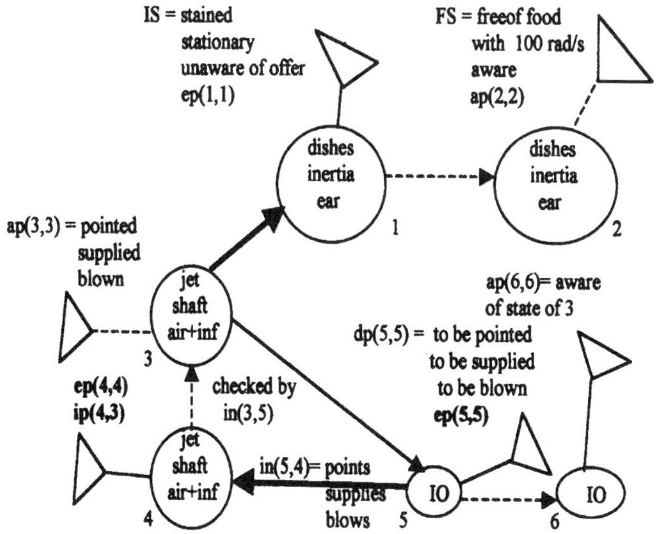

Figure 2. Diagram of change of state, product and interacting objects

<u>Seventh step</u>: Adverbial phrases for ip(4,3)

'Means is to exert force' ⊃ 'Jet of water is to impinge with force'

'Means is to be applied at an acute angle' ⊃ 'Jet of water is to impinge at an angle less than 90^{0}'

'Means is to operate with sufficient force to remove remainder of food ' ⊃ 'Jet of water is to impinge with force sufficiently large'

'Means is to operate gently' ⊃ 'Jet of water is to impinge with smoothly applied force'

'Means is to operate to reassure owner' ⊃ 'Jet of water is to impinge so as to be seen'

Adjectival phrases for ep(4,4)

'Means is to hold 30 dishes sized between 10 and 30 cm in diameter' ⊃ 'Housing for dishes is to be sufficient'

Adjectival phrases for ep(5,5)

'Means is to operate in removing remainder of food from the dishes without electricity' ⊃ 'IO is likely to be hand operated'

The 5th, 6th and 7th steps complete the semantic diagram in Fig.2. and enable :

1. A numerical measure of effectiveness of product to be worked out in relation to the requirements (Korn, 2001a),
2. Certainty factors to be assigned to ip(4,3), ep(4,4) and ep(5,5) (Durkin, 1994),

3. The inferential machine to be deduced leading to the evaluation of certainty of occurrence of outcome i.e. the *desired change of state*.

CONCLUSIONS

We have described a problem solving or design method for the design of products and 'systems' perhaps with sufficient details for its critical appreciation. The method is likely to be applicable to scenarios with 'hard' and/or 'soft' objects. Requirements for product and 'system' are deduced from a *change of state* through a number of steps and are needed for the selection of products and 'systems'. The change of state may be extracted from a scenario or given separately. Products are treated in a unified manner. The containment relation ($\supset$) between the generic and particular terms enables the effectiveness of product in bringing about a change of state to be assessed. Based on the steps detailed specifications can be drawn up.

REFERENCES

Burton, S.H. (1984). *Mastering English grammar*, Macmillan, London.

Checkland, P. (1981). *Systems thinking systems practice*, Wiley, Chichester.

Durkin, J. (1994). *Expert systems*, Macmillan, NY.

Flynn, D.J. (1998). *Information system requirement,* McGraw-Hill, London.

Hubka, W. and Eder, W.E. (1990). Design knowledge, *J of Engineering Design,* 1, pp.97-108.

Jackson, M.C. (2000). *Systems approaches to management,* Plenum, NY.

Korn, J. (1995). Theory of spontaneous processes, *Structural Eng.Review*, v7, n1.

Korn, J. (1996a). Domain-independent design theory, *J.Engineering Design,*v7,n3.

Korn, J. (1996b). Linguistic modelling of situations, *Systemist*, v18, n4.

Korn, J. (2000). Possibility of general systems theory, *44th Annual Meeting ISSS,* 16-22 July, Toronto, Ca.

Korn, J. (2001a). Design and delivery of information, *European J of Information Systems*, v10, n1.

Korn, J. (2001b). Theory and design of complex situations, *in Systems Science in Service to Humanity, Proceedings of the 45th Annual Meeting ISSS,* 8-13 July, Pacific Grove, USA.

Korn, J. (2001c). Linguistic modelling of information and IS, *6th Annual Conference of UKAIS,* 18-20 April, Portsmouth, UK.

Pahl, G. and Beitz, W. (1984). *Engineering Design,* The Design Council, London.

Roozenburg, N.F.M., and Eekels, J. (1995). *Product design: Fundamentals and methods,* Wiley, Chichester.

MODELLING ROLES IN BUSINESS SYSTEMS USING ROLE OBJECTS

Ying Liang

Division of Computing and Information Systems
School of Information and Communication Technology
University of Paisley, Paisley, U.K.
Email: lian-ci0@paisley.ac.uk

INTRODUCTION

Modelling techniques of information systems development have been created regarding data, information and knowledge representation. Data modelling techniques is concerned with data representation (Dewitz, 1996). Object modelling techniques pay attention to information representation (Sommerville, 1996). Now modelling techniques are moving towards knowledge representation in order to distinguish between information and knowledge in information systems. For example, Schoderbek *et al.*, (1990) suggested that information refers to inferentially intended material evaluated for a particular problem, for a specified individual, at a specific time, and for achieving a definite goal. Pressman (2000, p892) said "data is raw information-collections of facts that must be processed to be meaningful. Information is derived by associating facts within a given context. Knowledge associates information obtained in one context with other information obtained in a different context." If knowledge is regarded as different interpretations of the information, we need a specific approach to abstract and represent the knowledge in a business system. Currently many approaches can only represent information in business systems. For example, object-oriented modelling approaches use objects to represent information. However few approaches can represent the knowledge. In order to overcome the lack, I have created an approach that aims to use the role concept for abstracting knowledge from a business system and uses role objects to represent the knowledge. This paper will present the approach in details.

The role concept has been used in research on people's life and society for many years and found useful in examining the expectations that people hold of each other, and to consider each set of expectations as a separate role that one is required to play by virtue of one's position (Ruddock, 1969). Downie found that role-playing was a fact of social life and a role was a cluster of rights and duties with some sort of social function (Downie, 1971). Taylor (1995) found that almost everyone in an office plays multiple

Edited by Ragsdell *et al.*, Kluwer Academic/Plenum Publishers, New York, 2002.

roles, and a system that doesn't support that reality cannot adequately represent the workings of the business system. Multiple roles of a person/thing are identified from different contexts of a business system. Each of roles means a specific interpretation in a specific context, and has their own specific features and responsibilities for a specific goal. A collection of roles of a person/thing in a business system is the knowledge that we want to abstract and represent for that person/thing. Modelling multiple roles of a person/thing therefore is a critical step in designing any business model for empowering people to play multiple roles.

The importance of modelling roles in object-oriented analysis has also been indicated in other literatures such as Pernici (1990), Papazonglou (1991) and Martin and Odell (1992). The current object models, that handle job functions by subclassing a generic *employee* object by position such as *supervisor* and *machinist,* however cannot show that a supervisor could also be a machinist or become another kind of employee, as *supervisor* and *machinist* are *roles* that the employee plays, not *types* of *employee*. It is hardly practical to switch the object back and forth from one object to another each time the person changes roles (Taylor, 1995). Taylor showed an approach to analysing multiple roles of a person (i.e., the person can only play one role at a time) in a company. The approach however could not show the detail of a role. Oestereich (1997) pointed that objects had different roles in different situations and roles should be identified by object modelling. He however did not show how such identification could be done with the model. Gottlob (1996) suggested using role-tree to show roles of individual objects in implementation. The approach however did not support analysis. The industry standard, UML (the Unified Modelling Language) (Booch *et al.*, 1999), can represent single role of a person/thing in a context of a business system as shown in Figure 1 in which a person plays an *employee* role and a company plays an *employer* role in the working context. However, it cannot show multiple roles of that person/thing in different contexts of the same system.

Figure 1. A Class Diagram Containing Role Names

My approach represented in this paper was created in particular for abstracting knowledge by identifying multiple roles of a person/thing in different contexts of a business system at analysis stage. The approach has the following characteristics:

1. Focus of multiple roles of a person/thing in a business system.
2. Use of a role object to represent details of one of the multiple roles of the person/thing. The details include the knowledge the person/thing needs to know and the responsibilities the person/thing needs to take with that role.
3. Representation of trace of multiple roles of a person/thing in terms of a role network.
4. Provision of a role diagram and an extended class diagram for representing details of multiple roles of a person/thing in a business system.
5. Provision of a role modelling process for identifying and representing multiple roles.

This paper first defines role objects and their categories, and then defines the role diagram and the extended class diagram. It then shows and demonstrates the role modelling process, and finally concludes and summarizes the approach.

ROLE OBJECTS

I define a role object as follows:

- A *role object* represents one interpretation (i.e., a role) of a person/thing in a context of a business system, at a time. A role object has a name, which should be meaningful to the interpretation of the person/thing, and role features and role responsibilities. For example, the role object *student* represents an interpretation of a person in a learning context, that is, the person has registered in a course and has a matriculation number assigned by a university.
- *Roles features* represent the data that the interpretation includes and the role must hold in the context. For example, *matriculation number* is a role feature of *student*.
- *Role responsibilities* represent the services that the interpretation includes and that the role should provide in the context. The services are provided for all of the contracts the role supports (Wirfs-Block, *et al.*, 1990). For example, *select course* is a role responsibility of *student* in the context.

CATEGORIES OF ROLE OBJECTS

In order to trace roles of a person/thing in a business system, I classify the role objects in three categories as follows:

1. *Fundamental-role object,* which represents a primary interpretation of the person/thing (e.g., *Customer*) in all contexts of a business system, at all time.
2. *Initial-role object*, which represents a first extended interpretation of the primary interpretation in a different context. For example, *Non-credit customer* is an initial-role object that shows a first extended interpretation of *Customer* in a non-credit context (i.e., the customer has no credit as a new customer and must pay for the first purchase at once).
3. *Transfer-role object*, which represents a further extended interpretation of the primary interpretation in another different context. A transfer-role object should appear after either an initial-role object or another transfer-role object. For example, when *Non-credit customer* makes another request after completing the first purchase successfully its role is changed to a transfer-role object *Credit customer* in the credit context (i.e., a credit is assigned to the customer and allows to pay for the purchases within a month).

Role Network

A role network for a person/thing in a business system consists of a set of role objects and role transformations. The characteristics of role objects in a role network are as follows:

1. Each role object must have its own role features and/or role responsibilities in a context.
2. Initial-role objects and transfer-role objects may share role features and role responsibilities of the fundamental-role object in a role network. For example, *Non-credit customer* and *Credit customer* share *Customer*'s role features *customer name* and *address* and role responsibilities *make request* and *provide name and address*.
3. A transfer-role object may share role features and role responsibilities of the role object it is transferred from. For example, *Credit-restrict customer* may share *Credit customer*'s role features and role responsibilities.

I created a new role diagram and extended the class diagram in UML (Booch *et al.*, 1999) for representing the role network.

Role Diagram

A role diagram, as shown in Figure 2, shows a role network with roles and role transformations for a person/thing in a business system, at a high level.

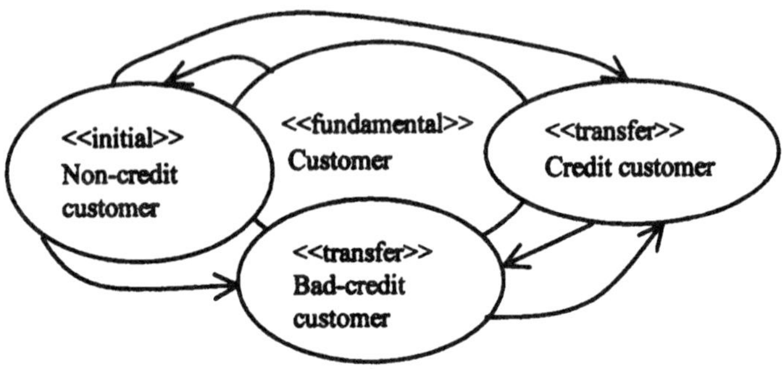

Figure 2. A Role Diagram

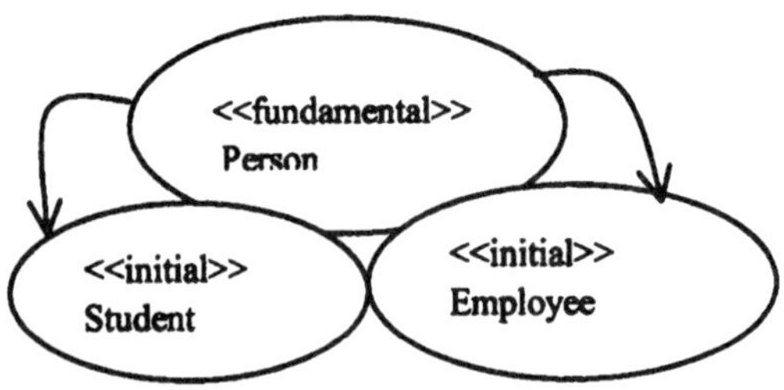

Figure 3. Parallel Roles *Student* and *Employee*

A *role diagram* consists of

1. A set of *roles*, each of which is represented by an oval containing a role name with a stereotype (either <<fundamental>>, or <<initial>>, or <<transfer>>). When a role shares the role features and/or role responsibilities of another role, its oval overlaps over the other's oval (e.g., *Credit customer* overlaps over *Customer*). A person/thing may play more than one role at the same time. For example, a person

who plays *Employee* in production line context may also play *Student* in a staff development context in a company in the same period. Such parallel roles are represented by two adjacent ovals, as shown in Figure 3.

2. A set of *role transformations*, each of which is shown by an arrow. A role transformation shows a change from one role in a context to another role in another context (e.g. from *Person* to *Student* in Figure 3).

Extended Class Diagram

I extend the class diagram in UML with role object and role transformation for representing roles including role features and role responsibilities, and role transformation, as shown in Figure 4:

1. *Role object*, represented by a rectangle box consisting of four sections. The top section contains the role name with a stereotype (either <<fundamental>>, <<initial>> or <<transfer>>) representing the role type. The second section contains the role names of shared role objects in *italic*. The third section contains the role features of the role object. The bottom section contains the role responsibilities of the role object.
2. *Role transformation*, represented by a *dark arrow* (e.g., from *Customer* to *Non-credit Customer* in Figure 4).

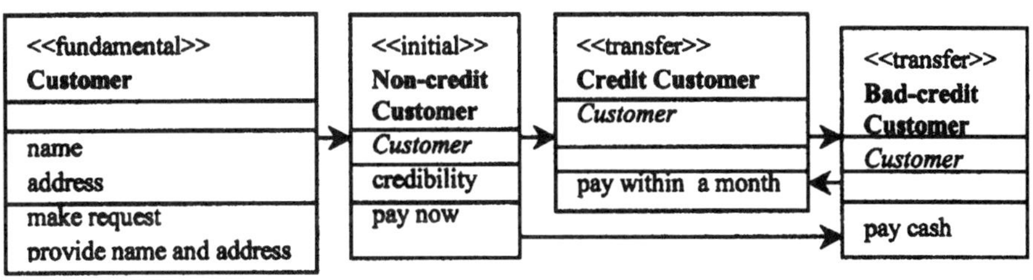

Figure 4. An Extended Class Diagram with Role Objects and Role Transformations.

The extended class diagram is drawn when specifying the details of roles using role objects.

PROCESS OF MODELLING ROLES WITH ROLE OBJECTS

The following steps are suggested for identifying and representing roles of people and things in a business system with role objects in terms of the role diagram and the extended class diagram.

Step 1. Identify people and things in the business system

In this step, the analyst contacts people (e.g., clerks, manager) working in a business and collects the details of the system (e.g., business processes, documents, customers, assistant staff) from them. All people and things that are used and involved in the system should be appreciated and captured.

Step 2. Decide the roles of people and things identified

In this step, the analyst needs to decide the roles of people and things identified in step 1 by considering different interpretations of them in different contexts of the business system. The analyst also needs to validate the interpretations with units/department of the organization and keeps the ones that are significant in the system. In general, a significant interpretation of a person/thing implies a role that the person/thing plays in the system. For example, in a video sales business system, a person sends a request to a video sales company for buying videos. In the sales department, the sales clerk interprets the request as an order when received. In the account department, the accountant interprets the request as an invoice when clearing payment. In the warehouse, the delivery clerk interprets the request as a receipt when delivering videos. There are therefore three different interpretations of the request in three different contexts of the business system. If all three interpretations are significant and useful in the system, they mean three roles of the request in the business: *order*, *invoice*, and *receipt*.

Step 3. Identify role transformations

In this step, the analyst identifies role transformations between the roles decided in Step 2. A role transformation is often needed when a person/thing is given a different interpretation in a different context. The type of each of roles should be also decided when role transformations are identified.

A role diagram such as Figure 5 is produced in this step for showing roles, roles types, and role transformations for each people/things in the system.

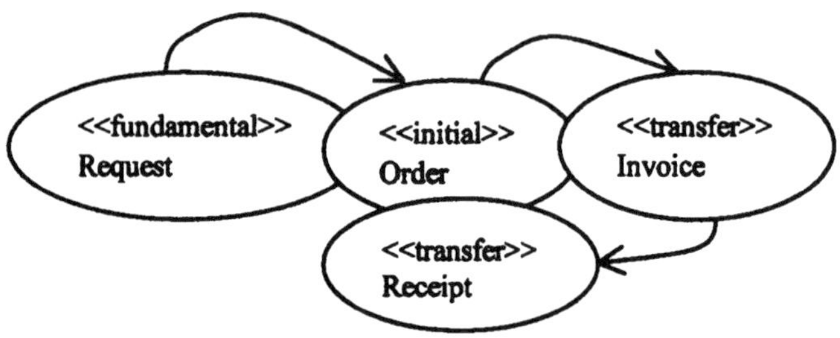

Figure 5. A Role Diagram for the Request in the Video Sales Business System

Step 4. Identify role features and role responsibilities

In this step, the analyst needs to consider which features and role responsibilities are part of an interpretation of a person/thing in a context and are important to that role. For example, *order number*, *date* and *order items* are part of interpretation of *order* in the video sales system and are important to an order. They are therefore the role features of *order*. Role responsibilities of a role can be identified in the similar way. For example, *provide order number*, *provide order date*, and *provide order items* are identified as the role responsibilities of *order*.

Sharable role features and role responsibilities of roles should be also identified and shown in the role diagrams in this step. For example, Figure 5 shows that *Order* shares the role features and role responsibilities of *Request* and that *Invoice* and *Receipt* shares the role features and role responsibilities of *Order* in the video sales business system.

Step 5. Specify the roles in terms of role objects

In this step, the analyst generates an extended class diagram for specifying role details identified in the above steps at a low level, as shown in Figure 6.

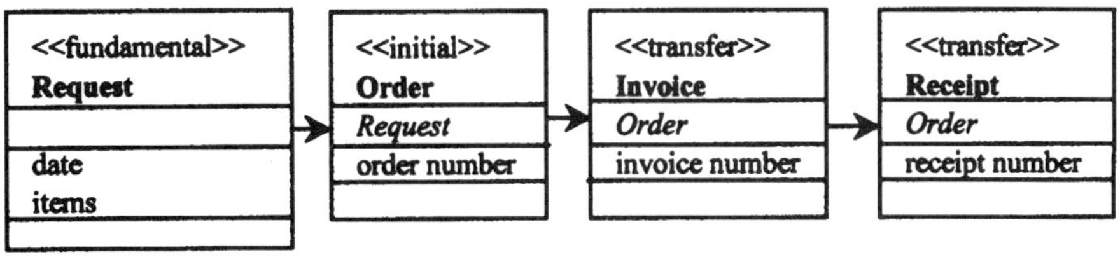

Figure 6. An Extended Class Diagram for the Video Sales Business

Step 6. Validate role diagrams and the extended class diagram

In this step, the analyst needs to check the role diagram and the extended class diagram against the interpretations of people and things in the business system. The roles, role features and role responsibilities defined must be consistent and complete with the interpretations. The two diagrams should be also reviewed by people who are working, or units/departments, in the organization in order to confirm that different interpretations of people and things in different contexts of the business system are fully, accurately and correctly represented.

CONCLUSIONS

The fundamental question of knowledge representation of whether there are primitive representational structures, richer than just objects, relationships, and so forth, that pervade human knowledge representation (Adeli, 1990). This paper presents a new approach to using a role network to represent knowledge on information (i.e., different interpretations of objects in different contexts) with the role concept and role objects. The role network is a presentation structure that is richer than objects and relationships.

In comparison with many current object modelling approaches, this approach has following differences and advantages:

1. It separates data, information and knowledge in a business system and represents them using different notations.
2. It abstracts knowledge from a business system, using the role concept. In particular, it models the roles of a general person/thing such as *customer*, rather than the roles of a specific person/thing such as *Jane Smith*, in a business system.
3. It represents roles using a role network instead of a role-tree so that iterative roles of a person/thing in a business system can be modelled.
4. It provides a role diagram and an extended class diagram for representing role objects in a business system at different levels.
5. It can identify role features and role responsibilities for a role and represent them explicitly in the extended class diagram.
6. It can abstract and represent common role features and role responsibilities based on roles instead of the same types of people/things in a business system. For

example, *invoice* and *receipt* share role features and role responsibilities of *order*, but three of them are different types of things in the video sales system.

Schank and Abelson (1977) indicated that information seems to be organized for everyday activity on the basis of goal-oriented tasks and strategies. The goals of an information system represent what the user requires from the system and they can be useful and helpful in identifying and selecting significant data, information and knowledge for the system (Liang, 2001). My future research will consider the impact of goals on roles in a business system and provide a way of modelling roles based on goals in terms of role objects by systems analysis.

REFERENCES

Adeli, H. (1990). *Knowledge Engineering, Vol.1 (Fundamentals),* McGraw-Hill, London.

Booch, G., Rumbaugh, J. and Jacobson, I. (1999). *The Unified Modelling Language User Guide*, Addison-Wesley, Reading, Massachusetts.

Dewitz, S.D. (1996). *System Analysis and Design and the Transition to Objects*, McGraw-Hill, London.

Downie, R.S. (1971). *Roles and Values: an introduction to social ethics*, Methuen, London.

Gottlob, G. (1996). "Extending Object-Oriented Systems with Roles," *ACM Transactions on Information Systems*. 14(3):268-296.

Liang, Y. (2001). "Actor-Led Object Modelling for Requirements and Systems Analysis," in *Proceedings of the 7th International Conference on Object-Oriented Information Systems*. Springer, London.

Martin, J. and Odell, J.J. (1992). *Object-oriented Analysis and Design*, Prentice Hall, Englewood Ciffs, New Jersey.

Oestereich, B. (1997). *Developing Software with UML-Object-Oriented Analysis and Design in Practice*, Addison-Wesley, Harlow, England.

Papazoglou, M.P. (1991). "A Methodology for representing multifaced objects," in *Proceedings of the International Conference on Database and Expert Systems Application*, ACM, New York.

Pernici, B. (1990). "Objects with Roles", in *Proceedings of the ACM Conference on Office Information Systems, ACM*, New York.

Pressman, R.S. (2000). *Software Engineering: A Practitioner's Approach* (5th edition), McGraw-Hill, London.

Ruddock, R. (1969). *Roles and Relationships*, Routledge and Kegan Paul Ltd., London.

Schank, R.C., and Abelson, R.P. (1977). *Scripts, Plans, and Understanding: An Inquiry into Human Knowledge Structures*, Hillsdale, New Jersey.

Schoderbek, C.G., Schoderbek, P.P., and Kefalas, A.G. (1990). *Management Systems: Conceptual Considerations* (4th edition), Business Publication, Dallas.

Sommerville, I. (1996). *Software Engineering* (5th edition), Addison-Wesley, Wokingham.

Taylor, D.A. (1995). *Business Engineering with Object Technology*, Wiley and Sons, New York.

Wirfs-Block, R., Wilkerson, B., and Wiener, L. (1990). *Designing Object-Oriented Software*, Prentice Hall, Englewood Ciffs, New Jersey.

TOWARDS A DEFINITION OF THE "INFORMATION BEARING CAPABILITY" OF A CONCEPTUAL DATA SCHEMA

Hongwei Xu and Junkang Feng

School of Information and Communications Technologies
University of Paisley
High Street, Paisley PA1 2BE, United Kingdom
hongwei.xu@paisley.ac.uk
jun-kang.feng@paisley.ac.uk

INTRODUCTION

A fundamental question in information systems (IS) and semantic theories of information is how data in an information system represents information. It would appear that little has been said in the IS literature on how the conditions for data to represent information should be formally defined. We will address this question by defining the Information Bearing Capability (IBC) of a data schema. In this paper, we will articulate this capability. This notion plays a central role in a semantic analysis of a conceptual data schema.

Our approach to defining the notion of IBC is to look at both syntactic and semantic levels by employing the ideas of semiotics. We will reveal the relationship between the two levels. To begin with, we will define the Principle of IBC, which will make use of a set of new notions that we will present, namely the *primary meaning*, *implied meaning*, *navigation uniqueness*, *relevant* and *irrelevant topological connection* that are involved in a conceptual data schema. In this paper, we will use the popular entity-relationship (ER) data model as a vehicle to present our ideas.

When the semantic level and relevant notions concerning meanings are considered, the ideas and methodological work of soft systems thinking are highly relevant. We will report our "real world" experience on this point briefly.

RELATED WORK

The closest work to ours in the literature is that concerning the notion of "information capacity" (Hull 1986; Miller *et al.*, 1993; Kwan and Fong, 1999). This work is mainly on schema transformation and integration by using the mathematical notion of mapping and focus on developing correctness criteria (Miller *et al.*, 1993), for example, verifying the transformed schemas without information loss for schema integration by means of denotational semantics (Kwan and Fong, 1999). This approach would seem to have concentrated on syntactic level of the problem.

Also, this approach does not seem to be able to address the issue of the *quantity of information* that a data construct represents. Moreover they do not seem to have provided a systematic semantic analysis on data schemata. We have not found so far work explicitly on the semantic correspondence between schemata.

From a philosophical point of view, talking about information capacity without a human and social setting as if it were absolutely objective can be misleading, to say the least. For example, two identical data schemata created by two different organisations can mean very differently. It would be absurd to declare that they had the same information capacity according to the rules that were given in the aforementioned references.

THE PRINCIPLE FOR DATA TO REPRESENT INFORMATION

Data represents (bears) information – The fundamental question to be addressed here is how a perceived real world fact, a relationship or a notion (more generally *a semantic relation*), say *f*, can be represented by a construct e.g., a path in an ER schema, in a data system, say *e*?

First of all, we should define *f* and *e* as follows:

f – is a semantic relation, which can be defined as a particular relationship between objects in the perceived real world or in imagination;

e – is a data construct, which can be defined as a node or a path in a graph that represents a data schema.

Here we need to introduce the notions of *primary meaning* and *implied meaning* of *e*. Before introducing those two notions, we should understand that what *topological connections* refer to. Using the ideas of *syntactics* and *semantics* of semiotics (Stamper 1997; Andersen 1997), it can be seen that there are two types of connections between entity instances. The connections that are made possible by the topological structure (i.e., a syntactic level formation of signs) of an ER schema can be termed *topological connections* regardless of their meaning (Feng and Crowe, 1999). The connections independent of a data system are termed *semantic relations*.

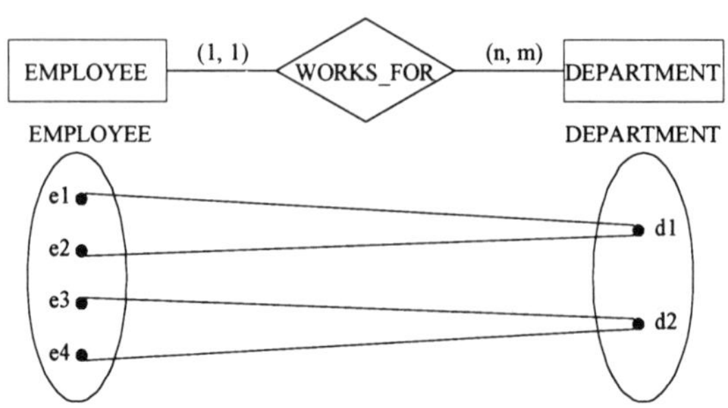

Figure 1. A Topological Connection vs. a Semantic Relation

For example, suppose that an employee works for a department is a true proposition, we say there is a semantic relation between the employee and the department. In database design, a data schema (see Figure 1) must enable this semantic relation to be represented by a topological connection. The topological connection between two entities exhibited in Figure 1 corresponds to the semantic relation.

Primary meaning and implied meaning – there are certain sets of semantic relation(s) that the topological connections of a path in an ER schema can always unconditionally and directly represent in a particular human and social setting, we will call these semantic relations *primary meaning* of *e*.

For example, the primary meaning of the path in Figure 2 is that a supplier supplies a part, and a project uses a part. These are the semantic relations that the topological connections in the path can always directly represent with no conditions.

Given certain conditions on both the topological level and the semantic level, a topological connection may represent a semantic relation that is beyond the primary meaning. We will call this semantic relation *implied meaning* of *e*. An implied meaning of the path could be a binary relationship. For example, a supplier supplies a project in Figure 2 if the following business rule is true, namely, if a supplier supplies a part and the part is used by a project, then the supplier supplies the project.

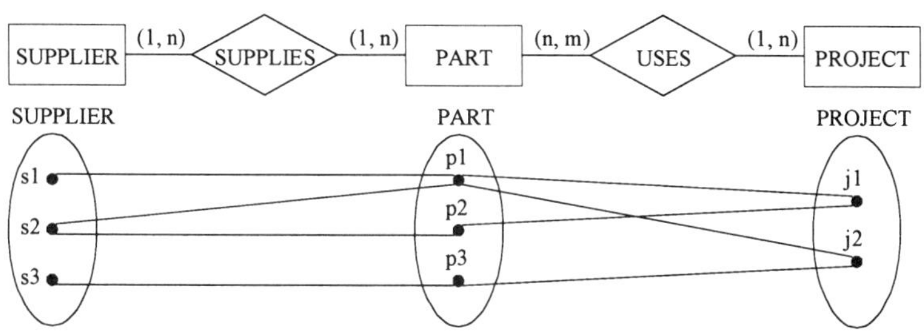

Figure 2. Primary Meaning vs. Implied Meaning of a Path

Conditions for Data to Represent Information

The following would seem to be necessary and sufficient conditions for *e* to represent *f*:

1. *e* must correspond to *f*.
2. *e* must be distinguishable from the rest of the data constructs by the only means available to the data system.

The first condition requires that *e* be looked at in isolation, that is to say, *e* is considered disregarding any other connections the entity instances in *e* have or may have, and furthermore either the *primary meaning* of *e* or one of its *implied meanings* equals *f*.

For the second condition, it is not sufficient for *e* to be distinguishable if we assume that we already know the identities of the entity instances that are involved in *e*. The question is actually whether we can reach all the entity instances that are involved in *e* with certainty. To answer this question, we propose to define a notion of *navigation uniqueness*.

For a path from node A to node B between which there may be other nodes, navigation from A to B through the path is unique if there is no "plurality of joins" (Codd, 1970) from A to B.

Navigation uniqueness holds if no *irrelevant* topological connections exist. The following two paragraphs will make it clear.

When one considers whether *e* is distinguishable, the notions of *irrelevant* and *relevant* topological connection would seem useful. Against a set (i.e., a type) of semantic relations, a topological connection is an irrelevant topological connection if it does not correspond to any semantic relation of this type. On the contrary, given a set of semantic relations, if it corresponds to an instance of the set, then a topological connection is a *relevant* topological connection with regard to the set.

For example, back to Figure 1, an accountant works for a finance department is a set of semantic relations. Assume that in the topological connection (e1, d1), e1 is a salesman, d1 is a sales department, then this connection does not correspond to any semantic relation of an accountant works for a finance department, so it is an irrelevant topological connection against the given set of semantic relations. If e3 is an accountant, and d2 is a finance department, then a topological connection (e3, d2) is able to correspond to an instance of the given set. So this topological connection can be called a relevant topological connection against the given set.

THE DEFINITION OF IBC

Based upon the Principle and the notions presented in the last section, we will now give our definition for the IBC. We will argue that the two aspects, namely *what* and *how*, should be looked at in order to define the IBC.

What Information can be Represented

The *what* aspect of the IBC can be further divided into the following:

1. Quantity - The *quantity* of information that a data schema can represent
2. Content - The *content* of information that a data schema can represent
3. Static IBC - The *static* quantity and content of information that a data schema can represent
4. Dynamic IBC - The *dynamic* quantity and content of information that a data schema can represent

Following Dretske (1981), when information is carried by a *signal* in the communication process of an information flow, the quantity of information that the *signal* carries, i.e., the amount of information transmitted, can be measured in a mathematical way. Moreover, the quantity of information that is carried by a signal imposes constraints on the content of the information that is carried by the signal.

For example, from an information transmission point of view, a *topological connection* should be viewed as the R, namely the *information receiver*, and the sign or signal may or may not carry information regarding a *real world relation*, which is the S, the *information source*. The amount of information transmitted from S to R can be denoted as Is(R); the amount of information that is created by or associated with a state of affairs can be denoted as I(S). To be able for a state of affairs at R to reflect faithfully what happens at S, one necessary condition is that the quantity of information that R receives from S is the same as the quantity of information that is created at S, namely Is(R) = I(S). But this condition is not sufficient for R to represent S.

The aforementioned Principle of the IBC was discovered for addressing the issue of the *content* of information that a data schema can represent, in addition to the issue of information quantity. Based upon the Principle, we can develop a set of rules for inferences on the two aforementioned conditions for data to represent information.

For example, inference rule $I(X) \supset Y, I(Y) \supset Z \models I(X) \supset Z$. This rule means that if the informational content of a state of affairs X includes Y and the informational content of a state of affairs Y includes Z, then the informational content of X includes Z. We will present a set of such inference rules in the near future.

Static IBC refers to the IBC of the data constructs considered at the *type* level. This means that we only consider whether a data construct *e* is capable of representing a semantic relation *f*, and that we do not consider what is captured by the change of values of *e*. For example, the change of the values of a woman's surname attribute (a field), which is a data construct, may represent the change of her marital status, which is a piece of information.

Dynamic IBC is an elaboration of the Static IBC, which refers to the information that is represented by the changes of the values of data constructs, as the example above shows. This idea came about due to the inspiration of the notion of "slowly changing dimensions" of a data warehousing model (Adamson and Venerable, 1998, p.12). This means that there exist value-based relationships between semantic relations and data constructs. This idea enables us to examine the IBC of a data schema for temporal databases.

How Well is the Information Represented?

The *how* aspect may be measured by whether it facilitates information storage through data and information derivation from data. This in turn may be defined in terms of:

1. The number of data elemental constructs involved
2. The length of a path that is involved
3. The operation types and numbers
4. How a path in a data schema is specified

Data elemental constructs (i.e., entities, attributes, and relationships) in totality should represent all *useful* or *valid* information for an application domain. The number or the size of data elemental constructs involved can be an issue depending on a certain database application requirements. From the point of view of performance, the designer should create a valid and minimal data schema for the given application requirements. To achieve that goal one should employ the Principle of the IBC.

For example, the structures of the data schemata in Figure 2 and Figure 3 are different. The relationship between a supplier and a project in figure 2 has no capability of telling us any useful information on which supplier supplies which project, because it does not satisfy the Principle of the IBC. More specifically, navigation (see Figure 2) from an instance of supplier s2 to an instance of project j1 is not unique. The ternary relationship in Figure 3 solves this problem, because all the conditions embodied by the

Principle for data to represent the information on which supplier supplies which parts to which projects are satisfied. Navigation between nodes is unique.

As for the length of a path that is involved in representing a semantic relation, consider the length of a path in those two schemata. The length of the path (supplies, uses) is great than one in Figure 2, and we might use an implied meaning (i.e., a suppler supplies a project) of the path to represent a semantic relation. In comparison, the data schema in Figure 3 is minimal for meeting all the requirements regarding *supplier-part-project* relationships. The length of the path is one and the numbers of the elemental constructs data required is also minimal compared with the schema in Figure 2.

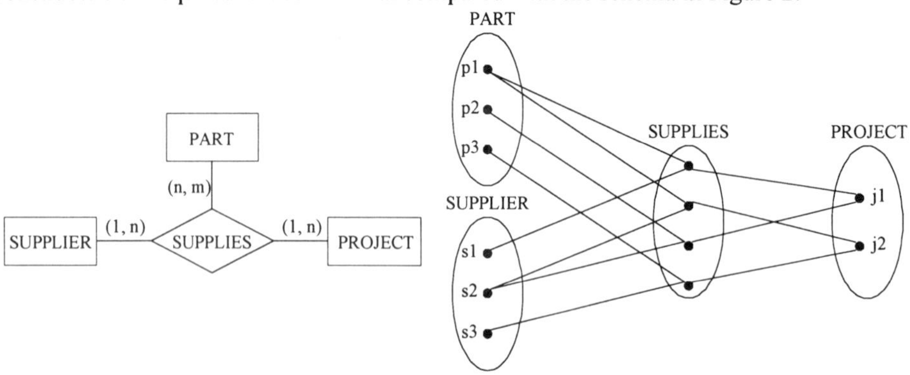

Figure 3. The Meaning of Path and Navigation Uniqueness

The Principle of the IBC helps specify the relationship between information and data. We can use it to arrive at a simplified or optimised data schema for bearing all the information required prior to the choice of database management systems (DBMS). For example, one consideration when normalizing relations is that the more tables to be created, the more likely it is that the *join* operation will be needed, which would worsen the performance of the database. The structure of a data schema determines the number of relations in certain normal forms to be created, which affects the operation types and numbers that would be required for deriving information from data.

One of the core tasks of our research is to develop some technique for specifying a path such that we can analyse both the *what* and *how* aspects of the IBC. That is to say, by specifying a path, we want to be able to accurately specify a set of topological connections (this point relates to the *what* question), and to be able to show how difficult to access this set of topological connections (this is concerned with the *how* question). This work is still on going.

THE RELEVANCE OF SOFT SYSTEMS THINKING TO ADDRESSING IBC

The development of the notions, arguments and reasoning above was made under a tacit assumption, namely the human and social setting was not an issue. If it were not the case, then we should preferably use soft systems thinking to look at all those notions, such as *semantic relation*, *primary meaning*, *implied meaning*, the scope of the *information source*, and that of the *information receiver*. While these notions may be generic, the instantiations of them for an ill-structured and non-trivial application domain could be a complex and difficult process, as our real world consultancy work has shown. Soft systems thinking leads us to anticipate that much work is needed in identifying semantic relations that the client concerns, and the correspondence between the semantic relations and those constructs in various data systems that they used to use and that we are designing with the client. We do spend a significant amount of time and make continuous efforts on resolving these issues throughout the projects. In the process, Soft Systems Methodology (Checkland and Scholes, 1990) is used to reach mutual understanding and consensus between the client and us.

THE BENEFITS OF THE NOTION OF IBC

Our theory about IBC should make a valuable contribution to schema minimisation, completeness, transformation, integration, and optimisation.

Based upon IBC, we can introduce the notion of *data construct equivalence*, which would be useful for comparing data schemata in the process of schema transformation and integration (Batini *et al.*, 1986). We can also look at the IBC of various data models.

The notion of IBC will help us study the specific rules for the structure, the constraints, and the operations of a data model upon which data schemata are based.

The theory lays the foundations for developing further concepts in handling heterogeneous data schemata. Those concepts include, for example, the notion of *IBC equivalence* in schema transformation and integration the notion of *minimal cover* in schema minimisation and optimisation, etc.

Following Dretske (1981), from the point of view of *informational relationship*, the data system should be seen as a *message system*, which is, in fact, the sates of affairs of the information receiver. These states of affairs must be discrete and mutually exclusive, otherwise there would be no question of *surprisal*. The notions defined in the second condition in the Principle of the IBC seem to be applicable to any further semantic analysis such as those that concern *informational content* (Dretske, 1981).

For practical database design, which considers the performance of a database, IBC would seem to be a good idea to offer, especially to query optimisation.

SUMMARY AND FUTURE WORK

We have defined the conditions for data to represent information in the form of the Principle. We have used the IBC concept to articulate the relationship between semantic relations and data constructs. We have presented a framework to capture the *what* and *how* aspects of the IBC. In relation to IBC, we also presented the notions of *primary* and *implied meaning*, *topological connection*, and *semantic relation*. Our future work on this topic will be mainly further elaborations of the ideas presented in this paper including the formulation of the notions, conditions, and rules. The work should be seen as an investigation into the fundamental question of how data should be organised in terms of structure rules, constraint rules, and operations by using computing and communications technologies from the viewpoint of bearing information.

REFERENCES

Adamson, C. and Venerable, M. (1998). *Data Warehouse Design Solutions*, John Wiley and Sons, New York.

Andersen, P. B. (1997). *Theory of Computer Semiotics*, Cambridge University Press, Cambridge.

Batini, C., Lenzerini, M. and Navathe, S. B. (1986). "A Comparative Analysis of Methodologies for Database Schema Integration", *ACM Computing Surveys*, 18(4):323-364.

Checkland, P. and Scholes, Jim. (1990). *Soft Systems Methodology in Action*, John Wiley & Sons, Chichester.

Codd, E.F. (1970). "A Relational Model of Data for Large Shared Data Banks", *Communications of the ACM*, 13(6):377-387.

Dretske, F. I. (1981). *Knowledge and the Flow of Information*, Basil Blackwell, Oxford.

Feng, J. and Crowe, M. (1999). "The Notion of 'Classes of a Path' in ER Schemas", in *Proceedings of Third East European Conference on Advances in Databases and Information Systems, ADBIS'99*, Springer, Berlin.

Hull, R. (1986). "Relative Information Capacity of Simple Relational Database Schemata", *SIAM Journal of Computing*, 15(3):856-886.

Kwan, I. and Fong, J. (1999). "Schema Integration Methodology and Its Verification by Use of Information Capacity", *Information Systems*, 24(5):355-376.

Miller, R. J., Ioannidis, Y. E., and Ramakrishnan, R. (1993). "The Use of Information Capacity in Schema Integration and Translation", in *Proceedings of the 19th International Conference on Very Large Data Base*, Morgan Kaufmann, San Francisco.

Mingers, J. (1995). "Information and Meaning: Foundations for an Intersubjective Account", *Infomation Systems Journal*, 5:285-306.

Stamper, R. (1997). "Organisational Semiotics", in *Information Systems: An Emerging Discipline?*, (J. Mingers and F.A. Stowell), McGraw-Hill, London.

Section Ten

Student Papers

ENACTING PLATFORMS TOWARDS CRITICAL SYSTEMS PRACTICE: REFLECTIONS FROM AN ORGANISATION-BASED ACTION RESEARCH

Alvaro Carrizosa

University of Lincolnshire and Humberside
Lincoln School of Management
Brayford Pool Lincoln, LN6 7TS, UK.
E-Mail: acarrizosa@lincoln.ac.uk

INTRODUCTION

This paper is the result of reflection and debate, which has been reciprocally enriched by theory and practice. The author of this paper entered into a real-world situation and aimed both at improving it and acquiring knowledge about the experience. He became, for a period of 3 years, involved in the flux of 'real-world problems' within an engineering company that invited him to do research by using systems ideas in practice. Thus, the research endeavour constituted an "organization-based action research" (Checkland and Scholes, 1999, pA40).

This organisation-based action research produced a series of findings contributing to the study of the process whereby different systems methodologies, methods, tools and techniques are used in combination. This process is known as Critical Systems Practice (CSP) (Jackson, 2000). The study yielded what Checkland and Holwell (1998, p16) refer to as "defensible generalisations" particularly regarding the enactment of 'platforms' for legitimising CSP. In responding to the complexity and heterogeneity of the problem situations at hand platforms adopt different 'forms'. This study supports the enactment of platforms to enhance the *process* of CSP and the *role of the researcher* in the process of intervention. In other words the findings of the research presented in this paper concentrate mainly on the "transferable problem solving capacity" (Jackson, 2000, p14), namely platforms.

In order to 'delineate' this notion of platforms three sections are presented. The first section presents the general research approach to set the conceptual and methodological background of the research. The second section provides a description of a series of characteristics attributed to platforms. And the third section discusses relevant aspects emerging from platforms in the light of the operationalisation of CSP.

Edited by Ragsdell *et al.*, Kluwer Academic/Plenum Publishers, New York, 2002.

GENERAL RESEARCH APPROACH

According to Checkland and Holwell (1998), the constitutive elements of any research are: an area of concern, a methodology to go about the study and a framework of epistemological principles in terms of which what counts as learning. "Particular linked ideas F are used in a methodology M to investigate an area of concern A. Using the methodology may then teach us not only about A but also about the adequacy of F and M" (ibid, p13). Therefore, they suggest a cycle and a process, which are aimed at generating a series of findings about the area of concern A, the framework of ideas F and the methodology M used. The A in this study consisted in testing and developing systems thinking ideas and methodologies in practice, tackling real-world management problems. Critical Systems Practice (CSP) was the methodology to be used and this carried the implication that the framework of ideas F supporting the intervention became Critical Systems Thinking (CST).

The research produced findings at these three different levels, however this paper concentrates on the findings generated at a methodological level. The study was not concerned with testing hypotheses, rather with generating research themes within which lessons could be sought. In this case the research themes relevant to M and A were: pluralism, improvement, modes of research and the role of the agent, which are identified by Jackson (2000) as relevant issues for CSP.

The crucial elements in a research approach of this kind according to Checkland and Holwell (1998, p12) are: (a) a collaborative process between researcher and those involved in the situation, (b) a process of critical inquiry, (c) a focus on social practice, and (d) a deliberate process of reflective learning. In this order of ideas, the general aim was to inquire in the area of concern, A, by using different methodologies, methods, tools and techniques as required being consequent with the methodology M (CSP) and the framework of ideas F (CST). Naturally, since this research was concerned with a real-world situation it continually evolved, becoming an ongoing cycle.

This cycle involves: first, stating the particular themes in which the researcher is interested; second, declaring in advance the framework of ideas F and the methodology M; third, entering into the real-world problem situation A; fourth, taking part in the change process in the situation; fifth, reflecting upon the experience; and finally, recognising what counts as learning derived from the findings and becoming new research themes. Naturally, this cycle is not a one-loop cycle; it is an ongoing process.

Table 1. Research Modes

Mode 1		Mode 2
Methodology-driven	Versus	Situation-driven
Intervention	Versus	Interaction
Sometimes sequential	Versus	Always interactive
Methodology as external recipe	Versus	Methodology an internalised model

(Adapted from Checkland and Scholes, 1999, pA36)

Another element to be considered as part of the general research approach is related to different ways of using methodology. The debate revolves around the distinction between two modes: Mode 1 and Mode 2 (Checkland and Scholes, 1990, 1999). This notion was initially used by Checkland and Scholes (1990) in Soft Systems Methodology and transferred by Jackson (2000) to CSP. The distinction between the two modes is summarised in Table 1.

The research was characterised by Mode 2 type of research. The emphasis was on the situation rather than on the methodologies; the interaction with those involved in the problem situation was considered of crucial importance; the methodologies, methods, tools and techniques were always to be used interactively rather than sequentially; and methodologies were internal models to be used during and after the intervention to help those involved reflect on what was happening.

Methodologies, methods, tools and techniques, in this context, were not considered relevant until they were effectively used, meaning they were worked out with actors in the problem situation at hand, defining a Mode 2 type of research. Thus the research did not contemplate external methodologies to be applied as guiding the research process, as would be the case in a Mode 1 research.

Bearing this research approach in mind the emergence of the notion of platforms constitutes the methodological finding of the research endeavour. The next section describes the emergence of these 'devices'.

THE EMERGENCE OF THE PLATFORMS

In this section defensible generalizations about the operationalisation of CSP are presented. The intention is not to create a new methodology or meta-methodology for the application of CSP but to 'delineate' instances that contribute to the operationalisation of CST by legitimising it.

A set of characteristics ascribed to platforms is presented. The different experiences, or projects, undertaken in the organisation-based action research, such as the Thinking Space (TS), the Book and the Wall Workshops (WW), which determined different platforms, in line with CSP have been described elsewhere (Carrizosa, 2000). Upon reflection it was possible to uncover five characteristics that account for the enactment of platforms and for which the enactment of platforms can be accountable. This relates to five constitutive intertwined forms of engagement: (a) engaging in doing organisation-based action research, more specifically, Mode 2 research; (b) engaging in 'problem-solving-setting' in contrast to 'problem-solving' as such; (c) engaging in 'creative explorations' where it is possible to question and change organisational 'ground rules'; (d) engaging in developing 'structured processes of thinking', that is structuring ideas for action; and (e) engaging in 'on-going' processes of enacting and re-enacting platforms.

Engagement In Doing Mode 2 Research

Mode 2 research was identified as a major characteristic of platforms. Members of the organisation were highly appreciative of the fact that the research was situation-driven, encouraged making sense of what was happening with reference to a critical systems approach, and reflected on the use of different methodologies, methods and tools accordingly. Indeed, throughout the research experience this type of engagement constituted a key factor for the emergence of platforms. Mode 2 research best supported this type of engagement since it emphasises the problem situation and not the methodology. This primarily implied an engagement with those involved and not only a concern with F and M.

Engagement with those involved meant the emphasis was on interaction with those people who were part of the problem situation, not on intervention. Consequently the researcher was not seen as an external expert but as an active participant. In practice the concern was always to keep the 'interaction' among participants active. This meant always being 'in touch' with the problem situations affecting the actors, an essential part of the TS, the Book and the WW. Thus, participants perceived research as a 'legitimate' problem-solving activity.

The process of interaction among actors and agents constituted the platforms, which were developmental and constantly responding to the problem situations at hand by engaging in doing research. In this context, nonetheless, the understanding of research was pragmatic, because in the organisation, actors were concerned with tackling problem situation and not necessarily with contributing to a particular body of knowledge or framework of ideas (F). However, there is no doubt that through the attention placed to the commitments of CSP, valuable insights occurred, which assisted local improvement.

Another feature of the research worth stressing is that within the platforms the difference between practice and research was not made apparent. Enhanced by Mode 2 research, the activities relating to 'research' were not divergent from the actual practice in which agents engaged. This approach provided opportunities to discuss what constituted the problem situations, to reflect on the unique cases that seemed to belong to none or all the actors, to play and have fun, 'observing' the problem situation from different rationales, to bring in new voices of actors affected but not heard, and to address conflicting values. By taking part in this research, participants were able to learn by collectively reflecting on their 'doing'. 'Knowledge form within' (Shotter, 1993) emerged as a result. In turn these interactions served to inform actions aimed at improving the problem situations. The focus of platforms was on the relationship between generating improvement in the problem situations and the creation of knowledge by doing research.

An essential part of this type of engagement relied on the fact that the researcher was not considered independent of his object of study. The "naïve subject/object dualism" referred by Midgley (2000, p43) was overcome by this Mode 2 approach. The researcher was considered an agent. The researcher was therefore not seen as an external expert but as an actor. The researcher could not be outside, as objective observer; platforms emerge from within the system as a result of interactions of the agent and those involved in the problem situations. They emerge through agents' disposition towards doing research, which is also translated into an engagement in 'problem-solving-setting'. This is further discussed below.

Engagement In Problem-Solving-Setting

It was recognised that platforms imply engagement in 'problem-solving-setting'. This meant that the agents (co-researchers) in the project on order to focus in 'problem-solving', had to generate the 'space' to engage in problem solving. This was accomplished by 'creating the organisational and intellectual space' (Whyte, 1991) within which new ideas could be effectively acted out. As Whyte pointed out, "The way we think about problems is shaped to a considerable extent by the social setting in which we find ourselves" (Whyte, 1991, p42).

In operationalising CSP this meant that platforms had to involve 'engaging in ongoing processes' where it was legitimate to contemplate alternative understandings of problem situations and thus generate various alternatives for intervention. This was initiated by inviting participants to 'do research'. The research was clearly "situation-driven", but within the TS, the Book and the WW, the distinctive feature was that actors became active 'observers' of the situation of which they were part. Platforms provided the space to enter into processes of reflection about the actions that had already been taken and were to be taken in the future. Thus the organisational social system became the focus for reflection; turning it backs on itself. Platforms 'materialise' the setting for the 'problem-solving' attempt. This was considered on its own a local improvement by members of the organisation.

Platforms were also seen as informed by reflective processes of research. This reflexivity distinguishes the 'realm' of platforms from other activities in the social system. This realm 'makes' the setting for the social system to become 'visible' for reflection. Platforms, then, allow for participants to 'enter' into their own knowledge about the organisation as a social system and about its transformation processes. In practice the organisation, as a 'human activity system' (Checkland, 1981), was not a research matter for the participants in their day-to-day activities; it was only from the TS, the Book and the WW that the organisation became a research matter 'observable' (to be reflected upon) by them.

Pursuing 'problem-solving-setting' thus facilitated 'entering' in 'problem-solving'. Creative exploration by the co-researchers inherently complemented this type of engagement. This practice is discussed below.

Engagement In Creative Explorations

The third characteristic of platforms is engaging in creative explorations. The research experience showed that creativity was intrinsic to the platforms enacted during the research process; this creativity was orientated towards questioning and/or changing organisational and intellectual ground rules and thus opened paths for creative solutions to new problem situations. Indeed, Whyte argued, "Creative solutions will depend upon the ability of the organisation to change the organisational and intellectual ground rules" (Whyte. 1991, p42).

Platforms, therefore, constituted the 'environment' (physical, and/or intellectual, and/or social, and/or visual) that 'launched' this creative process. Opening the TS, for example, facilitated engagement in creative problem solving TSI (Flood and Jackson, 1991) since bringing in alternative sets of understandings and proposals for improvement to be contemplated became legitimate within this setting. The creative process was enhanced further when it became possible to bring various different rationales into play as part of the TS, the Book and the WW so as to enrich participants' understanding about rationales for improvement. In this case the TS, the Book and the WW, as platforms, assumed the role of play-like learning activities (Stacey, 1996).

A number of different methods and tools were available to assist with this creative exploration. Rich pictures, metaphors and stories were all used in this research (Carrizosa, 1999).

Platforms 'launch' creativity, that is they provide a setting where it is legitimate to question the way we think and act about problems and contemplate and explore other views. Thus, engaging in the development of a 'structured way of thinking' is also a characteristic of platforms. This is discussed in the next section.

Engagement In Developing A Structured Way Of Thinking

Engagement in the development of a 'structured way of thinking' refers to structuring ideas for action. This process involves the building and re-building of 'ground rules' relevant to problem solving. In other words, engagement in developing structured way of thinking is about continuously defining 'frameworks' that provide criteria for approaching problem situations. These frameworks take the form of metaphors, conceptual maps, systems of rules, or image, among others.

In the particular case of the TS, this process resulted in defining a set of rules for structuring conversations and interactions. In the Book the process of writing each of the chapters demanded that a supporting conceptual framework be defined. In the case of the WW, the walls themselves constituted a visual structure in 'constant construction' for thinking and acting. Generally, platforms involved processes in which participants were encouraged to 'play' according to the 'rules' agreed collectively. Naturally, reconsidering the 'rules' was part of building a structured way of thinking and was always possible as part of the 'on going' process, presented below.

Engagement In 'On Going' Processes

Platforms exemplified engagement in 'on-going' processes. During engagement participants enacted and re-enacted platforms until these processes became a "permanent part of the continuous learning in the company beyond the implementation activities" (Jackson, 2000, p419). It was usual for an idea to arise within one platform thereby demonstrating the concept of 'stable' platforms giving birth to new, 'legitimate' ones. For instance, the reflection within TS progressively took the form of written documents, which resulted in authoring a book. This outcome in turn motivated the next platform. Through this type of engagement, new ideas were suggested and tested by facilitating participants' interactions.

As demonstrated by this research experience, platforms are indeed 'socially emergent processes' (Checkland, 1981). In the course of this research, these instances developed through 'interactions'; they arose when problem situations were tackled by engaging in collective reflection, i.e. by doing research. From this perspective the very way in which TS, Book and WW worked was not important. What mattered was that these platforms resulted from an on-going process made legitimate by the act of doing research; i.e. by problem-solving-setting, creative exploration and developing a structured way of thinking.

Thus, platforms contributed to improving the organisation by 'tuning' into the changing and transitory nature of the problem situations at hand. The knowledge derived from this learning process was always open-ended. What was of interest in these results was that the social system integrated by co-researchers and participants in the platforms was not merely a passive recipient of external or internal influences but was capable of taking action with reference to such influences. Participants were able to redefine their relations through the constant practice of enacting various platforms. This rendered the organisation's human activity system 'purposeful' in contrast to 'purposive' (Checkland and Scholes, 1990).

As pointed out in this research was not possible to reduce platforms to a series of prescriptions, requirements or conditions; they could only be described as a series of ever-flowing, ever-changing 'engagements'. Thus, results from this research project supported the operationalisation CSP. Platforms synthesised the 'transferable problem solving capability' resulting from this 'organisational-based action research' endeavour.

PLATFORMS TOWARDS CSP

Platforms provided a 'hook' for CSP. These instances represented engagement in interaction (purposeful activity), they were conducive to the legitimisation of CSP, supporting a post-modernist rationale.

As Jackson indicated when theoretically reflecting about this research process as a whole, "Post-modernism provides the theoretical support necessary to enable us to critically reflect on [the TS, the Book and the WW] with a view to improving the post modern input to critical systems practice in the future" (Jackson, 2000, p420). It was evident in the research that platforms synthesised aspects denoted by this rationale; researchers saw improvement based on local knowledge; they challenged power relations and encouraged diversity; they utilised different 'processes of facilitation' and 'modes of representation' (Taket and White, 2000). Their input to CSP was 'engagement' in the form of an enjoyable, fun, social, intellectually challenging, technical, an 'excuse' to embark in CSP.

The research process demonstrated that the legitimisation of CSP through platforms emanates from emphasising "the replacement of individual rationality by communal negotiation, the importance of social processes in the observational enterprise, the socio-practical function of language, and the significance of pluralistic cultural investments in the conception of true and good" (Gergen and Joseph, 1996, p368).

In addition, platforms provided a source of legitimacy for the research conducted in this organisation by portraying "a commitment to a range of empirically arbitrary and culturally embedded conceptualisations" (Gergen and Joseph, 1996, p368). Legitimisation in this context is about "introducing a realm where both expert and client can enter and into which both can also have access" (Taket and White, 2000, p81). Indeed legitimisation involved a process whereby the researcher and actors collaborated in the realisation of local knowledge; this realisation permeated the approach towards pluralism. Thus, platforms denoted purposeful human activity systems, where "it is people on whom the meaning of improvement depends first of all, for they posses the sense of purposefulness, the power, knowledge and sense of responsibility that together determine *what ought to count as 'improvement'*" (Ulrich, 1996, p20).

REFERENCES

Carrizosa, A, (1999). "Rich Pictures, Metaphors and Stories as Mechanisms to Improve Collective Actions," in *Synergy Matters: working with systems in the 21st century*, (A.M. Castell, A.J. Gregory, G.A. Hindle, M.E. James, and G. Ragsdell, eds), Plenum Publishers, New York.

Carrizosa, A, (2000). "Enacting Thinking Spaces towards Purposeful Actions: An action research project". World Congress of the Systems Science and the 44th Annual Meeting of the International Society for the Systems Science, Toronto, Canada.

Checkland, P. B. (1981). *Systems Thinking, Systems Practice*, Wiley and Sons, Chichester.

Checkland, P. B. and Holwell, S. (1998). "Action Research: Its Nature and Validity", *Systems Practice and Action Research.* 11(1):9-21.

Checkland, P. B. and Scholes, J. (1990). *Soft Systems Methodology in Action,* Wiley and Sons, Chichester.

Checkland, P. B. and Scholes, J. (1999). *Soft Systems Methodology: a 30-year retrospective*, Wiley and Sons, Chichester.

Flood, R.L. and Jackson, M.C. (1991). *Creative Problem Solving: Total Systems Intervention*, Wiley, Chichester.

Gergen, K.J. and Joseph, T. (1996). "Organizational Science in a Postmodern Context, *Journal of Applied Behavioral Science. 32, 356-378.*

Jackson, M.C. (2000). *Systems Approaches to Management*, Kluwer Academic/Plenum Publishers, New York.

Midgley, G. (2000). *Systemic Intervention: Philosophy and Practice*, Kluwer Academic/Plenum Publishers, New York.

Shotter, J. (1993). *Conversational Realities: Constructing Life through Language*, Sage, London.

Stacey, R.D. (1996). *Complexity and Creativity in Organizations*, Pitman Publishing, London.

Taket, A. and White L. (2000). *Partnership and Participation: Decision-making in the multiagency setting*, Wiley and Sons, Chichester.

Ulrich, W. (1996). *A Primer to Critical Systems Heuristics for Action Researchers*, The Centre for Systems Studies, The University of Hull.

Whyte, W.F. (1991). *Participatory Action Research*, Sage, CA.

A BASE FOR SIMULATING INFORMATION DISTRIBUTION

G. Egonsdotter and J. Palmius

Department of Information Technology and Media,
Mid Sweden University, S-83125 Östersund, Sweden.
Gunilla.Egonsdotter@itk.mh.se, Joel.Palmius@itk.mh.se

INTRODUCTION

A common problem with implementation of new communications hardware is that it is difficult to predict their effects on the information flow in an organisation. This leads to hardware investments, which are at best based on qualified guesses that they will improve the situation, and at worst based on half-pronounced promises from the local computer salesman. To help this problem it would be desirable to be able to simulate the information flow in an organisation in such a way that the effects of new hardware implementations could be studied. As the information flow in an organisation can be said to be a complex system, a Cellular Automata (CA) approach is appropriate to build such a simulation. CA approaches have been used for several simulations of complex situations before, as an example, models of combat (Dockery, and Woodcock, 1993) and models of traffic flow (Wahle, et al, 1999). In order to create a simulation, behaviour and information flows must be analysed and formalised using a computer-parsable notation language.

During early fall -01, the *Cellular Automata Based Simulation of Information Distribution* project (CASID) was started as a joint effort between Mid Sweden University (MSU) and Swedish National Defence College (SwNDC). The main participants were doctoral students from the centre for Research on Anticipatory Systems (RAS) at MSU, and from the AQUA group at SwNDC. The intent with the project is to construct software with the ability to simulate how information distributes in an organisation.

The whole project spanned over the entire fall -01, but the first part of the project consisted of charting, analysing and making operational observed information flows in two exercise organisations. In other words, the first part consisted of an exploratory phase with the purpose of deciding whether it was reasonable to think a simulation could be built. This paper describes the preliminary results of this first part.

The need to be able to study how information distributes through organisations has lately been a topic for some discussion in connection with ROLF 2010 and AQUA at

Systems Theory and Practice in the Knowledge Age
Edited by Ragsdell *et al.*, Kluwer Academic/Plenum Publishers, New York, 2002.

SwNDC (see below for explanations of ROLF and AQUA). It was felt that new solutions for information distribution were implemented even when the knowledge about their effects was incomplete or lacking. When formulating the project plan for CASID it was thought that one way to get a better base for decisions was through analysing information flows, behaviour and communication and build a simulation thereof.

Both the Swedish Defence Research Agency (FOI) and SwNDC have been studying information behaviour, and have even made formalised descriptions of communication (Wikberg and Modéer, 1999; Albinsson, 2000), but have as far as we know not made a computer-implementable description. Thus one major aim with the first exploratory phase of CASID to make an operational implementable description.

The project described in this paper is formally a part of AQUA, which in its turn was started as a complement to (or part of) ROLF 2010. The *Mobile Joint Command and Control System 2010* (ROLF 2010) has the purpose to: "test alternative Command and Control (C2) concepts, work forms, decision-making support and C2 organisations adapted to mobile or movable joint C2 systems for the year 2010" (Sundin, and Friman, 2000).

AQUA (which is a name, not an acronym) is a research and development project with the "Visionarium™", a specially designed room with technical infrastructure for supporting dialogue and decision-making, as its main product (see as an example Artman and Persson, 2000 for description). AQUA has over time developed into an umbrella covering several sub-projects concerning specific issues within the science of Command and Control. One of the planned applications of the results of the CASID project is to be able to study the effects of different configurations of information infrastructure, as an example the implementation of a Visionarium™.

THE CASE

SwNDC annually arranges exercises in strategic command as a part of the Programme for Advanced Command Studies. In these exercises, headquarters of command are formed in each of the two groups "battling" each other. The exercise tries to represent a crisis during a wartime situation. The staffs are modelled according to current North Atlantic Treaty (NATO) standards, something that makes analysing the organisation an easier task. This since the above type of organisation has clearly defined goals, clearly defined resources and is strictly hierarchical. To give a comprehensible description of this organisation here would require far too much space. However, a short summary could be as follows:

As said the staffs are organised strictly hierarchical. During the exercises there are operational and tactical command levels. The operational level is responsible for defining the long-term goals, while the tactical level implements the short-term decisions to reach this goal. Information is gathered "on the floor" and where it is sent upwards in the organisation for aggregation and analysis.

The exercises try implement the organisation during a situation of crisis, such as during a war, but in a somewhat smaller scale. As an example, sometimes a few persons will represent a whole staff.

The people participating in the exercise are a very homogeneous group. They have the same type of education, the same military language and they have good knowledge of the said hierarchical organisation. Thus they can easily understand and communicate with each other, something which might be different during a real situation.

The staffs communicate internally via an information system consisting of a combination of IT tools (email, phones...), meetings and manual routines (messengers, bulletin boards...). One important part of the exercises is to use and test new methods and tools for staff meetings.

THEORY AND WORLD VIEW

During the last decades, several attempts have been made to simulate human behaviour, and many of these with a specific Artificial Intelligence approach (Vickery and Vickery, 1987). While trying to simulate behaviour and strategies for finding information it has, however, not been the intention of this project to claim that it will implement AI in the traditional sense. The focus has rather been on the information distribution *per se*, within which the human information-moving behaviours claim a significant role.

It should be noted that when the term "information system" is used in this paper, the humans using the physical infrastructure are considered part of that system. The authors believe that only concentrating on the technical infrastructure of an information system is too limited, since significant parts of the information distribution in an organisation takes places during unplanned encounters in corridors and coffee rooms.

One of the fundamental background theories used in the project is Shannon's Information Theory. It was decided that the traditional model with transmitter, channel and receiver would well describe what the project was trying to simulate. The participants of the project are well aware that it is virtually impossible to handle all the problems of semantics, technology and efficiency described within Information Theory. The military communications are, however, strictly formal, explicit and redundant, properties which might somewhat balance these problems (Fiske, 1997).

Another fundamental theory, or rather technology, which is the base for the whole project, is Cellular Automata (CA). The choice of CA was made since it had previously been applied successfully in battle simulations (Dockery and Woodcock, 1993).

The use of CA necessarily limits the possible variation in worldview when analysing the case situation. First, it is inherent in the technology that systems are described as being very mechanical. Secondly, it is assumed that everything that should be described can be analysed down to explicits without losing anything important on the way: Synergy and relational effects are described as properties of the "cells" rather than modelled separately. Thus, as an example, a group is never described *per se*. Instead, the individual behaviour when being in the presence of other automatons is modelled.

While this "all contained in the individual" approach might be discussed in relation to its relevance for descriptions of human behaviour, it is thought in the project that it will be a practical approach to construct a good representation of the said behaviour. It might not be entirely exact, but it could paint a usable picture.

METHOD

Scope

Falling outside the scope of the project (and therefore outside this paper), is the actual content of information. To model the significance and consequences of information content would prove far too complex to be possible to handle. Therefore, all material has been studied with the intention, as far as possible, to exclude all information content. The authors realise that this delimitation is a drastic reduction of the situation.

Another aspect is the question about how general the results can be said to be. As mentioned, the modelled situation is a military organisation in crisis (in this case a war situation). The authors cannot, because of this, claim that the results are representative for "normal" organisations.

Methodological Steps

The first step in the project was to do an inventory of existing material about the behaviour and organisation of the said exercises, and in the practise about what the exercises were built to emulate. As SwNDC has significant amounts of documentation, it was necessary to limit the search somewhat.

While having formulated a project vision, a purpose and a goal during the project planning, the first actual step of the project was to formulate an operative picture in the sense of Löwgren and Stolterman (Löwgren and Stolterman, 1998). In practise this consisted of an exploratory modelling of the different entities identified within the situation. This could be viewed as a kind of exploratory research in the sense of Patton (Patton, 1990).

The information points and the behaviour of the observed individuals in the organisation (this material having been fetched primarily from videotapes of the exercises), were analysed with a base in Shannon's Information Theory.

After having constructed a basic operative picture of the situation, the third step consisted of complementing the material with structured deep interviews of exercise leaders, and back-comparisons with the existing documentation. Further, the results have been continually discussed with a senior representative from SwNDC in order to ensure basic reasonableness.

The results of the back-comparison were fed into a new modelling, which were then compared to the documentation, to form iterative steps in an evolutionary design of the models.

MAKING AN OPERATIONAL DESCRIPTION

Based on the initial exploratory modelling of the situation, it was agreed that descriptions of four entities could provide a sufficient basis for simulating an information system. These were the physical structure (walls, doors, areas...), the information points (bulletin boards, phones, computers...), the person automatons (with their roles and behaviour) and the actual information.

Information Points

The first and most obvious part of an information system is the physical infrastructure, the points through which the information spreads. In this study the information points are defined as "physical objects with the ability to receive, store and transmit information packets".

The information points were found to be possible to describe using a few general properties. The most notable of these were "send rate", "receive rate", "storage capacity", "medium" and "area of influence".

The send and receive rates are measured in packets per second and define the speed of information transmission between the point and a person automaton. These rates, as well as many of the other properties, are described as a mean and a standard deviation. After discussions with statisticians at MSU, it was in the project group decided that a randomisation from a normal distribution would better describe the values than if the values were chosen from a square distribution. The randomisation function taking the mean and the standard deviation as parameters, and in which r_1 = [0,1] and r_2 = [0,1], is described as:

$$Nf(\mu,\sigma) = \mu * \sigma * \sqrt{-1 * \ln(r_1)} * \cos(2\pi * r_2)$$

The storage capacity decides how many information packets the point can contain before being full. The medium is which type of information packets the point can handle. As an example, a bulletin board can accept pictures and text, but not sound.

The area of influence is two values: one describing how close an automaton must be to receive an information packet from the point, and one how close it must be to add an information packet to the storage of the point.

The Physical Structure

The physical structure describes the "arena" within which the person automatons move and operate. It was observed that a significant part of the information distribution was done in chance encounters within corridors and in the coffee room. Therefore it was decided that the simulation needed to recapture the movements of the individuals, even when they did not explicitly work with information distribution.

In order to provide a basis for the behaviour schemata (described below), key areas were implemented. These areas represent localities where the individuals within the organisation spend time, such as "lunch room", "personal office", "corridor" and "bathroom".

The description of the arena will (when these descriptions are implemented) be modelled in a special editor. In practise this work will consist of trying to represent a plan sketch of the modelled building.

Within the arena, the information points described above will be placed according to the modelled situation.

Information Packets

The definition of an information package is in this study "a discreet quantity of data". While in the simulation being completely devoid of actual content, they represent "information" in the modelled situation. Therefore the term is "information package" rather than "data package".

The information packages have four main parameters: A serial number, a type, a priority and a class. The serial number is plainly a unique identifier to enable tracing of specific packages. The type is the media through which the information can be spread, at this stage "visual", "text" or "audio". This decides through which information points the package can be spread. The priority is "normal", "medium" or "high", and provides a base for selection when the automatons have several information packages to choose between.

The class is primarily a way of grouping a stream of information packages, in order to provide a base for statistics after the simulation run. The class also decides whether the information package exists when the simulation starts, or if it has to be constructed through an aggregation of other information packages.

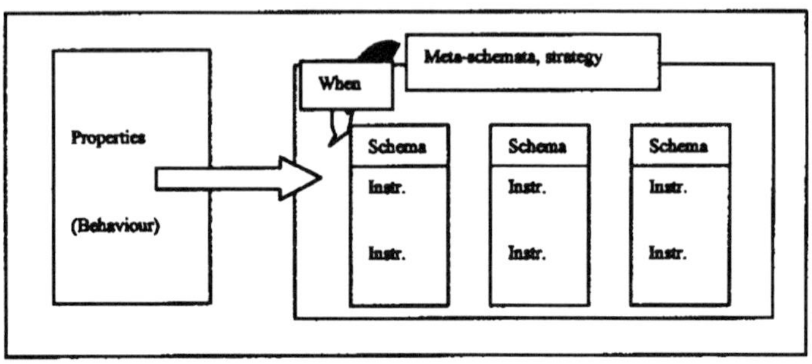

Figure 1. The behaviour of a person automaton consists of schemata and a strategy controlling when the schemata are triggered.

Person Automatons and their Behaviour

The by far most complex modelling was that of the behaviour of the person automatons. At the time of writing this modelling is only partly finished, and therefore what is offered below is the preliminary results of this modelling.

The person automatons represent the individuals working within the organisation. As a simulation might contain several hundred person-automatons, it was early decided that it would be impossible to describe all the represented automatons individually. Instead, the preliminary modelling was aimed at grouping behavioural patterns into a number of distinct classes. As these groups do not necessarily equate with a formal function within the organisation, the term "behaviour group" was chosen rather than "role" in order to avoid confusion when discussing with personnel at SwNDC.

A person automaton consists of several properties and a general behaviour. Most of the properties, and the whole behaviour, are inherited from the definition of the behaviour group to which the automaton belongs.

The behaviour is an aggregate of a number of schemata (such as a schema for "take coffee break"), and a meta-schema controlling the overall strategy of the automaton. The schemata will be triggered through events in the environment ("interact with automaton met in corridor"), according to schedule ("take lunch break", "sleep") and according to the overall strategy ("look for information through email in personal office").

The schemata consist of "instructions" or in other words a low-level code language with atomic functions such as "go to area", "wait" and "interact with point". At the moment of writing, this language is at best at a conceptual stage. Apart from the behaviour, the automatons will have properties such as "position in organisation" and "home area".

XML:ifying the Entities

As a basis for a simulation the models are pointless unless they are possible to implement in code. Therefore, all the modelling were made with the possible ways for implementation as a constant points of reference. Parallel to the conceptual modelling, the project members have worked with describing the results of the models in XML code.

At the moment of writing, preliminary document type definitions (DTD) exist for the physical structure and for the information points. Initial attempts at describing behaviour and schemata have been made.

SUMMARY AND CONCLUSIONS

The above material describes the preliminary results of the first exploratory phase of the CASID project, a material which will serve as a base for the further work. As such, the models and operational descriptions can be seen as somewhat sketchy. However the purpose with the described phase was to collect information and do preliminary exploratory modelling in order to decide whether it could be believed that a simulation of the mentioned kind could be built. With the existing material, the authors of this paper are quite confident that believing so is reasonable.

The further work will be concentrated at polishing the models, and to implement them as Java-based software. Work with formulating DTDs for the description language used in the applications has, as mentioned, already begun. As material of this kind start to mature, it will be made available on the project homepage (Egonsdotter and Palmius, 2001).

The initial attempts at trying to describe information distribution through modelling individual behaviour and modelling the structure the individuals interact with, indicates interesting possibilities. The approach to describe human behaviour as a number of schemata constructed by low-level instructions, and aggregating into an overall behaviour, seems promising at this stage.

In spite of the above operational description being a drastic reduction of human behaviour, it is the belief of the authors that a simulation based on the material can serve as useful tool for studying information distribution, and for being a decision support for implementations of new technological structures.

REFERENCES

Albinsson, P.A. (2000). *Visualisering av loggfiler – ett hjälpmedel för utvärdering av verksamhetsmodeller*. FOA, Linköping, Sweden.

Artman, H. and Persson, M. (2000). "Old Practices – new technology: Observation of how established practices meet new technology". In *ROLF 2010 the way Ahead and The First Step* (C. Sundin and H. Friman, eds.) The Swedish National Defence Collage, Department of Operational Studies and the Centre for Command and Control Science, Stockholm, Sweden.

Dockery, J. and Woodcock, A. (1993). *The Military Landscape - Mathematical Models of Combat*. Woodhead Publishing Limited, England.

Egonsdotter, G. and Palmius, J. (2001) *CASID - Homepage*. Available: Http://gathering.itm.mh.se/casid.

Fiske, J. (1997). *Kommunikationsteorier: en introduktion*. Central tryckeriet AB, Borås, Sweden.

Löwgren, J. and Stolterman, E. (1998). *Design av Informationsteknik - Materialet Utan Egenskaper*. Studentlitteratur, Lund, Sweden.

Patton, M.Q. (1990). *Qualitative Evaluation And Research Methods*. SAGE Publications, Inc.,USA

Sundin, C. and Friman, H. (2000). *ROLF 2010 the way Ahead and The First Step*. The Swedish National Defence Collage, Department of Operational Studies and the Centre for Command and Control Science, Stockholm, Sweden.

Vickery, B.C. and Vickery, A. (1978). *Information Science in Theory and Practice*. Bowker-Saur, London, England.

Wahle, J., Neubert, L., Esser, J. and Schreckenberg, M. (1999). "A Cellular Automaton Traffic Flow Model for Online Simulation of Traffic". *Parallel Computing*, 27: 719-735.

Wikberg, P. and Modeér, B. (1999). *Modellering av operativ ledningsenhet – Visuell modellering som metod för problemanalys och operationalisering av begrepp i komplexa system*. FOA, Linköping, Sweden.

SYSTEMS THINKING AND THE KNOWLEDGE AGE: REVISITING KUHN

Hilary C. M. Kane

School of Information and Communication Technologies
University of Paisley
Paisley, PA1 2BE, UK

INTRODUCTION

At the start of the 21st century, it may seem unusual to look back at the work of Thomas Kuhn (1970). Kuhn's approach offered a new insight in to how 'mature' science and discovery were inextricably intertwined. This insight suggested that the process of scientific discovery was in essence cyclical and that much of the mature scientific community worked within the confines of, what he termed "paradigms", as a means of framing their work. With this in mind, the reason for revisiting Kuhn (1970), is that what he has to say about a cyclical process and paradigms has a great deal of resonance within the systems thinking movement, both in the way that it has evolved and in light of the interest that is being shown in knowledge and related terms such as knowledge age, knowledge economy, knowledge management.

The reason for using Kuhn (1970) as a frame of reference is due, in large part, to the initial framing of a PhD proposal. Reading his work (1970) raised ideas about parallels that may be drawn between Kuhn's exposition of science and writings concerning knowledge acquisition, transfer and dissemination. When considering knowledge one of the central features appears to have been how to take what is inside, i.e. tacit knowledge, and put it outside, i.e. explicit so that others can utilise it. There seem to be parallels here with his (1970) work, when he talks about the apparently cyclical process of science. Also, in the area of knowledge there is a clear distinction between those writers who produce work in the field and those who write from the position of academia whilst conducting field work. The former are much more focused on practice whilst writers such as Polanyi (1966) and Popper (1959) tend to be in the realms of philosophy and theory with less apparent practical application. All this provides for interesting dichotomies. The final reason for consideration of Kuhn (1970) is that he is not without his critics. Feyerabend, Popper and Masterman all offer criticism of Kuhn. Masterman (1965) suggests that his definition of a paradigm is too loose, Feyerband (1965) believes that the thinking upon which he bases his book would inhibit the

Edited by Ragsdell *et al.*, Kluwer Academic/Plenum Publishers, New York, 2002.

advancement of knowledge and Popper (1965) thinks that Kuhn is misguided about the central idea of what constitutes normal science. The essence of this is that each writer has a view on Kuhn, whether positive or negative, and this suggests that Kuhn still has relevance today.

Of the three areas that will be considered in more detail, the first is concerned with the apparently cyclical process of scientific discovery, from the emergence of a paradigm, its refinement, through to its ultimate destruction and the emergence of a new paradigm. This cyclical process will be overlaid with the movement between tacit and codified knowledge to elicit any similarities. Secondly, the presence of any parallels between the evolution of systems thinking and scientific knowledge will be examined. Finally, what Kuhn says about paradigms will be explored in terms of systems thinking in the knowledge age.

THE STRUCTURE OF SCIENTIFIC REVOLUTIONS

Having given a brief outline of Kuhn's (1970) work, the specific areas deemed worthy of further investigation and reasons why this work has been chosen in preference to others, it is worthwhile at this stage to actually look at "The Structure of Scientific Revolutions" (1970) in greater detail. He (1970) attempts to explain the nature and background to scientific discovery from which emerges a clear cyclical process. He places this in the context of paradigms, which he defines as

> "...some accepted examples of scientific practice...[which] provide models from which spring particular coherent traditions of scientific research." (ibid, p. 10).

Work undertaken under the guise of science, prior to the emergence of a paradigm, is actually a great deal more random as it lacks the structure and boundaries of a paradigm (1970). As a paradigm 'triumphs', the scientific community are able to direct their research efforts more precisely. Kuhn (1970) does not believe that a paradigm is the whole answer to a problem. There still remain areas remaining requiring further exploration. However, the boundaries are set by the paradigm. This enables greater analysis of an area than might otherwise have taken place. It is this, which he (1970) refers to as 'paradigm-based research'.

During this 'paradigm-based research' enquiry is steered towards marrying fact with theory and, thereby, enabling articulation and refinement of the paradigm. It may, in turn, alter theory. This research is seen as pivotal in scientific revolutions. Although Kuhn (1970) acknowledges a swathe of rules, equipment and methods for the conduct of this paradigmatic research, he makes it clear that they are subordinate to a paradigm which can channel research "in the absence of rules" (ibid, p. 42).

A paradigm's hold on the scientific community is weakened upon discovery of an anomaly and the community's desire to explain it. Before a paradigm is rejected, an alternative paradigm must be placed before the community for its acceptance. Competing schools may arise, as in the pre-paradigm era, prior to the acceptance of the new paradigm and thus the process starts anew.

This greatly synthesised account of Kuhn's (1970) work is intended to extract the main points felt relevant and to set these alongside thinking in terms of knowledge, systems thinking and what paradigms might be emergent, if any, within this framework.

CYCLES

The first area that emerges from Kuhn (1970) is the cyclical nature of science. Cycles appear in all areas of research, from the quality cycle propounded by Deming (1986) to the crystallisation of codified law into statutes from older common law and the further refinement of these with interpretations by the judiciary. There appears to be a cycle evident in Kuhn (1970) from the first signs of 'casual collection' of data to the emergence of a paradigm, through the process of refinement of the paradigm to its destruction and the emergence of a new one. Thus cycles would appear to have played an important part in the way that research has depicted movement in a particular field.

Within the field of knowledge, cycles also appear to have played an important role. Where this may have the most relevance is in the area of tacit knowledge and its interaction with codified knowledge.

Polanyi (1966), considered one of the foremost writers in the field of knowledge, in his work "The Tacit Dimension", specifically examines tacit knowledge and how scientists find new theories. His (Polanyi, 1966) argument is that it is the scientist's application of tacit knowledge that generates new theories. This is based on his belief that people "know more than they.. tell" (ibid, p. 18). This use of tacit knowledge to enable the creation of new theories suggests that tacit knowing can be transformed into something explicit that others can acquire; thus what is tacit is made tangible. Taking this a stage further, other scientists may acquire this explicit theorising and add to their own research. In turn, this is assimilated with 'what they already know', thereby extending their own tacit knowledge. This cycle is continued with the creation of further explicit knowledge available for dissemination and acquisition.

Nonaka (1995) is particularly focused on the creation and dissemination of knowledge with his development of the 'knowledge spiral'. His view is that knowledge can be created moving from tacit or explicit to tacit or explicit. He describes explicit knowledge as things that are easy to communicate and tacit knowledge as things that are hard to explain or communicate, placing this within a cyclical concept. In a sense this mirrors the extension of Polanyi's (1966) views although Polanyi was more concerned with what was internal to the individual.

This apparent evidence of cycles in both Polanyi (1966) and Nonaka (1995) seems very similar to the use of a cycle evident in Kuhn (1970). Other writing appears to acknowledge that to promote the extension of knowledge or its transfer, specifically within an organisational context, it is necessary to move from tacit to codified knowledge. In Stewart's (1997) work in "Intellectual Capital" there is this recognition, but only within the framework of a strategy. Both Davenport & Prusak (1998) and Stewart (1997) recognise that making tacit knowledge explicit is inherently difficult.

What does this mean for systems thinking in the knowledge age? It is posited that cycles play an important part of the evolution of scientific knowledge, from the emergence of a paradigm through to its ultimate destruction and the emergence of a new paradigm (Kuhn, 1970). Parallels can be drawn between the writings. In the field of knowledge one of the main concerns appears to focus on the ability to move from tacit to codified knowledge, or, put another way, its evolution. Cycles play an important part in describing how that evolution takes place and may also provide a metaphor for analysing or making conjectures about how research in this area might proceed. Although he (1970) does not specifically state that science is cyclical in nature, he describes a process that appears to be this and provides valuable insight into writers in the field of knowledge who have, perhaps unwittingly, adopted an approach mirroring Kuhn's exposition of knowledge evolution in science.

SYSTEMS THINKING

Another way to consider the evolutionary nature of scientific discovery, described by Kuhn (1970), is to examine whether it echoes an evolution of systems thinking. There may be similarities between the concept of competing schools prior to the acceptance of a paradigm, followed by a period of stability, which provides the basis for the paradigm's destruction and replacement and the way that systems thinking has evolved.

The starting point for considering whether similarities exist might be to look at the roots of systems thinking. Historically research in organisations has centred around solutions to perceived 'problem situations'. To solve these 'problems' the traditional approach has been characterised by the use of quantitative methods to analyse a problem situation and attempt to reduce factors to statistics or hard facts. Such analysis has tended to exclude consideration of such variables as human input, as these are not perceived to form part of the problem. Broadly, this response might be termed a hard systems approach. Although, it is reductionist in approach, it still operates within the context of a systemic methodology.

This approach has been criticised. Rosenhead (1989) believed that use of a hard systems approach rested on certain assumptions, which included the use of this approach to analyse the suitability of various solutions in comparison with each other and select the optimal solution this way. A soft systems approach, he thought, was different or less obviously based on a strategy of solution comparison. Rather the focus was placed more on analysis of the process of how a decision emerged, and this then used as an aid to achieve better decisions. Rosenhead (1989) went further as he attempted to look at a method or variety of methods that might be used to construct a framework for making decisions and dealing with problems.

Over time, there has been some movement away from this hard systems approach, with its tendency to rely on quantitative measures, and a move towards what is perceived as a more holistic approach through the use of soft systems methodology. Adherents to the latter methodology point to a perceived failure on the part of hard systems advocates to acknowledge people as part of any 'problem situation'. The movement towards soft systems was viewed as a way to redress the balance.

As Kuhn (1970) states, during a period of normal science, research is directed to 'puzzle-solving'. The use of the word puzzle is important as puzzles must have a solution. Although it may be arrived at in a number of limited ways, there is a solution. Whilst the hard approach was in its ascendancy, it appeared to provide the answers that allowed the resolution of problems and thus might be said to reflect a period of 'normal' science. With the recognition that not all puzzles had been solved and the growth in the belief of a new approach, the necessary ingredients were available to pave the way for a 'revolution'. Alternatively, with the emergence of a soft approach, it may represent the pre-paradigmatic position before the acceptance of a new paradigm. The reductionist approach of the hard systems paradigm failed to acknowledge the importance of the whole situation, as it existed interconnected through all its sub parts. The soft systems paradigm could be regarded as a way to overcome the deficiencies in the hard paradigm, by allowing recognition of the whole, its interrelations and people as part of any given problem situation. Another way to look at it is that hard systems failed to acknowledge a central aspect of any problem (i.e. that people can and are part of the problem), thus loosening the hard systems paradigm, whilst soft systems was present to take up the challenge as a replacement.

Checkland (1981) views the hard systems paradigm of as "a special case of the soft"(ibid, p. 191). This provides an alternative standpoint that does not fit neatly into the suggested evolutionary approach that has been described above. To accept this view within what has already been stated would place the soft systems paradigm in the role of 'normal' science, relegating hard systems to a pre-paradigmatic position. However again there is still a parallel that can be drawn in the way Kuhn (1970) talks about the evolution of scientific knowledge and systems thinking. What has not been considered is the position of critical systems thinking and whether there its place represents another stage within the evolutionary process of systems thinking.

A further alternative, if one accepts that broadly the term systems thinking might be described as a way of thinking of systems, then both hard and soft approaches fit together more easily. It could be that systems thinking exists at a higher level, i.e. it is the paradigm, with hard and soft approaches relegated to, what Kuhn (1970) describes as 'rules for puzzle solving'. In essence, we can conceive everything within a system, even problems; and the tools, such as a hard or soft systems approach are merely that; tools. He (1970) refers to the apparatus that enables problem solving as a body of rules, equipment and procedures that practitioners use. This concept of rules encompasses "established viewpoints or preconceptions" (ibid, p. 39). Given this potentially broad analogy, a point acknowledged by Kuhn (1970), it is possible for the community of systems thinkers to agree on a shared view of the paradigm, i.e. what constitutes systems thinking, whilst allowing disagreement about the methods that might be used to solve problem situations, thus allowing the encapsulation of soft and hard systems methods. Practitioners learn methods through the use of prior example, e.g. learning about qualitative and quantitative methods of research, about hard and soft systems methodologies, and can look at problem situations (i.e. puzzles to be solved) using the tools and techniques that they have learned.

Given Kuhn's alternative definition of a paradigm as representing:

> "..the entire constellation of beliefs, values and techniques...shared by the members of a community" (Kuhn, 1970),

it is not unreasonable to suggest that systems thinking could represent a paradigm for the systems thinking community.

PARADIGMS IN SYSTEMS THINKING AND THE KNOWLEDGE AGE

The final area and perhaps one that Kuhn acknowledges as quite contentious, is that of paradigms. He states that paradigms are:

> "..some examples of scientific practice..[which] provide models from which spring particular coherent tradition of scientific research" (ibid, p. 10),

although he later concedes that he provides a variety of meanings for this term. Using this as a definition of a paradigm, consideration can be given to what paradigms exist in the 'knowledge age' and how this may help development of systems thinking in this realm. In order to do this, it is necessary to develop an understanding of what is meant by the 'knowledge age'.

As has been commented earlier, there has been a good deal of writing in the area of knowledge even when not explicitly dealing with this subject such as Popper's (1959) work "The Logic of Scientific Discovery". The terminology in this area has talked variously about the knowledge age, knowledge working and intellectual capital all in an attempt to enunciate methods of managing knowledge under the banner of knowledge management. Davenport and Prusak (1998) describe knowledge as a "fluid mix of framed experience, values, contextual information and insight" (p. 5). They distinguish knowledge from information on the basis of a process that adds value, a distinction rejected by Stewart (1997) who regards this as unhelpful as "one man's knowledge is another man's data" (p. 69). Fahey and Prusak (1998) refer to knowledge as 'what a knower knows' which is slightly different from Polanyi's (1966) explanation of "we can know more than we can tell", albeit this is specifically directed towards tacit knowledge.

From these definitions of knowledge it is clear that writers place different interpretations on their understanding of what knowledge is and this is carried through in discussions of knowledge management. The term has been characterised by phrases such as:

> "putting information and people's competence to work together" (Kalseth, 1997),

> "a discipline that promotes an integrated approach to identifying, managing and sharing all of an enterprise's information assets" (Corrall, 1999)

or

> "the practice of harnessing and exploiting intellectual capital to gain competitive advantage and customer commitment through efficiency, innovation and faster, more effective decision-making" (Barth, 2000).

From these descriptions there would seem to be, as yet, no coherent definition of knowledge, its subdivisions, i.e. tacit and explicit, and by extension the phrase 'knowledge age' lacks precision. The interpretation that is considered most illuminating is that of Davenport and Prusak's (1998), which considers that there must be some additional process to change information in to something, termed 'knowledge'.

Given the lack of coherency around the term knowledge it may be difficult to ascertain what paradigms exist in a 'knowledge age'. Kuhn has implied that the essence of a paradigm is that it provides a model upon which research can be guided. Burrell and Morgan (1979) see a paradigm as a conceptual framework for seeing and making sense of the world so that being positioned in a particular paradigm leads to viewing the world in a particular way. This would appear to broaden Kuhn's (1970) conception of a paradigm. Senge (1992) describes it as "a discipline for seeing wholes, recognising patterns and interrelationships" and learning how to structure those relationships in more effective, efficient ways. There is an acknowledgement of the interrelations of subsystems or processes that constitute a whole. Thus, if Kuhn (1970) looks at a paradigm as a guide, Burrell and Morgan (1979) see a framework, then Senge's (1992) interpretation appears to fulfil the characteristics of a paradigm for systems thinking.

Corrall (1999) sees knowledge management as "an evolutionary...development" as it entails "taking a more holistic view of information..[and] requires a strategic focus" (p.3). Taking this description a similarity can be drawn between the fields of knowledge management and systems thinking. Kuhn sees the paradigm as a framework within which scientists can conduct research, a reference point, a touchstone against which their efforts can be measured. Within the field of systems and systems thinking the articulation of this has been somewhat broadened by Burrell and Morgan (1979) as they seek to encapsulate what might be described as a 'world view'. This is not so far removed from Kuhn, as he believes that the paradigm shapes the view of a scientist and their endeavours. Taking this a stage further, Senge's (1992) comments merely seek to elucidate in a sense Burrell and Morgan (1979) whilst Corrall's (1999) opinion pulls in Senge's view to encapsulate knowledge management. Thus, it might be argued that research in the field of knowledge or the knowledge age should be rooted in the paradigm of systems thinking.

Within the context of the practical application of knowledge management, much of the literature stems from the work of Karl Sveiby. Sveiby (1996) talks about the 'knowledge organisation'. He talks in terms of 'strategic paradigms' that organisations must address if they are to move from service company to a knowledge organisation. However much of this style of writing suffers from a number of problems. Although written in a style readily accessible to the business community, there is a lack of rigour with case studies used as illustration without an underlying methodological analysis. Whilst Sveiby offers definitions of knowledge, knowledge management and other terminology under this umbrella, it is not necessarily in accord with any other definitions put forward, raising again issues of coherence within the field of knowledge.

There is little evidence to suggest a clear paradigm to support a 'tradition of research', even though there are various strands of thought on how knowledge is to be managed. In addition there appears to be a gulf between practice and amongst academics, which would strengthen any potential argument for returning to Senge's (1992) view on systems thinking as a jumping off point to consider the facets of the knowledge age.

CONCLUSION

Kuhn has much to offer in looking at systems thinking specifically with reference to knowledge. In the same way that he hints at cycles occurring in the evolution of scientific knowledge, so too, does the writing in the field of knowledge. He offers a filter through which the field of systems thinking can be analysed and illustrates the difficulty of identifying paradigms in the knowledge age. Worthy of further examination as it has only been touched upon in this paper is to consider the position of critical systems thinking in light of its position as a potential meta-framework for systems thinking with particular reference to the 21st century.

REFERENCES

Barth, S. (2000). Defining Knowledge Management, *CRM Magazine*, 4 July 2000.

Burrell, G. and Morgan, G. (1979). *Sociological Paradigms and Organisational Analysis: Elements of the Sociology of Corporate Life*, Ashgate, Aldershot.

Checkland, P. (1981). Systems Thinking, Systems Practice, J. Wiley & Sons, Chichester.

Corrall, S. (1999). *Are We in the Knowledge Management Business* at http://www.ariadne.ac.uk/issue18/knowledge-mgt/

Davenport, T. and Prusak, L. (1998). *Working Knowledge*, Harvard Business School Press, Boston, MA.

Deming, W. E. (1986). *Out of Crisis: Quality, Productivity and Competitive Position*, Cambridge University Press, Cambridge.

Fahey, L. and Prusak, L. (1998). The Eleven Deadliest Sins of Knowledge Management, *California Management Review*, 40, (3), pp 265 – 76.

Feyerabend, P. (1965). Consolations for the Specialist, in Lakatos, I. and Musgrave, A. (eds.), *Criticism and the Growth of Knowledge*, Proceedings of International Colloquium, The Philosophy of Science, 4, Cambridge University Press, Amsterdam.

Kalseth, K. (1997). Knowledge Management: Putting Information and People's Competence to Work Together, *FID News Bulletin*, 47 (7/8), pp 191 – 192.

Kuhn, T. (1970). *The Structure of Scientific Revolutions*, University of Chicago Press, Chicago.

Masterman, M. (1965). The Nature of a Paradigm, in Lakatos, I. and Musgrave, A. (eds.), *Criticism and the Growth of Knowledge*, Proceedings of International Colloquium, The Philosophy of Science, 4, Cambridge University Press, Amsterdam.

Nonaka, I. (1995). *Knowledge Creating Company: How Japanese Companies Create the Dynamics of Innovation*, Oxford University Press, Oxford.

Polanyi, M. (1966). The Tacit Dimension, Routledge & Kegan Paul, London.

Popper, K. (1959). The Logic of Scientific Discovery, Hutchinson, London.

Popper, K. (1965). Normal Science and its Dangers, in Lakatos, I. and Musgrave, A. (eds.), *Criticism and the Growth of Knowledge*, Proceedings of International Colloquium, The Philosophy of Science, 4, Cambridge University Press, Amsterdam.

Rosenhead, J. (1989). *Rational Analysis for a Problematic World*, J. Wiley & Sons, Chichester.

Senge, P. M. (1992). *The Fifth Discipline*, Random House, London.

Stewart, T. A. (1997). *Intellectual Capital*, Nicholas Brealey Publishing, London.

Sveiby, K. (1996). *The Knowledge Organisation* at http://www.sveiby.com/KOS12.html

CHAOS AND COMPLEXITY THEORY IN MANAGEMENT: AN EXPLORATION FROM A CRITICAL SYSTEMS THINKING PERSPECTIVE

Maria Carolina Ortegón-Monroy

Lincoln School of Management
University of Lincoln
Brayford Pool, Lincoln LN6 7TS, UK
E-mail: mortegon@lincoln.ac.uk

INTRODUCTION

This paper presents the theoretical and practical findings emerging from an exploratory study of Chaos and Complexity Theory (C&CT) as an alternative approach to propitiate new understandings about the nature of management in organisations. They show that C&CT is not theoretically coherent enough to be applied in practice.

The paper involves a theoretical and a practical part. The theoretical exploration is conducted from a Critical Systems Thinking (CST) perspective. It involves a theoretical inquiry particularly of Stacey's account of C&CT, presented in: *Complexity and Creativity in Organisations*. This is presently considered to be one of the fullest and probably the most widely read management account on the subject. Jackson's framework of CST, used to interrogate systems approaches constitutes the selected approach to conduct this theoretical exploration. From this theoretical exploration a series of findings were produced. They reveal incoherencies, which were further explored through the findings emerging from the exploration at the practical level. The practical exploration was conducted in the Humberside Training and Enterprise Council (H-TEC), which introduced C&CT ideas as part of its process of becoming a learning organisation.

Based on these findings, this paper thus presents a critique and suggests ways forward. In order to do this, the research model used to conduct the exploratory study supporting this critique is described first. This sets the scene to give an account of the findings, that is the theoretical incoherencies and their manifestations in the practice. The paper concludes with a reflection about the implications of the theoretical incoherencies, indicating what C&CT can deliver and suggesting how it can be used in the practice.

RESEARCH MODEL

The model of the research became the cyclic process described in the following diagram.

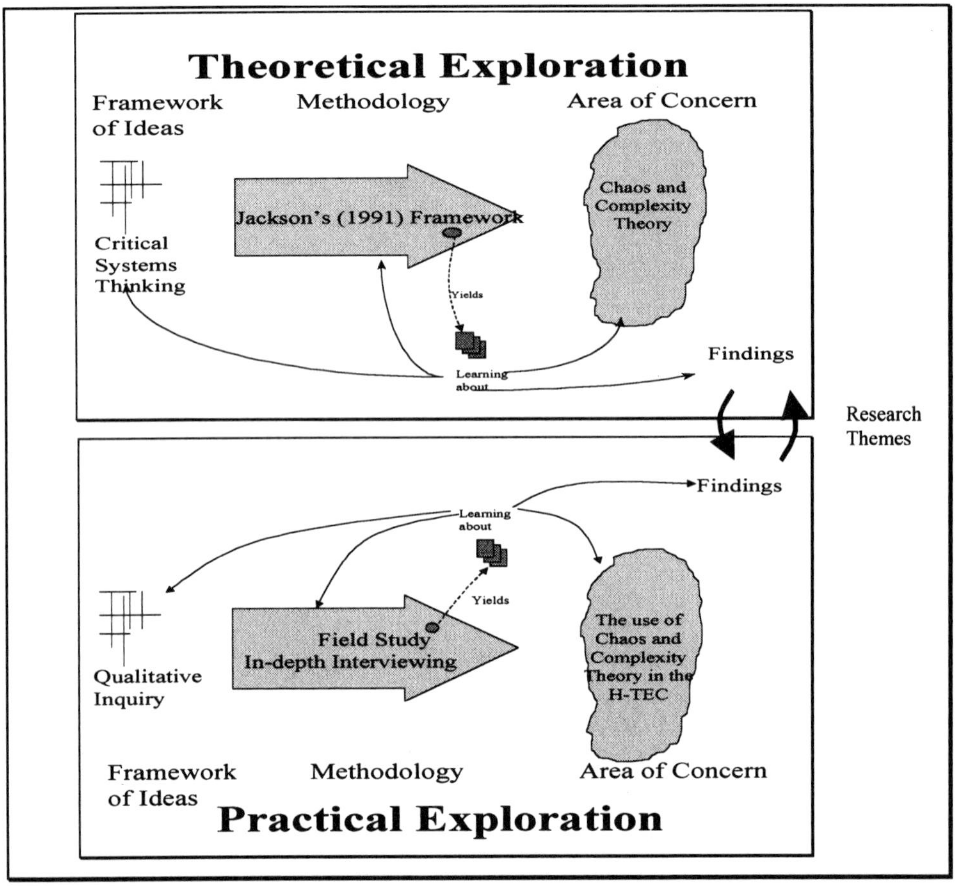

Figure 1. Research Model using the Framework of Checkland and Holwell (1998)

A framework developed by Jackson (1991) to interrogate systems methodologies is used to develop the critique. This framework consists of an examination of systems approaches at a number of levels: (1) Habermas' theory of human cognitive interests (1972, 1974); (2) Burrell and Morgan's sociological paradigms (1979); (3) Morgan's metaphors of organisation (1986); and (4) Jackson and Keys' System of Systems Methodologies (1984). In the case of this study, this framework is used to interrogate Stacey's version of Chaos and Complexity Theory (C&CT). These outcomes are used to inform the practical exploration.

At a practical level this exploration was conducted in the Investors in People directorate of the H-TEC, concretely the first self-managed team of consultants belonging to this area. The researcher was invited by the Managing Director of the organisation to conduct her research in this area of the organisation. The situation of the team was explored along whatever path it took through a period of two years. The aim was to inquire about the use of C&CT ideas. The study focused on determining about individuals' perceptions and experiences as related to the approach adopted by the Managing Director and the Human Resources Team to introduce and operationalise C&CT; that is the 'framework of organisational design principles' (FODP, see Storr (1997)). Qualitative, informal, in-depth interviewing (Taylor and Bodgan, 1984) ('exploratory dialogues') was the methodology applied to investigate this area of interest. The findings emerging form these processes constituted a series of recurrent themes that reflected the common understandings and experiences of the self-managed team members about their use of C&CT. They were used to test the theoretical findings.

THEORETICAL INCOHERENCIES AND THEIR MANIFESTATION IN THE PRACTICE

This section demonstrates how the theoretical incoherencies of C&CT come to the fore and affect the practice. Each sub-section introduces one incoherence; discusses it in terms of the underlying assumptions made in the approach used to incorporate these ideas in the Humberside TEC; presents these incoherencies as experienced by the self-managed team members; and comments about their implications.

Abandoning the Technical Interest without Doing So: A Pretension

The theoretical incoherence discussed here is that the view supporting the descriptions of the C&CT approach demands abandoning the technical interest while the view supporting the intervention encourages determining regularities in behaviour to identify underlying structures that account for learning to occur and thus accomplish the central management task of 'driving the organisation to the edge of chaos' (Stacey, 1996) or 'nudging systems into desired trajectories' (Morgan, 1986).

The above incoherence was effectively experienced by the MD of the H-TEC as a dilemma: "How to remove command and control without using command and control?" This situation was present at the level of the self-managed team by acknowledging the ownership of the process of incorporation of these ideas to the MD, but at the same time experiencing a modus operandi in which they were free to start their own initiatives and flexibly design their tasks, that is to self-organise.

Thus the pretension of abandoning the technical interest is not realised within this approach; managers are faced with the tension between the need for control and allowing self-organisation. At the level of the self-managed team this false pretension generated lack of trust and ownership of the process of incorporation of these ideas. Clearly, the approach contemplates other more subtle forms of pursuing the technical interest, indicating that it is by no means abandoned.

Making the Practical Interest Technical: A Fallacy

The theoretical incoherence discussed here is that the view supporting the descriptions of the C&CT approach assumes that no one can be in control but the view supporting intervention accepts the regulation of behaviour to propitiate conditions for mutual understanding and learning to take place in the organisation.

The above was evidenced in the approach used at the H-TEC as a whole by promoting that the organisation is a self-organising system and control is disperse throughout the organisation to the extent that strategy is an emergent phenomenon, but aiming at regulating behaviour through the FODP to build a learning culture in order to reach a better fit with the environment.

At the level of the self-managed team this incoherence was present through the MD's management style promoting the assumption that "control is only an illusion" but imposing their modus operandi, and directly attempting to influence and control the culture and their learning process. As a consequence, a counter culture of individualism predominated within the self-managed team, which did not feel very identified with the process of incorporation of these ideas.

The approach used at the H-TEC to incorporate C&CT ideas shows that pursuit of the practical interest (learning) is subservient to technical purposes (adapting to the environment) confirming the fallacy of the C&CT approach in portraying the practical interest as predominant when it is the technical interest that is most favoured.

Making the Emancipatory Interest Technical: Instrumentalising Learning

The theoretical incoherence discussed here is that the view supporting the descriptions of the C&CT approach considers changes in schemas (double loop learning) as liberating but the view supporting the intervention is enslaved to one schema or mindset: operating far from equilibrium and co-evolving with the forces of the market.

The above was evidenced in the approach used at the H-TEC incorporate C&CT as a whole on the one hand, by emphasising that a key feature of a complex-evolving system is its capacity for both single and double loop learning. Individuals were thus seen as being self-responsible for processes of changes of schema to take place in the organisation. But on the other hand, by accentuating the view of organisations as complex evolving systems where learning is central to build the capacity to deal with the unknown and co-evolve in a deforming landscape, confirming that responding to the forces of the market is more important than emancipation.

At the level of the self-managed team this incoherence was experienced as a freedom to design their job and develop initiatives that was severely constrained by a structure and a series of controls that rendered them powerless. This re-confirms the predominant schema operating in the organisation.

This reveals that learning, contemplated as changes in schemas, was seen predominantly as instrumental, and not as emancipatory within an approach where adapting to the environment is more important, and not emancipation in any normal sense of the word.

Making One Schema Pervade:using its Own Tools to Remove It

The theoretical incoherence discussed here is that the view supporting intervention in the C&CT approach is based on the pervasiveness of a dominant schema (learning, cybernetic, mechanistic). While the view supporting the descriptions is based on understanding change in organisations as a dynamic of changes in schemas.

The above was shown in the approach propitiated in the H-TEC by contemplating that the organisation can adapt its behaviour to respond to its environment and it can also change its schema or the rules by which it operates. But at the same time favouring the learning organisation view, which underpinned and embraced the whole approach.

At the level of the self-managed team this was experienced through a management style that promoted to "think out of the box" and "be comfortable with the uncomfortable" but at the same time using a command and control system to make them part of this process of becoming a learning organisation. The schema through which change became operative was the command and control one, determining its predominance for self-managed team members.

The above corroborates that using mechanistic tools to remove a mechanistic approach just makes it more pervasive. The C&CT approach encourages managers to influence control parameters to propitiate the conditions for learning and creativity and innovation to take place. This is an undeniable indication that in spite of contemplating the eventual dominance of any view, mechanistic views to management are predominant. Despite Stacey's claims to the contrary this approach is trapped in this one view. This considerably hinders the adoption of pluralist perspectives and the propitiation of double loop learning.

Unitary Assumption behind Propitiating a Pluralist Context

The theoretical incoherence discussed here is that the view supporting the descriptions in the C&CT approach considers complex-pluralist contexts as normal, and as the contexts to be propitiated but the view supporting intervention makes complex unitary assumptions.

The above was evidenced in the approach employed at the H-TEC as a whole through its predominant command and control orientation used to incorporate these ideas in this organisation while at the same time presupposing the view of the organisation as a learning system. At the level of the self-managed team this was experienced as a sudden imposition of their present modus operandi while at the same time encouraging learning and self-responsibility.

Instead of propitiating a pluralist (edge of chaos) context in the pursuit of creativity and innovation, the tension generated has resulted in the enforcement of a context where unitary, mechanistic assumptions were being made; self-managed team members complied with it in the pursuit of improving self-performance.

The C&CT approach apparently seems to indicate that it is not a 'problem-solving' approach but a 'context-propitiating' one. However on propitiating a pluralist context, unitary assumptions are made (the end is to drive the organisation to the edge of chaos by means of manipulating control parameters). The approach seems to overlook that propitiating a pluralist context involves making a unitary assumption.

The Structuralist Project behind Social Constructivist View: Theoretical Un-Reflectiveness

The theoretical incoherence discussed here is that the C&CT approach adopts a structuralist view from which it is expected that managers embark on the instauration of a social constructivist view, which conforms to its descriptions about organisations and change.

The above was evidenced in the approach used to incorporate these ideas in the H-TEC through the propitiation of the edge of chaos as an ideal context for learning in which members of the organisation could become engaged in "working with multiple realities", while at the same time introducing the FODP to establish in familiar terms the underlying conditions for learning and innovation to occur in the organisation for their utilisation.

At the level of the self managed team this incoherence was present as the influence of the MD in identifying underlying patterns and thus in propitiating the conditions and supportive structures for learning to take place within the team. But at the same time being faced with a situation in which they had to co-construct their new 'environment' through their own initiatives and definition of their own roles. However, meeting targets became central and determined the approach to managing their new situation.

The above indicates that the structuralist project of building a social constructivist view is flawed with theoretical un-reflectiveness (Jackson, 2000). In theory the C&CT approach unreflectively makes use of two currents of thought, which generates a lot of tension and renders the approach self-contradictory. We see a lot of tension between the voluntaristic ideas supporting the social-constructivist view underlying these descriptions, whereby individuals play an active role in "constructing " themselves and their own world through their interactions (they can re-alize new schemas or worldviews); and deterministic explanations (human systems exposing the properties of complex evolving systems) indicating how to influence the underlying structures or systems of relationships (control parameters) that generate surface phenomena, as defined by the structuralist approach to intervening.

In the practice the above discussion inevitably leads to reflect on the need of generating critical awareness about the different organisational approaches within the spirit of 'pluralism' as in Jackson (1991, 1999, 2000). Indeed when an organisation is anchored in one paradigm, which "specifies what types of entities exist, what forms of access [members] have to them, how [members] should generate knowledge, and how [members] might thereby act" (Mingers and Gill, 1997, p.243), unreflectively introducing approaches reflecting other paradigms becomes counterproductive. The experience of the self-managed team seems to confirm this. The table below synthesises the incoherencies presented in this section.

CONTRIBUTION OF C&CT APPROACH TO MANAGEMENT: WAYS FORWARD

The main contribution of Stacey's approach to management is in terms of description. It generates an *alternative* appreciation of processes of change and evolution in organisations. It does this by providing a framework to articulate various dimensions of organisational reality. In other words, it tries to integrate various existing views in the management literature about organisations to account for the underlying structures and systems that generate change. This denotes a structuralist position.

Table 1. Incoherencies and their Implications

Incoherence	View supporting descriptions	View supporting Intervention	Implications for practice
Abandoning the technical interest without doing so: seeking order in chaos	-Abandon the technical interest: "accept the lack of foresight and control" (Stacey, 1996, p.242).	-Determine regularities in behaviour: there are underlying structures accounting for innovation and learning to occur in organisations.	-There are still controllers who seek order in chaos. -Tension between the need for control and allowing self-organisation.
Making the Practical Interest Technical: Driving The Organisation to The Edge of Chaos	-No one can be in control: "no amount of human intelligence, self-awareness and determ-ination can alter the fundamental dynamics of non-linear feedback" (Stacey, 1996 p.218).	-Regulate behaviour so as to propitiate conditions for mutual understanding: control parameters are to be influenced to drive the organisation to the edge of chaos.	-Fallacy: the technical interest is predominant; the attempt is to manipulate behaviour to propitiate creativity and innovation.
Making the Emancipatory Interest Technical: Co-evolution and the Forces of the Market	-Changes in schemas are liberating: Individuals are free to take their own authority, self-organising and thus producing creative new directions.	- One schema is favoured: the mindset is coevolving with the forces of the market, operating far from equilibrium.	-Responding to the forces of the market is more important, not emancipation in any normal sense of the word.
The pervasiveness of one dominant schema vs. the dynamic of changes in schemas	-The dynamic of change in organisations is described as a process of changes in schemas.	-One schema is favoured: "...organisations are systems that learn in complex ways" (Stacey, 1996, p.242).	-The approach constrains under-standing about organisation from a pluralist perspective and also constrains double loop learning.
Assuming complex-pluralist problem contexts as normal but acting according to complex-unitary problem contexts	-Complex-pluralist assumptions are made: at the edge of chaos: there is high divergence in terms of values and beliefs; and disagreement about ends and means.	-Complex-unitary assumptions are made: Creativity and innov-ation are the purposes potentially realised only by driving the organisation to the edge of chaos.	- Mechanistic, unitary contexts are enforced: propitiating a pluralist context involves making unitary assumptions.
Social Construcivism and Structuralism: influencing control parameters	-A social constructivist view is advocated: individuals co-construct themselves and their world as they interact.	-A structuralist approach is presupposed: underlying structures generating surface phenomena are to be uncovered and understood in order to be influenced.	-Control is central to the management of organisations.

This articulation of one view by using C&CT terms and concepts to integrate various existing views in the management literature is fruitful in that it can enhance understanding, precisely by providing an alternative view. The particular 'assembly' of these ideas and concepts (metaphors) justifies its claim of being novel. However, as comprehensive as this view might seem in terms of integrating other views, it presents a particular arrangement of them, namely, the organisations as complex evolving systems view, confirming its structuralist position. Awareness of this position, points to a series of contradictions which seriously limit the possibilities of the approach in terms of its claims on intervention.

In terms of intervention, the approach does not propose any methodology, but recommends influencing a series of 'control parameters' to drive the organisation to the 'edge'. This means propitiating an 'ideal' problem context, but it is precisely here that it is deemed to fail. In effect, the approach cannot escape one paradigm by using the paradigm's own tools to do so. This is its major pretension and also its major limitation in terms of its contribution to the field.

Thus the main contribution of this approach is in terms of using C&CT concepts to integrate available ideas and concepts in the management literature, generating an alternative understanding about the processes of change and evolution in organisations. It creates a framework that 'assembles' different existing views on organisations under the umbrella of the organisations as complex evolving systems view. However, on intervening, it is evident that the approach cannot take all the views of its interpretations. Consequently in using the very tools of the position (functionalist) it tries to escape, it inevitably traps itself, limiting its own claims in terms of intervention.

REFERENCES

Burrell, G., and Morgan, G. (1979). *Sociological Paradigms and Organisational Analysis*, Heinemann, London.

Checkland, P.B. and Holwell, S. (1998). "Action Research: Its Nature and Validity", *Systems Practice and Action Research.* 11(1): 9-21.

Habermas, J. (1972). *Knowledge and Human Interests*, Heinemann, London.

Habermas, J. (1974). *Theory and Practice,* Heinemann, London.

Jackson, M.C. (1991). *Systems Methodology for the Management Sciences*, Plenum, New York.

Jackson, M.C. (1999). "Towards Coherent Pluralism in Management Science". *Journal of the Operational Research Society.* 50(11):12-22.

Jackson, M.C. (2000). *Systems Approaches to Management*, Kluwer Academic/Plenum Publishers, New York.

Jackson, M.C. and Keys, P. (1984). "Towards a System of Systems Methodologies", *Journal of the Operational Research Society.* 35:473-86.

Mingers J. and Gill A. (1997) *Multimethodology: the theory and practice of combining management science methodologies*, Wiley, Chichester.

Morgan, G. (1986). *Images of Organisation,* Sage Press, Beverly Hills, California.

Morgan, G. (1993). *Imaginization,* Sage, New York.

Stacey, R. (1996). *Complexity and Creativity in Organisations*, Berret-Koehler, San Francisco.

Storr, F. (1997). *On Becoming a Learning Company.* Humberside Training and Enterprise Council, UK: Internal Publication.

Taylor, S. and Bodgan, R. (1984). *Introduction to Qualitative Research Methods: the search for meanings.* 2nd Edition. Wiley, New York.

Author Index

Index